LITCHI : Preventive Practices and Curative Measures

LITCHI
Preventive Practices and Curative Measures

By:

Hari Shankar Singh
Vishal Nath
Smt. Abha Singh
Shesh Dhar Pandey

SATISH SERIAL PUBLISHING HOUSE
403, Express Tower, Commercial Complex
Azadpur, Delhi-110033 (India)
Phone : 011-27672852, Fax : 91-11-27672046
e-mail : info@satishserial.com, hkjain1975@yahoo.com
Website : www.satishserial.com

Published by :

SATISH SERIAL PUBLISHING HOUSE
403, Express Tower, Commercial Complex, Azadpur, Delhi-110033 (INDIA)
Phone : 011-27672852 Fax : 91-11-27672046
E-mail : info@satishserial.com, hkjain1975@yahoo.com

Print Again, 2024

ISBN : 978-93-81226-04-9

Composed, Designed & Printed in India

About the Authors

Dr. Hari Shankar Singh did M.Sc. (Ag) and obtained Doctorate degree in Agricultural Entomology from N.D.University of Agriculture and Technology, Faizabad. He served in Central Silk Board in Jammu and Kashmir as Senior Research Associate, Indira Gandhi Krishi Vishwa Vidyalaya, Raipur, Chattisgarh as scientist, Vivekanand Parvatiya Krishi Anusandhan Sansthan- ICAR, Almora-Uttarakhand and Horticulture and Agro-forestry Programme, Ranchi-Jharkhand as Sr. Scientist. He has nearly 200 publications including 2 books. Having good experience of participatory works, he has developed various IPM and field technologies. He has handled and worked in the projects funded by NHB, APCESS, DFID, NATP and DBT in the field of insect pest management in fruits and vegetable crops. He is deeply involved in tribal horticulture development in KBK region of Orissa, India. At present, he is serving as Principal Scientist and Head at Central Horticultural Experiment Station, Bhubaneswar-Orissa, under Indian Institute of Horticultural Research-ICAR.

Prof. Vishal Nath, has completed his Master's degree from N. D. University of Agricultural and Technology, Faizabad and Ph.D. from Kanpur. He worked as scientist (Horticulture) at Lucknow and Bikaner. He worked as Senior Scientist at, HARP, Ranchi, as Professor (Horticulture) at Birsa Agricultural University, Ranchi and as Head, Central Horticultural Experiment Station, Bhubaneswar-Orissa. Presently, he is working as Director, NRC for Litchi, Muzaffarpur, Bihar. Apart from his contribution to National Agricultural Research System (NARS) at various capacities, he has wider exposure to Hitech Horticulture being practiced in developed countries like Israel. Prof. Vishal Nath has exclusively worked on fruits of major and minor importance being grown in tropical, sub-tropical, arid and semi-arid regions and under waste land conditions. He has developed many technologies and varieties of different fruit crops suitable for resource poor areas. He has written several technical bulletins, extension folders and information brochures on these aspects. Prof. Vishal Nath has authored very important books on arid horticulture, 'Fruits for Future', Litchi and Fruit Production Technology for Plateau regions. He has more than 225 publications on different fruits which include research papers, popular articles, technical bulletins, research manuals and information bulletins in his credit.

Smt. Abha Singh obtained her master degree from Banaras Hindu University and joined ICAR as scientist. Basically she is involved in research and training in the field of food and nutrition with special reference to fruit and vegetable preservation. She has developed low cost vitamin A rich sweet potato based weaning mix of children. She has published 15 research papers, 18 articles, in addition to abstracts and popular articles and 2 books. Presently she is working as scientist in Directorate of Research for Women in Agriculture, Bhubaneswar-Orissa.

Dr. Shesh Dhar Pandey did his M.Sc. and Ph.D. in Horticulture from N. D. University of Agricultural and Technology, Faizabad. Initially he worked as Asstt. Professor in Indira Gandhi University of Agriculture and Technology, Raipur, M.P. and then joined NRC for banana at Trichy. He has 107 publications including 3 books. He has specialization in production technology of horticultural crops including integrated water and nutrient management, high density planting and organic farming. Presently, he is working as Principal scientist Horticulture at NRC for Litchi, Muzaffarpur, Bihar.

Preface

India is the second largest producer of litchi. Bihar, West Bengal, Tripura, Assam, Uttarakhand, Jharkhand and Punjab are the major litchi producing states. Litchi area is further expected to increase in many other states with the advancement of entrepreneurship among the farmers. Shahi, China, Culcuttia, Bedana, Late Bedana and Longia are being considered among the most superior types of litchi in India. Most of the commercial cultivars of litchi are location specific with long gestation period. Commercial litchi farming is done by large farmers, however, the scenario is changing gradually due to participation of small and marginal farmers. The productivity of litchi is low for many reasons including poor initial planning, inferior planting material, poor orchard management practices, unscientific post harvest practices and natural calamities. When the tree is full of vigour, vegetative dominance prevails over reproductive phase in certain varieties and locations, which adversely affects flowering, fruit set and fruit growth. High fruit drop and post harvest losses along with poor management of pests, diseases and disorders and poor upkeep of orchard affect the economic production of litchi. In order to avoid loss and to increase profitability, it is necessary to take steps from initial stage so that the vigour, growth and fruiting capacity of the plant is maintained.

There are two types of losses in litchi cultivation. First, arising due to grower's negligence/ mistakes in planning, management of orchard and harvest and the second, arising due to natural factors such as, pests, diseases, disorders and adverse climatic factors. Mistakes may occur in setting up of an orchard, selecting varieties, planting, fertilizing, drainage, irrigation systems, shaping, weed control, mulching, pruning and canopy management, use of growth regulators, harvesting, packing, marketing and quality management. Natural problems such as, pests, diseases and fruit disorders may occur by their own and may aggravate further due to negligence of farmer. Number of books and information bulletin on litchi are available presently, however, most of them are reference based which fulfil the need of academicians, researchers and students. The present book has been prepared to educate farmers and orchardists so that they can take proper step from very beginning to avoid problems in litchi cultivation. This book has been prepared in action mode as "loss preventive practices and curative options" in such a way as the farmers can easily understand the technology and initiate appropriate and timely action. The manuscript is based on the research and experiences of the authors' vis-à-vis the available published literature on the subject. We hope that large number of farmers who are engaged in litchi cultivation and beginners will be immensely benefited from this book. We acknowledge the litchi workers whose published reports/works/research papers/book/monograph in India and abroad have been referred / consulted for the preparation of this book. Special thank are due to workers at RAU, Pusa, Samastipur, GBPUAT, Pantnagar, FAO and NRC, Litchi, Muzaffarpur for their valuable help to bring out this publication. Thanks are also due to Shri K. Gouri Sankara Rao for helping in preparation of this manuscript.

Authors

Contents

Chapter 1

Introduction

Litchi (*Litchi chinensis* Sonn.: Sapindaceae), a sub-tropical evergreen fruit crop, has highly specific climatic requirements. Probably due to this reason, its economic cultivation is restricted to only a few subtropical countries in the world. The major litchi growing countries are China, Israel, Australia, Thailand, Taiwan, India, Vietnam, parts of Africa and at higher elevations in Mexico and Central and South America. World production of litchi is estimated to be around 2.11 million tons, with more than 95% of the area and production share of Asia. A relatively small amount of litchi is produced in the United States, Mexico, and Central and South America. The top five world litchi producing countries are China, India, Taiwan, Thailand, and Vietnam. India and China account for 91 per cent of the world litchi production but it is mainly marketed locally. India enjoys a prominent position in the litchi map of the world both, in terms of production and productivity. Over the years, India has recorded significant growth in production and productivity of litchi, though the trend is not consistent. Both, the Indian and the world markets are expanding fast for litchi.

China is the largest producer of litchi where production season extends from mid May to mid-August. India, where litchi was introduced in the 18th century, is the second largest producer with relatively higher productivity compared to other growing regions having production period from the first week of May to the first week of July. India accounts for about one-fifth of the global production. Taiwan is the third largest litchi producer with a production period extending from June to August. Thailand is the fourth largest producer, with a production season running from April to June. Vietnam is next in the order having harvesting period from May to June. Other notable producers include South Africa, Australia, Mexico, Florida, Hawaii and California. Country wise major litchi cultivars are presented in Table 1.

Table 1: Major litchi producing countries and major cultivars

S.No.	Country	Major cultivars
1	China	Bah Lup, Baitang-ying, Fay Zee Siu, Haak Yip, Kwai May (Red), Lanzhu, No Mai Chee and Wai Chee
2	Australia	Fay Zee Siu, Kwai May Pink, Salathiel, Souey Tung, Tai So and Wai Chee
3	India	Bedana, China, Culcuttia, Late Bedana, Longia and Shahi
4	Indonesia	Local selection
5	Israel	Mauritius
6	Madagascar	Madras and Mauritius
7	Philippines	Sinco, Tai So and ULPB Red
8	South Africa	Mauritius, McLean's Red
9	Thailand	Chacapat, Haak Yip, Kom, Tai So, Wai Chee
10	USA	Brewster, Haak Yip, Kwai Wai, No Mai Chee, Shan Chi
11	Vietnam	Vaithieu

Source: The litchi crop in Asia and the Pacific, Regional Office for Asia and the Pacific, FAO corporate document repository. http://www.fao.org/docrep/

In India, litchi ranks 10th in area and 11th in production among fruit crops (Table 2), but in value terms, it ranks sixth. At national level, banana and mango are the most important fruit but in Bihar, litchi is considered to be the most important fruit as it contributes significantly to its total fruit production.

There has been substantial increase in area and production of litchi in the last 50 years. Area has increased from 9,400 hectares in 1949-50 to 69200 hectares in 2007-08. The contribution of litchi to total area under fruit has increased from 0.75 to 1.30 per cent. Increase in area between 1991-92 and 2007-08 (16 years) has been 40.30 per cent, while production increased during the same period is to the tune of 71.67 per cent. Productivity also recorded an increase of 22.44 per cent during the same period. Evidently, production and productivity of litchi is constantly increasing in the country (Table 2).

Table 2 : Crop-wise area (000 ha), production (000 MT) and productivity (MT/ha) of major fruit crops in India during 2006-07 and 2007-08

Crop	2006-07			2007-08		
	Area	Production	Productivity	Area	Production	Productivity
Banana	604	20998	34.80	647	23205	35.90
Mango	2154	13734	6.40	2205	13792	6.30
Citrus	798	7145	9.00	843	7574	9.00
Papaya	72	2482	34.50	80	2686	33.40
Apple	252	1624	6.50	264	2002	7.60
Guava	176	1830	10.40	178	1975	11.10
Grapes	65	1685	25.90	64	1677	26.00
Sapota	149	1216	8.20	150	1238	8.20
Pineapple	86	1362	15.80	80	1216	15.10
Pomegranate	117	840	7.20	122	858	7.00
Litchi	65	403	6.20	69	418	6.00
Others	1016	6244	6.10	1071	6862	6.40
Total	**5554**	**59563**	**0.70**	**5775**	**63503**	**11.0**

Source: Indian Horticulture Database, 2009

Being specific in climatic and soil requirements, litchi has limited distribution. It is grown in the states of Bihar, Tripura, West Bengal, Uttar Pradesh, Punjab and Haryana. Of the total production of litchi in India, 74 per cent is contributed by Bihar. The second largest litchi producing state is West Bengal followed by Assam and Tripura (Table 3&4). Productivity is highest in Punjab followed by West Bengal and Bihar.

Table 3 : State wise area, production and productivity of litchi

State	Area (000 ha)			Production (000MT)			Productivity (MT/ha)		
	05-06	06-07	07-08	05-06	06-07	07-08	05-06	06-07	07-08
Bihar	28.40	28.80	29.60	200.10	211.90	223.20	7.00	7.40	7.50
W.Bengal	8.10	8.10	8.10	74.90	77.20	77.50	9.30	9.50	9.60
Assam	4.70	4.70	4.80	34.90	33.30	34.00	7.40	7.00	7.00
Punjab	1.40	1.50	1.50	13.80	14.60	19.30	10.00	10.00	12.80
Jharkhand	1.40	1.40	3.30	16.50	16.50	16.50	12.00	12.00	5.10
Uttaranchal	7.90	8.40	8.80	13.60	14.40	15.10	1.70	1.70	1.70
Tripura	2.30	2.50	2.60	14.30	14.40	14.40	6.20	5.60	5.40
Orissa	3.90	3.90	4.00	11.90	12.10	12.80	3.00	3.10	3.20
H. P.	3.50	3.80	3.90	3.70	2.80	2.50	1.00	0.70	0.60
U.P.	0.20	0.20	0.30	0.50	0.80	0.90	2.00	3.30	3.40
Others	1.30	1.70	2.10	7.60	4.50	1.80	5.70	2.60	0.90
Total/ Average	63.20	64.90	69.20	391.80	402.60	418.40	6.20	6.20	6.00

Source: Indian Horticulture Database, 2009

Table 4 : Area, Production and Productivity of litchi in India

Year	Area (000 Ha)	% of total fruit area	Production (000MT)	% total fruit production	Productivity (MT/Ha)
1991-92	49.30	1.70	243.80	0.90	4.90
2001-02	58.10	1.40	355.90	0.80	6.10
2002-03	54.10	1.40	476.40	1.10	8.80
2003-04	53.70	1.10	478.50	1.00	8.90
2004-05	60.00	1.20	368.60	0.70	6.10
2005-06	63.20	1.20	391.10	0.70	6.20
2006-07	64.90	1.20	402.60	0.70	6.20
2007-08	69.20	1.30	418.40	0.70	6.00

Source: Indian Horticulture Database, 2009

Cultivation intensity of litchi varies even within state. Based on cultivation intensity, the place specific distribution of litchi is given in Table 5.

Table 5 : Cultivation intensity of litchi in different parts of India

Cultivation	Production areas (District/State)
Concentrated	Muzaffarpur, Samstipur, Vaishali, Sitamarhi and Motihari in Bihar, Dehradun, Udham Singh Nagar and valley of Pithoragarh in Uttarakhand, Saharanpur and Kushinagar in U.P., Ranchi and Hazaribagh in Jharkhand, Murshidabad, Maldah, 24 Parganas and Coochbehar in West Bengal.
Medium concentrated	Darbhanga, Shivapuri and Begusarai in Bihar, West Tripura in Tripura; Ghorakhpur and Basti in Uttar Pradesh, Gumla and Lohardaga in Jharkhand, Sarguja in Chattisgarh, Kangara valley in Himachal Pradesh.
Sporadic	Gurudaspur in Punjab, Kachhar in Assam, Garhwal in Uttaranchal, Ootty and Nilgiri in Tamil Nadu; North and South Tripura in Tripura, Nadia in West Bengal, Panchmarhi in Madhya Pradesh, Angul in Orissa, Ambikapur in Chattisgarh

Source: Litchi: M.Rai,Vishal Nath and P.Day (editors), CHES-IIHR,ICAR,Plandu, Ranchi

In India, litchi maturity commences first in Tripura followed by West Bengal and then in Bihar. The first and second week of May are the time for harvest in the eastern region, while litchi of Bihar matures in the 3rd - 4th week of May and continues up to the first week of June. Litchi in Uttar Pradesh and Punjab is ready for harvest during the 2nd - 3rd week of June. In Himachal Pradesh, litchi of the same cultivar is harvested in the last week of June.

Litchi is known for its taste and flavour and rules the domestic market as "Queen of fruits" during the season. However, set of fruits like mango and grapes dominate the domestic and export trade and have strong competition with litchi fruits during the season. This can be attributed to numerous factors including research and development efforts, production technologies, infrastructure and institutional support extended by the government policy and cooperation among cultivators for earlier crops. Absence or underplay of similar institutional and infrastructure facilities for a number of other fruits having similar production advantage is responsible for impeding the diversification of the Indian export fruit basket. Litchi is one such fruit, which is widely grown and has a vast untapped potential in the domestic and global market. It offers immense scope to boost exports through appropriate institutional support and has major demand in Middle East countries. A limited quantity of litchi is also exported to European market. Less than 5% of the world's litchi production or approximately 100,000 tons enters into world trade on an annual basis which is expected to increase in the coming years. During the Indian litchi season (May to June), good quality litchi is not available from other parts of the world except from Thailand (May and June) and Israel (July). In spite of these advantages, India has negligible share (less than 1%) in the world trade.

The fresh fruit market dominates the trade, followed by dried and canned fruit. The main importing countries are the European Union, the United States, Hong Kong, Singapore, Japan, and Canada. The main exporting countries are China, Taiwan, Thailand, Madagascar, South Africa, Australia, and Mexico.

Table 6 : Export of litchi from India (Value in lakh and Qty in tones)

Country name	Quantity (2005-06)	Value (2005-06)	Quantity (2006-07)	Value (2006-07)	Quantity (2007-08)	Value (2007-08)
Nepal	143.98	18.98	33.83	9.67	142.74	45.71
U.A.E.	10.40	6.55	17.30	12.69	16.10	13.78
U.K.	0.00	0.00	0.03	0.01	1.22	3.06
Bahamas	0.00	0.00	0.00	0.00	0.84	0.51
Bahrain	13.83	5.68	0.00	0.00	0.42	0.25
Canada	0.00	0.00	0.00	0.00	0.20	0.12
Oman	0.00	0.00	0.00	0.00	0.08	0.05
Saudi Arabia	1.10	0.34	0.00	0.00	0.00	0.00
Vietnam	28.00	17.28	0.00	0.00	0.00	0.00
Germany	0.65	0.08	0.00	0.00	0.00	0.00
Others	520.88	4.45	1609.98	141.99	0.00	0.00

Source: APEDA, Website, 7.1.2009

Litchi is a delicious fruit having aromatic pulp with sweet and acid taste. Litchi is largely preferred as table fruit. It is also used for canning, which has made its impact in international trade. 'Litchi Nut' a dried product of whole litchi fruit is also potential product world wide. High quality flavoured squashes can be prepared from litchi. Litchi is a non-climacteric fruit hence, it is not expected to increase in soluble solids once it has been picked. Depending upon variety and climate, the fruits contain 60% juice, 8% rag, 19% seed, and 13% peel. The moisture varies from 77-83 per cent with variable level of TSS (15.90- 20.10^{0} Brix). The acid content varies from 0.2 per cent to 0. 64 per cent. Litchi is not a significant dietary source of protein (only 0.8-0.9 per cent) and poor source of fat (0.30-0.50 per cent). It is not a good source of thiamine, riboflavin, calcium, phosphorus and iron. It lacks provitamin A. It is a good source of ascorbic acid with a level of 27.8 mg/100 g. The daily vitamin C requirement for the average adult can be met by consuming 14-17 litchis. The consumption of a litchi fruit would meet 2-4% of the dietary reference intakes (DRI) for P, K, Mg, Fe, Zn and Mn and provide 22% of the DRI for Cu. Most of the energy in a litchi is in the form of carbohydrate (sugar). Litchis are high in heart-healthy polyphenols, containing 15% more than grapes, a fruit commonly referenced as high in polyphenols. A new pharmaceutical composition and processes have been prepared for treating and preventing diabetes. The pharmaceutical composition of litchi is composed of three ingredients i.e. Oleanolic acid, Saponins and Kuguasu which is nontoxic. It appears that the high ascorbic acid and sugar content are the chief nutritional constituents of litchi as presented in Table 7.

Table 7 : Nutritional components of Litchi aril (Food Value Per 100 g of Edible Portion*)

Constituents	Fresh aril	Dried aril
Calories	63.00-64.00	277.00
Moisture	81.90-84.83%	17.90-22.30%
Protein	0.68-1.0 0g	2.90-3.80 g
Fat	0.30-0.58 g	0.20-1.20 g
Carbohydrates	13.31-16.40 g	70.70-77.50 g
Fiber	0.23-0.40 g	1.40 g
Ash	0.37-0.50 g	1.50-2.00 g
Calcium	8.00-10.00 mg	33.00 mg
Phosphorus	30.00-42.00 mg	-
Iron	0.40 mg	1.70 mg
Sodium	3.00mg	3.00 mg
Potassium	170.00 mg	1,100 mg
Thiamine	28.00 mcg	-
Nicotinic Acid	0.40 mg	-
Riboflavin	0.05 mg	0.05 mg
Ascorbic Acid	24.00-60 .00mg	42.00 mg

*According to analyses made in China, India and the Philippines.

Litchi is said to relieve coughing and to have a beneficial effect on gastralgia, tumors and enlargements of the glands. Chinese people believe that excessive consumption of raw litchi causes fever and nosebleed. In India, the seeds are powdered and administered in intestinal troubles. Decoctions of the root, bark and flowers are gargled to alleviate ailments of the throat. Litchi roots have shown activity against one type of tumor in experimental animals in the United States Department of Agriculture/National Cancer Institute Cancer Chemotherapy Screening Program.

Now a days, many farmers in north India, wish to invest in litchi as this crop has ready market, high yield and demand in local as well as export market and can be easily grown in suitable climate. Today, the litchi cultivating areas are being notified as litchi belt for commercial production. In these areas, regulatory norms are being developed to provide safe environment and logistic support system to have better production and marketing as special zone. Litchi is sensitive to the growing environment; therefore, any change in environmental factors severely affects the productivity and fruit quality. In some areas, litchi is the livelihood for many people as it provides both on-farm and off-farm employment. Small and marginal farmers get additional income from litchi plants in their homesteads. Thus, litchi cultivation is the livelihood security for a large population, especially in the state of Bihar. Commercial litchi plantation creates 'a source of job opportunity' for the people associated with growing and managing orchards, harvesting and post harvest handling, packing, transportation, export and value addition. Commercial litchi production is attaining the status of an industry in certain pockets with forward and backward linkages.

In the production side, the major problems are poor plant establishment, less fruiting span, low and irregular yields due to poor flowering and fruit set. Likewise, fruit cracking, browning and rotting of fruits, poor shelf life and lack of suitable varieties with early and late maturity and good quality fruits are some of the factors hindering the growth of litchi industry at commercial scale.

A profitable litchi industry depends on the production capacity of orchards. Lack of proper planning at initial stage and orchard management practices at later part may lead to various kinds of production related problems in the established litchi orchard. Trees must be in position to produce sufficient quantity of fruits every year for market balance. Therefore, it is important that the trees are managed with scientific tree management system. Besides, the crop growing conditions, the cultural practices and the cultivar selection determine the quality of litchi at a particular location. Mineral deficiencies, improper water and nutrient management, poor pest management and cultural neglect eventually affect the yield and quality.

The present level of litchi production with increased quantum of post harvest losses due to fast deterioration, quality produce is unable to meet the demand as well as economic return to the grower due to many production and post harvest constraints such as:

- Slow rate of adoption of improved techniques
- Poor orchard management
- Tendency of imbalance bearing in few leading cultivars
- Incidence of pests and diseases
- Loss due to physiological disorders
- Difficulty in harvesting as poor status of mechanization
- Low yield potential and decline in quality production
- Greater post harvest losses

The following factors may cause future problem and loss to any farmer having a litchi plantation or planning to venture into litchi industry.

- Poor understanding of litchi growth pattern and behavior leading to unscientific plant, pest and produce management.
- Litchi cultivation under unsuitable geographical location and unsuitable cultivars under poor growing conditions.
- Poor orchard planning, such as deciding appropriate spacing based on requirement like high density or regular density, methods of layout, pit preparation, pit mixture and pit filling, improper production or procurement of planting material, odd time of planting, post planting neglect, faulty tree frame and fruiting arms, etc.
- Non adoption of proper orchard management techniques like tree basin and orchard floor management, pruning, weeding, methods for ensuring regular bearing and its further impact on tree health, water and nutrient management methods and schedule.

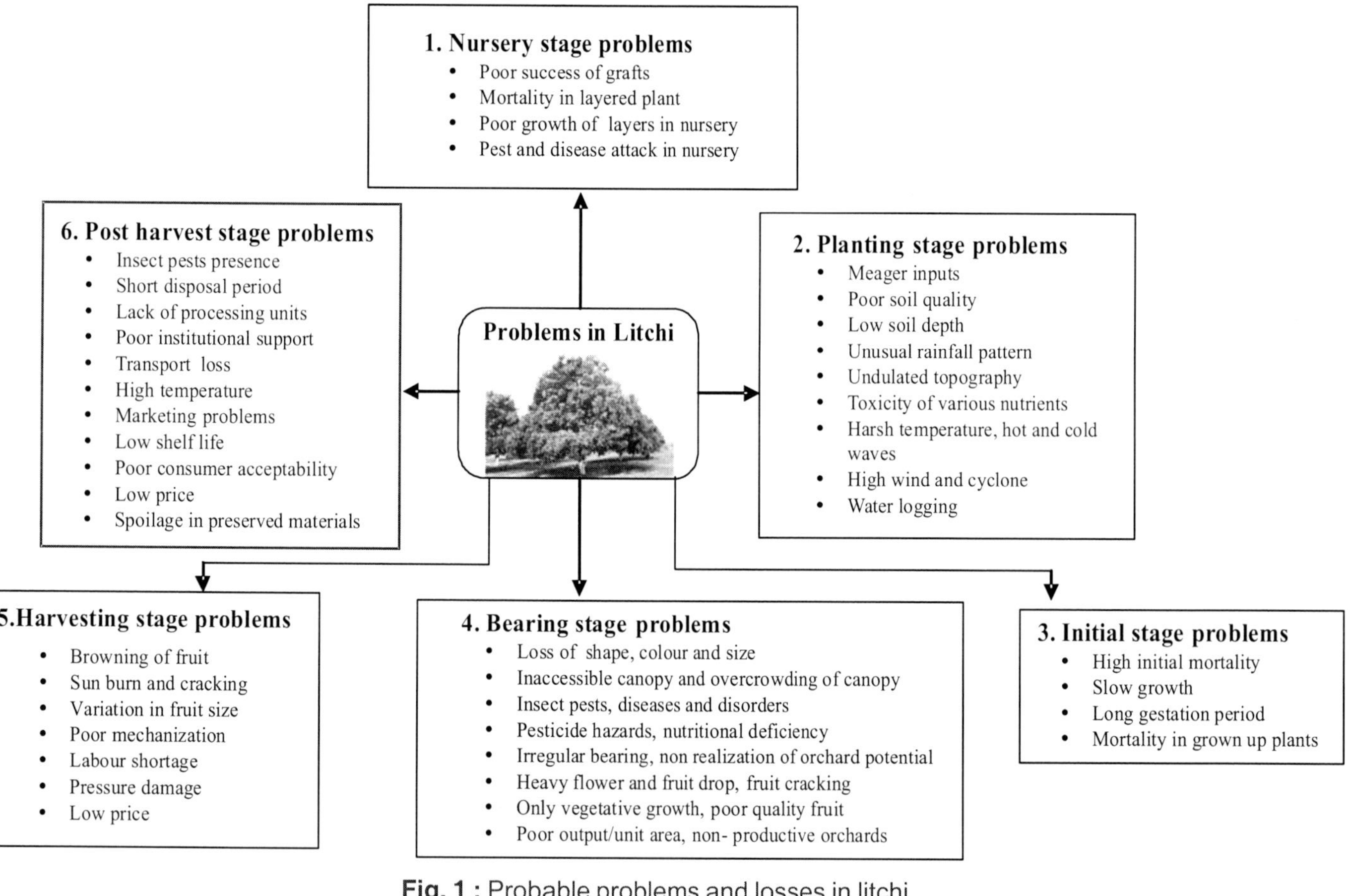

Fig. 1 : Probable problems and losses in litchi

- Poor skill of pest and disease identification and their management.
- Poor knowledge about maturity standard, harvesting time and method, anticipated production volume at particular time and prices at farm gate, wholesale and retail markets. Unscientific post harvest handlings at field level and at packing house.
- Poor knowledge about methods to withstand the post-harvest abuses of the marketing and distribution process as markets can accept 'only limited quantities'.
- Unawareness of pest clinic information services, directory information services, policy information services and market information services.
- Poor market and transport infrastructure site of production to place of disposal
- Mounting pressure of climate change and related impact on litchi.

Considering the importance of this fruit crop in the country, efforts have been made to provide technological support through research and promoting production, post-harvest management and marketing, including export, through development programmes and schemes. Based on up to date knowledge, an effort has been made to develop this book perceiving the "good horticultural practices as tools for prevention of problems and losses in later stage" in litchi production. Attempts have also been made to enlighten the readers regarding preventive practices that can be adopted to avoid the losses and curative options that can mitigate the problems. The information furnished in the manuscript is based on the working experience, survey of available literature and applied aspects of core research.

❑❑❑

Chapter 2

Understand the Growth and Behaviour of Litchi Plant

The final result of the entire growth and reproductive process is fruit yield which is comparatively low and irregular in litchi production, therefore, a perfect understanding of the litchi is necessary so that adequate cropping methods may be developed to reach the final objective. Before undertaking commercial plantation of litchi, one should understand that how litchi plant grows and yields fruit over the years because this basic information makes strong base for initial planning and orchard management. The orchard management practices differ with the age and stage of litchi plant as the practices for juvenile plant (non bearing young plant) may be different from the mature plant. Broadly, based on progressive age, the litchi plant has 6 distinct phases of growth and development. Based upon these stages, its behaviour may vary and so the management practices.

(A) Growth stages of litchi

(i) Nursery stage

The very first stage of litchi plant i.e. up to 1 year of plant age is known as nursery stage. During this stage, the plants are maintained by expert nursery operators.

(ii) Establishment and non-bearing stage

The second stage which lies between 1-6 years is called the young non-bearing stage when plant tries to attain the desired frame. During this stage, plant has accelerated growth to develop scaffolds and main limbs and produces vegetative flushes 3-5 times/ year. In this phase, the orchardist should start efforts to develop uniform frame of the tree. Irrigation and fertilizer requirement of this stage vary from other stages. Even in subtropical climates, where temperature are sufficiently low to induce floral initiation, vegetatively propagated litchi takes several years before it begins to flower. This is a clear indication that litchi has longer juvenile period with respect to floral initiation. In seedling litchi, this juvenile period may last from 6 to 10 years, while in layered plants it lasts 5 to 6 years. It has nevertheless been reported that layered trees obtained from senile branches frequently flower in the year of planting. Data in subtropical Africa have revealed that grafted trees flower earlier than layered trees due to the fact that sap movement is restricted with grafting.

Management and upkeep of young non-bearing litchi orchard is essential which accelerates the growth of the young plant and minimizes the juvenile period. Up to 3 years of age, the main aim for the litchi orchard management is to accelerate the plant growth to develop better tree frame which makes platform for higher production throughout the tree life span. Generally, young litchi plant sprouts 5 times in a year to form canopy with a certain quantity of vegetative growth. Therefore, newly planted young saplings should be properly nourished to promote better vegetative growth in this period. It has been anticipated that after attaining the age of 5-6 years, the plant develops good canopy width and height (1.5-2m) and becomes ready for bearing some fruits. Other management practices like weeding, hoeing of plant basins, plant protection; water management, etc. should be done as per requirement.

(iii) Junior bearing stage

The third stage of litchi plant between 6-10 years of age is known as junior bearing stage. This is a delicate stage when a major transformation from vegetative to reproductive growth takes place. During this stage, the plant keeps enlarging its canopy and produces fruits also. At this stage, it is essential to maintain a balance between vegetative growth and fruiting since the young plants tends to have more vegetative growth. Some times, the fruit production is suppressed or reduced due to excessive vegetative flushing particularly in spring season during this stage.

(iv) Young bearing orchards

This is the very delicate stage of orchard management where more visionary approach should be followed in skillful manner. During young bearing stage between 10-25 years, plant needs more nutrients for proper growth and bearing. At this stage, the tree keeps enlarging canopy and produces economic fruit yield. Therefore, the management should be aimed at balance both vegetative and reproductive growth. It is important that at this stage, there must be optimum balance between vegetative growth and fruit production. Excessive vegetative growth adversely affects fruit setting while excessive fruit load reduces development of tree canopy. Skillful orchard management at this stage includes maintenance of optimum number of bearing branches as per the nutritional status and health of plant. Young trees tends to have more vegetative growth, therefore, it has been observed that sometimes fruit production is suppressed by excessive vegetative flushing particularly during autumn. Therefore, effective measures like ringing, girdling, fostering of winter shoots, use of paclobutrazol, etc. can be applied for proportional crop production.

(v) Adult bearing orchards

After attaining the age of 25 years, litchi tree enters into the phase where vegetative growth and fruiting are relatively balanced and the phase is called as adult bearing phase. The stage of litchi tree between 25 to 40 years of age is also known as full bearing stage. The main features of this stage are well distributed tertiary branches and fruiting terminals which bear the fruits and vegetative flush. In this phase, tree reaches its full yield potential. The main aim of orchard management at this stage is to produce maximum yield with maximum feeding. At this stage, different fertilizers should be applied at different periods of cycle. Foliar sprays of nutrients and plant growth regulators are also essential to ensure higher production. Timely plant protection, short treatment, intercultural operations, water and canopy management for higher production should be judiciously followed. To have regular yields, the vegetative flushing pattern, its time and health should be monitored and attempts should be made to check excessive vegetative growth through chemical and cultural manipulations.

(vi) Old bearing (Senile) orchards

This stage comes 40 years onward when symptoms of ageing are evident. The plant becomes old with big canopy, restricted flush growth and low fruiting. Such tree need special management plan like reiterative pruning, re-canopy building and promotion of new shoots etc. Litchi tree has strong renewal ability therefore, the old and unproductive orchards can be rejuvenated for further higher production. Heavy regenerative pruning followed by optimum fertilizer and moisture management near feeding root zone are the main management practices for such orchards.

(B) Plant characteristics of litchi

(i) Plant

Litchi is a subtropical evergreen tree having uniform growth and dense canopy. It does not grow very tall (up to 12m), but some vigorous plants can go up to 20m at 25-30 years of age. The tree is attractive having a dense, rounded, conical and symmetrical canopy of dark green foliage and can grow to be very large tree having tendency to produce layered canopy with horizontally low branches. Growth habit differs considerably among cultivars. Some genotypes have vigourous vegetative flush and are more erect. Litchi is slow growing tree to reach full size. Stem is thick, uneven, and short with off white coloured bark. The branches tend to form V-shaped crotches in the young plants which are easily broken off by the wind. In grown up plants, however, the thick limbs develop wider crotches which sustains wind pressure. It produces a number of annual shoots which give it a very attractive appearance in all seasons. Its ornamental qualities are further enhanced at harvest time when bunches of bright red fruits hang from the periphery of the tree. It has good longevity for more than 100 years.

(ii) Shoot and branch

Terminal shoots in litchi give out leafy out growth and reproductive bud appears on the tip of the vegetative shoot. Vegetative growth of shoot in form of flushes continues for some time followed by appearance of reproductive growth in form of panicle but in some cases the emergence of new 2-3 vegetative flushes and the inflorescence grows simultaneously in form of mixed panicles. Branches droop as the tree grows. Some times multiple trunks are also found in air layered plants where branches are allowed at low level. Current year twigs are green or pink and thin. Spread of full grownup tree is 6-8 m round and spreading.

(iii) Leaf

Litchi leaves are compound with 7-12 lamella, alternate, coppery red or red to pale green or even creamy when young and bright, dark green when matured. They may be lanceolate, oblong-lanceolate or even elliptic in shape, measuring 7.5 to 20cm in length, 2.5 to 6cm in width (petiole and rachis) and smooth but coriaceous in texture. The base of the leaf blade is rounded. A leaf consists of ten to twelve pairs of leaflets. They may be arranged along the rachis directly or obliquely opposite to each other, each on its own small stem (petiole) and are between 3 and 25mm in length. The leaflets are glossy with shorter internodal length, dark green on the upper side and dull and waxy on the underside. Small-leaved varieties with shorter internodal length are usually less vigorous. The leaves in the young growth flushes tend to have beautiful reddish bronze colour in some genotypes.

(iv) Root system

The root system of the litchi plant varies depending on the method of propagation used. Trees, grown from seed, develop fairly strong tap-roots and a well-distributed secondary and tertiary root system. However, most of the trees in commercial orchards are propagated by air layering. Layered plants are characterized by a very shallow root system where the tap-root is virtually non-existent. The absorbing roots start at soil level and descend to a depth of approximately one meter. Majority of litchi plants developed through air layering have adventitious root system whereas; grafted plant can have tap root system. Feeder roots in litchi are present 25-30 cm below the soil surface, which are mainly responsible for the absorption of mineral nutrients and water.

(v) Inflorescence

The flowering period varies with the genotype and environmental condition. Generally, the flower bud differentiation in litchi starts in December and completed by the end of January. Subsequently emergence of flower panicle starts in January and is continued up to the end of February. Prior to floral bud initiation, the twig first begins to give out vegetative growth and then just at the terminal end of the vegetative shoot, a reproductive bud in form of recemose panicle appears.

The litchi inflorescence is determinate and composed of several panicles produced on current season's wood. The panicles are normally produced terminally in clusters of 10 or more, but in some trees high percentage of sub-terminal panicles or axillaries are also produced. Panicles are generally in mixed form with the lowest buds producing leaves only, the middle buds producing floral buds in the axils of the leaves and the topmost buds producing only floral branches and sometimes very small leaves which do not persist. Usually the terminal inflorescences are more robust than the auxiliary.

The number and percentage of the different types of flowers in litchi vary greatly with cultivar, environmental conditions, tree and panicle within a tree. The proportion of functional female flowers varies between 10 and 60 per cent. Each panicle produces hundreds of small, white, greenish or yellowish flowers which emit a not-too-pleasant smell when the tree is in full flower.

(vi) Flower

Flower is terminal or auxiliary arising from the axil of first or second upper most leaf of the new shoot. It bears tiny, greenish white or yellowish flowers and is branched racemosely. The inflorescence varies from 10 to 35 cm in length and 5 to 30 cm in spread. Litchi flowers measure 3 to 6mm in width when fully open and rest on pedicels measuring approximately 1.5mm. They possess a cup-shaped calyx with 4-5 short, serrated sepals from which petals are totally absent. Flowers are usually produced in later winter or early spring and there are three types of flowers which open in succession on the same panicle. The flowers vary in terms of degree of sexual development, that is, length and functionality of the stamens and development and functionality of the pistil. In litchi three types of flowers are found:

Fig. 2 : Litchi flower

i) Male or staminate flowers (Type-I or M_1)

These flowers are lacking pistil. The position is occupied by a structure which contains a pink, pubescent protuberance which is very rudimentary or aborted form of pistil lacking both, stigma and style. The pistil is surrounded by 4-12 stamens (usually 8) with hairy filaments. The filament varies 5-6mm in

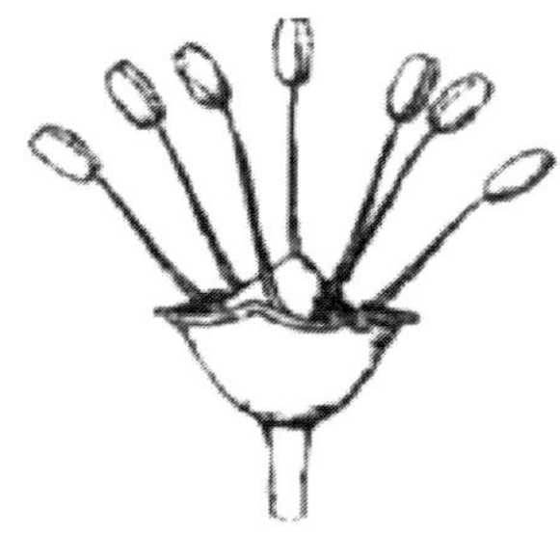

Fig. 3 : M_1 of I Type

length. In some flowers, the filaments are 2-3 times longer than the anther. While in some, they are equal to length of the anthers. Type I is defined as non-functional male, which opens for 10 days. Stamens are long, thin creamy white filaments which are deflexed and inserted into an angular glabrous nectary disc. Anthers are small, 2 celled, elliptical, emarginated at apex, fixed, and longitudinally dehiscent, initially pale white and become brownish at maturity. During the time that an individual flower remains open, the pollen sacs mature and the anther dehisce longitudinally one after the other, not simultaneously and discharge the pollen. Pollen grains are small, triangular having nipple like angles with three germinating pores. Ovary is very small, reddish white and abortive. The flowers possess nectar discs at their base, but these are not very well developed.

(ii) Hermaphrodite female flower (Type-II or F)

These flowers resemble the hermaphrodite male flower with short filament except that the lobes of the stigma open down the vertical cleft. They are called Type II or F flowers which appear and remain functional only for 2 days. These flowers have small but completely developed pistil attached to a short peduncle. The ovary has 2 to 4 carpals, each containing an ovule. The surface of the ovary is pubescent with protuberances which persist and give the fruit its rough surface. The ovary is long with a short style whose apex is split into two lobes. When these are open, the white, sticky surfaces of the stigmas are exposed. The nectar disc is found at the base of the ovary and produces an abundant secretion which attracts insects. Usually, only one of the lobes of the ovary develops into a fruit, the others are aborting. Occasionally, however, two lobes may develop, producing 2 fruits each containing a seed, which remain embedded in their bases. The pistil is usually surrounded by 5-8 stamens with very short filaments (less than 1.5mm in length). The anthers do not normally dehisce and contain little viable pollen which creates the situation of cross pollination.

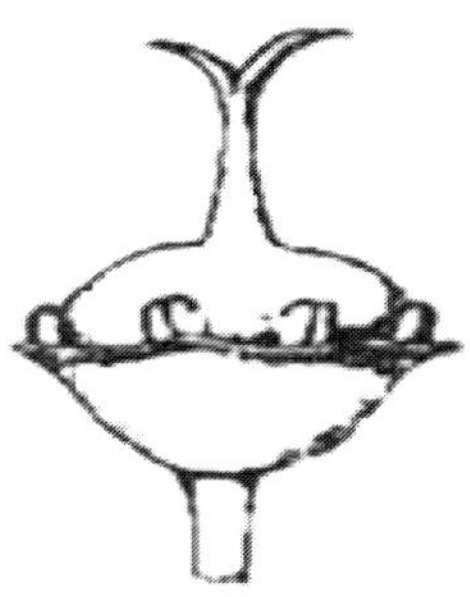

Fig. 4. : F or II Type

(iii) Hermaphrodite male flower (Type-III or M_2)

Type III is a functional male flower but with more hermaphrodite features than Type I flower. Stamens and pistil are present but the lobes of stigma don't open to permit the entrance of pollen. The filaments vary in length. Stamens are in two sets of length out of which, those with long filaments are fertile. Pistil consists of short undivided style and stigmatic branches which is nonfunctional. Type III flowers have an ovary which has certain resemblances to both the above-mentioned types. Type-III flowers open for 7-10 days. It is surrounded by 6 to 10 stamens which are similar to those of Type I flowers, has a rudimentary style and stigma and also a nectar disc. Pollination does not appear to occur in these flowers since the stigmatic lobes do not open.

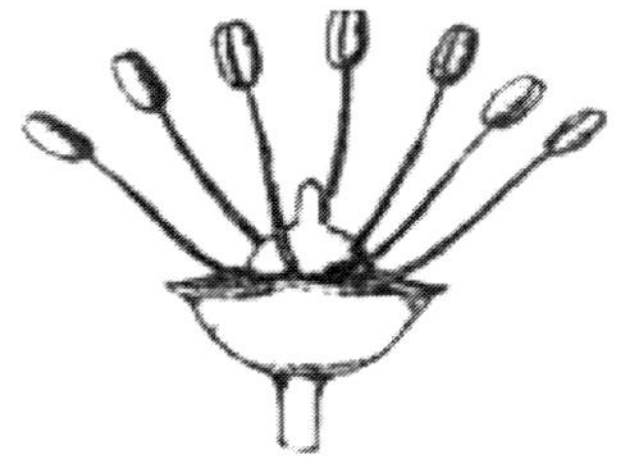

Fig. 5 : M_2 or III Type

The three types of flowers occur in succession on the same tree, usually on different panicle or branches. However, the first flowers to appear on any given tree are may not necessarily be Type-I. There seems to be a tendency in young trees for Type II flowers to open first.

The duration of flowering (anthesis to pollination) in litchi is between 20 and 45 days, depending on the cultivar, season and the location. Floral anthesis occurs both during day and night, with peak opening in the early hours of the morning (6.00 hrs). Temperatures below 8^0C suppress flower opening, but this occurs in the rainy and dry seasons. Under very dry conditions, the young flowers may dehydrate and fail to develop. High night temperatures (approx. 21^0C) reduce the duration of female flower anthesis.

Dehiscence of the anthers begins about one day after anthesis and continues for up to three days, with not all the anthers in a flower dehiscing simultaneously. This takes place continuously, occurring more frequently between 08.00 and 10.00hrs, with no apparent environmental, cultural or genetic effect. The pollen of Type-III flowers are more viable than that of Type-I flowers which are in turn, more viable than that of Type-II flowers. The stigma becomes receptive as soon as it begins to divide into lobes and remain receptive up to 3 days after anthesis. Maximum receptivity (75%) occurs one day after anthesis. When the stigmas cease its receptiveness, they become dry and lose their glossy appearance. Anthers dehisce longitudinally. Pollen produced by the M_2 flowers is most abundant and viable. Estimated pollen viability ranges from 4-40 per cent at the time of pollen release and decreases rapidly thereafter.

(vii) Fruit

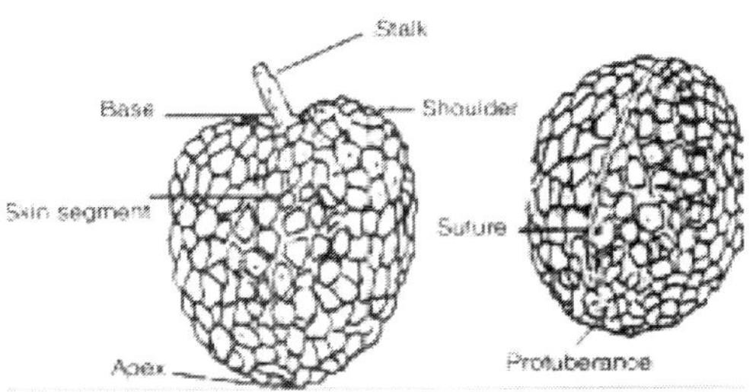

Fig. 6 : External parts of a litchi fruit

The litchi fruit is botanically a drupe, which is pendant in a loose cluster (up to 0.6m long) of several dozen fruits. It may be round, ovoid, heart-shaped or even kidney-shaped. It varies in size depending on the cultivar, but can measure 5cm in length and up to 4cm in diameter. The fruit has shoulder, suture, apex and skin segments which are seen from out side. Different parts of the fruits are as follows.

(i) Epicarp

The rind of the fruit (pericarp) is thin, tough, hard, brittle and frangible. In the immature fruit, it is green in colour but in many commercial varieties it becomes bright red when ripe. Less bright shades of red may also occur and in some cases, the fruit may even be yellow or green. The skin has sharp protuberances which are much less marked than in related species such as rambutan and pulusan. When the fruit begins to dry, the skin changes to a dirty brown colour and becomes much more brittle. This occurs rapidly, even when the fruit is still perfectly edible. Four distinct types of tissues have been noted.

1. The epidermis is composed of rectangular cells with slightly thickened out cells.
2. Hypodermal sclerotic cells have characteristic transverse striations. These cells occur in one to three layers and show the presence of tannin inside.
3. Ordinary parenchymatous cells with slightly lignified walls and dense cell contents similar to those mentioned above.
4. Thin walled parenchymatous cells occurring at the base of the tubercle. Special cells containing reginous or latex-secretion occur mostly at the sides.

(ii) Mesocarp

The mesocarp tissue is somewhat loose in consistency. Any line of demarcation between the epicarp and the mesocarp does not appear as such. In the initial stages, this tissue is composed of ordinary

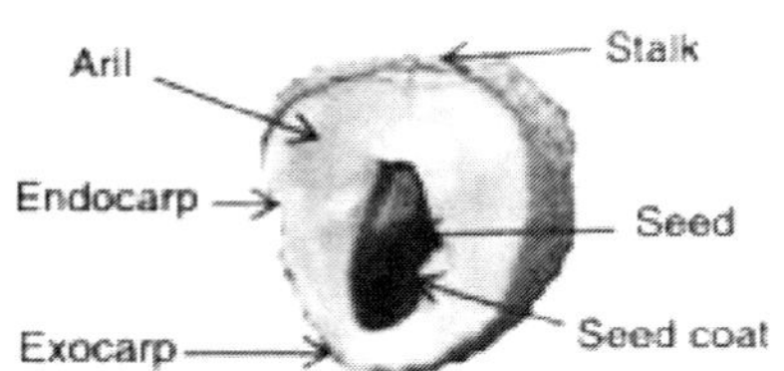

Fig. 7 : Internal part of a litchi fruit

parenchymatous cells which, later becomes separated on account of the development of large intercellular spaces. In botanical terminology, the edible part of the fruit is called aril. When the fruit is developing, the aril grows continuously around the seed from which, it separates as early as it does from the pericarp. The aril contains little material from the cell membranes. The pulp or aril is an outgrowth of the outer cells of the seed coat (outer integument) and in good cultivars; it may comprise 80 per cent of fruit weight. The aril is generally translucent usually white, in some cases verging on pink, translucent, slightly acid, juicy and sweet, with a faint and pleasant aroma. Many cultivars of litchi can be distinguished by their flavour and aroma.

(iii) Endocarp

Two distinct tissues have been noted in endocarp. (i) the outmost one, i.e. next to the aril, is composed of two layers of rectangular cells which are considerably thickened and lignified, (ii) the tissue above it is composed of thick celled elongated parenchymatous cells with tapering walls.

(viii) The Seed

The fruit contains a single dark brown seed measuring 6 to 12 mm wide and 10 to 23 mm long. Some cultivars have a high proportion of aborted seeds (chicken tongued seeds) and thus, a high flesh recovery. The seed is surrounded by creamy and pulpy edible aril. It is cylindrical, compressed piano-convex or concavo convex, ex-albuminous and chocolate in colour. In the seedless cultivars, seed is small, shriveled and non-viable.

Fig. 8(a) : Normal seed **Fig. 8(b) :** Chicken tongued seed

(C) Growth, flowering and pollination in litchi

Litchi is an evergreen plant which produces continuous flushes. The details of growth and flowering in litchi are as under.

(i) Flushing or emergence of new shoots

Flushing refers to the emergence of new shoots on the comparatively older shoots. The flushing enlarges tree size and produces leaves for utilizing sun light for carbohydrates synthesis which supports fruit development in the following season. The intraplant variation in flushing and shoot growth pattern influences the overall floriferousness of the litchi plants. All the shoots with second flush emerging during August and the third flush, emerging during November bear panicle in the month of February. A higher shoot girth and larger number of leaves per unit shoot length in the second flush are crucial for flower bud differentiation. Cessation of growth of second flush before the emergence of third flush results in panicle emergence. The carbohydrate content of previous season flush appears to contribute towards the emergence of third flush and ultimately the panicle as the content decreases during the third flushing. The phenol content of all the flushes increases during the initiation of new flush.

Shoot growth in litchi is not continuous. Generally, there is a rapid period of shoot elongation and leaf expansion followed by a period of leaf maturation, before the next period of shoot growth. In litchi, the duration and interval of growth are strongly related to environmental conditions. Combination of low temperature and high vegetative flush maturity is necessary for floral initiation to occur. Exposure to high temperatures invariably results in the production of vegetative shoots, irrespective of flush maturity. Strong floral initiation is marked by the emergence of terminal panicles and accompanying axillary panicles. A decrease in vegetative flush maturity or increase in temperature results in a decrease in axillary shoot formation and the production of several intermediate shoot structures. These include leafy panicles, stunted panicles, partially emerged buds and non-emergent swollen buds, often produced on the same tree. Both temperature and flush maturity can influence subsequent shoot structure of litchi. Relationship between leaf flushing (*viz.* early, mid and late season) and yield of litchi indicates that out of the three flushes, *viz.* early (after harvest), mid (August to October) and late (after November) season; the early and mid-season flushing influence the yield, whereas, the late season flushing does not have any contribution towards yield. The mid-season flush (appearing in August-October) is of more significance in cultivars Bedana, Bombai and Deshi, whereas, the early season flush (appearing in July) is the desirable vegetative flush in the rest of the cultivars with respect to yield.

(ii) Dormancy or rest period

Generally, the litchi trees undergo a period of rest or dormancy. It is a period without any externally observable growth activity in mature shoots for about 3 - 5 months depending on the variety. This is a pre condition required for the successful flowering. This dormancy is naturally induced by low temperatures in subtropical environments and soil moisture deficits under tropical conditions or may be by both ways under some situations. Low nitrogen supply also promotes this rest in mature shoots. Though, the term dormancy or rest is used to explain this stage of growth, however the plants are not physiologically in a dormant stage. During this period plants photosynthesize very actively and produce carbohydrates for storage which is used for further physiological control of flower bud differentiation and flowering. High temperature, high soil moisture and high nitrogen levels in the soil or in plant tissues might break this rest and force the plant to flush again and again. Induced and untimely flushing affects the physiology of trees by changing the food reserve base and leads to a further vegetative growth. Plants under such conditions produce poor crops during the ensuing fruiting season. This generally happens due to planting of a cultivar which is not suitable for a particular climate. When a variety, with a tendency to have rest under low temperature led climatic conditions is planted under high temperature environments, it does not undergo rest and continues to produce vegetative flushes which hamper its performance in the changed environment. Although much research is required before it can be precisely defined how physiological processes and environment determine floral initiation, it is generally believed that litchi needs a period of vegetative dormancy to initiate floral buds. Apparently this dormancy can be induced by several factors including low temperatures, moisture stress, nutritional factors, cincturing and auxin sprays.

(iii) Flowering

In Indian condition, irregular bearing is seen in China group of cultivars whereas, Shahi group of cultivars are regular bearer and are most preferred and thus, occupy the maximum area. The problem of low and irregular bearing in some of the cultivars like China and Purbi is a major obstacle for its commercial cultivation. This is generally due to failure of plant to initiate flowers, although sometimes the flowers and fruit may be dropped due to adverse conditions. It is therefore, essential to understand the phenomenon and the conditions influencing the flowering behaviour to make the litchi a profitable fruit crop.

The litchi flowers are produced on the present season's wood. The wood arises from buds of shoots which failed to produce flowers in the previous season and from those branches which flowered, but failed to set fruit. Under conditions prevailing in India, all shoots emerging in any month are capable of producing panicles in spring of the following year. Cool temperatures are known to induce flowering, but other factors such as the age of the stems since the previous flush also influence the response. The highest flowering occurs only in those shoots which are 15 to 16 weeks old. Both the timing of cold weather and intensity of cool night temperatures affect the flowering response.

Emergence of the panicle is the first obvious sign of flowering in litchi. After initiation, the primary axis of the inflorescence elongates and primordial of the secondary branches appear. The secondary branches develop similarly, with the development of the different floral parts taking place in axils of bracts of the main axis and branches. Each central primordium enlarges and differentiates into 2-4 hairy calyx lobes, with primordial protuberances develop from the central primordium into the carpel. The vegetative bud, which is typically dome-shaped, flattens and broadens with rapid elevations on both sides of the apex. The development of the different floral parts takes place in acropetal succession, without a dormant period between floral initiation and anthesis.

The future panicle structure, number and type of flowers and their spatial distribution in litchi is set during panicle and floral bud differentiation. Panicles generally appear in two or three flushes, with the terminal branches developing first. Panicles are generally in mixed form and the number of flowers per panicle seems to increase in subtropical climates where low temperatures encourage dormancy. Growth of the panicle is sigmoidal and is usually completed in about 5-8 weeks, with considerable variation in the growth of individual panicles within a tree. Differences exist between cultivars with respect to the length and spread of the panicle and number of floral branches. Some general factors affecting flowering in litchi are:

1. Total duration of relatively low temperature is more important than frequency during the critical period preceding flowering.

2. Moisture stress conditions which influence floral initiation should be present during the preceding autumn and winter and fertilizer (nitrogen) should be withheld during the same period.
3. The response to cincturing depends on several factors including plant vigour, time of cincturing, cultivar and environmental conditions.
4. Exogenous auxins restrict vegetative flushing and induce floral initiation, but results have not always been consistent. The most widely used auxin is sodium naphthylene acetic acid (Na-NAA) at 100-200 mg/l is applied once or more. There is, nevertheless, a limit to the ability of exogenous auxins to inhibit vegetative flushing under constant warm and wet conditions. Growth retardants to restrict vegetative growth tested at different concentrations (1.0-4.0 g/l) of SADH (dimethyl amino succinamic acid), P_{333} (Paclobutrazole) have significant effect on vegetative flushing or flowering but needs confirmation under Indian conditions.
5. An effect of crop competition has been reported where flowering is reduced in trees which have flowered abundantly in the previous year. However, this effect has not always been observed.
6. Genotypic effect on floral initiation has been observed in India, where early variety like Shahi is showing regular trend and late variety China, shows the alternate or heavy crop followed by low crop trend.
7. The above-mentioned factors do not act in isolation. They interact to control the pattern of vegetative development and floral initiation. Many studies have shown that floral initiation is always preceded by a period of induced dormancy in the terminal branches, but whether dormancy is the primary factor, is not known. Both carbohydrates and plant hormones have been proposed as factors controlling floral differentiation in tree crops, but a causal relationship between flowering and starch content has not been established.
8. It has been suggested that floral initiation in litchi may depend on a balance between a growth promoter (flowering inhibitor) produced by young, expanding leaves and a flowering substance (growth inhibitor) produced by mature leaves. This assumption is supported by the fact that vegetative dormancy in the litchi appears to reduce the number of young leaves on terminal branches.

(iv) Pollination

Litchi is a highly cross pollinated crop. As a matter of fact, self-pollination in litchi can occur since functional male and female flowers are present at the same time, on the same or different inflorescences on a single tree but it is generally recognized that litchi flowers are self-sterile and therefore, require insects for pollination. An individual flower is self sterile, but possesses nectar which attracts insects like honey bees, flies, ants and wasp leading to entomophilic cross pollination. A nectary occurs on every flower as a large fleshy crenulate gland within the calyx to which the stamens and pistils are inserted. Nectar, secreted only in the morning, is highly attractive to honey bees and flies. At the time of flowering, installation of honey bee boxes in litchi orchard increases the yield up to 30-40 per cent. Beside that, growers can also obtain honey and can get extra income. Pollinators forage primarily between 06.00 and 12.00 hrs although foraging continues later in the day at much lower levels. Other pollinators are screw-worm, ants, wasps, Coleopterans, Hemipterans, Homopterans and Lepidopterans but it is considered that bees are the most outstanding beneficial insects on litchi. The value of the honey bee is obvious in the setting of litchi fruit. When litchi trees are plentiful, honey bees gather immense stores of high-quality honey. Pollinators' activity is reduced during extremes of temperature, cloud cover, heavy rain, and strong winds and in the presence of insecticides.

Although no specific number of colonies per unit of litchi has been recommended, it has been emphasized that supplying honey bees in litchi orchards is an important and practical recommendation for assuring adequate pollination and fruit-set. Bees should be present continuously throughout bloom. Colonies should be provided at a minimal rate of four per acre.

(D) Fruiting in litchi

(i) Fruit-set

Fruit set in litchi is climate dependent and profoundly affected by temperature and humidity. Reproductive failure is common and not always explained. In some years, certain cultivars produce few or only male flowers and as a result, there is little or no fruit set. This problem can be minimized through the use of better adapted varieties and management practices to retard growth and induce flowering. Winter/spring temperature extremes affecting bloom phenology and unsettled weather limiting bee movement during bloom have been identified as other causes of reproductive failure. Fruit set occurs when pollen, primarily from M_2 flowers, is transferred to female flowers. The respective stages of bloom overlap between panicles and trees, but rarely overlap within individual panicles. There is considerable variability in the number of female flowers per panicle. Cultivars with the greatest number of F flowers per panicle produce the highest yields. Female flowers comprise 10-60 per cent of all flowers produced which depend upon the age of tree, panicle and environmental conditions.

Most of the flowers are either male or pseudo hermaphrodite. Thus, a small proportion of female flowers are produced and a limited numbers of flowers have successful pollination and fertilization. As a result of successful pollination and fertilization, fruit development starts and fruit set is observed on a panicle. With the successful pollination of the flower, a burst of growth of the erstwhile ovary occurs and development of a fruit begins. However, it has been observed that the fruit-set in litchi ranges from 1 to 48 per cent depending upon the cultivar, tree vigour and environment condition. It is also believed that 20 fruits per panicle finally produce a good crop. Except under extreme environmental conditions, pollen production does not seem to be a serious limitation to fruit set, provided there is an adequate pollinating insect activity.

(ii) Fruit growth

Only one of the two ovaries of the female flower normally develops into a fruit. Very rarely, two lobes develop, with the mature fruit superficially resembling two fruit adhering to each other at their bases, each containing a seed. Fruit growth is normally faster when it occurs late in the season during warmer weather. All parts of the fruit do not develop at the same time. During the first seven to eight weeks after fertilization, the fruit skin, the embryo and the seed coat are formed. At the end of this stage, the aril or flesh is only a negligible portion of the fruit. During the next two to three weeks, the cotyledons (or seed leaves) that comprise most of the seed are formed and the development of the aril begins. At the end of this stage, the aril is about third of fruit fresh weight. The final period of fruit growth is dominated by rapid growth of the aril (seed development also continues). At fruit maturity, the aril is about 65 to 75 per cent of fruit weight. Depending on the season and cultivar, fruit take about 12 to 16 weeks to mature.

Maximum fruit weight occurs about two to three weeks before the fruit mature. In most cultivars, the colour changes from green to yellow-red to red with advancing maturity. This change is associated with a flattening of the skin segments and protuberances, and an increase in sugar/acid ratio and eating quality. Fruit weight is related to weather and tree culture, and ranges from 15 to 35 g for different cultivars. Cultivars that have a high proportion of chicken tongue seed normally produce smaller fruit. Some of these cultivars may also produce nearly seedless fruit. These normally weigh only 8 or 10 g. High leaf nitrogen and potassium concentrations and regular irrigation are essential for good fruit yields. High temperatures often accelerate fruit development at the expense of fruit weight. However, at very low temperatures, photosynthesis is reduced. Growth curves are typically sigmoidal for the pericarp, seed and aril as well as for the fruit itself, with final size depending on the cultivar and the environmental factor. Growth regulators (exogenous auxins and gibberellins) have shown positive influence on fruit growth.

(iii) Fruit drop

In litchi, more female flowers are produced than fruit set and far more fruits are set than develop to maturity. Large numbers of fruits are set initially but only few remain up to maturity. Typically, in premature fruit, abscission commences soon after anthesis and continues till fruit maturity, with most fruit abscising in the first two to six weeks. This varies greatly with location, cultivar, weather and culture. In some cases, all of the fruits are dropped before maturity. The initial abscission is mainly due to failure of fertilization. Fruit can also fall due to embryo abortion. The later abscission of fruit is mainly due to competition for assimilates. Girdling at this stage often reduces fruit drop, while drought, shade and leaf removal increase it. Fruit thinning at this time also increases the retention of the remaining crop. The major fruit drop period occurs before the peak demand in carbohydrates by the developing fruit. The young green fruit can photosynthesis, however, most of the carbohydrates for the fruit come from current assimilation in the leaves behind the fruit clusters. Reserves in the branches can also be used. Young leaves do not induce fruit abscission unless they develop directly behind the fruit cluster. This generally occurs when fruit set is poor. Fruit-sucking bugs and fruit borers induce fruit abscission in many orchards. In some areas, they can account for more than 90 per cent of green fruit drop.

Nutritional and hormonal imbalances have been implicated in premature fruit abscission. Experiments have shown that fruit retention can be improved by applying auxins when the fruit weigh about 1 to 2 g. Earlier or later applications are ineffective. Some of these growth regulators can also increase fruit size. Treatment with 2,4, 5-TP in order to prevent abscission and obtain larger fruit succeeded in producing fruits of an excellent colour, with chicken-tongued seeds, but the skins are prone to cracking and the fruits weigh not more than 10g . As with other fruit crops, the causes of fruit drop are numerous which includes, poor fertilization or its failure, non survival of embryo, internal nutritional factors, hormonal imbalance, moisture stress at critical stage, high wind velocity, sudden increase in temperature, in appropriate cultural factors and genetic factors of plant.

Application of micronutrients such as boron and zinc alone or in combination with growth regulators has been found effective for fruit retention up to some extent. There have been no experiments to define the role of environmental factors in premature fruit drop, although it is thought that high temperature, strong winds, extremes of soil moisture and humidity accelerates the process of fruit drop. Little work has been done on the relationship between fruit growth rate and abscission or between abscission and fruit load. The relative contribution of tree reserves and assimilate production to fruiting has not been studied either, although it seems logical to suppose that abscission of the reproductive structures is accelerated by the aforementioned environmental conditions which reduce assimilate production or favour rapid vegetative growth (over-irrigation or fertilization, especially in combination with warm temperatures).

(iv) Fruit yield

The crop load which a tree can bear is primarily determined by the tree reserve level. Excess fruit drop may happen at various stages of fruit development. Even if trees have sufficient reserves, lack of nutrients or soil moisture deficit during fruit development phase or pest and disease attacks on flowers and developing fruits may act as limiting factors that control the quantum of the final crop. Therefore, at this stage trees have to be managed very well providing all necessary inputs correctly. Litchi yields are commonly unreliable and erratic and rarely approach the capacity of the tree. Abscission of fruitlets resulting from self-pollination may occur at high rates in some cultivars, resulting in high levels of hybrid fruit (76-95%) at maturity. Moreover, hybrid fruit are heavier and have larger seeds, although the latter is not necessarily desirable. Well managed mature litchi orchards with usually about 100 trees/ha, can produce as much as 10-15 tones of fruit/ha in a good year. Orchards are usually planted at a much higher density and later thinned to have high yield in litchi. Average yields per tree ranges from 10-80 kg. A 6-year-old tree may produce (0.9-1.4 kg of fruit, while a 12-year-old tree may yield 140 kg. In general, 60-70 kg fruit/adult tree/year is considered an acceptable yield and 120-130 kg/tree on excellent one. These acceptable and excellent yields correspond to a level of 12-13 tones productivity per hectare with normal spacing.

❑❑❑

Chapter 3

Cultivate the Litchi Where Soil and Climate Favour it

Establishment of litchi orchard is a long-term investment. As litchi trees continue to produce fruit for longer period than many other fruit crops, it deserves careful and well thought planning. It has been seen that genetical as well as environmental factors decide the success of crop and the final harvest; however, it requires proper and suitable management practices for its successful performance. Thus, while selecting the site, care should be taken that environmental factors and soil is supporting the litchi crop because any mistake made at initial planning is likely to put the growers to heavy losses throughout the life span of the orchard.

Genetical factors can be looked by selecting the appropriate variety eg. slow-growing, late-maturing cultivars but very little can be done in the way of environmental conditions, except by choosing locations. Therefore, the planting of litchi requires proper planning and careful selection of site. Soil conditions and cropping practices also play an important part in the success of plant, its growth and development and final yield. At a particular place, the available natural resource based planning should be done to cultivate litchi. Therefore, the following care should be taken while selecting the area for litchi orchard establishment.

(1) Suitable site and location

Litchi is a subtropical fruit and needs specific climate, hence, the selection of appropriate location for a litchi orchard is very important. Litchi plantation can only be made commercially viable if it is planted at suitable site. In selecting the site for litchi, pre-requisite factors like soil type, soil depth, soil fertility, drainage, water table, etc. must be carefully examined. Among the factors determining the desirability of a piece of land for litchi growing is the land value as the cost of land contributes towards cost of production. A cheap land may prove expensive in the long run, if the probable expenses in making the land suitable for planting is not considered.

Successful orcharding consist not only production, but efficient disposal of the produce. Location viz. proximity to road, market, transport, storage facilities, cheap source of electricity, water supply, availability of processing plant etc. are important from economic point of view. Almost 25 to 35 per cent of the total cost of the fruit growing is incurred on packing, transportation and other handling charges. The margins of profit can be increased if costs on these items are reduced. The site has direct impact on the growth, bearing and health of trees and location mainly concerns the economic disposal of the produce. Locations away from forest are preferred to avoid menace of wild animals.

The contour of the land affects the system of planting of orchard. It is especially important from the view point of erosion, frost, freezing and desiccation. A slope more than 15 per cent makes the land uncultivable whereas, 1-4 per cent slope may be kept clean without soil loss. Uneven land involves heavy expenditure in leveling, introduces other unfavourable situations and cultural difficulties. It is a strong belief that land slopes are chief cause of success or failure of the orchards. Southern slopes are generally warmer and affect the plants in a manner different from northern slops because of the difference in the quantum of sunshine that the plants receive.

While selecting the site, well developed litchi growing areas popular by name as traditional litchi belt should be of first choice, however, developing areas, potential areas and demand driven areas should also be taken into consideration. It is presumed that traditional litchi belts have all the basic requirements for litchi cultivation and its trade whereas, in newer areas, the selected site should be near to main road and market. It should be situated as close as possible to roads and other transport facilities to allow quick forwarding of fruit to local or foreign markets. Any mistake in selection of site can not be altered after planting while modifications in other factors are possible.

In selecting the location, it is advantageous to have a place where fruit growing is not new since the experience of others, the knowledge of the cultivation practices and availability of skilled labour are really useful. Location with a heavy output attracts large buyers and the transport problem disappears and thus, the profits are greatly increased

Some of the states are well equipped with adequate infrastructure of meteorological data enabling farmers to study climatic conditions in each and every zone. In any case, it is necessary to choose areas which meet best of the conditions ideal for litchi cultivation. There should be easy access to tree canopy from all directions to facilitate sprays and harvesting operations. For the same economic reasons, flat or slightly sloping land is preferred.

(2) Soil and climatic needs

Soil plays a crucial role in nutrient uptake phenomenon of litchi plants. Soil creates a buffer of nutrients which support the plant growth and fruit development. Litchi is very much particular to soil type and good drainage. It can be grown on a range of soils, including alluvial sands, loams and soils with a high content of organic matter and pH. From the physical viewpoint, litchi does not require as deep soils as do mangoes. It can be adequately grown in soils with 40cm depth, and even in rocky and calcareous soils. There can also be problems on very sandy soils that dry out during hot weather and on calcareous soils with potential iron, zinc or manganese deficiencies. These soils need to be carefully managed. Trees perform best on well drained clay loams of medium to high fertility with a minimum one meter of well drained topsoil. In general, litchi prefers deep alluvial soils, containing sufficient amount of organic matter but it could be detrimental to flowering in the adult phase. Litchi nevertheless adapts to soils of different textures like sandy, loamy, clayey-sand, clayey-loam and even soils with 40 per cent clay content, but the heavier soils requires better drainage.

Slopes more than 15 per cent are avoided as they do not allow the safe use of machinery and prone to erosion. When trees are grown on terraces, the planting site is generally filled with quality loam and organic matter to improve soil structure and fertility. Soil type influences total root density and feeder root distribution (depth of the soil where 80 per cent of roots are located). Though some roots are found below 300 cm in a deep calcareous sandy loam, most roots are located in the top 45 cm. A tremendous network of roots up to one meter has been observed in sandy soils while trees growing in clay soils have a shallow root system. The deep roots are, however, capable of absorbing enough water during the dry season to support a large crop. Soils with a low water retention capacity and a low fertility level are important where commercial production is concerned since, in such soils, vegetative growth is easily checked which has a favourable effect on flowering and fruit production. This is why light soils are preferred for very vigorous varieties but are not so essential for other less vigorous varieties which can produce adequately under a wide range of soil conditions. From the chemical viewpoint, litchi does not require vary fertile soil. For commercial production, it requires a high nitrogen level and a moderate level of phosphorous, potassium and calcium. Many poor soils contain adequate quantities of magnesium and micro-elements which are suitable for litchi.

Soil should be examined for suitability in respect of depth, drainage and compacted layers. It should preferably be 1 to 2 m deep. Once it has been determined that the soil is suitable from this point of view, a soil analysis should be done six months prior to planting to determine the quantities of fertilizer which will have to be applied sufficiently in advance. Preparation of the soil according to the results of the soil analysis is needed, especially when large quantities of lime are required. For successful litchi production, some soil related issues such as pH, wetting and water table are important.

(i) High wetting is harmful

A constant water table is ideal for good growth and development of litchi crop. Litchi can withstand up to 14 days of immersion, provided the water does not become stagnant. It may die on heavy clay soils if water becomes logged. Trees subjected to continue flooding found to be dwarfer than those on better drained soils. Poor drainage in heavy clays can increase the incidence of collar rots and root diseases. Hilling of the soil along the rows to give ridges of 0.5 m high is recommended in wet sites. The addition of drainage can also assist growth in wet soils.

(ii) Avoid abnormal soil pH

There is little information on the response of litchi to excess salts as data on salt tolerances are scarce and contradictory. Some reports indicate that litchi tolerates the presence of salts in its root area well, even better than mango whereas, others state that salt tolerance is practically nil. However, litchi appears to be less sensitive than other fruit crops, still it is in the low tolerance class of plants. It is recommended that trees should not be irrigated with water having an electrical conductivity greater than 0.5 dsm^{-1} m or about 500 mg soluble salts per litre. Sometimes, plant may get damaged during dry weather under high salt condition especially when young trees are over-fertilized. The tips and margins of the old leaves may burn due to toxicity.

Acid soils with pH of between 5.5 and 6.5 are generally considered as ideal but litchi tolerates soils with a pH of less than 5.5 easily. They may also be planted in alkaline soils with a pH of up to 8.5 subject to the adequate supply of micronutrients externally. Nutrition has to be carefully managed on these soils to avoid deficiencies of micronutrients such as iron.

Most growers aim at a pH between 5.5 and 6.0, although lower pH is probably acceptable with slight pit management. Nutrition management, especially the application of micronutrients needs to be modified at extremes of soil pH. No more than 5 tones of lime per ha should be applied in a single application on sandy soils. Where more lime is required, a second amount should be applied three months later. Dolomite can be used instead of the lime, if soil magnesium concentrations are low.

(3) Suitable temperature and moisture

Litchi is found in a narrow range of climates in comparison to many other sub-tropical fruit such as citrus and guava, etc. There are few critical studies on the response of the litchi to weather. These have indicated that under field conditions, sunshine, temperature and soil moisture content are often correlated with the plant performance. Temperature, soil moisture and atmospheric humidity for different stages of litchi have been described. The main commercial plantings of litchi are found at low elevation in the sub-tropics. These areas have cool to cold winters and warm to hot summers. Rainfall is highest in summer and least in winter or spring. The climatic factors which encourage floral initiation in litchi are low temperatures and moisture stress before flowering. However, these two factors do not alone determine floral initiation and there have been cases of profuse flowering where they were not present. Other factors, particularly nutritional and hormonal factors also play an essential role in flowering of litchi. On the other hand, high relative humidity and high soil moisture content are other vital factors during the flowering and fruit setting stages, but rain or fog at this time can cause serious damage by reducing pollination. Low humidity can have a number of adverse effects such as dehydration of young flowers, incomplete pollen germination and abortion of the embryo. Temperatures at flowering must be moderate so as not to jeopardize anthesis. Low and high temperatures both during flowering and fruit set can cause the young fruits to drop and the embryo to abort.

(i) For growth and flowering

Adult trees suffer serious damage and may be killed at -4^{0}C even though they may have withstood temperatures of -6^{0}C for a few hours in the resting state with only slight damage. Freezing temperatures below-1^{0}C can cause serious damage to tender shoot and young trees are usually died. It has also been found that layered plants are more susceptible to cold in their early years than plants grown from seed. It seems that the shallower root system of layered plants is more easily damaged than the deeper root system of those grown from seed. Below freezing temperatures and hot, dry winds kill inflorescences. When the terminal buds are frosted just prior to panicle emergence, a large number of axillary panicles develop. When the terminal panicle is destroyed by physical or chemical means, even after fruit set, a second axillary flowering occurs, provided that temperatures are low so as to stimulate floral initiation.

Litchi also tolerates relatively high temperatures during growth and flowering. The normal day temperature limit during the most sensitive growing phase is 40^{0}C whereas, it grows satisfactory well under temperatures range from 20^{0}C to 35^{0}C; the optimum being 30^{0}C. Below 20^{0}C, growth is slowed down significantly and below 15^{0}C or 16^{0}C, vegetative activity ceases. Finally, widely fluctuating day temperatures in sub-tropical climates seem to contribute to vigorous vegetative growth.

Litchi flowers best with day temperature below 20°C. Temperature governs the flowering time intensity and duration, sex ratio, fruit set, fruit growth and development and ripening time of litchi fruits. Total duration of relatively low temperatures has been considered more important than frequency. Therefore, it is concluded that a cold, dry winter favours flowering. It has been demonstrated that winter and autumn temperature of between 0^0C and 13^0C coupled with a moisture deficit induce more profuse flowering in seasons where there have been below 13^0C temperature for about 200 hours.

High temperatures reduce both the duration of flushing and the interval between flushes. Temperature also affects the rate of reproductive development with panicles emerging earlier but taking longer time to reach anthesis. Higher numbers of female flowers are associated with an average temperature of 18^0C during early flower development and the lower numbers of female flowers at the average temperature of 23^0C. In contrast, the rate of flower opening is related to the number of flowers per panicle. Therefore, it has been concluded that the areas with winter temperature maximum of above 25°C are not well suited for litchi culture.

Temperature has been shown to have strong effects on pollination, but these responses do not necessarily translate into better fruit production. Maximum fertilization occurs when the number of pollen tubes per ovary does not increase with time after fertilization. Pollination is optimum at 19^0 to 22^0C, with maximum fertilization obtained after seven days. At 15^0C, pollen tube elongation is strongly inhibited. However, from 15^0 to 27^0C, at least 10 per cent of ovules contain pollen tubes indicating that they are fertilized. Such a level of fertilization appears sufficient for most cultivars to produce a high yield, although at 33^0C, all female flowers abscise, suggesting a limitation for good yields when days are above 30^0C for long periods.

(ii) For fruit development

The relationship between fruit set and weather is not well understood. However, continuous hot, dry conditions may reduce yields. Bagging can improve fruit quality, possibly due to cooler temperatures and higher humidity. Moderately high temperatures and humidity are important during the fruit development. Moisture stress causes cracking and reduces the final weight and quality of the fruit. In some cultivars, it prevents the seed from being totally covered by flesh. If moisture stress is accompanied by high temperatures, sun-burning results and the fruit is totally ruined. It is also worth mentioning that excess rainfall or humidity during the ripening phase can cause cracking.

(4) Drought has adverse impact on plant and fruit development

Drought during fruit development generally reduces production but it induces flowering. Shoot growth is very sensitive to changes in tree water status as its growth decreases as the level and duration of drought increases. A period of drought before flower induction may assist flowering by delaying early shoot growth until winter. Once flower panicles are initiated, best fruit set is achieved when plants are well watered. Most of the flowers abscise prematurely in drought hit plants and the few flowers that reach anthesis become male. It indicates that trees should be irrigated from panicle emergence to prevent water deficits reducing fruit set, although threshold level of moisture below which production is affected has not been worked out precisely.

There is very little information on the response to irrigation during fruiting. The results indicate that there may be different effects on fruit production depending on the level and timing of the water shortage. Drought reduces the number of fruits per tree, average fruit weight, aril recovery and yield. Fruit splitting or skin cracking led by drought or temperature fluctuation at fruit development stage is a serious problem where some times 50 per cent or more of the crop may be lost.

(5) Intensity of solar radiation has impact on plant development

Litchi originated as one of the dominant species in sub-tropical rain forests of Asia. However, as with many crops, the original environment may not be ideal for its commercial production. Solar radiation is the source of photosynthesis and its intensity varies with the location of orchard and plant density. Both flowering and fruiting are reduced once adjacent trees start to touch each other and therefore at this stage, thinning/pruning becomes necessary. The decline of yield in crowded orchards begins when a portion of the canopy are shaded for most of the day. Light may limit flower development although high sunshine hours are expected to be correlated with higher temperatures and therefore early anthesis. The reduction in fruit set during low sunshine hours or cloudy weather could be due to lack of assimilates for flower development, but is more likely to be related to a direct effect of rain on the anthers or stigmas. Overcast weather may reduce bee activity. Early flowering fails because of cool, overcast weather during fruit set. It has been reported that more than 75 per cent of terminal branches flower, even if the plants are shaded several months before flowering. Average seasonal changes in light would not be expected to strongly influence flowering, unless overcast weather persists for several weeks. Heavy shade for one week increases fruit drop. Thus, it is inferred that, solar radiation is essential for litchi.

The photoperiod also has influence on floral initiation. As it reduces from 16 to 8 hours, there is an increase in Type-II flower anthesis. Although it is generally thought that bright light is essential for commercial production but sometimes 9 to 15 hours of sunlight per day has no effect on production. Shady conditions for young plants nevertheless seem to favour initial growth and it has been common practice in Israel to grow litchi under banana plants at least for the first two years. It is quite obvious that in addition to shade, the banana plant also provide protection from strong wind which is harmful to litchi.

(6) Rain fall pattern affects the tree productivity

Litchi needs relatively more water. It is generally observed that abundant rainfall or irrigation and the resulting high soil moisture level encourages vegetative flushing. The optimum rainfall level lies between 1250 and 1700mm. In India, there are areas where annual rainfall is less than 1000mm but litchis are cultivated on a commercial basis. The adult plant of litchi is drought-resistant and can survive 4 to 12 weeks without water. Even though a high rate of relative humidity favours growth, it has been shown that except in extreme conditions, this is not always an important factor provided that there is an adequate supply of water for irrigation. On the other hand, recent research in Australia has shown that even under high soil humidity conditions, litchi appears to suffer moisture stress on hot, dry, windy days of low relative humidity.

(7) Frost prone areas are harmful

Frost is common feature in northern and hilly areas of India. The temperature below 1.1^{0}C in the night, bright sunshine during day and sharp drop in evening temperature with no wind are the pre-condition of frost. The frost affects litchi plants adversely. The extent of damage by frost depends upon age of the tree, moisture content of the soil, condition of growth, actual timing of frost occurrence, severity and duration of the frost. Well grown up trees with mature wood are less affected than young trees which are immature and in active growth phase. Level of soil moisture affects the extent of frost injury. Moist soil or irrigated soil raises the soil temperature and provides protection against mild frosts therefore, trees of such growth and age growing under dry soils conditions are severely damaged by frost than to those growing in wet land.

(8) Hail storms destroy the crop and plant

The hailstorms which often occur sporadically are main disasters during flowering and fruiting in litchi. The ice pallets destroy the panicle, developing flowers and small fruits. The damage caused to larger fruits is by physical hitting on the fruit surface which leads to cracks in skin, rupture of aril and rotting of pulp. Secondary infection by fungus starts some time after the physical damage. Affected fruits are unfit for consumption. Netting of individual plant is only method to protect hail storm damage.

(9) Orchards in high wind velocity areas need special attention

Hot/cool and strong winds are harmful to litchi crop. In fact, high velocity of wind affects litchi trees in several ways. High wind velocity leads to quick evaporation of water from the soil and thus, reduces moisture availability which is very much necessary for ideal growth and development of the plant. Litchi trees normally assume a dome shaped canopy and are comparatively less prone to damage of limbs and branches due to wind however, owing to week attachment of peduncle, a large amount of fruit drop takes place due to wind. Strong winds (more than 120 km/hour), gales or storms blow away the branches, leaves or panicles and sometimes fully grown trees also get toppled. It is more so in the areas adjacent to open areas that are not fully protected with windbreaks. It prevents flowering on branches which are exposed to constant buffeting, burns inflorescences, causes flower and fruit shed and its drying effect leads to skin-cracking.

Wind damages young shoots by way of burning especially when it is accompanied by high temperatures, breaks branches and uproots trees. Wind also leads to a loss of vigour, and consequently, slower growth. Fruit drop at initial stages of its growth can be noticed due to high velocity of wind. Various measures may be taken to limit the damage but the most recommendable is perhaps to tie the main branches with thick, rubber insulated wire and the raising of thick and strong wind breaks around the litchi orchard.

(10) Ideal conditions for commercial litchi production

The key factors to consider when assessing the potential of different areas for litchi are temperatures in winter that affect flower initiation, temperatures and light levels in spring which affect fruit set and reliability of rainfall which affects fruit development. Normally temperatures below 20°C induce flowers, while flowering is irregular at higher temperatures. A short drought period in winter assist flowering, especially in the more tropical cultivars, but is not essential. Annual rainfall of 1,200 to 1,500 mm is probably required in the absence of irrigation. The other critical part of the crop cycle is fruit set that is reduced when temperature falls below 20°C for extended periods during flowering. Persistent cloud cover at flowering time can also be a problem. The ideal condition for litchi cultivation is:

- A frost-free climate.
- Absence of strong wind.
- A cold (minimum temperatures between 8°C and 14°C), dry period prior to flowering (autumn and early winter).
- Moderate rainfall and temperature during flowering.
- High temperature and humidity during fruit setting.
- Moderate temperature and humidity during maturation.
- Deep, well-drained, non-saline calcareous soils with proper texture, fertility and high organic matter.

❑❑❑

Chapter 4

Produce and Use the Quality Planting Material

The cost of self propagated sapling is less than purchased one. Further, self propagated planting material ensures the genuineness and truthfulness of the variety. A farmer, if wish to reduce the cost on sapling, can prepare plants by himself provided that he has expertise. Nursery business is a very lucrative aspect. However, a nursery operator should know the basic and practical aspects to be able to produce good quality planting materials as it is the key to economic viability of commercial nursery. Important factors considered for establishing a nursery are agro-climatic conditions, soil types, soil pH, location, area, irrigation facilities, communication, market demand, availability of mother plant and skilled persons, etc.

1. Knowledge of the nursery Act

Most of the States have now enacted the Nursery Act. This act varies from state to state. Nurserymen should know about it. First, nurseryman has to obtain license from State Government then he should procure mother plants from ICAR institutions/Agricultural University or Govt. nurseries. He has to label mother plants and grafts layers as these nurseries are to be inspected by Government agency.

2. Have propagation structures if large volume is to be prepared

Reducing or managing mortality is a key to the nursery business. Like other business, nursery also suffers from business risk. This includes competition, variable input costs and uncertain returns. An extra problem is environmental risk causing uncertain levels of production, difficulty in forecasting when one will have product to sell and even period when prices are likely to be higher. In order to produce more and maintaining the stock, some structures are needed.

Polyhouse is a transparent or partially transparent material supported by a structure to enclose an area for the propagation of plants. Specifically, where the covering material is glass, the structure may be referred to as a glasshouse. A green house or polyhouse refers to the use of plastic films or plastic sheeting. When the enclosing material is woven or otherwise constructed to allow sunlight, moisture and air to pass through the gaps, the structure is known as a shade house. A better quality product can be achieved by eliminating adverse environmental conditions using a polyhouse to provide an optimum growing environment for nursery plants.

Classification of a polyhouse is done according to its basic shape. Types include Gable, Flat arch, Raised dome, Saw tooth, Skillion, Tunnel etc. There are categories based on investment and budget.

Low cost polyhouses are less than 3 meters in total height. Tunnel houses are the most common type. They do not have vertical walls. They have poor ventilation. This type of structure is relatively inexpensive and easy to erect. Little or no automation is used. While this sort of structure provides basic advantages over open field nursery, its potential is still limited by the growing environment and management is relatively difficult. Low level poly houses generally result in a suboptimal growing environmen and does little to reduce the incidence of pests and diseases. Low technology polyhouses have significant environmental limitations, but they offer a cost effective structure to nurserymen.

Medium technology polyhouses offer a compromise between cost and productivity. Medium level polyhouses are typically characterised by vertical walls more than 2m but less than 4 meters tall and a total height usually less than 5.5 meters. They may have roof or side wall ventilation or both. Medium level polyhouses are usually clad with either single or double skin plastic film or glass and use varying degrees of automation. Medium level polyhouses represent a reasonable economic and environmental basis for the nursery. Plant propagation in medium level polyhouses can be more efficient than field production. There is greater opportunity to use non-chemical pest and disease management strategies.

High technology polyhouse offer superior nursery performance. High level polyhouses have a wall height of at least 4 meters, with the roof peak being up to 8 meters above ground level These structures offer superior environmental performance. High technology structures have roof ventilation and may also have side wall vents. Cladding may be plastic film (single or double), polycarbonate sheeting or glass. Environmental controls are almost always automated. These structures offer enormous opportunities for economic and environmental sustainability. Use of pesticides can be significantly reduced. High technology structures provide a generally impressive sight and internationally are increasingly being used in agribusiness opportunities.

Although these polyhouses are capital intensive, they offer a highly productive and environmentally sustainable opportunity. Polyhouses provide a shelter in which a suitable environment is maintained for plants. Solar energy from the sun provides sunlight and some heat, but it should be provided a system to regulate the environment. This is done by using fans and cooling pads.

Ventilation is the exchange of inside air for outside air to control temperature, remove moisture, or replenish carbon dioxide (CO2). Several ventilation systems can be used. Natural ventilation uses roof vents on the ridge line with side inlet vents. Warm air rises on convective currents to escape through the top, drawing cool air in through the sides. Mechanical ventilation uses an exhaust fan to move air out one end of the polyhouses while outside air enters the other end through motorized inlet louvers. Exhaust fans should be sized to exchange the total volume of air in the polyhouses each minute.

Air movement by ventilation alone may not be adequate in the middle of the summer; the air temperature may need to be lowered with evaporative cooling. Also, the light intensity may be too great for the plants. During the summer, evaporative cooling, shade cloth or paint may be necessary. Shade materials include roll-up screens of wood or aluminum, vinyl netting and paint.

Automatic control is essential to maintain a reasonable environment in the polyhouses. On a winter day with varying amounts of sunlight and clouds, the temperature can fluctuate greatly; close supervision would be required if a manual ventilation system are in use. Therefore, unless close monitoring is possible, commercial operators should have automated systems.

A water supply is essential. Hand watering is acceptable if someone is available when the task needs to be done. A variety of automatic watering systems is available to help to do the task over short periods of time. The small polyhouse is likely to have a variety of plant materials, containers and soil mixes that need different amounts of water. Time clocks can be used to control automatic watering systems. Mist sprays can be used to create humidity or to moisten seedlings.

Shade net houses

Shade houses are structures which are covered in woven or otherwise constructed materials to allow sunlight, moisture and air to pass through the gaps. The covering material is used to provide a particular environmental modification, such as reduced light or protection from severe weather conditions. The height of the structure may vary according to the type of crop being produced and may be as high as 4 meters.

3. Selection of site for nursery

The site selected for establishing a nursery should preferably be located near the production and marketing centers for the convenience of publicity and transportation of the plants with minimum or no damage. The site should be convenient enough for transportation of input materials also. It is important to have or develop a perennial source of water inside the nursery. If need arises, wind breaks of tall and useful plants like neem, tamarind, aonla, seedling mango, etc. may be planted to provide necessary shade and protection.

4. Selection of varieties

The varietal choice primarily depends on the market demand in nearby areas. For wider market coverage, the choice may be dependent on market studies in the desired areas.

5. Establishment of structure / materials and training to workers

The required structures and facilities include a nursery shed, input storage room, potting area, a compost pit, a water system which involves a tube well, a pump, pressure tank and fittings. The work shed with thatch roofs and locally available materials like bamboo, wood, etc. may be constructed. For potting during rainy season, the soil and compost should be stored during hot or summer months as in rainy season, collection of dried soil and manure are very difficult and without these, the seedlings cannot be raised. The newly cut air layers are planted in 350 ml root trainers for 45-60 days followed by well compensated and filled poly bags for better growth or hardening of the plants. In general, this is done under partial shade. A store-cum-office of 6.0m x 4.5 m constructed with locally available materials may also be needed. Fencing is needed to protect plants from animals.

Equipments and tools such as pruning secateurs, garden rakes, khurpa (pvc and wooden grip), bill hook, garden saw, garden sword, sprinklers with heavy duty stand and garden tool kit, pickaxe, watering can, crowbar, mugs, trowels, large funnel wheel barrows, shovels, pruning shears, grafting knives and knapsack sprayers, root trainers with stand are needed for success of nursery. The other consumables materials include black plastic bags, urea, insecticide, fungicide, river sand, bamboo poles, plastic bags, garden soil, compost and poly feeds.

Labour force is needed for land clearing and leveling, layering operations, plastic bagging of plants, watering and applying fertilizer, weeding, spraying, removing from plastic wrap and covering living graft/layers. Hence, they should have some basic idea of nursery activities. A well trained Mali can take care of technical aspects of litchi nursery.

Adopt proper method of propagation

1. Propagation by seed

Litchi can be propagated by seed and vegetative means, however, for commercial orcharding, vegetative methods (layering and grafting) are preferred. Seeds are used for raising root stocks to raise the grafted plants.

Litchi seed is recalcitrant in nature and therefore, losses its viability in a very short period if it is exposed to the air in the shade under normal humidity conditions. The seed begins to shrivel within 24 hours and in five days it becomes incapable of germinating. Seeds can be preserved for up to 8 weeks between 2-2.5cm thick layers of wet sphagnum moss or, for somewhat shorter periods, if wrapped in peat in a refrigerator. They may also be stored for at least a month in the shade, in closed patri dishes dusted with captan at temperatures of 15^{0}C to 25^{0}C, without losing their germinating capacity. Another method of preservation is to leave the seeds in the fruit. This prevents them from drying and preserves their viability for at least 3 to 4 weeks. Eighty per cent of fresh seed germinate after three weeks, provided soil water and aeration are adequate. Large seeds germinate better than small seeds and also have stronger growth initially. In contrast, chicken-tongue seeds are not viable. Growth is usually better with organic mixes, acid pH and inoculation of mycorrhiza.

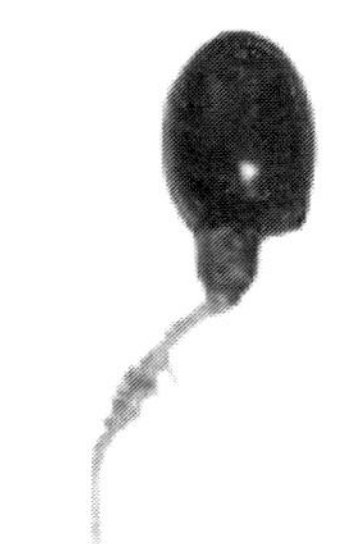

Fig. 9 : Germinating litchi seed

After separating from the fruit and cleaning properly, the seeds must be sown horizontally to a depth of 1-2.5 cm in a well-drained sowing medium in partly shade, well irrigated location. The sowing medium may be either sand or various mixtures of sand, vermiculite, soil and compost provided that they offer good aeration which is considered essential for seed germination. Incorporation of soil from old plantations into the sowing medium usually encourages germination because of the positive effect of the mycorhizal presence. This practice is not always recommended since the soil may contain pathogens, nematodes, etc., which could attack the root system. The compulsory disinfection has to be done in such cases. Germination should commence within three days, under normal conditions. Seeds should be sown in trays and transplanted into individual bags when the young plants have reached a height of 10-15cm. Only mature seeds must be used for propagation purposes. For obvious reasons, small seeds in which the embryo has aborted must never be used for sowing. To facilitate transplanting, the seeds should be sown at a distance of not less than double their length and depth double to their width. The seedlings should also be watered regularly and protected from biotic and abiotic stresses. Temperatures from 25°C to 30°C with high humidity are ideal for litchi seed to germinate.

Litchi propagation from seed is unsatisfactory, not usually recommended, because varieties do not reproduce true-to type from seed and have poor to average fruit quality. Seedling trees often take more than 10 years to come into bearing. This type of propagation is, therefore, used only in research work or to produce rootstocks. New cultivars can only be developed from the selection of seedlings with improved characteristics. New cultivars might bear more regularly, earlier or later than existing cultivars. They might also have larger fruit, brighter skin or smaller seeds. Seedlings that are to be evaluated as potential new cultivars can be planted out after a year. They are usually planted closer than in commercial orchards preferably, one meter apart.

2. Propagation by cuttings

Spring and autumn season are the best time for planting the cuttings. The semi-hard to hard wood cuttings with two leaves, measuring approximately 13-20cm in length and 15mm in diameter are taken from young and vigorous trees. Different growth regulators may be used to promote rooting. Good response in cuttings can be obtained by over night dipping in 100-200ppm IBA. 2000 to 5000ppm of indolebutyric acid (IBA), indoleacetic acid (IAA) and nephtalenacetic acid (NAA) through quick dip methods are the most successful growth regulators. The basal ends of the cuttings are usually dipped in the hormonal powder or in a solution of auxins for a few seconds. In some case, a fungicide (bavistan) should also be added to the growth regulator.

Once the cuttings have been prepared, they must be placed in a propagation bed, under mist. The medium should be as loose as possible and preferably temperature should be maintained between 30-32^{0}C. Because litchi rooting is a slow process, the cuttings must remain in the rooting bed for 3 to 4 months. Only when a good root system has been formed, they can be transferred to bags. Once transplanted, they must remain in a greenhouse for 15-16 months before finally being planted in the field. If cuttings are few in number or are being prepared for experimental purposes, it is advisable to place each one in a small, transparent plastic bag so that progress can be easily observed.

Attention must be drawn to the fact that genetics plays an important role in rooting success of cuttings. Because of the slow development of callus tissue, the slow-growing cultivars do not root well. Success also depends on selecting the correct type of wood, misting or fogging, and good temperature control in the propagation house. The rooting media also needs to be free draining. Some cultivars can provide 80 per cent success rate. Young plants of about 50 to 60 cm height are recommended for planting because smaller plants often die in the field.

Sixty to eighty per cent of semi-hardwood and hardwood cuttings get rooted with shoots collected from older wood behind the soft tips prior to flowering. Soft terminal cuttings are unsuccessful therefore, should be avoided for cuttings. Better results are obtained if the shoots are girdled a few months before taking the cuttings because of accumulated Carbohydrates in the above girdle portion.

Although, on some occasion, rooting has been 100 per cent successful under 100 per cent humidity conditions but litchi propagation by cuttings does not always have the desired results. The method calls for expensive, complex installations and the plants produced are weaker and root systems usually less developed than those obtained by layering. Because of this, propagation by cutting is used for experimental purposes, but not commercially.

3. Propagation by layering

i) Air layering

Air-layering or 'gootee' is widely accepted method of litchi propagation. This is the most widely used propagation method and the one, which gives the most satisfactory results. Its major advantages are that it is simple to use and genetically identical plants are produced. The most serious drawback of air layering is the damage to the parent plant if a large number of layers are required and poor survival in the nursery after shifting of layers. Plants prepared by this method are delicate and difficult to transport.

Air-layering is done when leaves of the previous growth flush have proper maturity. Even though, it has been reported that the better the branch used, the better the root system obtained. Excellent results can be obtained with branches of 10-25 mm diameter and 46-60cm length. These shoots produce rooting rate above 90 per cent and damage to the mother plant is also minimal.

Air layering can be done at any time of years as long as there is sufficient moisture, but best results are obtained in rainy and spring season. Branches should be selected on the periphery of the trees, so that they can easily be worked on. The selected branch should consist of a single stem and other stems must be removed. Preference should be given to branches that are erect and in satisfactory physiological conditions i.e. the last vegetative flush must be well advanced.

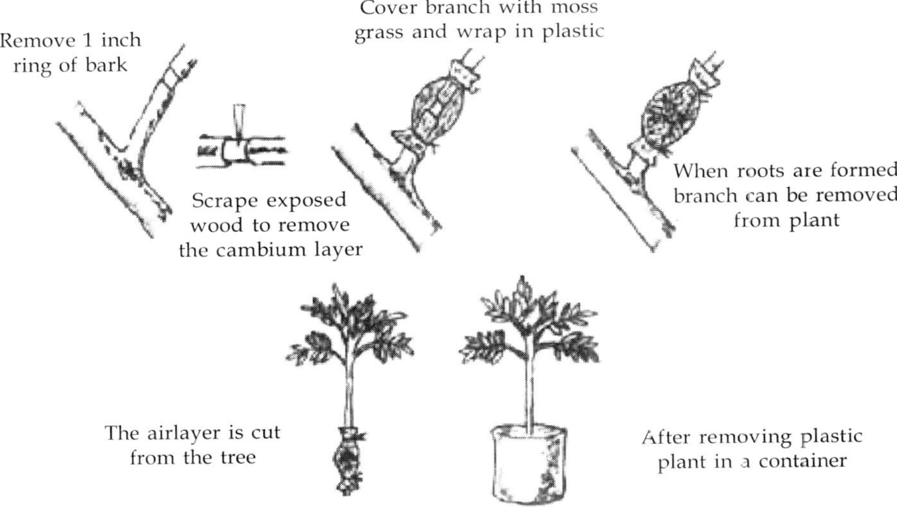

Fig. 10 : Method of air layering

A ring of bark measuring 2.5 cm width is removed from the branch at the point where root formation is desired. The thin cambium layer beneath the bark is scraped away. For best results, this ringed area should be left exposed for few weeks until callus tissue begins to form at the end toward the branch tip. If air-layering is done too soon after branches are ringed, rooting often fails. This is because freshly exposed cambium cells may continue to divide and overgrow the ringed area, thus inhibiting rooting. An ordinary knife or clippers may be used for this operation. The ringed area should be surrounded by a layer of moss of about 2.5cm thick and 10cm long. The traditional method used soil, organic matter, sawdust and woodchips wrapped in cloth to enclose the rings. However, moist peat moss and polyethylene bags are now used in many areas (Fig. 10). The use of the plastic alleviates the problem of daily hand watering. A medium consisting of 100 per cent peat moss and limed to a pH of 6 is ideal. Auxins sometimes improve rooting, but are not essential.The plastic wrap ends are tied tightly around the branch with string and sealed with plastic tape to keep rainwater out.

Under adequate conditions (temperatures 25-30^{0}C), two to three months after layering several roots are formed and are visible changing colour from white to creamy brown through the plastic film. The air-layered branch is then cut off immediately below the plastic. The wrapping is then removed and the rooted air-layer planted in a suitable container. Special care must be taken to avoid damage to the young root system while transplanting. Use of root trainer and organic matter rich rooting media has been found to promote the secondary and tertiary root in the litchi air layers which improves the establishment. Application of super phosphate in rooting medium of root trainer also improves root growth. The planting must not be done in dry weather. It is also necessary to reduce the foliage on the plant by about half. The young plants must be placed in individual bags of 18x15x35cm filled with a well drained, dense mixture, which must be carefully packed around the bag containing the roots. The trunk and rooted area should not be buried more than 2.5-5cm below the level of the soil in the container. Containers with newly transplanted air-layered plants should be placed in a shaded area for about two weeks with a lightweight plastic bag placed over the plant to retain humidity until the plant begins to put out new growth. Later on, the plants can be gradually exposed to full sun to "harden." Two or three vegetative flushes must appear before the plants can be hardened off in the open air prior to final planting. The whole process, from the start of layering operations to transplanting into the field takes between 12 and 18 months. Transplanting to the field is best done during a rainy season, but if this is not possible, the plants should be watered every two to three days until well established. Removing about half of the leaves at the time of planting in the container and again when transplanting to the ground will prevent excessive moisture loss. Major bottleneck associated with this method of propagation is the high mortality of layers after cutting them from the mother plants and establishment in the nursery on their own root systems. Time of layering plays an important role in rooting and production of quality layers, because root induction during layering may be associated with the particular physiological conditions in the stem.

Year Round Production of Litchi Plants under Polyhouse

The following steps should be followed for the year round production of litchi under polyhouse.

- Establishment of mother block under controlled conditions at 2 x 2 m spacing accommodating 125 plants in 500 m^2 polyhouse.

- Layering operations are started on atleast two year old mother plants.
- This operation will increase the no. of shoots for further layering and after each operation, number of shoots will be available in multiples.
- Then air layers are detached from mother plants.
- After removal of polythene strip, air-layers are shifted to 300/ 500 cc. root trainers in shade net or polyhouse.
- Soil, FYM and sand should be in equal ratio for filling of root trainers.
- Then after 2 months, these plants are transferred in 1 kg capacity polybag and kept in polyhouse.
- Air-layered plants are ready in 4-6 months for planting in polyhouse.
- To harden, these newly developed plants are put in net house for 15 days.
- For year round propagation of litchi plants, the above steps are employed two times in a year. In this way under polyhouse, higher number of plants is produced regularly throughout the year in comparison to open field on per unit area basis.

Economic Assessment of Polyhouse (500 m^2) for Multiplication of Litchi Plant/Year

A. Fixed cost

i.	Rental value of land for 500 m^2 @ 10,000/acre/year	=	Rs. 1250.00
ii.	Depreciation on polyhouse (cost Rs. 8,00,000)	=	Rs. 16,000.00
iii.	Interest on fixed capital @ 12%	=	Rs. 96,000.00
	Total fixed cost (I + II + III)	=	Rs. 1,14,500.00

Table 8 : Operational cost/year (litchi air-layers)

Operation	No. of plants	Cost involved Rs.
Air layering	80,000	---
Rooted air layer obtained	70,000	---
Air layer raising (plastic sheet + sphagnum moss + sutali + labour)	70,000	2,80,000
Rooted layer shifted in root trainer	70,000	70,000
No. of survival layers in root trainer	68,500	---
No. of survival layers in poly bag and maintenance	68,500	1,37,000
Labour charges	—	25,000
Electricity charges	—	25,000
Total cost of operation	—	5,48,000

Total cost involved/year = A + B = Rs. 1,14,500 + Rs. 5,48,000 = Rs. 6,62,500
Total income generated from 500 m2 polyhouse in a year from litchi air-layers.
A. Total income from sale of 68,500 litchi plants @ Rs. 15 / plant = Rs. 10,27,000
B. Total annual income from litchi air-layers = Rs. 10,27,500
C. Total net income from 500 m^2 polyhouse
= C — (A+B) = Rs. 10,27,500 — 6,62,500 = Rs. 3,65,000

ii) Pot layering

Some nurseries practice pot layering wherein a lower branch of mature wood is cinctured and the cut surface is buried in a pot or container filled with rooting medium. The pot is watered regularly. The roots develop in the cinctured portion of the branch in about 2 months. Then the branch is detached from the main plant by giving sharp cut, preferably in 2-3 stages. No repotting is required before transplanting in the field. Application of IBA (2000-5000 ppm) improves rooting and survival of the layers.

iii) Stool layering

It involves cutting of a 2-3 years old well established air layered litchi plant in the month of February at 25cm from the ground level. After a month 6-8 side shoots appear. Except one shoot all the other shoots are selected for stooling in the month of June. A ring of bark measuring 3cm width below 20cm from the tip of the shoots is removed. A paste of IBA (2500ppm) is applied to the ringed area. Ten days later, soil is mounded around the base of the newly developed shoots so as to cover 10-15 cm of the stem above the ring to encourage adventitious roots. This causes the shoots to root profusely in 2 months. The rooted shoots are separated from the mother plant in the month of September and immediately planted in nursery beds or pots. It is reported that the transplanted shoots have a survival rate of 81-82% as compared with 40-50% in air-layers.

4. Propagation by grafting

Grafting is the act of joining two plant parts i.e. rootstock and scion together. The upper part of the graft (the scion) makes the top of the plant; the lower portion (the rootstock) makes the root system and part of the trunk. Litchi is a plant which does not graft easily. This is basically due to the fact that cambium is active over only about one-third of its circumference at any given time except when the plant is very young, During a given growth period, tissue development takes place on one side of the matrix and during a subsequent vegetative flush, another part of the cambium becomes active. The whole process is regulated so that a fairly symmetrical cylinder of wood is formed. Successful grafting therefore depends basically on cambium activity occurring in areas adjacent to point of graft operation, the matrix. It has recently been suggested that some litchi cultivars may be incompatible. Grafting is used in some nurseries in China and Viet Nam. The Chinese grafted and budded new cultivars several centuries ago, but they relied mainly on layers for commercial propagation. It is only recently that this method has been evaluated in high-density orchards. Little thought has been given for using rootstocks to control tree growth, productivity and fruit quality as in apple and stone fruit. Related species and ecotypes have been suggested as potential rootstocks, but no commercial orchards have been developed on these systems.

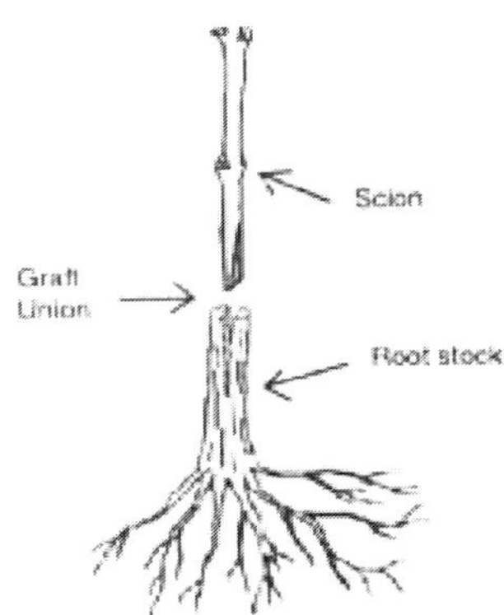

Fig. 11 : Grafting process

Budding and grafting often fail because of incompatibility between the scion and stock. Poor grafting technique, grafting at the wrong physiological stage and poor care in the nursery can also cause problems. Incompatibility has been reported in some countries. A number of grafting methods have been tried which have given varying success.

(i) Approach grafting

The distinguishing feature of approach grafting (Inarching) is that two independently growing, self-sustaining plants are grafted together. This self-sustaining characteristic of both plants which are to be grafted insures survival of both even if the grafting attempt is not successful. However, the chance of being successful is greatly enhanced because of the active growing condition of both plants involved. For inarching, plants of an established seedling are planted close to the base of the mother plant, without extensively damaging the root structure of the established plant. From both plants, closely position shoots which are at least of pencil diameter or preferably close to the same size are selected. At the point where the union is to occur, a slice of 3.0-5.0cm bark along with some wood is removed from both stems uniformly. The both surfaces are then bound tightly together with bond or electrical tape. Completely wrapping is done around the joint area. Some of the top portion of the foliage is removed from the adapted variety 15-20cm above the graft union. This encourages a more rapid healing of the grafted union.

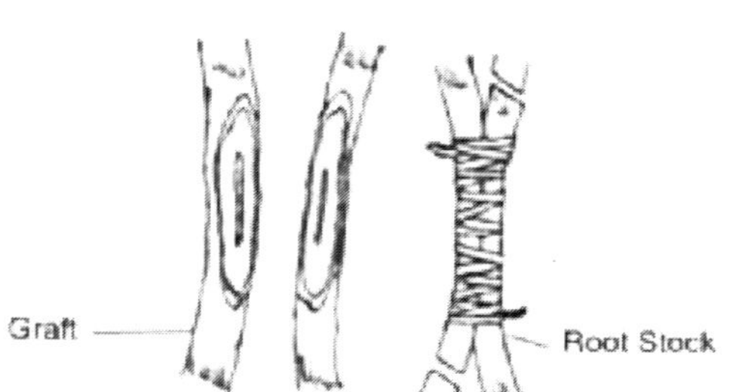

Fig. 12 : Approach grafting

The union should be complete in four weeks. This type of grafting is most successful if performed during growth season. After the parts are well united (4 weeks or more), the remainder of the top of the rootstock plant and base of the scion variety can be cut off immediately. Immediately after the portion of each plant is removed, it may be necessary to reduce the leaf area of the top.

A high proportion of graft success is obtained with this method, but its use in commercial nurseries is rather limited since adult, established plants of the variety to be grafted, must be available so that potted plants grown from seed may be placed next to them. The method does not overcome the difficultly of introducing cultivars into new areas although plants obtained in this way may later be distributed elsewhere.

(ii) Side grafting

The scion should have two to three buds with a length of about 8-10 cm long. It is usually recommended to girdle the branches from where the scions are taken before 21 days of grafting. For autumn grafts short (3-4 buds), round scions of green with completely mature wood (just before the opening of the new shoots) are preferred, while for spring grafts, hard and semi-hard wood is chosen. In any case, the wood must be relatively mature which is about to begin a new vegetative flush of growth. A wedge is made at the end of the scion slightly thicker than the other. It is not necessary to make the cuts more than 2.5-3cm long. Cuts must be made straight and smooth, with a single movement of a sharp knife. A smooth area 15-20cm above the base of the rootstock is selected and a slanting cut is made into it. The cut should angle downward and extend about halfway through the branch. The upper part of the stock back is pulled to open the cut. The scion is inserted into the open cut with the slightly thicker side lying along the cambium. The scion is set at a slight angle to give maximum contact. When the top is released, the scion should be held in place, so that no wrapping is necessary however, some people prefer to wrap the union. It is done carefully, so the cambial alignment is not disturbed. The root stock branch should then be cut off 5 to 6 inches beyond the graft. Any lateral branches on the rootstock that might cover the graft are removed, as it begins to grow. After several weeks, when the scion has started growth, the remaining shoot of the stock should be carefully cut closer to the graft.

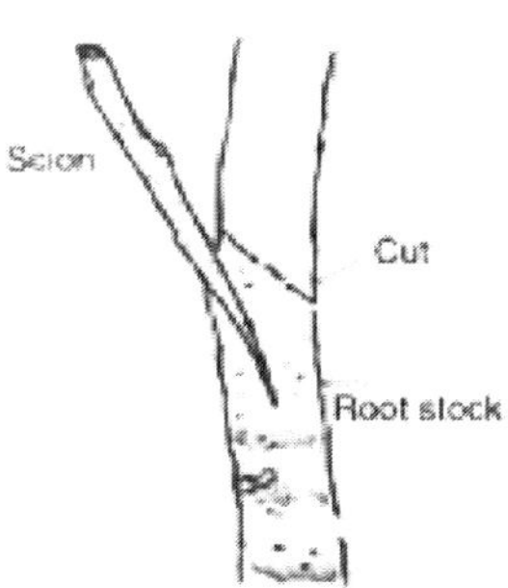

Fig. 13 : Side grafting

Under favourable temperature conditions, buds begin to sprout within 3-4 weeks and the tops of the stocks may be cut off two to three weeks after the scion has been inserted. The new trees may be planted out once they have hardened off sufficiently and reached a height of 50-100cm, but not later than one year after grafting. To reduce costs, it is always convenient to graft plants in small bags (4kg capacity), at the start of the growing season and then into larger bags (about 13 kg).

It has been seen that with side grafting, the stock contributes more significantly to callus formation than does the cultivar, while with approach grafting, the faster growing symbiont makes the greater contribution. This means that fast-growing plants which can be selected form nurseries prior to grafting should be preferred as rootstock and, in principle; seeds from the more vigorous varieties should be used. Seeds sown in early summer are ready for grafting the following autumn or spring.

(iii) Bark grafting

Bark grafting is used for repairing the established plant of litchi. For this purpose, a smooth and knot-free root stock is selected. The branch is saw off at a right angle smoothly avoiding tearing or splitting the bark. If the saw cut is not smooth, a knife may be used to remove the rough edges. The bark must be tight to form a successful graft. A grafting knife may be tapped with a mallet to drive the blade into the stub to split the stock through the center so a split extends about 5cm into the branch (Fig-14). The scion should be 10 to 12cm long with two to three buds. The base of the scion is prepared by cutting inward 3 to 5cm from the base then downward, forming a shoulder and long smooth cut. The long cut should extend about one-third through the twig, keeping its base strong enough to insert. On opposite side of the long cut, a short cut is made to give the base of the scion a wedge shape for easier insertion. A knife may be used to lift the bark at the top of the slit. The scion is pushed down and centered in the slit or between both slits if the double slit method is used. Insert the scion until the shoulder rests on the stub. If the scion is large enough, one or two small nails may be used to tighten the scion to the root stock. Some prefer to use electrical tape or masking tape to pull the surfaces tight. As in the case of side grafting, the cuts must be long so as to obtain the best possible areas of contact. The plants produced by this method do not grow in the normal shape. A number of different versions of this method of propagation have been tried on litchi with relative success.

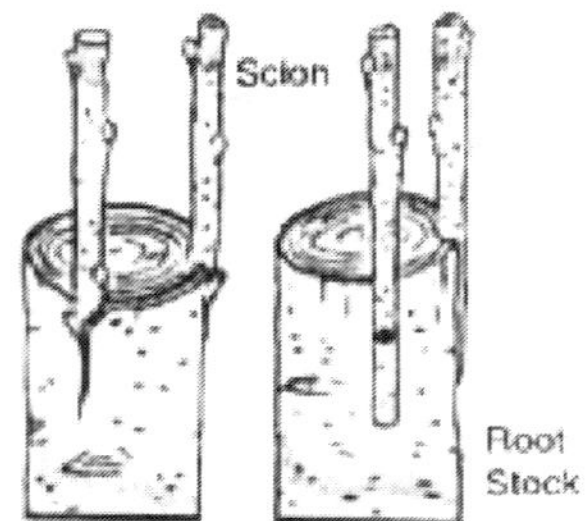

Fig. 14 : Bark grafting

(iv) Whip grafting

The whip graft is fairly easy and heals rapidly. It works best when the root stock and scion are of similar diameter preferably between 0.7 and 1.25cm. The scion is first prepared by girdling branches of

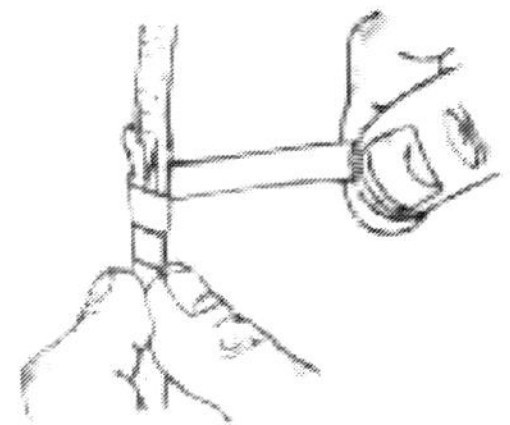
Fig. 15 : Whip grafting

new growth 21 days prior to grafting. Girdling involves removing a strip of bark approximately 6mm in width, situated some 0.5m from the apex, over the entire circumference of the branch. Scions are taken only from young trees which had a period of vigorous growth. Scion material up to 6mm thickness may be used. The scion must be approximately 10 cm long and contain at least 2-4 buds. It is preferable to use wood from the pervious year rather than young terminal wood.

A long, slanting cut is made in the rootstock 30cm above the ground at a point where the diameter of the stock is the same as that of the scion. This cut must not be less than 2.5cm long, 4mm thick and in stock and somewhat longer if the stock is wider in diameter. It is nevertheless not recommended to use stock more than 1cm thick.

The advantage of grafting over other methods of propagation for commercial purposes is sufficiently well-known. A deep, well-formed root system is obtained in this system in which fruit production commences earlier and grafting operations are conducted in the nursery. Another advantage is that desired characteristics are reproduced. This is, furthermore, the only practical and simple method of introducing new cultivars from other countries into a given area. The only drawback with grafting is that use of different rootstocks is likely to lead to heterogeneity in a commercial plantation and the poor success of graft. Because of repeated failures to obtain successful unions and good growth, this method is not used on a commercial basis in litchi nurseries. Many a times the grafting fails because of following reasons.

- Stock and scion were not compatible to each other.
- The cambiums were not meeting properly.
- Scions were upside down.
- Grafting was done at the wrong time.
- Under stock or scion were not healthy.
- Scions were dried out or injured by cold.
- Scions were not dormant.
- The graft was not properly covered with grafting wax.
- The scion was displaced by storm, birds or other means.

- Excessive growth of wild sprouts from rootstock portion which has weaken the scion.
- The graft was shaded too much by other growth.
- The graft was attacked by insects or disease.
- The graft union was girdled because tape was not cut or released in time.

Relative Merits of Propagation Methods

Various methods of propagation have been attempted to experimental level but the air layering has been found most successful and being practiced commercially in India and abroad for litchi propagation. The comparative information on propagation method has been given in following table.

Table 9 : Relative merits of different propagation methods

Parameter	Seedling	Cutting	Air layering	Stooling	Grafting
Uniformity	Highly variable	Uniform	Uniform	Uniform	Uniform
Juvenility period	Long	Short	Short	Short	Short
Root system	Very good	Poor	Fair	Good	Good
Size of new plant	Small	Small	Large	Large	Variable
Time to plant out (months)	12-24	12-24	6-12	6-12	Variable
Multiplication rate	High	High	Low	Low	Low
Destruction of parent tree	Low	High	High	Low	Low
Simplicity	Simple	Simple	Simple	Complex	Difficult
Cost	Cheap	Cheap	Cheap	Expensive	Costly
Labour	Low	Low	High	High	High

Modified from K.S. Chauhan (2001). Litchi: Boatny, production and utilization, Kalayani Publishers New Delhi

Manuring, watering, drainage and plant protection of nursery plant

Fertilizers can be applied once the young plants begin to produce new growth. Fertilizer application is to be done very carefully. Vigorous growth of plant is always attractive to the buyer. Heavy manuring is not beneficial for storage of plants. Light manuring, watering is also important. Watering is done according to need of the plant. The nursery should have a water source of its own. Digging a well (12 m deep x 3 m dia) and installation of a 2.0 HP pump set with accessories are considered. Sprinkler system of irrigation is not advisable at the beginning.

For sufficient vegetative and reproductive growth of plants, good drainage system must be developed in between the beds and around the nursery. Adequately gentle slope in the pot bed surface is also desirable. It is extremely important to ensure that water logging does not occur in and around the pots and beds. Keen observation on attack of different pests and diseases is required. If the mother plants are infected, the propagated plants will also be infected therefore, necessary control measures in mother plants as well as in nursery plants should be taken immediately on observation.

❑❑❑

Chapter 5

Select the Suitable Variety

Establishment of litchi orchard is a long-term investment because litchi trees continue to produce fruit for more than 60 years. The productivity of litchi basically depends on variety and climate. The climate, sometimes leads to greater fluctuation in yield of a litchi orchard but by and large it is the variety which governs the potential yield of the orchard. Therefore, a right variety should be selected for plantation at a particular place. Although, all the litchi varieties have a wide range of adaptability, but yield and quality performance are the area, region and location specific.

Systematic classification of the cultivars of subtropical and tropical fruits, such as mango, citrus, bananas, avocados and pineapples has been done, however no serious study on these lines has been conducted with respect to litchi. Most of the commercial varieties in the various litchi producing countries originate from China. Litchi varieties grown in India are highly variable under different climate and soil conditions. When distinguishing the cultivar, the shape of skin segments and protuberances are the reliable and stable genetic characteristics. Fruit size, shape and taste are also variables but are influenced by other than genetic factors. Indian cultivars vary greatly in vegetative flushing pattern, flush colour and flowering ability. Based on these characteristics, cultivars were classified in five groups. Group A, which has 7 cultivars is the early group, B and C groups are mid-season, and group D is the late group. Only one cultivar, which is very late, is under group E and its cultivation is confined to Muzaffarpur. Yield and physicochemical characteristics of some important Indian cultivars are given in Table 10. These cultivars also show variation in yield, cracking, and physico-chemical quality.

Table 10 : Physicochemical characteristics of important Indian litchi cultivars

Varieties	1	2	3	4	5	6	7	8	9
Ajhauli	27.58	72.42	3.48	1.15	15.29	63.13	20.22	0.3	12.53
Bedana	0.24	32.75	3.32	0.94	16.33	70.68	19.33	0.28	10.20
China	0.33	95.33	3.40	1.10	14.77	58.46	20.22	0.29	10.95
Dehra Dun	14.27	67.00	2.82	1.17	16.77	71.74	20.75	0.41	11.87
Dehra Rose	10.01	74.87	3.34	1.13	19.63	74.12	21.27	0.37	12.19
Deshi	13.63	83.03	3.50	1.18	15.94	66.65	22.82	0.37	13.48
Green	24.44	84.72	3.57	1.27	16.70	61.35	21.42	0.33	11.55
Kasba	0.00	37.00	3.78	1.12	25.93	72.23	20.23	0.40	11.24
Late Bedana	0.00	54.22	3.36	1.06	16.70	75.08	18.17	0.27	10.38
Longia	0.00	63.85	3.17	1.16	13.84	67.51	19.27	0.28	11.50
Purbi	0.95	80.65	3.31	1.07	20.00	71.73	20.10	0.29	11.04
Rose Scented	8.85	88.05	3.63	1.16	19.39	69.05	20.37	0.37	12.98
Shahi	12.25	100.30	3.40	1.21	19.47	74.78	22.30	0.38	12.97
Trikolia	12.03	42.37	3.37	1.14	17.35	70.83	22.43	0.36	12.70

1.Cracking (%), 2.Estimated Fruit Yield (Kg/plant), 3. Length (cm), 4. Shape Index, 5. Fruit Weight (g), 6. Pulp (%), 7.TSS (^{0}Brix), 8. Acidity (%), 9. Total Sugar (g/100g)

Source:Litchi:M.Rai,Vishal Nath and P.Day (editors),CHES-IIHR,ICAR,Plandu, Ranchi

Criteria for selection of litchi varieties

None of the existing commercial varieties of litchi can be called as an ideal, as these lack one desirable character or the other. In general, the varietal selection is frequently made on the basis of yield potential, disease resistance, suitability for local conditions and availability of planting material in a nearby place and often without sufficient consideration of market preferences and post-harvest behaviour particularly the keeping quality. In order to meet the challenge of the present-day requirements, some criteria should be considered while selecting a litchi variety.

No existing cultivar possesses all these desirable characteristics, some of which may even be strongly influenced by environmental conditions, that is why cultivar selection must be carried out in every region. It is worth pointing out that some characteristics may be more desirable in some areas than in others. The most desirable characteristics of a litchi cultivar are set out in following text.

Desirable characteristics of a litchi cultivar

Fruit characters

- *Weight*: Light weight fruit is an undesirable feature which affects profitability. Any varietal selection for cultivation must include the heaviest weight possibly not less than 25g.
- *Shape*: Round shape is the most efficient shape as it allows for easy packaging and reduces transport cost and it should therefore be given preference, however, it is not an absolutely essential characteristic. Varieties of other shapes, oval, or even a slight heart shape may also be selected.
- *Post harvest quality*: Fruit should lend itself well to preservation by freezing so that it can be consumed over a fairly long time after harvesting. This is an important factor to be kept in mind while selecting the variety for cultivation.
- *Ripening*: Uniform ripening pattern and short ripening period is desirable so that all the fruits on the same trees ripe at the same time. This makes harvesting easier and keeps harvesting cost down.
- *Season of fruit availability*: Litchi fruits are harvested in the summer, when there is an abundance of all other fruit. Early or late varieties would therefore, be more desirable.
- *Shelf life*: The litchi fruit get deteriorated with in short period after harvesting, hence, selected variety should have longer shelf life of fruit.
- *Fruit colour*: Bright red fruit colour is the desirable feature as it is eye pleasing. Fast skin browning is a demerit. A litchi cultivar should have slow darkening of skin colour after harvesting. This will reduce the need of refrigeration of the fruits immediately after harvesting and would keep export cost down.

- *Skin texture*: Skin should have acceptable thickness and texture. It provides adequate protection in the period between harvesting and consumption and allow for easier peeling.
- *Resistance for biotic and abiotic stress*: Pest and disease resistant varieties are of prime importance. It is an important factor for the tree as a whole. Fruit should be resistant to cracking and sun-burning. Since the fruits are particularly sensitive to cracking, which is caused by extremes of humidity and temperature, it would be highly desirable to have a variety free from this defect as far as possible.
- *Seed*: Small seed is an essential characteristic for litchi fruit. Of existing cultivars, only those in which the seed accounts for less than 30 per cent of the weight of the fruit (practically more than 70 per cent pulp) should be selected. Since litchi with chicken-tongued seeds do not suffer any loss of quality, this characteristic is very much appreciated. It is influenced to some extent by environment. However, whatever the circumstances, the higher the proportion of this type of fruit obtained the better is the variety. Small seeds are not absolutely essential in large fruits.
- *Taste*: Pulp should be of pleasant taste and texture. Fruits that are slightly fragrant with a sweet or slightly bitter sweet taste even prior to maturity, are preferred. The pulp should be crisp in texture and not too watery.

Tree characters

- Early bearing is an important tree character for litchi. It is very desirable to have early varieties so that income may be obtained as soon as possible to recover establishment costs.
- Variety should give abundant and regular harvests. It is very important to obtain cultivars which flower and bear abundantly without the presence of special temperature and humidity conditions. This character will make variety wider adoptable to extend cultivation to other areas.
- Selected variety must have branches with wide crotch angle preferably 120-130^{0}angle.
- The plant type or variety should be resistant to extremes of temperature which will enhance production in relatively disadvantages areas. Although the species is relatively resistant, but still, it is better to be more resistant.

- Saline soils and irrigation water with a high salt concentration are common in many parts of the litchi growing belts. Therefore, salt resistance or tolerance is desirable characteristic.
- Litchi plant is moisture stress tolerant however; the fruit suffer a lot due to moisture stress. Some varieties are relatively more tolerant to the same and therefore, these should be selected for cultivation.
- Cultivars which present a wide range of adaptability are to be preferred over those requiring light soils for commercial cultivation. It is safe to select a variety which can withstand the adverse soil conditions.
- It has been noted that moderately to slightly vigorous varieties contribute to some extent to flowering and should generally be preferred.
- Spreading growth habit makes harvesting easier and increases wind resistance, hence, should be preferred.

How to identify litchi variety

Based on leaf: Leaf colour along with shape and size of the leaves are of great importance in varietal identification. The leaf of Rose Scented and Shahi is boat-shaped while China has a distinctive twist along the length curved upward from the midrib and down along its length. Small leaflets of Bedana are oval shaped.

Based on fruit shape: The fruit shape of the litchi is very distinguishing. The round shape of Bedana is distinguished from the oblong shape of China or Shahi. The fruit is smooth and pulp is even or uneven. The apex of the fruit can be round, obtuse, blunt as in Shahi, or pointed as in China (Fig. 16).

Colour of new flush: The varieties can also be distinguished depending upon the colour of the new flush and season of flushing. Shahi produces very light coloured flush while China has pinkish flush. Bedana and Swarna Roopa have very dark pink flush. Emergence of the panicle and its shape also differs.

Based on panicle: Panicle shape and compactness are important features. Shahi has long panicles while Bedana produces short and compact panicles.

Colour of fruit: The colour of the litchi fruit is pinkish brown or dark red depending on the cultivar. Colour of the skin varies and is also influenced by growing conditions.

Peel characters: Peel thickness depends on the cultivars. Bedana, Swarna Roopa and China have very thick skin. Peel surface at maturity also varies being smooth, swelling, sharp and pointed. Protuberances of the pericarp can be smooth and sharply pointed. Bedana has very smooth protuberances while Shahi has very distinct protuberances.

Seed characters: The presence and absence of seed as well as structure and size of seeds also vary from cultivar to cultivar, although it is influenced by environmental conditions. In Rose Scented and Bedana, a high proportion of chicken-tongued seeds (aborted seeds) are observed while China has bold seeds.

Season of maturity: Although litchi has short duration not exceeding 30-40 days, cultivars can be distinguished based on season and maturity, provided they are grown at same location. Accordingly, cultivars can be grouped into three categories namely, early, medium and late season. Fruit maturity also varies slightly and relatively from year to year depending upon prevailing weather conditions specially the Growing Degree Days (Heat Unit Submission). The maturity period of the fruit also varies depending upon agro-climatic regions. Some of the varieties in West Bengal come to maturity earlier, while the same varieties can be late in Uttar Pradesh. Due to the production of litchi in varying agro-climatic conditions, maturity, fruit colour, shape and size are reported to be varying. Thus, there has been much confusion in the names of varieties. As a result, the same variety is called by different names at different locations. The key characteristics of some of the litchi varieties are presented in Table 11.

At present, none of the commercial varieties of litchi has all these characteristics combined into one. As a matter of fact, all these characters may not be available in a single variety or if at all it is available, the variety may not perform well in all the litchi growing areas. Therefore, it is essential to choose a wider adaptable variety, because the success of any litchi plantation depends on tree health and its management. Adaptable variety and its compatible environment are the foundation on which tree management systems are built up. When the right variety is in place, tree management would facilitate enhancing the tree productivity and orchard profitability. However, when incompatible varieties are grown in an unmatched environment, any tree management system would hardly make any significant improvements in fruit yield. If growers at a particular place are interested for new varieties, especially the exotic which have good market demands, it is better to first analyze the growing environment of those varieties with that of the local environment where the orchard is going to be established.

Table 11 : Key to Important Cultivars of Litchi

Character	Cultivar
1. Flush pink, leaf boat-shaped, dark green, panicle long, fruits oblong with round apex.	
Colour of fruit deep pink	Shahi and Tirkolia
Rose flavour	Rose scented
Colour of fruit light and greenish	Green
High cracking and big seed	Ajhauli
Late in Maturity	Dehradun
2. Deep pink flush, leaf with twist along the length, curved upward from the midrib and down along their length, panicle long, fruit oblong with pointed apex	
Color of fruit pink	China
Fruits deep pink	Purbi/Manraji
Fruits in bunches	Bombai/Calcuttia
Early maturity	CHES-2
3. Dark pink flush, oval shaped leaves, compact and small panicles, Fruit round, smooth, morechicken tongue seed (aborted seed)	
Early maturing	Early Bedana or Early Seedless
Late maturing	Late Seedless/Late Bedana
Deep pink colour	Swarna Roopa
Mid season maturity	Swarna Roopa
4. Deep pink flush, boat-shaped and dark green long leaves, panicle long, largest size fruit	
Deep in colour	Kasba
5. Small elongated leaves, light green in colour, panicle compact, fruit medium in size, very late maturity.	
Pulp sweet and excellent flavour	Longia
Pulp sour	Kaselia and Khatti

Source : H.P.Singh and S. Babita (2002) Lychee Production in the Asia-Pacific Region

Some litchi varieties are region specific

Selection of a cultivar should aim at its suitability to every location which is rather difficult but not impossible. Commercial cultivars of litchi are geographically adapted to certain particular locations and give their best possible performance in these reasons only (Table 12).

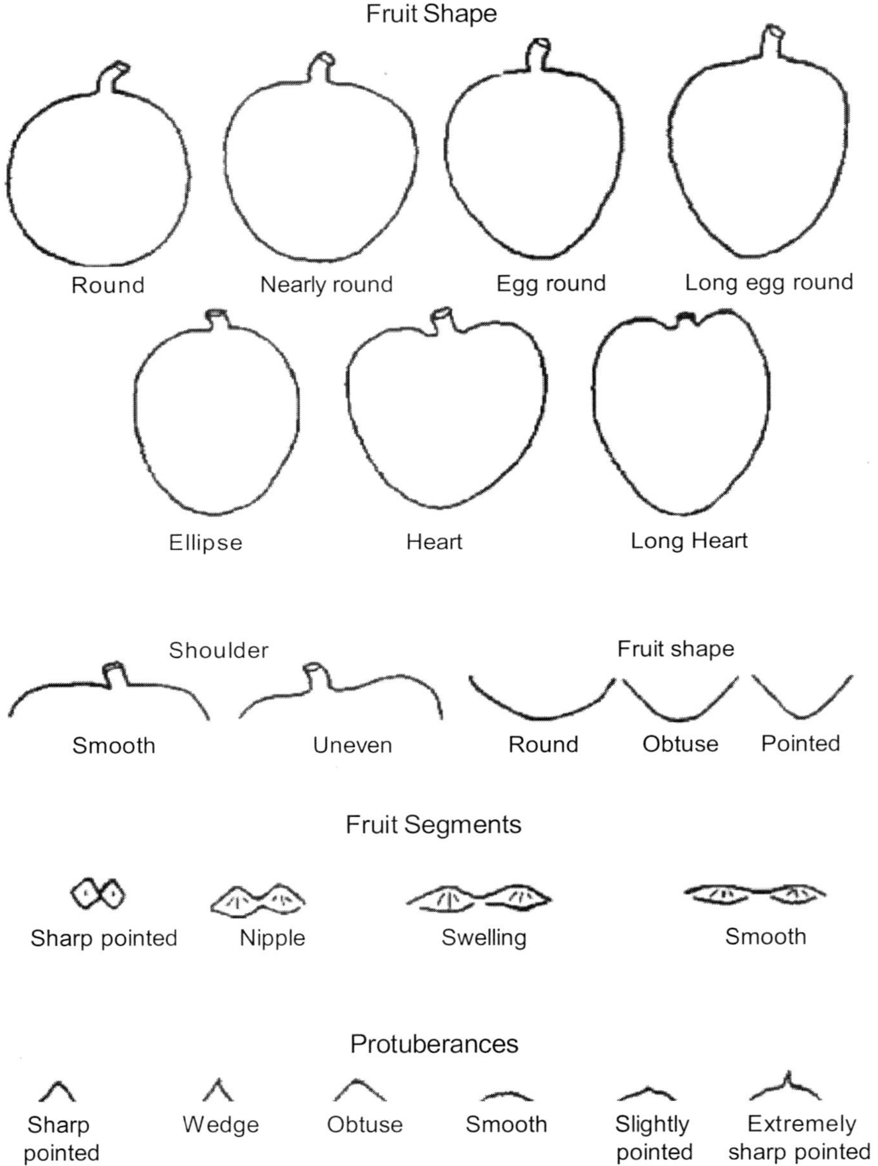

Fig. 16 : Fruit characteristics used to describe different cultivars

(Source :The lychee crop in Asia and the Pacific, FAO corporate document repository Regional Office for Asia and the Pacific)

Table 12 : Commercial cultivars of litchi in different Indian States

State	Total No.	Cultivar
Bihar	19	Deshi, Purbi, China, Dehra Rose, Shahi, Kasba, Ajhauli, Madrai, Maclean, Longia, Kaselia, Late Bedana, Early Bedana, Muzaffarpur, Trikolia, Swarna Roopa, Dehradun
Uttar Pradesh	13	Early Large Red, Early Bedana, Late Large Red, Late Bedana, Shahi, Rose Scented, Culcuttia, Extra Early Green, Gulabi, Pickling, Khatti, Dehradun, Piazi
West Bengal	8	Bombai, Ellaichi, Early Ellaichi, Late China, Deshi, Purbi, Kasba, Muzaffarpur
Punjab and Haryana	11	Saharanpur, Dehradun, Muzaffarpur, Late Seedless, Early Seedless, Rose Scented, Early Large Red, Late Large Red, Culcuttia, Khatti, Gulabi

Source: Litchi: M.Rai, V. Nath and P.Day (editors), CHES-IIHR, ICAR, Plandu, Ranchi

Select a few among many varieties available in the country

Chinese consider that the litchi has more cultivars than any other fruit. Although over 100 cultivars are grown out of which only about 8 are of commercially important. It is extremely difficult to maintain the yield and fruit quality of highly prized litchi cultivars under soil or climatic conditions other than those of the trees in their native area.

There are thirty three cultivars of litchi which are grown in India. Most of these cultivars are seedling selection from Chinese cultivars. The majority of litchi production comes from only six cultivars namely Saharanpur, Dehradun, Muzaffarpur, Rose Scented, Seedless and Calcuttia. As a matter of fact, each district has its own major cultivars. Based on maturity period in Jharkhand litchi varieties are grouped as under:

Table 13 : Maturity period of litchi cultivars

S.No.	Period	Cultivars
1	Early	Ajhauli, Green, Dehra rose, Trikolia, Shahi, Rose Scented, Deshi, Mujaffarpur,Swarna Roopa
2.	Mid	China, Purbi, Culcuttia, Bombai, Bedana, Dehradun, Kasba
3	Late	Late Bedana, Kaselia, Longia

Source: Litchi: M.Rai, V. Nath and P. Day (editors), CHES-IIHR, ICAR, Plandu, Ranchi

In Bihar, which has the largest area under litchi in India, Shahi, Rose Scented and China are of most commercial varieties in Muzaffarpur while Kasba and Purbi are the choicest cultivars of eastern part of Bihar. The other cultivars being grown in Bihar on small scale are Early Bedana, Late Bedana, Desi, Mandaraji, Longia and Kaselia. Among eight cultivars being grown in West Bengal, the cultivars Bombai, Ellaichi Early, Ellaichi Late and China are considered to be more promising both from quality and production point of view. Other cultivars of this area are Deshi, Purbi, Muzaffarpur and Kasba. Late Seedless and Culcuttia are partly resistant to hot and desiccating winds and bear good quality fruits profusely.

Litchi varieties vary in their shape, size, colour, taste, regional suitability, yield, and response to post harvest management practices, commercial use and pest and disease problems. For reference to growers the important features of some of the conventional varieties of litchi in the country has been enumerated as follows:

1. *Ajhauli*: This is an early maturing variety selected from Ajhauli village. It yields about 80-100 kg fruit from a full grown up sixteen years old tree. Fruits are red in colour, weighing 15 to 18 g and have big seeds. It cannot be distinguished from Shahi on vegetative characteristics as it has many similarities but it has pink floral disc which is distinct from Shahi. This variety is highly prone to cracking but under irrigated condition cracking is minimized.

2. *Bombai*: This is an important commercial variety cultivated for table purpose in West Bengal. It ripens during first to second week of May. The trees are vigorous and attain an average height of 6-7 m and spread up to a diameter of 7- 8 m. Average yield is 80-90 kg/tree. Fruits are large in size (3.5 cm long and 3.2 cm diameter); obliquely heart shaped and weigh 15-20 g. The colour of ripe fruit has attractive carmine red with uranium green skin background. In this variety, each developed fruit has another tiny under-developed fruit attached to the fruit stalk. Pulp is grayish-white, sweet, soft, and juicy with 17^{0}B T.S.S., 11 per cent total sugar and 0.45 per cent acidity. The elongated, smooth and shining seed of light chocolate colour is 2.3 cm long, 1.6 cm in diameter and weighs 3.4 g.

3. *CHES-2*: This is an early maturing selection from the population of the China group. It bears fruits both at the outer and inner canopy, which helps in reducing the sunburn as well as fruit

cracking. The fruits are deep red, conical shaped and appear in a cluster of about 15-20. The fruit has an average weight of 21.3 g containing 3.8 g seed and 16.1 g pulp. Vegetative characteristics are similar to China however, flowering and fruiting is earlier.

4. *Calcuttia*: It is the best of all the varieties grown in northern parts of India for table purpose. It can be successfully cultivated even in hotter areas provided there is protection from strong hot winds and provision for plenty of water for irrigation. The average yield is 80-100 kg/tree. It is a late season variety and the fruits ripes in the last week of June. Trees of this variety attain an average height of 4m and spread of 6 m. Fruits are large in size, oblong, with tyrian rose colour and dark tubercles at maturity. Pulp is creamy white, soft, juicy, very sweet with 18.2 ^{0} Brix and T.S.S. 18.9%.

5. *China*: The origin of this cultivar is not known but the name indicates that it was selected for its superiority and named after the major litchi producing country 'China' because of high yield potential. It is tolerant to hot waves and fluctuations in soil moisture, which cause fruit cracking. It is similar to Purbi, Calcuttia, Bengalia, Bombaiya and Manaraji of different regions for few external characters. This is a medium-late season cultivar. Fruits ripe during the end of May in West Bengal, the first week of June in Jharkhand and North Bihar and the third week of June in Uttar Pradesh. Trees are comparatively dwarf and high yielders but it is prone to alternate bearing. Rains at the time of fruit bud differentiation cause emergence of vegetative flush resulting in loss of crop. It bears fruits in cluster of 12-18. In some cases, more than 30 fruits per cluster are also recorded. The plants bear less fruit in eastern and southern directions. Fruits are large in size, medium-heavy in weight, oblong in shape and tyrant rose in colour with dark tubercles at maturity. The aril is creamy-white, soft, juicy, sweet having 17 to 18^{0}Brix TSS, 11 per cent total sugar and 0.43 per cent titratable acidity. Seeds are glaucous, dark chocolate in colour, oblong to concave or planoconvex in shape, medium in size (2.9 cm length and 1.5 cm diameter), and average in weight (3.49 g/seed). The ratio of rind: pulp: seed by weight is 16.42: 69.22: 14.36. The flavour of the pulp is not pleasant like Shahi, but owing to its high yield and non cracking, this cultivar is popular. This cultivar cannot be distinguished from Mandraji and Purbi grown in the eastern part of Bihar state.

6. *Dehradun*: This variety is mainly grown for table purpose in Uttar Pradesh, Uttaranchal and Punjab. In Bihar, it is cultivated under the name of 'Dehra-Rose'. The fruits start ripening by the third week of June in Uttar Pradesh but in Jharkhand it matures with Shahi. Fruit yield is 80-90 kg/tree. It is a medium vigorous tree (5 m height and 7 m spread) which produces medium to high yield. Fruits are medium in size, measuring 3.7 cm in length, 3.5 cm in diameter, weighing 15.2 g and having oblique-heart to conical shape with bright rose pink colour. The pulp of this cultivar is grayish-white, soft, and moderately juicy with 18^0 Brix TSS, 10.4 per cent sugar and 0.44 per cent acidity. This variety is highly susceptible to sunburn and cracking especially under rainfed conditions. The name of the cultivar suggests that it is a selection made in Dehradun.
7. *Early Seedless (Syn. Early Bedana):* It is known as Early Seedless in Punjab because of its early ripening and small seeds. This cultivar has distinguishing leaf and flower characters. The cultivar is very much popular in Uttar Pradesh and Punjab. It is a medium to poor-yielding variety (50-60 kg/tree) but bears regularly. Trees are medium, attaining an average height of 5.0 m and spread of 6.2 m. Fruits are medium in size and weight ranges from 15 to 18 g having oval or heart shape, rough surface with uranium green skin covered with carmine red tubercles at maturity. Aril is creamy white, soft, juicy and sweet containing 17 to 20^0 Brix TSS. Seed is very small, shrunken, glamorous and dirty chocolate in colour. The overall fruit quality of the cultivar is good.
8. *Ellaichi*: This is an important cultivar of West Bengal having brighter prospects of commercialization. The trees are moderately vigourous, attaining an average height of 5-6 m and spread of 6-7 m. It is a mid season cultivar which ripens in the first week of June. The cultivar yields 50-60 kg fruit annually. Fruits are conical, marigold orange red with an average weight of 12-15g. Pulp creamy white in colour, sweet, soft juicy with agreeable flavor. The cultivar has 18^0 Brix TSS, 11.50% total sugar 0.45% acidity and 6.91:1 pulp: stone ratio at ripening. Seeds are relatively small, shining, weighing 1.50-2.00g. Fruits are less susceptible to sunburn and cracking.
9. *Gulabi*: This is another late maturing cultivar of North India in which, ripening takes place by the fourth week of June. Trees

attain a height of 6.0 m and spread of 7.0 m. Early rain hampers the quality of fruits. The medium vigorous tree bears profusely and regularly with medium to large sized fruits giving 90-100 kg fruits/tree. The shape of the fruit is variable from oblong-oval to heart shaped whereas; the rind colour at ripening varies from shrimp red to carmine red with mandarin red tubercles. The pulp is firm, grayish white, sweet with 18.2^0 Brix TSS, 10.7 per cent total sugar and 0.49 per cent titratable acidity. The seed is rather big, heavy, and oblong-cylindrical in shape with a shining chocolate seed coat.

10. *Kasba*: This is an important cultivar selected from Kasba village in West Bengal having brighter prospects for commercialization. The trees are moderately vigorous, attaining an average height of 5-6 m and spread of 6-7 m. It is a mid season cultivar which ripens in the first week of June. The variety is cultivated for table purpose with an average yield of 50-60 kg/tree. Fruits are conical, marigold-orange red in colour, fruit weighs between 23-27g, perhaps the heaviest fruit among the known varieties, but the number of fruit is less. The pulp is creamy white in colour, sweet, soft and juicy with agreeable flavour. The cultivar has 18^0 Brix TSS, 11.5 per cent total sugar and 0.45 per cent acidity, and 6.91:1 pulp: stone ratio at ripening. Seeds are relatively small, shining, and weigh 1.5-2.0 g. Fruits are less susceptible to sunburn and cracking. Interestingly, the cultivar performs better in marginal soils as it has the capacity to absorb more nutrients however; this cultivar has not assumed commercial success.

11. *Kaselia*: This is later maturing cultivar found growing in isolation. The tree is medium in size and fruits mature very late. Fruit attain a pink red colour. The pulp content is comparatively low and the seed is big. This cultivar is also known as 'Khatti' or 'Pickling'. The cultivar has not assumed commercial success.

12. *Late Large Red (Syn. Muzaffarpur):* It is one of the important litchi varieties mainly grown in Bihar and its adjacent states. This variety bears profusely every year with an average yield of 80-100 kg/tree. Fruits are large in size, oval or oblong conical with crimson red tubercles. Pulp is grayish white, soft, moderately juicy with a T.S.S. of 20.3^0 Brix.

13. *Late Seedless (Syn. Late Bedana):* This is a late maturing cultivar, which usually ripens in the last week of June in Uttaranchal, the

end of May in Jharkhand and in the first week of June in Muzaffarpur. The new flush is dark pink in colour and its leaf can be distinguished from other cultivars. The panicle is compact. This variety is not completely devoid of seed but it is rather shriveled and small in size. It can be successfully cultivated even in hotter areas provided there is protection from strong hot winds and provision for plenty of water for irrigation. The trees of this variety are vigorous and attain an average height of 5.5 m and spread of 7.0m. The average yield is 80-100 kg/tree. Fruits are medium to large in size, conical in shape, colour at maturity vermilion to carmine with dark-blackish brown tubercles. Pulp is creamy-white, soft, juicy with $18\text{-}20^0$ Brix T.S.S., but acidity is low. Seeds are very small, shrunken, glamorous, and chocolate in colour with fusiform shape. Overall quality is very good. It is cultivated for table and processing purpose.

14. *Longia*: This cultivar is well distributed in North Bihar and is preferred for late maturity. The tree is medium in size, leaves are small and light in colour and it has compact panicles. Fruits are medium in size and the aril has an excellent aroma. Due to shy bearing habit, there is a declining preference for this cultivar.

15. *Purbi*: It is mostly grown for table purpose in eastern part of Bihar. Fruits are medium-large, oblong-conical in shape which ripen at the end of May or first week of June. At maturity, red tubercles appear on pinkish brown background. The average yield is 90-100 kg/tree.

16. *Rose Scented*: It is commercially cultivated for table purpose in Uttarkhand and Muzaffarpur area of Bihar. Besides high fruit quality, it is famous for distinct rose aroma and hence called Rose Scented. It is a mid-season variety that starts ripening in the first week of June. Average yield is around 80-90 kg/tree. Fruits are medium to large in shape mostly oval or heart shaped and deep rose pink in colour. Pulp is grayish white soft moderately juicy with T.S.S. 20^0 Brix.

17. *Shahi*: This is the most popular cultivar grown in North Bihar, Jharkhand, Uttarakhand and Uttar Pradesh. Besides having high quality fruit it has a distinct aroma. The vegetative flush of this cultivar is light, and fruit weight ranges from 20-25 g. This cultivar is earliest in maturity and ripens during the second week of May to the first week of June at various locations. It matures on 12-15

May in Jharkhand, the 25th May in North Bihar and by the first week of June in the Tarai region of Uttarakhand. Trees of this cultivar are very vigorous and produce fruits ranging from 100-150 kg per plant. Mature fruits are prone to cracking in zones with low humidity and poor moisture content in the soil. Fruits are globous-heart or obtuse in shape having rose madder and fuchsia purple background with red tubercles at ripening. Pulp is grayish-white, soft, moderately juicy and sweet, and TSS ranges from 19-22^{0}Brix. Seed size varies. On the same plant larger fruits have big seeds while seeds in small fruits are shrunken. The fruits are known for excellent aroma and quality. This cultivar occupies a major area under litchi in India.

18. *Swarna Roopa*: It is a selection in litchi identified and released by HARP, Ranchi with high T.S.S and resistant to fruit cracking. This is the first variety developed and systematically released in India. This is a late maturing, cracking resistant cultivar of litchi. The fruits are of attractive red colour with small seed and high aril percentage. Leaves are similar to Bedana in shape and size. New flush is pink and mature leaves are dark in colour. The cultivar has 18.5-22.5 cm long, compact panicles. Fruits are medium in size weighing 12-15 g and have high pulp content. The pulp contains high TSS and low acidity. Total sugar content in the fruit is 13 per cent out of which 8.5 per cent is reducing sugar. The cultivar is suitable for extended harvest as it matures after China and is prized for its attractive fruit colour. This cultivar is recommended for commercial production.

In addition, Ambika litchi-1 and Indira litchi-2 are the litchi varieties released by IGKV, Raipur which are grown in Sarguja area of Chattisgarh.

❑❑❑

Chapter 6

Adopt Scientific Orchard Establishment Practices

The success of a litchi orchard mainly depends on the initial establishment of plants. A well established orchard itself avoids many problems related to plant growth and performance. Therefore, one can prevent many risks through appropriate orchard establishment practices. This includes:

1. Land preparation

Poor land preparation and inappropriate lay out posses recurring problem. Therefore, selected fields should be deeply ploughed and then harrowed to root out the perennial weeds, roots of perennial bushes and heavy clods. If the litchi plantation has to be under taken on bushy/ minor land, clearing of bushes and forest stuff is essential before deep ploughing. Under hillocks and steep sloppy land (30-40% slope), contour trenching instead of deep ploughing should be done. This operation provides congenial tilth to young roots for their healthy development. After harrowing, proper leveling of land is done and a gentle slope is provided in one direction to facilitate irrigation as well as drainage of excess water during rains. The soils which have drainage problems should be provided with adequate trenching from the very beginning to avoid serious damage to young plants due to water stagnation. On sloping land with a gradient up to 10 per cent, it is advisable to plant the trees on mounds or in trenches. This facilitates management operations and improves drainage. Where the land is more steeply sloping, terraces should be established, although this makes ploughing considerably more difficult. It is also advisable to have grass or stone-lined irrigation ditches to prevent soil erosion.

Deep ploughing of the land in dry weather prior to planting will ensure uniform growth and is essential in previously cultivated soils where a compact layer has usually been formed. It is desirable, but not essential to use cover crops prior to planting and continuing the first 3 or 4 years. They will prevent weed growth, act as a windbreak for the young trees and provide a source of *"in situ"* mulching.

2. Lime application

In normal as well as alkali soil, plants do not require liming however, if the soil analysis show the lower pH than required (5.6-6.0), lime should be applied and if there is magnesium deficiency, dolomite, should be used to improve the soil pH. If the soil is very acidic, heavy lime applications may be necessary. Two-third of the recommended quantity of lime must be scattered over the planting area, mixed with the top soil and then ploughed as deep as possible, at least 9 to 12 months before planting. Calcium (lime) moves very slowly downwards into the soil and must therefore be applied in to the depth of the root zone. If a lighter lime application (2-4 t/ha) is required, the lime can be applied into the soil at least three months before planting. The lime should only be applied when the pH is lower than the target soil (water) pH.

Table 14 : Lime requirement (at 10 cm depth) for soils with different pH

Soil buffer pH	Lime required (t/ha) to bring the soil pH (water eq.) to	
	5.5	6.5
4.5	8.6	18
4.7	7.1	15
4.9	5.7	12
5.1	4.4	10
5.3	3.2	7.5
5.5	2.2	5.7
5.7	1.3	4.1
5.9	0.7	2.8

Source : Phil Moody and Bob Aitken, Queensland Department of Natural Resources and Mines.

In alkaline soils with a pH of 7-8, repeated applications of sulphur may be made at the rate of 500 kg/ha, until pH falls to 5.6. This is not indispensable, however, litchi can be grown in highly alkaline soils provided that micro-elements are available in sufficient quantity.

3. Field layout

Proper layout of a litchi orchard is necessary. After field preparation, one should decide the layout pattern of orchard. Similar to selection of site, any mistake committed during layout cannot be rectified at later stage. Therefore, the aim and objective of plantation should be very much clear before the actual layout in the field. The system of layout and distance of planting is decided according to needs. Laying out an orchard on level land is a simple matter of establishing a straight baseline, usually next to a fence or roadway. Then, lines at right angles to the baseline are established at both ends of the plot and one or two places in the middle. An easy way to establish these angles is to use three ropes whose lengths are in 3:4:5 proportions (based on the Pythagorean Theorem). Put the 4m rope along the baseline, then place the 3m rope at approximately a right angle, and finally, close the triangle with the 5m rope (Fig. 17). Adjust the 3m segment in either direction so that it just touches the end of the 5m piece. Ensure that the 3m section is at a right angle to the baseline. Next, place stakes along the baseline and the right angle line for sighting to extend these lines. From this point on, any desired row and tree spacing can be established using a tape measure or knotted rope to measure at the proper intervals.

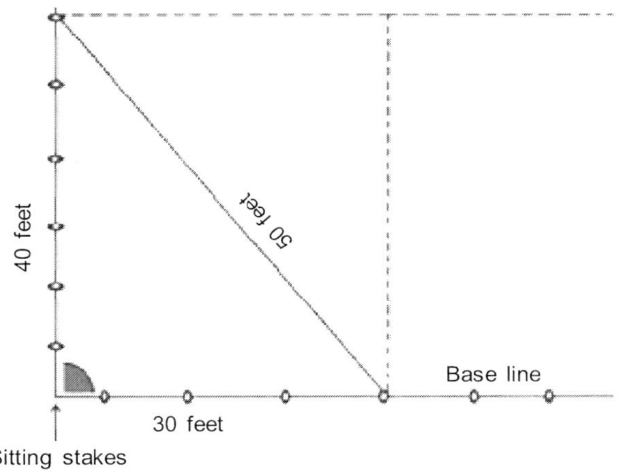

Fig.17 : Orchard lay out based on a triangle

Orchard layout

When boundary lines are drawn, it becomes easy to divide the whole orchard area into squares or rectangles with their sides equal to the planting distance. For planting under the hexagonal pattern, two chains equal in length to the planting distance are used. These chains are joined together by a ring. Two rings are also attached to the other two ends. Thus, the connected chain is equal in length to double the planting distance with one ring at centre and two rings at either end. First of all pegs or stakes are placed on the base line at appropriate distance. Then, the two rings at the ends are fixed in the two adjacent pegs on the base line and the chain is pulled tight holding the centre ring in hand. This tightened position is held and a stake or peg is fixed in the centre ring. This way another row of trees will be marked out and this can be used as base line for marking out a third now. For triangular planting, the markings are set as in the square or rectangular systems, except that those in the even numbered rows are mid-way between, instead of opposition in old pattern.

Contour layout for sloppy land requires the use of a surveyor's level and rod. The first row is at the highest elevation, and is staked out on the level (i.e. all points on the line are at the same elevation). Next, the steepest slope along the first row is found and the distance that has been selected as the minimum distance between rows is measured down the slope. From that point the next row is laid out on a level line as before. Moving away from this steepest slope to less-steep slopes, the rows will be wider apart. Wherever, the distance between two adjacent rows becomes twice the minimum distance, short contour row is laid out between them that point to the end of the plot. While planting litchi on contour, staggered method of planting is preferred between the two contours which utilizes the additional natural resources for production.

4. Planting methods

Among the prevalent systems of layout Square, Rectangular, Quincunx, Hexagonal and Contour are quite prevalent. However, for high density planting single hedge row system (Rectangular method) or double hedge row system can be adopted. For normal orcharding square system is very much common.

(i) Square Planting

In this method, the orchard space is divided into squares and the trees are planted at the four corners of the square in straight rows running at right angles. It is easy to layout (Fig. 18). All cultural operations in the orchard can be carried out in either direction.

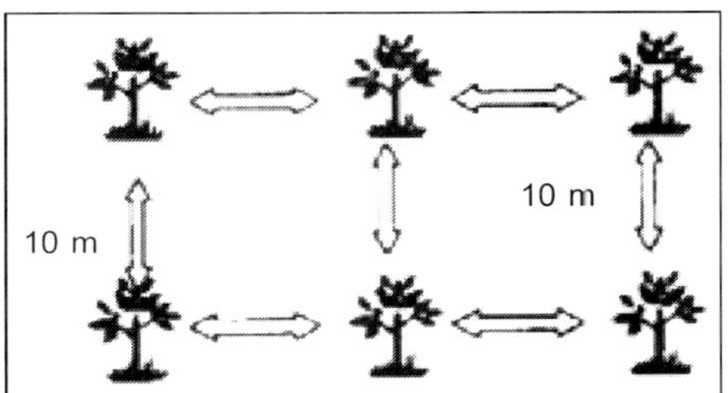

Fig. 18 : Square planting

(ii) Rectangular Planting

In this case, planting trees are set out in straight rows that line up at right angles to one another with spacing between them in a row somewhat less than that between rows forming a rectangle. Like square planting, this is easy to layout and manage with respect to cultural and harvesting operations. This layout is often adopted for high density planting system in single or double hedge row methods. The advantage of a rectangular pattern over a square pattern is that, by removing alternate trees in every row, additional width is obtained in the drive row when fillers are removed. This can only be accomplished, however, when the in-row spacing is greater than half the between-row-spacing. Such a plan allows for more space for the trees as they mature and also adequate space for power driven equipment (tractor/harvester).

(iii) Quincunx (Five spot)

In this, planting is done in the same way as in the square or rectangular pattern except that a fifth tree is set in the centre of the square or rectangle (Fig. 19). The central tree is usually the filler which is kept for a short period or till the main trees develop to full canopy size. Alternatively, this space can be utilized for growing perennial short statured plants. If the main trees become crowded, the middle tree can be headed back for few years and finally removed. This arrangement permits more plants than the main square or rectangular pattern.

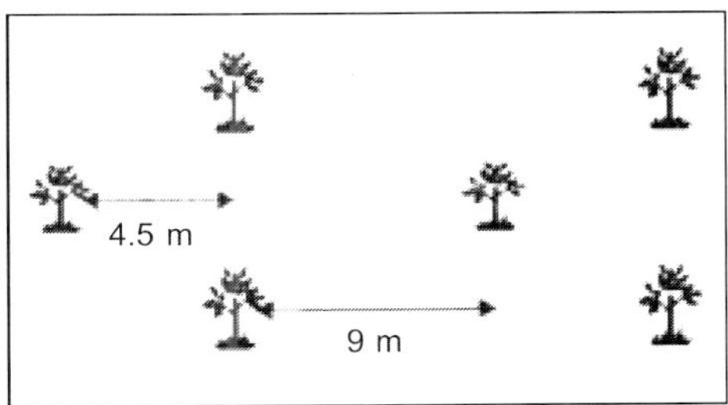

Fig.19 : Quincunx planting

(iv) Hexagonal

A pattern in which plants in alternate rows are spaced in the form of an equilateral triangle. The distance between the rows is equal to $(3/2)^2$ x a where 'a' is the length of one side of the triangle (Fig.20). Thus, though the trees are equidistant from each other, the row to row distance is less than plant to plant distance. This allows maximum use of the land with about 15% more plants per unit area, compared with a square pattern of the same dimensions. This is widely used where the land is expensive and fertile.

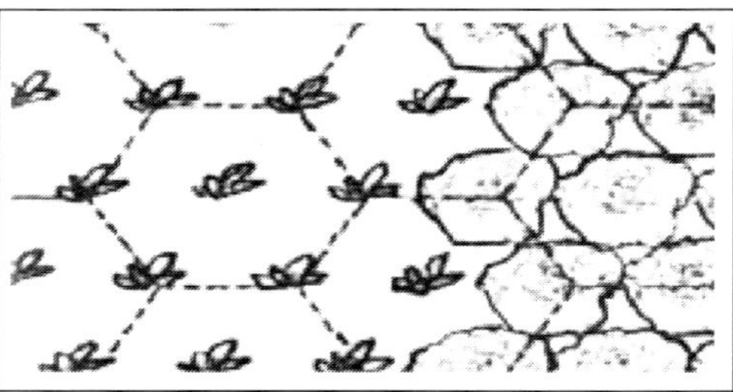

Fig. 20 : Hexagonal planting

(v) Triangular planting

A pattern in which plants in alternate rows offset half the space between plants in a row (Fig. 21). The distance between the rows is the same or more than that in a row. Thus a series of iso-scales triangles (two sides equal, instead of three as in an equilateral triangle) are formed. This is easier to layout than the hexagonal pattern but results in 9% fewer plants than the equivalent square or rectangle.

Fig. 21 : Triangular planting

(vi) Alternate (Double-Hedge row system)

This refers to the practice of spacing litchi trees in a row at about half the usual distance or adding additional rows to a conventional square or rectangular system. This arrangement is used for better use of the land during the early years of tree growth but does necessitate removal of the extra plants to keep them in bounds when they begin to crowd. After removal of the trees, the planting will become a square or rectangular one with original spacing. In double hedge row system, the additional row of the same crop and variety is permitted whereas; in hexagonal or triangular system other plants can be adjusted. The pattern in this system is consists of two closely spaced rows with a wide middle between the rows.

(vii) Hedgerow planting

In this system the litchi plants are spaced closely in rows. The distance between trees in a row is usually one half-one-third the distance between rows (Fig. 22). This is practiced only for high density plantings of litchi. Such plantings are still at experimental stage. The yield potential and longevity of the orchards under hedge row system are yet to be ascertained. Initial results have shown that such plantings are more remunerative during early fruiting years.

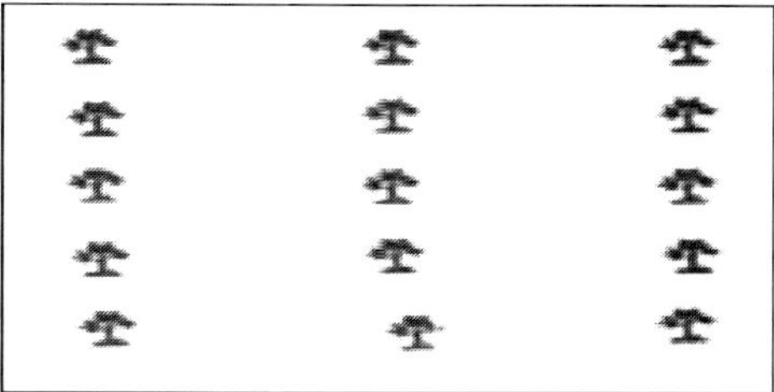

Fig. 22 : Hedgerow planting

(viii) Contour planting

The contour planting is used on rolling slopes or hillsides where some terracing may be needed. This planting arrangement permits production from sloppy waste land that otherwise could not be utilized. Considerable care must be taken to stop erosion by heavy rains or by irrigation by diverting the water to run among the tree rows rather than straight down slope. Any of the regular patterns described above can be used for planting trees on the sloping terrain. If the gradient of the slope is high, terraces are constructed following the natural contour as much as possible (Fig. 23). The edges of the terraces are slightly higher than the side against the hill so that the irrigation water remains in the terraces. Drip irrigation and sod culture greatly minimize the problem of soil erosion. While planting the trees, marking for the tree points is from lowest level to the top is needed. Rows or group of rows are laid out along the contour.

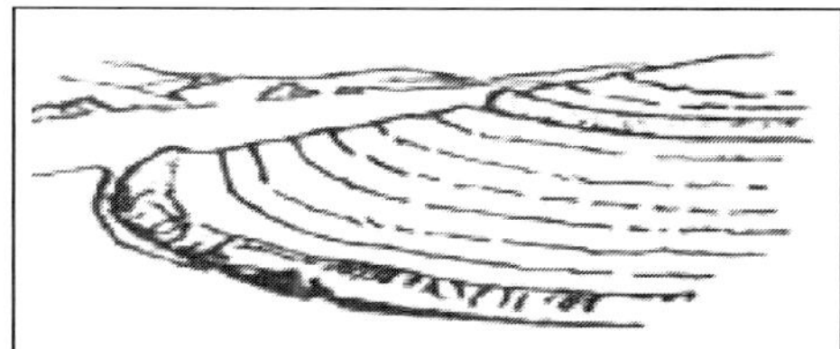

Fig. 23 : Contour planting

5. Dig the pits based on the soil type and other factors

Once the planting layout has been chosen and the land prepared, the pits in which the saplings are to be planted must be dug. This should be done several months prior to planting. The main purpose of digging and filling the pit is to provide congenial conditions for plant growth and development, especially during the establishment phase. The pits should be invariably opened during summer to expose its inner surface to heat and sun light. Pits should be filled before the rainy season for maximum settling of loose soil. To prepare the pit certain steps has to be followed.

In loose soils 60 x 60 x 60 cm planting pits are sufficient for litchi while in hard soils 90 x 90 x 90 cm pits should be prepared. In murrum soil conditions, where good soil is lacking, pit size may be increased up to 200 x 150 cm and pit soil is replaced by good soil. In locations, where the soil is loamy and deep, pits of 50 x 50 x 50m may be dug at desired distances. Whereas, at hill slopes, the pits should be of 1 x 1 x 1 m size along the contour. At sloppy places, the pits are prepared in form of trench by cutting the upper side of slope and leveling the lower side of the slope. In this way a leveled platform is developed near the pit. Pit should be made in proper alignment so that saplings can be planted in the centre of the pit.

When the soil is taken out of the pit, the upper half soil is kept on one side and the lower half on other, under normal soil conditions; however, in murram or problematic soil the whole soil of pit is kept aside and top soil is filled. Dug out soil and open pit are left as such to weather for 2-4 weeks during summer months to destroy any type of infection under sunlight and heat. In stony soils, it is better to separate all the stones from the excavated materials and left over soils should be mixed with manure and scrapped soil available in the field. Using a JCB machine for pit digging specially in hard and murrum soil is advisable because the pressure exerted by bucket of JCB cracks the hard pan which allows the litchi roots to penetrate.

6. Fill the pits with appropriate substrates

Initial establishment and vigour of the litchi plant is governed by the available substrate in the pit. Therefore, the pits should be filled with good quality substrate. Substrates mainly include farm yard manures, cakes, vermi-compost, *Trichoderma* rich manures and bio-fertilizers along with soil from litchi root rizosphere and VAM culture. In rocky and murram soils, which are almost zero in their inherent nutrient capacity, the initial substrate supplied in the pit governs the performance of litchi plant. As a common dose, 30-40kg *Trichoderma* enriched FYM, 2 kg neem/karanj cake, 250g SSP or bone meal is sufficient for one pit in normal soil. In problematic soil however, pit soil should also be replaced with good soil. In highly acidic situation, application of 1-2 kg slaked lime, 250g each of Zinc and Boron (Agriculture grade) is helpful to restore initial vigour of plant. Where magnesium deficiency is thought to occur, 500g of dolomite should also be incorporated in pit filling mixture. Where soils are very sandy or infertile, the super phosphate may be replaced by a 12-12-17 (46) or similar type fertilizer mixture not less than two weeks prior to planting and, if possible, three to five months in advance. Fertilizers carefully mixed with topsoil should be placed at the bottom of the pit. If the soil is having infestation of white ants, fenvulrate dust is mixed with upper as well as lower soil of the pit. The upper soil of pit along with substrate is filled first followed by lower soil mixture. High quality soil mixture particularly the soil beneath the well grownup litchi tree with probable mychorrhiza culture improves initial establishment and growth of litchi saplings. The main purpose of filling rich mixture in the bottom of the pit is that the plant will first proliferate their roots in $2/3^{rd}$ lower portion of the pit. The upper level of pit is kept 15cm above from the field level. After filling, if rain is not there, the pits should be irrigated to settle down the soil.

6. Apply mycorrhiza

It has been observed that litchi plant produces better growth if mycorrhizal fungi grows on its roots. Mycorrhizae are symbiotic associations that form between the roots of litchi plant and fungi. These symbiosis are characterized by bi-directional movement of nutrients where, carbon flows to the fungus and inorganic nutrients move to the plant, thereby providing a critical linkage between the plant root and soil. In infertile soils, nutrients taken up by the mycorrhizal fungi can lead to improved plant growth and reproduction. As a result, mycorrhizal plants are often more competitive and better able to tolerate environmental stresses than are non-mycorrhizal plants. Fungi make root tubercles in plants and play vital role in nutrient absorption. The mycorrhiza, *Rhizophagus litchi* belongs to the vasicular-arbuscular group of phytomycetous endophytes popularly known as VAM. The VAM is endophyte and cannot be cultured on artificial media therefore, the presence of living roots are necessary for its survival. Mycorrhiza is only found on short-lived sublateral roots. The fungi penetrate the roots through the epidermal cells into the cortex, whereas the root hairs, endodermis and vascular tissue are free of infection. Plants with the tubercles are larger and have more roots than plants without the fungi. Many workers have suggested that litchi requires mycorrhiza to grow satisfactorily, although healthy plants have been examined which were completely devoid of tubercles. Nevertheless, there are several reports that confirm the existence of the fungus in the litchi roots. The poor growth of litchi under alkaline pH has been attributed to micro-nutrient toxicity/deficiency or lack of mycorrhiza. It has been observed that the fungi grow better in acid compared to alkaline soils.

In majority of the cases, improved phosphate uptake is the primary cause of growth and yield enhancement. Highly significant positive correlation between the incidence of mycorrhizal infection and phosphorus and potash contents of litchi leaves has been observed. It is also seen that infection with the VAM increases uptake of P and K into the shoot and increased root weight, trunk diameter and fruit yield of litchi. Therefore, it is suggested that new plants be grown in soil taken from the vicinity of old trees to introduce the mycorrhiza.

7. Select healthy and genuine saplings

The establishment of a litchi orchard is a long-term venture, so it must be undertaken with great care. True-to-type, healthy and vigorous planting materials have a direct effect on the success of the litchi orchard. Healthy planting material of genuine quality makes the base of orchard efficiency and profitability. The quality of trees grown in the nursery will determine their performance when planted in the field. Therefore, the orchard must be developed with a good quality, genuine planting material.

Planting materials should be purchased from a registered nursery to guarantee the authenticity of the variety and hygiene of plants. Plants can also be produced by the growers adopting the standard procedures well in advance so that planting materials attain appropriate stage by the time of planting in the orchard. Under such conditions, reliability of layer must be ensured. Selection of genuine mother plant has great impact on orchard efficiency therefore, every care must be taken to select the mother plant for nursery or orchard establishment. The grafts prepared from poor, stale or diseased mother plant give poor health resulting into lower production and poor quality. On the other hand, litchi grafts of desired cultivar procured from genuine sources show better orchard efficiency and production.

Buyers guide for litchi saplings

The buyer must visit the nursery and examine the layers closely to see if they are of a good standard. Select good quality plants by examining the following :

- *Inspect mother orchard for the health and genuineness of layers*: Before purchase of planting material one should inspect the mother orchard for its health and genuineness. Saplings should not be purchased from nurseries which do not maintain the high yielding and healthy mother plants for propagation. It will be still better to select own mother plants and prepare grafts from them. Where attack by the litchi mite (*Aceria litchi* Kiefer) is feared, all plants must be treated several months prior to planting.

- *Don't purchase over aged saplings*: Saplings which have been in the nursery only for 1-2 months not over aged (young-less than one year or old saplings-more than 2 years) should not be purchased, as there are fair chances of their mortality. A good sapling should be straight and established for a year in the nursery.
- *Sapling height*: Height should be 40-60cm.
- *Leaves give clue of sapling quality*: A first-grade plant should have a few older leaves and followed by new flush sufficient for photosynthesis. The leaves of a nursery should have a glossy, green colour giving a general appearance of health. New leaves may have a coppery -red colour and these should not be mistaken for a deficiency. If the leaves are deformed or discoloured then there is something wrong with the plant. Such plants may be stunted due to incorrect nursery practices, leading to damage by pests, diseases or due to mineral deficiencies.
- *Shoot length and appearance*: Average size of shoot length is a good indication of a vigorous and healthy plant. Avoid saplings with compressed internodes and a rosette of leaves at the shoot apex. Plants having compressed shoot are normally deficient in nutrients.
- *Potting mixture in the sapling bag*: The soil mixture plays an important role in the health of plant. Mixture with high clay content limits drainage and results in poor root respiration and growth. A mixture without essential nutrients especially phosphorus, results in poor root development. A best soil mixture should have sand FYM and soil in right proportion to ensure good aeration and drainage in the polythene bag or pot and also to ensure good plant growth.
- *Seedling bag size and root system*: Small containers result in limited root growth. The plants may appear healthy, but restricted root growth in the nursery will greatly affect growth in the orchard. Polythene bag / pots of at least 20 x 15 x 45 cm are recommended for use in litchi nurseries. The suitability of the container and root system can be tested by purchasing a saplings trees and removing them from their containers. The root should be strong and there should be secondary and tertiary absorbing roots.

8. Harden the sapling before planting

Use only vigorous plants for field planting. Minimize the stress during field planting by hardening plants exposing to direct sunlight and with less water application before planting. This hardening improves of the success rate of field establishment.

9. Start planting at appropriate time

Litchi may be planted at any time of year, except in very hot, cold or windy weather. Planting is best done on cloudy, very humid days, which usually occur in spring, late summer, early monsoon and early autumn. Since litchi requires abundant water during its establishment phase, therefore if irrigation facilities are not available, the onset of monsoon is usually preferred. If the plantation is in an area which is prone to frosts, spring would be the best time for planting.

The best time for planting all over India is during the monsoon when there is sufficient moisture in the atmosphere. However, under North Indian conditions, litchi planting should be done preferably during rainy season (July to August). Planting can also be done during spring season (February to March) where assured irrigation is available. The spring season is short and followed by a hot dry period of low humidity and hence, high percentage of mortality of young plants is observed in the field. The best time of planting litchi in the heavy rainfall zone is the end of the rainy season. In dry areas, the rainfall is less; the planting can be done in the early part of the monsoon for better establishment. If it is planted during bright sun shine hours of day which turns out to be unusually hot and dry, the plants may wither due to excessive loss of water. If the sky is cloudy or sun light is hazy, planting can be done during day time also. However, it is better to plant the saplings in evening hours when the high humidity prevails in the atmosphere.

The plants should be watered abundantly and frequently after planting for the first fortnight. Care should be taken to ensure that the root ball and the soil around the roots are well moistened. It must be kept in mind that if the layers in the nursery are raised in a mixture containing large amounts of peat and vermiculite (the porous material) it may dry out rather quickly, hence watering should be judicious and need based. Mulching is desirable for a period of 2-3 years to assist with weed and moisture control and to have a steady supply of nutrients.

10. Plant the saplings at proper distance

Prior to preparing the land, spacing and layout must be finalized. The main purpose of planning and layout of an orchard is to provide adequate space to the plant for normal development, proper intercultural operation and easy passage of air and sunlight and orchard sanitation. Distance of planting in litchi is need based and dependent on various factors like nature and fertility level of soil, vigour of the cultivar, general growth conditions of the area and purpose of orchard establishment. In poor soils, litchi plants makes slow growth, so require less space while in heavy soils plant grow vigourusly and therefore require more spacing. A 25-year old litchi tree may attain a canopy diameter of 12m and will require adequate aeration and light around its entire circumference in order to produce a good yield. Minimum recommended distance between main crop trees is therefore 10-12 metres with a plant population of 70-100 trees/ha. However, distances may differ depending on the cultivar. For slow-growing varieties, in cold locations or in poor soils, trees may be planted at distances of between 9m x 9m (123 plants/ha) and 6m x 6m (278 plants/ha). Since litchi is a relatively slow-growing species, it may be advisable to have a higher-density plantation. Indeed, trees may be planted at distances of 5 to 8m, with about half of them subsequently being removed (and then being used as propagation material) as their branches begin to intermingle. For high density planting, 5 x 5m spacing is the best with almost all the cultivars except that of vigrours. As matter of fact, the vigorous varieties are not suitable for high density plantations. A compromise can be arrived at by planting litchi at distances of 7m to 9m and overcrowding is prevented by means of hand or mechanical pruning. Climatic factors of a place should always be taken into consideration while deciding the distance and planting of litchi.

Table 15 : Sapling requirement at different planting density

Sl.No	Planting distance (mxm)	No of plant/ha	No of sapling to be procured *
1	9 x 6	185	203
2	7.5 x 7.5	178	196
3	8 x 8	156	172
4	10 x 10	100	110
5	12 x 12	69	76

* 10 per cent more sapling are needed to meet out mortality

11. Ensure minimum root disturbance of saplings

While transporting the plants, care should be taken that roots of the plants are not damaged and also ensure that the root ball in the bag is intact. Litchi do not like having their roots exposed or damaged during transplanting. To get good soil contact, the loose soil must be thoroughly consolidated around the root ball. Now a days, mostly the litchi plants are propagated in poly bags and plant becomes ready for planting within one year. These saplings do not have much disturbance in root system. However, plants after detachment from the mother trees, face great problem of root damage. Plants in poly bag are easy to carry and transport with minimum mortality, however, gooties raised in nursery beds should be dug out carefully with minimum injury to the roots and transported safely up to field. During digging it should be kept in mind that earth ball is clean, intact, medium in size having about 1.5-2.00 kg weight and packed properly with grass or straw. Heavier and bigger earth ball is likely to break and require much space for transit. The wrapping of earth ball plays important role in transportation. If loosely wrapped, chances of breaking of ball are more pronounced as compared to tightly wrapped. The earth ball should be tied at collar portion as well as in middle of the ball. Such type of tying avoids breaking of ball during transit. For transportation, layers should be packed in medium size baskets accommodating 10-12 layers in each basket. Before packing, the dried grasses are put at the bottom of basket which works as cushion and saves ball from jerk. Plants can also be packed in plastic crates. After making cushion, the so saplings are put in basket and they are tied together with string. Finally, the gunny bag piece is placed over the earth ball portion of the basket. Such packed litchi layers are transported to any distance by road or train without any damage to the plants. This is for plants with earth ball whereas, the plants raised in bags can be easily transported in the plastic crates.

If the distance is long and more time is required in transit, the layers are sprinkled with water on the way and are handled gently during loading and unloading. After receiving the layers at the particular destination, it should be kept under the shade and sprayed with water before opening of packing. Layers with sound and intact earth ball should be planted in the field and damaged ones should be put in the nursery for hardening. Generally, 5-10 per cent more plants is needed than the actual requirement. These additional plants are kept reserved. Plants should preferably be planted before the root system out grows the containers. This will ensure the presence of strong root system that can require lower levels for water and nutrients.

12. Adopt right method of planting of saplings

Under no circumstances the plants should be carried by the stem. Dig a slightly bigger hole than the size of earth ball at the already marked point to accommodate the root-ball. Before taking out the polythene bags from root ball, it should be moist with water for 5 minutes which helps compact earth ball. Thereafter, cut around the bottom edge of the polythene to remove the bottom polythene. Put the plant with its intact ball of earth in the hole in such a manner that the plant remains straight and collar portion of plant is in the level or slightly higher than the ground level. Then remove the polyethylene bag with two longitudinal cuts from bottom up. After removing the polyethene cover and placing the plant in the hole, fill the planting hole with soil and slightly tighten the soil by pressing the moist soil of the pit all around the root ball so that the plant is set firmly in the field. These steps help to reduce root damage due to breaking and splitting of potting media block. The right place of planting saplings in the pit can be done with the help of a planting board.

While planting, one should be careful that the earth ball does not break during pressing the soil. A small basin is then made immediately after planting for irrigation. The frequency of irrigation in the first month depends on the weather. Planting should not be done so deep as to bury the major portion in the soil or so high as to expose the upper roots. Allow the plant to grow directly up. Use a stick closer to the plant and tighten it into the stick carefully. Regular checking of plants is essential to detect the faults like sinking of soil, tilting of plant and cracking in basin of plant after planting the grafts. Sufficient moisture should be maintained till the plants do not set or start new growth in the field.

13. Arrange sufficient water for irrigation

After planting, watering of plant is a must. Water is a crucial input for establishment of litchi orchard. Assured availability of water and appropriate irrigation system is the most important aspect in a commercial litchi orchard. The type of irrigation system can be decided on the basis of water availability and capital budget. Flood, basin, check basin and furrow are the common methods; however drip method can also be used for litchi orchard establishment. Under tree, drip or micro sprinkler system is the best on account of water saving but the requirement of power, technical expertise and high initial cost makes the system most expensive as well. In the upland, with less irrigation facilities, litchi plants can be raised by providing life saving irrigation from natural water collected in plastic lined pots/Jal kundas/open ponds. Under these situations, practices of *in situ* water harvesting and moisture conservation near plant root zone should be practiced.

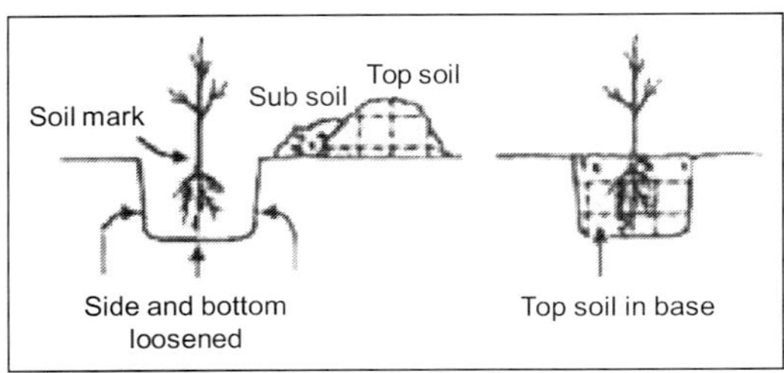

Fig. 24 : Right method of planting

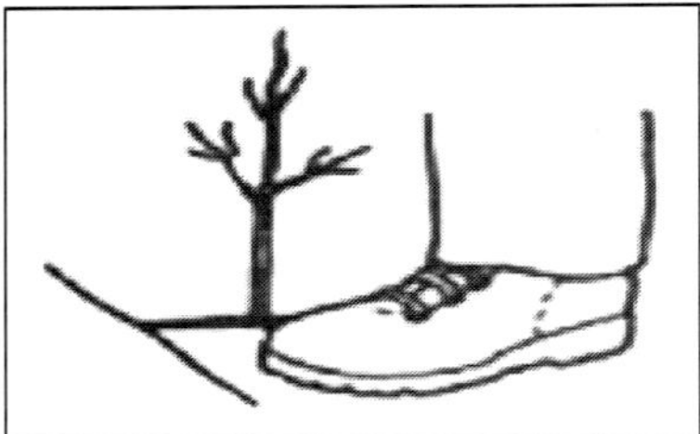

Fig. 25 : Soil pressing

14.Aarrange tree guard and staking

Newly planted litchi trees should have a tree guard of bamboo wooven round or rectangle or square shape installed at the time of planting. It ensures protection from winds, bright sunlight and frost burn of main stem and branches. It also helps partially protection from small ruminants. The tree guard is placed in a perimeter about 50cm from the tree to a height of about 1.5m. However, within one year, this tree guard should be removed.

Fig. 26 : Tree guard

Plants may be staked, at least for the first year until they are well established, to prevent damage by occasional winds. Stakes and ties add to the cost of the plant establishment and the maintenance requirement. Trees need to be staked against movement by the wind, but only a short stake should be used with a fairly loose tie. The reason is that the flexing of the plant in the wind stimulates root growth to naturally brace the plant, and also stimulates growth in stem diameter at the base of the trunk. If the plant is so tightly staked that it cannot way, this natural bracing does not develop, and it is then very likely to snap or be blown over when the stake and tie are removed. Don't use a taller stakes. Stakes pressure-treated with preservative are not required, as the stake only has to last about two seasons.

Push the stake into the bottom of the pit on the windward side of the tree as close as possible to the root ball without damaging it. The stake should finish at a height no more than one third of the height of the tree. Loosely attach the tree to the stake, using a rope. Rope should form a figure of eight or have a space between the plant and the stake to prevent chafing. In many cases, ropes need nailing to the stake to stop them slipping, unless they can be fitted just above a branch. Check the plant and stake frequently and remove the rope at the beginning of the second growing season, by which time the plant should be securely anchored by its roots. If it is not, then the plant is not growing sufficiently well and other steps will need to be taken to encourage the plant into growth. Double stakes can also be used as shown in the diagram. These are more reliable and less likely to cause chafing. They also act as protectors against careless mowing, which is a significant cause of damage to young trees. It could be done by planting four stakes one metre high in a square at one metre intervals and attaching some synthetic cloth of the same height to them of the same height as the plants and suitably intertwined around them.

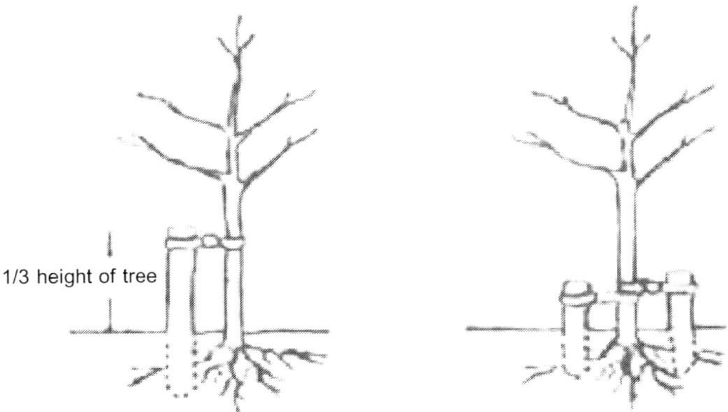

Fig. 27 : Method of staking

15. Apply mulch and provide shade

Use mulch around the plant using easily available mulching material such as dry grasses. Mulching helps reduce soil temperature in the root zone, controls weeds reduces drying of soil and soil erosion. Provide shade appropriately to protect plants from heavy sunlight.

16. Replanting if failed

In spite of all possible care some litchi plants die during first two years of transplanting. The extent of such deaths does not exceed 5% or maximum to 10% of the total population if proper care is taken. These gaps are filled by replanting fresh saplings of the same cultivar. The replanting can be taken up during the planting season or off season except in severe winter or summer months. The most common faults observed in establishing the newly planted trees are improper irrigation and lack of weed control. These two faults cause more loss of plants than any other factor during the initial years. Proper attention must be paid to irrigation and mulching.

When an old litchi orchard is terminated, it is not replanted again. This is simply because, owing to long life of the orchard, the soil gets impoverished and the pests and diseases associated with the crop are fully established. Thus, if the same site is replanted, the success of the second establishment becomes doubtful.

17. Make provision to drain excess water

Drainage of excess water is a problem in litchi orchards which is accumulated in heavy soils or low lying flood prone areas. In these areas, poor provisions are reflected by the poor growth, sickly appearance or death of the young plants. Building up of a persistent water table may damage root growth and the functioning of roots is affected. High water tables and high soil moisture at critical stage of plant establishment and production cycle have been found harmful. Water logging affects growth of young plants and flowering in grown up trees and subsequent productivity of orchard. The main objectives of a proper drainage system is to remove the excess water from the active root zone during rainy months and as far as possible to keep the water table in the orchard below the active root zone, (1.80-2.40 m below the ground level). Soils with too much of slope are prone to erosion and too much of drainage in such soils also get moisture depletion quickly and require frequent irrigations.

18. Prepare live fence and shelter belt as wind break

Encroachment of trespassers, theft and animal grazing are the matter of common concern in litchi orchard. Fencing is done in two ways; fencing by using cement pole and wire or by erecting boundary wall and live fencing by growing of specific plants species. The first method of fencing is costly and can be afforded by rich farmers only. Live fencing is an age old and traditional practice. The fences act as wind break also if grown by using live plants particularly in the areas where cold and hot winds are the problem. The main purpose of fencing is to protect the plants from stray animals and harmful effects of cold and hot waves. Windbreaks are practically indispensable even on hillsides, since they serve to promise high humidity. Besides protection, the main function of live fences, to provide fuel wood, fodder and food, and enriches the soil. Therefore, fencing is an important aspect in litchi cultivation.

Apart from high loss of water from leaf and soil surface during windy days, the wind causes fruit drop in litchi. Establishing wind breaks such as artificial structures or fast-growing trees on the upwind side of prevailing winds helps in reducing the wind damages. Windbreak trees are commonly tall having dense foliage and keep the surrounding atmosphere humid. These trees help in minimizing the wind velocity, low or high temperatures which affect adversely the young plants and immature fall of the fruits.

Traditional farmers have vast knowledge about the practice of live fencing and the species used. Planting live fence with tall growing trees as first row and small, bushy thorny plants in staggered manner as second row along the boundary of litchi orchard is very much important. In sloppy lands, wind breaks should be planted along with contour in ascending and descending order to protect the plants even at higher points. At such places, the live fencing can not be combined with wind break and must be planted along the boundary. Having a live fence around the farm has multiple benefits. It does however; take two to three years to develop. The species suitable for live fence should be thorny, inedible and non-grazable for cattle and goats, hardy and relatively maintenance free (other than pruning /lopping), adaptable to the local conditions, fast-growing with ability to produce something that can generate some revenue. Live fences can be divided into two basic categories on the basis of their growth habit and use.

Live fence posts

Fig. 28 : Wind break

Fast growing hardy plants species can be grown as fence post at a spacing of 4-5m. These plants commonly grown by seeds or maintained from natural vegetations growing along the boundary to have strong root system. Farmers can also plant stakes of easy to root species such as *Gliricidia sepium, Moringa,* Flame tree etc. The live fence posts are more durable than traditional wooden poles as they are less prone to attack by termites and decay fungi. When grazing animals are part of the farming system, the only way to establish live fence posts and a living fence, is to start with a conventional wire fence supported by wooden pole and to gradually establish live fence plants to substitute for the decaying poles. *Gliricidia* live posts fence can be established by planting a few large (1.5-2.0 m) stakes in the existing conventional wire fence. These stakes normally take root within a month or so. After the first pruning, subsequent pruning can be carried out every 4 to 8 months. Shoot pruning at intervals of 6 to 8 months result in woody sprouts that are suitable for use as stakes. These pruned twigs can be used to multiply live stakes for fence posts within a year or two after establishing the first live fence posts. Tar, Babool, Simaruba, Jamun, Jangal gilebee, etc. can also be planted as fence posts.

Live barriers or hedges: Hedges are thicker, more densely spaced fences that generally include a number of thick growing thorny species which do not require support of barbed wire. To minimize the initial cost of litchi orchard establishment, it is advisable to establish dense and thorny hedges to protect the orchard. Another alternative often used by farmers is the combination of easy to establish live fence plants that are poisonous and unpalatable species. If well established, these natural barriers can deter both animal and human trespassers from entering into the orchard. A live fence should ideally be planted just before the monsoons and watered regularly after the rainy season is over to ensure optimum growth. Agave, Sisal, Boradi, Karonda and Mehandi etc. are the promising plant species which can be planted as live hedge along the litchi orchard boundary.

The trees, which are generally used as windbreak, are seedling mango, jamun, mulberry, moringa, jackfruit, carambola, shisham, teak, gamhar and bamboo. These can also be used as live fencing posts. These trees are usually planted close to each other to provide an effective shield. Tall growing trees like shisham and jamun are planted 6m apart while low headed trees like mulberry and carambola are planted at 7m distance. Normally, plants grown as wind break may extract nutrients from litchi plants. Therefore, a trench of one meter depth should be dug between litchi row and wind break.

❑❑❑

Chapter 7

Care and Rear the Developing Plants

Litchi saplings planted in June-July take full advantage of monsoon season for their establishment and initial flushing. Up to October, plant grows well except the attack of some pests and put forth one or two new extension flush. Litchi plants are supposed to bear fruits 6-7 years after planting. By this time, plant attains a definite framework and becomes physiologically sound to bear the fruits. Therefore, tree management system of developing plants should be followed until the management system is switched over to that for bearing trees. In the first two years, plants need some special care for better growth and establishment. Young orchards in particular, suffer readily from poor management and hence delay in bearing of commercial crops. Old trees, though more tolerant because of their greater reserve of food materials, will nevertheless, respond readily to cultural management. Moreover, it is not necessary and often not desirable to follow a single system to manage the orchard throughout its life. An orchard is handled in one way while the trees are young, and in different way when the trees are mature. Following care is taken initially in establishing litchi orchard to avoid future problems.

Do not neglect plant basins

Proper aeration near root zone and good sanitary conditions in the basins of newly planted litchi sapling is a must for keeping the plants healthy and disease free. Proper growth of young plants can be ensured by timely hoeing and weeding of the basin which is considered to be the essential and should always receive first priority. If intercropping is being practiced, the litchi plants should not be kept in neglected condition during pre-bearing stage. After harvest of every intercrops, the plant basins should be properly cultivated and essential nutrients should be applied. It should also be seen that the pest and diseases of inter crop are not causing damage to litchi plants. Any such neglect of young orchards proves costly as fruiting is delayed and longevity and productivity of the plants are reduced.

Protect plants from cold and hot air

Newly planted litchi saplings face first winter from November-February under North Indian conditions. For the successful establishment and growth during winter, it is necessary to protect the newly planted sapling from cold. In North India, some times, winter becomes more severe leading to occurrence of frost. The frost damages the young leaves and growing tips and finally the plant may die. Therefore, the protection of litchi saplings for at least one winter or in some cases two seasons is very essential. It is advisable to establish litchi orchards in such a manner that the rows run diagonally to the prevailing wind direction to avoid creating a tunnel effect. Before planting of litchi orchard, it is essential to reserve some place for planting of wind breaks at the border sides of orchard from which hot and high winds and frost are expected. To protect the plants from frost, thatching of individual plant is essential. They can be thatched with any type of dried grass or dried stalk of maize or *bajra* along with paddy straw. The plant is covered from 3 sides leaving a gap towards east to allow entrance of sunlight and air. For protection against frosts, the young litchi plants should be covered properly well in time and the thatching should be removed only when the danger is over. Some growers remove the covers too early, with the first warming up of temperatures after winter; but the trees are likely to suffer if late frosts occur. Basin irrigation, mulching and spray of water before anticipated frost date may be helpful in reducing the extent of loss. Occurrence of hailstorms is a natural phenomenon which causes partial or total loss of litchi crop. Likewise, it must be protected in summer to avoid scorching from hot winds (Loo) and intense heat of the summer months.

It is necessary to provide support to the plant by staking in the areas where in high winds is more prevalent. The staking is helpful to keep plant stem straight and save the plant from damage. When litchi have grown taller, the stem may be white washed with lime to protect from sunscald.

Follow proper water management practices

Young litchi plants must be frequently irrigated during the drought spell in rainy season, winter and summer months for at least one year. When the plants are well established in the field, irrigation may be given as per need preferably on the basis of soil moisture status. During vegetative growth period, every effort should be made to maintain sufficient moisture near root zone of the plant. Generally, irrigation at weekly interval during summer and fortnightly interval in winter along with mulching should be done in the first two years of plant establishment. However, the quantity and frequency may be modified depending on the soil type, climatic condition and method of raising orchard. Lighter soils will have to be irrigated more frequently than the clay soils. Irrigation frequency may be adjusted according to rainfall and evapo-transpiration during each season. During rainy season, water should not be allowed to stagnate around the plant as it will affect the root activity adversely. Drainage may be provided near root zone by making slope at lower side. In marginally water logged area, planting of litchi at raised platform can also be done. For more details, chapter-8 may be referred.

Apply balanced fertilizers

The litchi plant needs more nitrogen at young stage to promote plant growth, therefore during the establishment and growth phase, higher proportion of nitrogenous fertilizer should be given than phosphatic and potassic fertilizers. As the plant grows, the requirement of fertilizer increases and therefore, the doses should be increased annually until the plant attains full growth. For efficient use of applied fertilizer, it is always better to split the annual quantity of fertilizers in 2-3 parts and apply in the plant basin. The loss of fertilizer can further be minimized by avoiding its application in the peak rainy season. As a matter of fact, fertilizer in young litchi plants should be applied during October and March whereas, the manures should be applied during pre monsoon period (June-July). Fertilizers should be placed in a band spread in circle approximately 30cm away from the base of the plants and should be incorporated to the soil mechanically without disturbing the roots. The distance of band should increase as the plant grows. Sufficient soil moisture is necessary after application of the fertilizer mixture. Soil incorporation is not necessary if under tree sprinklers are used for irrigation of trees as fertilizers can be effectively incorporated by irrigation water. For more details, chapter-10 may be referred.

Keep orchard weed free

Weeds should be completely removed from the root zone of litchi plants. Weeds, not only compete for water and nutrients applied to plants but also shelters many insects and diseases. Use of deep cultivation tools are not recommended in young litchi orchard as it disturbs the soil and damage active roots. However, alternative methods to suppress or control weeds should be applied. Uses of systemic herbicides and / or mulching with plastics or organic wastes have been found to control weeds satisfactorily without disturbing the roots. Chopping of the weeds is more beneficial than doing clean weeding around the base of plants. Weeds in between plants and between rows should be controlled mechanically, using a tractor mounted or a hand operated grass cutters or by manual slashing. Dry weed can be used as mulches around the plants. Which ever chemical weedicides is used, it is better to use it around the trees. Care should be taken to avoid direct contact of weedicides with the green tissues of plants. If weeds are dried, mow the weeds before rain and apply chemicals when weeds are in active growth stage to maximize the efficiency of chemicals and effective weed control. For more details, chapter-8 may be referred.

Apply mulch in plant basins

Mulching the trees is an important and very useful practice. Mulches are effective in conserving soil moisture, maintaining root zone temperature, controlling weed and soil erosion. Most of the organic materials including farm waste and other bi-products can be used as mulching material but hard to decay materials with high C:N ratio are better. Materials like rice straw, rice husk, dried weeds, saw dust etc. can be used for mulching purposes. When mulches are used, care should be taken to inspect trees regularly for any termite activity to take appropriate control measures. Termite activity may be increased with mulch application. For more details, chapter-8 may be referred.

Initial canopy architecture

Providing a better frame is the first step to make plant prepared for heavy yields expected under good management. Almost one year after planting, when plants are established, every care should be taken to maximize the plant growth so that the plant can be pruned and trained to provide good framework with well spaced spreading branches and open center canopy. For a proper framework, do not allow branches at the base of the main stem. Such branches should be removed / pinched off in the early stages of growth to leave a standard main trunk without any branch up to a height of 70-80cm. Between this height, if 3-4 side branches are there, allow them to grow as limb but if branches are not there force the plant to produce heading back or twisting the main trunk at approximate height and ensure 4 well spaced shoots in each direction. These will develop into limbs. After 60-70cm, each limb should have 2-3 primary branches. Each of these primary branches should culminate into 2-3 secondary and further into 2-4 tertiary branches. To train branches as spreading limbs, staking of individual branch can be done to avoid breaking. This way, plants can be provided with better canopy architecture. The pruning and training process must be continued for 2-3 years because all these can not be done in a single operation or two. Therefore, growers must regularly visit the plants and train them as and when necessary. Pruning should be commenced in well established plants before any anticipated major growth flush. For more details, chapter-11 may be referred.

Grow inter crop and cover crop

Clean cultivation throughout the year is not desirable in litchi plantation. Therefore, vacant space between the trees can be utilized for growing suitable intercrops. The selection of crops should be purely area based and as per need of the farmer. Preferably, the leguminous crops should be used. For more details, chapter-9 may be referred.

Maintain bio-fencing and wind break

Bio-fencing should be maintained on sides of orchard to save the young orchard from cattle and wild animal. The method of bio-fencing, chapter-6 may be referred for the detailed information.

Protect plant from insect pests and fire

During dry periods, termites may damage underground parts of the plant. To overcome this problem apply a band of engine oil on the base of the stem up to about 45 cm height. Similarly, the measures to control the pests and diseases should be taken time and time to save the developing litchi plants. For more detail refer the chapter on insect pest management.

In the dry zone, heavy winds blow during May-June. Most cultivated lands are damaged by fire at this time of the year. To protect the litchi plantation by spreading fire, have a 5-10 m wide fire belt around the orchard plough and harrowed to remove dry grass. Also during the dry months of the year, keep an area of about 3 m around the base of trees weed free. This might protect trees even if a fire spreads over the orchard.

❐❐❐

Chapter 8

Avoid Moisture Stress and Weed Excess

Water is one of the most important factors limiting the production of fruit crops. Lack of water can lead to declining tree health, lower yields and poor quality fruits. In fact, every aspect of tree growth is affected by water stress or drought. The long term viability of orchards, therefore, depends on avoiding water stress, particularly during critical stages of tree's production cycle.

To obtain the maximum yield of litchi, tree must not be allowed to be deficient of water at any time during its life. In commercial practice, this ideal situation can rarely be achieved. Shortage of water, labour or equipment can result in period of crop water stress. It is, therefore, important for the orchardist to know the response of his litchi trees and to manipulate these to his best advantages in endeavouring to obtain maximum profit with minimum expenditure on water. Under some circumstances, maximum yield can be obtained if the crop is permitted to suffer water stress only between moisture sensitive stages, but in most of the cases this will result in yield reduction. What matters, however, is the profit in term of return per unit of cash spent and this can be obtained by accepting a total yield less than the maximum but expending each liter of water in a way by which it produces its maximum response. This usually means making the best use of natural precipitation and soil moisture reserves and applying artificial irrigation only at known moisture sensitive growth stages. Irrigation in litchi is required for:

- For establishment and growth of young plants
- Maintenance of tree health & vigour
- Reducing fruit drop & fluctuation in yield
- Production of quality fruits
- Reduction in cracking of fruits

The litchi is very sensitive to water. The rapid growth of the plants during the first five years can be achieved by high rates of irrigation and adequate nutrition. The timing and rate of irrigation depend on the normal rainfall pattern which varies from place to place and year to year. The amount of water required, will also vary with soil type, root depth, age of tree, intensity of transpiration, season, topographical location, wind fluctuation, etc. Thus, it is not easy to determine the amount of water required for optimum growth and production of litchi trees. In fact, to determine irrigation requirements, one must know the evapotranspiration (ET) losses. This is because only a small fraction (1%) of water is used in the plant metabolism (photosynthesis and other bio-chemical reactions). Most of the absorbed water (often as much as 99%) escapes as vapour in the process of transpiration by the plant canopies. Therefore, if the rate of water loss through leaves and the plant and evaporation from the soil are estimated precisely, it would be possible to predict the water requirements of the crop more accurately. Unfortunately, limited information is available on the transpiration rate of litchi trees under different agro-climatic situations. Since the rate of transpiration is greatly dependent on the atmospheric condition of a particular place and time, it is important to work out the transpiration rate for the particular area or locality.

Fig. 29 : Healthy litchi plant

When preparing a litchi irrigation programme, a number of general factors, such as climate, the plant and soil must be kept in mind. Although litchi prefers light textured soils, where the water-holding capacity is much lower than that of heavy soils and therefore, more frequent but lighter irrigation is needed. It is also worth mentioning that shallow soils have a lower water retention capacity and water must be applied at shorter intervals. The water retention capacity of soils covered with mulching or planted with cover crops (desirable for litchis) is higher than that of bare soils.

Climatic factors are also extremely important. Evapotranspiration for instance will be higher under conditions of high temperature, low moisture and high wind and with a high leaf surface index, which means that differences may exist depending on the cultivar and / or foliar density. Although litchi can tolerate flooding for short periods, a good drainage system is particularly important especially on hill sites where the subsoil is usually less permeable than the topsoil. Since the effects of soil salinization on litchi has not been adequately studied, in areas where this is likely to occur, it would be advisable to leach the salts out of the soil. In the absence of any comparative studies dealing with other crops, leaching with double the quantity of water used for normal irrigation may be done at six week intervals during the irrigation period. It is of course unnecessary to do this in high rainfall areas.

Prior to planning a litchi irrigation schedule it would be worth to remember some of the points enlisted below.

- Litchi trees need regular watering and therefore, it is essential that enough water must be available from the flowering stage after the February/March flush till the crop the harvest.
- Because the edible portion of the litchi fruit has a water content of 86%, the availability of water remains important during the fruit development period.
- A water shortage will delay development of the fruit and adversely affect the size, mass and quality of the litchis.
- Need based irrigation must continue after harvesting to ensure a normal growth flush.
- After the final vegetative growths flush (October-November), the irrigation must be withheld or reduced till the fruit set.
- Young plants that are not producing yet are irrigated throughout the year.

- Litchi requires abundant water particularly during its development phase, when optimum rainfall level in the region is below 1000 mm / year.
- Abundant rainfall, mist and excessive irrigation during flowering may reduce fruit set.
- Too high or too low moisture conditions during fruit maturation may cause skin-cracking.

Stress and non-stress periods

Litchi trees require adequate soil moisture during fruit setting and subsequent development of the fruits till full maturity. The high soil moisture level during maturity reduces peel cracking. Lack of moisture during blooming is also found to be detrimental, as it reduces the amount of fruit set. The effect of drought on litchi tree has been studied in detail and inferred that it can cause reduced photosynthesis, leaf wilting and dropping, poor flowering, small and cracked fruits, excessive fruit drop and even tree death.

However, ecological and experimental evidences seem to indicate that floral initiation in litchi is promoted by water stress during the preceding autumn and winter. Several workers are of the opinion that high soil moisture prior to floral initiation promotes vegetative growth and suppresses flowering, while low soil moisture restricts vegetative growth and promotes flowering. In contrast, in some cases there has been little or no correlation between soil moisture and flowering of litchi, probably because the trees were deep rooted or because the range of soil moisture within different plantings was too narrow or because of the interaction between soil moisture and other factors.

Wetting depth

Since the major portion of the active root zone of litchi trees is located in the upper (0.6m) soil profile, it is believed that water application to deeper layers is not necessary. Moreover, it is questionable whether the entire root zone must be constantly supplied with water. If the volume of wet root could be reduced without affecting the yield, a saving of irrigation water might be achieved. The wet root volume can be decreased either by reducing the volume of water applied to the entire surface area, reducing the depth of soil wetted or by reducing the proportion of the surface area wetted without changing the wetting depth. The partial wetting of the soil volume reduces the total water use, but increases the frequency of irrigation required to maintain equal levels of available moisture in soil. However, in some fruit crops like orange, the increased water stress symptoms have been noted by the partial wetting. Therefore, it is believed that partial wetting causes desiccation of new roots developing in the non-irrigated soil.

In the past 2-3 decades, impact of partial wetting of the root zone has been assessed in detail. The partial wetting of the root zone by sprinkling and tendency to apply insufficient water give a low financial return. In developed countries, when sprinkler system came into general use, the amount of water applied often is inadequate for good yield. But the failure of partial wetting by sprinkling has been found to occur mainly in regions where winter rainfall is insufficient to wet the entire potential zone. Thus, use of the partial wetting method must be restricted to those localities where natural precipitation is high in winter and where the root zone moisture is close to field capacity. Sprinklers below the litchi tree canopy improves the micro-climate and causes cooling affect during fruit development also.

Table 16 : Root depth and water extraction pattern

Soil depth (m)	Water extraction (%)
0.0 – 0.3	28
0.3 – 0.6	23
0.6 – 0.9	17
0.9 – 1.2	19
1.2 – 1.5	13

Source: Menzel *et al.* (1995)

Determining irrigation requirements

There are different approaches for determination of irrigation requirements in litchi.

1. The actual measurement of soil moisture and subsequent calculation of available moisture.
2. Calculation of actual water requirement from meteorological data like solar radiation, temperature, wind velocity, precipitation and evaporation.
3. Water status in plant tissues which may give an indication of the water requirement and the timing of irrigation for crops.

4. The integration of plant and soil based measurements for deciding the irrigation

Plant and soil based measurements may also be integrated with atmosphere based measurements or any analogous system may be followed for a better prediction. Under certain conditions, practically all these approaches might provide a sound basis for irrigation scheduling. In fact, based on various methods, different techniques of irrigation scheduling have been developed. However, to have better understanding of these methods, one must be conversant with methods of soil-water measurements used for its expression and appropriate interpretation.

(i) Irrigation based on soil moisture measurement

There are many ways of measuring soil water content viz. gravimetric analysis, by using a tensiometer or a resistance block or a neutron moisture probe. All these methods have their own limitations and are widely used depending upon the user's choice and access to them.

There are traditional method of determining when and how much to irrigate. The idea is to observe the moisture reserve of the root zone which gradually diminishes following each irrigation, so as to know when that reserve has been depleted to a predetermined level to serve as the minimum allowable level. At the point the irrigation is to apply, the volume of water calculated to replenish the soil reservoir of the root zone to its "full" level i.e. field capacity level.

A precondition to effective management of root zone soil moisture in litchi, is to establish its roots at a specific depth which of course varies with tree age. Root density or the soil volume occupied by roots usually increases when the plant grows. Roots of fruit tree, in general, extend down to 1.5m. Litchi trees, have been found to have a shallow root distribution when grown on a range of acid soils in subtropical climate. Although some roots are found down to a depth of 100cm, most roots are located in the top 0.20 to 0.60cm of the soil. The roots which are found at greater depths, their contribution to the feeder root system is less than 3%. The highest feeder root density lies in the 30cm soil layer. The depth of root is more in fine textured soil compared with coarse textured soil but the total amount of roots in the profile showed the reverse trend. In a deep calcareous sandy loam soil of subtropical India, most roots are located in top 45cm. The deep root system is capable of absorbing enough water from the water table during dry season to support a large crop. Maximum root growth is observed prior to flowering and after harvest. Being a shallow rooted crop in most of the cases, it has been observed that maintaining a grass-covered sward has a detrimental effect on both roots and shoot growth in litchi trees, because of competition for water and nutrients.

Irrigation scheduling in litchi must be controlled by the use of 2 tensiometers, 1 installed in the middle of the root zone and the other below the root zone. The top tensiometer indicates when irrigation is to commence, while the lower tensiometer is used to indicate when there is adequate reserves of soil moisture at depth. Irrigation in litchi orchard must take place when a 300mm tensiometer rises to 30cb on a clay soil and 10cb on a sandy loam. Since most (80-85%) of the available soil moisture is drawn from 0-60cm layer, the amount of water applied must replenish the deficit from field capacity in this layer for all practical purposes.

Irrigation timings in litchi is very important, therefore, one must keep in mind that for effective utilization, it should be applied before soil moisture becomes limiting to litchi trees. As a thumb rule, water is applied when 50% of the available water (i.e. 50% of the water between field capacity and the wilting point) in the root zone (0-60cm) has been depleted or when water tension in the zone of rapid water exceeds 4 bars. If further depletion is allowed, the plants may be subjected to a level of stress that might cause an appreciable reduction in yield. Though the highest water use efficiency is observed with irrigation at 45% ASM (available soil moisture) depletion, the best yield could be harvested with only 30% ASM depletion. It has been reported that the yield is not seriously affected if one allows ASM to deplete up to 45%.

(ii) Plant based measurement

In addition or as an alternative to monitoring the soil, it is possible to monitor the water status of the plants. This can be done visually, as well as instrumentally, to detect early sings of thirst (incipient stress) very well in time to irrigate the plants and prevent any significant reduction in yield. Numerous methods have been proposed to monitor the physiological state of the water in the plant. Important among these are techniques to estimate transpiration using excised leaves, determination of leaf tissue hydration (relative water content) with punched disks or intact leaves, observations of stomatal aperture, monitoring stem diameter, pressure cell and psychrometric measurements of leaf water potential, determination of free prolin or ABA accumulation in the leaves and more. Perhaps the most comprehensive measurements are of total plant transpiration and photosynthesis, using portable tents with transparent (plastic) walls. Nearly all of these techniques require specialized instrumentation and trained personnel and are difficult to carry out on a routine basis in fields. Hence, there is need to develop a gadget based on these technique which can be used in field for ordinary irrigation management.

Since perennial fruit trees, owing to their large canopies have a marked capacity for passive and active adjustment of cell solute concentration in response to varying water status in the soil and air, leaf water potential is often not a good measure indicating the exact nature and degree of stress experienced by the plants. Significant osmotic adjustment has been reported in many temperate and subtropical fruit trees.

Still the most common way to monitor the crop is by the well tried and tested method of direct visual inspection. An experienced orchardist can detect early sings of thirst by the appearance of the foliage especially during the period of peak transpiration demand (generally at mid day). Young leaves are the most sensitive organ in this regard. They begin to curl or become flaccid or change their angle of orientation. When that happens, an irrigation is indeed needed, so it is good to irrigate the tree before such sings are too evident, that is to say before the stress becomes severe enough to reduce the yield potential.

(iii) Meteorological observations

These methods are more or less directly based on the evaluation of the so-called evaporation power of the atmosphere and aim at inferring from it and eventually other parameters like the time and depth of water application. The idea here is to follow the meteorologically imposed evapotranspirational demand as it varies over time and to set the quantity of irrigation accordingly. Conceptually, it is supposed that there ought to be a definable rate of ET for the special case, in which the field is maintained perpetually wet and this rate should depend only on the meteorological regime. The concept of "potential evapotranspiration" (PET) is an attempt to characterize the climatic environment in terms of its evaporative power i.e. the maximal evaporation rate which the atmosphere is capable of extracting from a well watered field under given climatic conditions. PET is thus said to represent the externally imposed "evaporative demand".

Soil moisture determination also helps in estimating crop evapotranspiration through water balance technique. In this, an account of all quantities of water added to, subtracted from and stored within the root zone during a given period of time is maintained. ET is then calculated as ET = (P + I + U) – (R + D) ± Ä ASM; where P = precipitation, I = irrigation, U = upward capillary flow into the root zone from below, R =runoff, D = downward drainage out of the root zone and ± Ä ASM = change in soil moisture over a given time interval. Adopting the technique crop ET, water requirements of many fruit crops have been estimated, but no work has been done litchi on these lines.

(iv) Soil-plant-atmosphere continum

In the last two decades, new concepts of irrigation management have been evolved. A fundamental change has taken place in the old concept of soil-plant-water relations, leading to more dynamics and holistic approach. The field is not perceived to be unified system in which all processes are inter-dependently linked as a chain. This unified system has been called the SPAC, for "Soil-Plant-Atmosphere Continuum". Accordingly, the availability of soil moisture is not a property of the soil alone but indeed a combined function of the plant, the soil and climate. Now, several mathematical models exist which attempt to describe almost all the aspects of SPAC. With the help of a computer, it is possible to correlate or combine more than one hundred characters of the three components of SPAC to reach a more precise and better controlled irrigation schedule for a particular crop. This has greatly helped in attaining maximum water use efficiency (i.e. maximum yield per unit of water transpired).

Methods of irrigation

There are four main methods to apply water to the litchi plants : (1) surface irrigation, (2) sprinkle irrigation, (3) drip irrigation and (4) sub-irrigation. Each system has advantages and disadvantages. Physical factors involved in the system selection include soils, climate, topography, water quality and availability, water table depth, orchard size, system performance, maintenance and repair etc. Human and economic factors involved are labour and management skills, availability and cost as well as capital and energy costs in relation to expectable returns. Not all of the relevant factors can be defined or weighed quantitatively in each case, so often the decision as to which system to select rests in part on personal judgment and subjective preference rather than on completely objective analysis.

1. Surface irrigation

Surface irrigation has four main techniques e.g. (i) normal flooding, (ii) basin irrigation (iii) border irrigation and (iv) furrow irrigation. All these methods are employed to irrigate litchi orchards. This type of irrigation is recommendable only in those regions where low-cost labour and abundant water are available or where rainfall distribution is adequate to meet litchi requirements and only occasional irrigation is necessary. It is suited to land that is almost flat.

Normal flooding is the least controlled of all surface irrigation techniques. Water is delivered to the orchard and allowed to spread over the entire orchard surface in a continuous sheet. Since the degree of control over the flowing water is minimal, the resulting distribution of water is usually highly uneven. Despite its obvious disadvantages, flooding is still prevalent in many areas for irrigation of litchi orchards. In basin irrigation, low earth dikes are made around each tree or a group of trees, within which water can be impounded for irrigation. Young litchi plants are usually irrigated by basin irrigation because at this stage the plant are irrigated individually through a channel connecting the basins made around their trunks. Old trees can also be irrigated by this method in those area where the natural land surface is nearly level. In sloppy land, basin irrigation can be carried out in conjunction with terracing. Border irrigation is followed in large leveled litchi orchards where the slope of the surface does not exceed 0.5%. Here the entire orchard area is divided into elongated plots confined to low earth bunds. Depending upon the slope and the size of the land, the number of border strips can be decided. In furrow irrigation, the water runs in small furrows between the tree rows. The furrows vary from 30 to 60cm in depth, 20 to 60cm in width and 200 to 300cm in length and follow the contours of the land. Slopes along the furrow may range from 0.2 to 2% but some times may be as high as 5%. This system is more suitable for irrigating young litchi orchards as one furrow running along side of each tree row may suffice to supply water to the limited zone of rooting. When the trees develop, several furrows are needed between the rows in order to wet the greater volume of soil occupied by the extensive root system.

2. High frequency irrigation

Under continuously favourable moisture conditions, improved cultivars can attain their high yield potential and can respond to great amounts of fertilizer and more intensive management practices. The advent of new irrigation system (sprinklers, drippers or tricklers, micro-sprayers, bubblers etc.) has made it possible to establish and maintain soil moisture conditions at a more nearly optimal level. Since a high frequency irrigation system can be adjusted to supply water at very nearly the exact rate required by the crop, one no longer needs to depend on the soil's own ability to store water during long interval between irrigations. The consequences of this fact are far-reaching. New lands, until recently considered totally unsuited for irrigation, can now be brought into production. One outstanding example is the case of coarse sand and gravels, where moisture storage capacity is minimal and where the conveyance and spreading of water by surface flooding would involve inordinate losses by excess and non-uniform seepage. Such soils can now be irrigated quite readily even on sloping around without expensive leveling by means of drip or micro-sprayer systems.

With high frequency irrigation, the orchardist need no longer worry about when available soil moisture is depleted or when plants begin to suffer stress. Field capacity and permanent wilting point indeed loose their relevance in the present context of high-tech irrigation. It has been suggested that the best the pressurized irrigation systems is almost continuous irrigation during the period of peak water demand because it costs little more to use a system once it is permanently installed.

(i) Sprinkler irrigation

This type of irrigation is highly recommended for use in frost-prone or low-moisture areas in order to offset the very unfavourable effects of this type of climate. In sprinkler irrigation, there is usually one sprinkler for every tree. This is because in order to obtain good cover an overlap of more than 200 per cent is necessary. Sprinkler irrigation, also called overhead irrigation is the application and distribution of water on the trees in the form of a spray or a jet which breaks into drops or droplets, created by expelling water under pressure from a nozzle. In effect, it is a simulated series or range of drop sizes. If run-off occurs, this means that the system has not been correctly set up and it will be necessary to check that the right type of sprinkler is being used. Rate of discharge and sprinkler size are suitable / or whether some adjustment to the sprinklers is required. A system of sprinkler irrigation consists of water source, pumping unit, a pressurized conveyance system (pipe network) and a set of nozzles to eject the water into the air and spray it on the litchi trees. There are many different systems of sprinklers. They may be classified as over-tree or under tree sprinkler and fixed or portable sprinkler system. Where cost of water is expensive and wind conditions are severe, it is preferable to use the fixed under tree type sprinkler. Where atmospheric humidity is low and wind speed is usually normal, the productivity of litchi trees can be improved by overhead intermittent misting or sprinkling as it decreases leaf air water vapour concentration gradient. In recent years, increasing number of litchi growers in developed countries have abandoned the old-fashioned surface irrigation in favour of overhead sprinklers. Some farmers, however, do not favour this system as it causes some injury to plant parts, flowers and fruit. This happens probably due to poor installation of the irrigation system.

(ii) Drip irrigation

Drip irrigation is the slow localized application of water drop by drop at a point or grid of points on or just below the soil surface. Water is delivered to the plants via a set of plastic lateral tubes laid along the ground or buried just beneath it for protection. The lateral lines are connected to a buried main line that receives water from a head source. The trickling rate, generally in the range of 4-8 litres / hour per emitter, must not exceed the soil's infilterability if run off is to be avoided. The operating water pressure is usually in the range of 1 to 3 atmospheres. This system is specially successful in the regions where water supply is scarce. Although it is highly efficient, it presents certain management problems, particularly when the trees reach adulthood, because a large number of drippers are needed per tree, especially in light soils. In soils with a low water infiltration capacity, problems relating to salt concentration in the topsoil could arise. This system, like the micro-sprinkler method, is very suitable for both flat and steeply sloping land.

A general problem which arises with drip irrigation systems is that the pipes and drippers are likely to become blocked by mineral or organic particles. It is therefore indispensable to have a good filtering system, especially where drip irrigation is concerned. There is also the risk that pipes exposed on the soil surface may be damaged by rodents and dogs. Added advantage of this system is that plant nutrients can also be applied to the trees through irrigation water (fertigation). Some of the problems of drip or sprinkler irrigation include the high cost of installation and difficulties in keeping equipment operational.

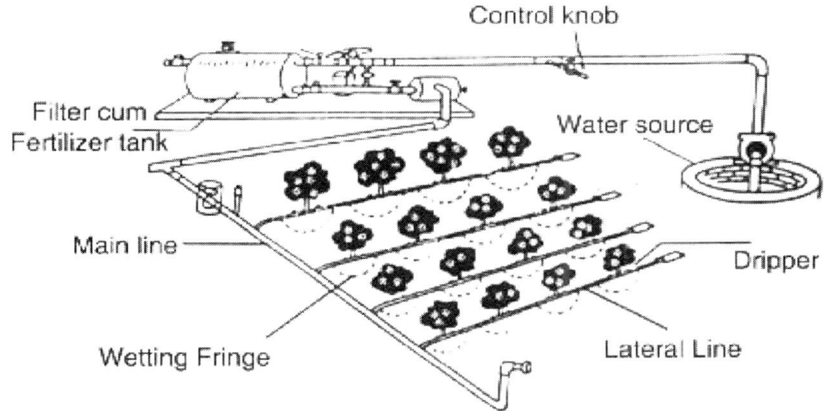

Fig. 30 : Drip irrigation system for litchi orchard

As is mentioned with respect to sprinkler irrigation, special attention must be paid to the design of drip irrigation systems based on the soil type, emitters / dripper characteristics and tree development patterns and each individual case must be examined by qualified technical staff. For uniform distribution of water to the entire root zone of litchi plants hydrogols can be attached with laterals which has flexibility of adjustment as the root zone is expanding.

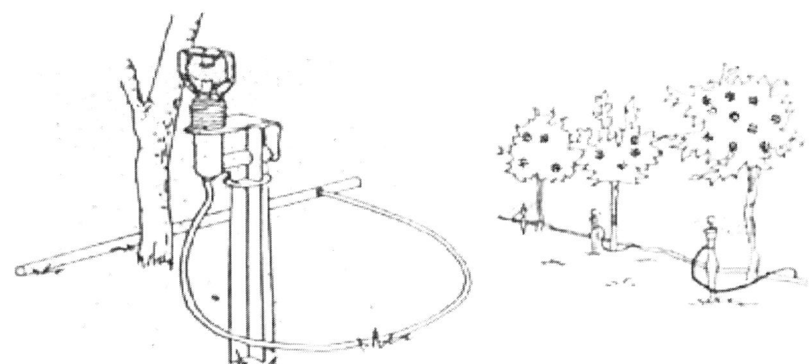

Fig. 31 : Tree Mini Sprinklers in Litchi

Some farmers prefer both drip and micro-sprinkler below the canopy to increase the humidity in the orchard at the fruiting time. This reduces fruit cracking. Recently, overhead sprinklers with fine droplets or raingun has been found effective to maintain required humidity in the orchard.

Determination of water requirement through drip irrigation

Water requirement for litchi crop (W/R), is a function of surface area covered by plants, evaporation rate and infiltration capacity of soil. The irrigation water requirement for each plant is calculated first and thereafter for the whole plot of 1 ha based on plant population for the different seasons. The maximum discharge required during any one of the three seasons is adopted for design purposes. The daily water requirement for fully grown plants can be calculated as under.

WR = A X B X C X D X E

Where :

WR = Water requirement (lpd /plant)
A = Open Pan evaporation (mm/ day)
B = Pan factor (0.7) - This may differ area wise
C = Spacing of plant (m^2)
D = Crop factor (factor depends on plant growth-for fully grown plants = 1)
E = Wetted Area (0.3 for widely spaced crops)

The total water requirement of the farm plot would be WR x No.of Plants.

The irrigation water requirement is determined using average season wise pan evaporation data for the area.

Table 17 : The daily water requirement of plants using above equation can be worked out as under.

S.No	Season	Evaporation mm/day (Varies from area to area)	Water requirement Lpd/plant	M^3/ day/ha
1	Kharif	4.15	41.83	8.31
2	Rabi	3.55	35.78	7.44
3	Summer	6.45	65.07	13.53

Adopted from :http://planning.up.nic.in/innovations/mi/litchi

The drip irrigation system may be designed depending upon E.T. as per Table 17, with maximum requirement of 65.07 liter /day/plant during the summer season and for this, the water required would be 13.53 M^3/ day/ha of plantation. If the average working hour of pump set is taken as 4 hours per day, the discharge required would be as, pumping rate per hectare = 13.53 m^3 /day/ha = 3.38 m^3 /hr/ha = 0.94 LPS or say 1 LPS.

As required discharge is only 13.53 m^3/day/ha, it can be pumped for one hour only from a well giving a discharge of 5-6 lps. For the estimated water requirement of 1 lps only, an arrangement to divert excess water to irrigate other crops may be provided, especially during Kharif and Rabi periods. Alternatively, a tank of 14 m^3 capacity can be provided so that uninterrupted irrigation may continue even in areas where power shut down are frequent.

Cost

The unit cost of the drip irrigation system to be installed on a 1 ha plot of litchi garden is given below.

Table 18 : Cost estimate for drip irrigation in one ha.

S. No	Item	Quantity	Rate (Rs./m)	Cost (Rs.)
1	PVC Mainline 62. 50 mm dia (4kgf/cm^2)	100 m	48.5	4,850
2	PVC Lateral Line 50 mm dia (6kgf/cm^2)	120 m	47.5	5,700
3	Drip Lateral line 16 mm dia (2.5 kgf/cm^2)	1350 m	7.05	9,870
4	Lateral connectors	13 nos	10 each	130
5	Sub main flush valve 50x65 mm	2 No	89.5	179
6	Gate valve	1 No	710	710
7	Drippers (4lph)	832 No	2 each	2,300
8	Micro tube (6mm.) @ 4 m per plant	832 m.	2/m	1,664
9	Tees, end-plugs.	one	LS	1,500
10	Bypass assembly 1.5" x 1.5"	one	848	848
11	Super Screen filter (15 m^3x1.5")	one	Lump sum	1,860
12	Ventury assembly plastic 1.5"	one	Lump sum	1,500
13	Fitting and accessories	one	Lump sum	2,000
	Sub total	(a)	Say	33,000
	Cost of 14 m^3 storage water tank	(b)	Lump sum	25,000
	Creation of water source (variable)	(c)	Lump sum	30,000
	Total	(a+b+c)		88,000

Adopted from :http://planning.up.nic.in/innovations/inno3/mi/litchi.htm.

Irrigation recommendations

When the ET of litchi is worked out for a particular set of climatic conditions, the task of an irrigator becomes easy. The irrigator needs to make up the ET losses through irrigation allowing 20-30% more water to compensate for deep drainage and conveyance losses. But this will happen in area where water is plentiful and cheap in comparison with other production inputs. The orchardist is free to select any method of irrigation or any time interval to satisfy the crop's need. He must, however, be aware of the dangers of excessive irrigation. In areas where irrigation water is limited, the question of its efficient use arises. How the water is to be distributed on the entire orchard area. Should the litchi orchard have part of its area irrigated, leaving other parts un-irrigated? Should partial wetting of the root zone be sufficient enough? Some other similar questions may also boggle the mind of a litchi grower. Answer to these questions lies in allowing flexibility in irrigation scheduling. For example, in areas where water unavailability is unpredictable, a safety factor or buffer of several days is needed to protect the crop.

Since soil and climate of different places are highly diverse, there is a wide variation in plant growth or fruit yield vis-à-vis the number of irrigation applied to the litchi trees. Where annual rainfall is more than 125cm and is well distributed throughout the year, litchi can be grown without irrigation. The young plant neither establishes well nor make optimum growth if there is a shortage of water. Adequate and frequent watering is also essential for the bearing tree as the atmospheric humidity is very low during the course of fruit development.

Table 19 : Water requirement of litchi/tree in different season

Year	Canopy diameter (m)	Water requirement in litre/irrigation*			
		Winter	Spring	Autumn	Summer
1	0.50	3	6	4.50	7.50
2	1.00	12	24	18	30
3	1.50	30	60	45	75
4	2.00	60	120	90	150
5 onward	-	-	-	-	160

*Irrigate two to three times a week in sands and one to two times a week in heavy clays.

Weed management

Weeds are a serious threat in the initial stage of nursery plant establishment where they competes for nutrients and act as shelter for pests and diseases, but in full grownup plants weed do not have that impact except that of unpleasant look and shelter for pests. In litchi orchards, however, owing to shallow root system, it is necessary to keep the weeds under check at least twice a year which otherwise may prove harmful to the growth and productivity of litchi trees. First 1-3 years after transplanting are crucial for the growth of litchi plants. Application of organic manure and inorganic fertilizers, irrigation at regular intervals and good sunshine results in the growth of a broad spectrum of grassy and broad leaf weeds in litchi orchards. These weeds should be removed to have better growth and performance, failing which the ultimate normal growth of litchi trees will be hampered.

Depending on the duration of growth, weeds may be classified into three groups as annual, biennial and perennial. Annual weeds multiply by seeds and complete their life cycle in one season. The competition offered by weeds is due to the combination of their prolific reproductive capacity and vigorous growth. Annual weeds produce large quantities of seeds. Prolific seed production ensures that many seedlings conditions are favourable. Perennial weeds usually last for two seasons. These weeds grow from seeds, rhizome, runners and bulbs, etc., and flower and fruit every year. If not suppressed, they continue to live for a number of years. They are very harmful to young trees. The common perennial weeds of the are : *Cynodon dactylon, Cyperus rotundus, doddar* or *swamalata* (*Cascuta reflexa*) causes considerable damage to trees. It is very difficult to eradicate *Cascuta* unless systematic hand picking is resorted to.

If weeds are not controlled, they interfere in the cultural operation like irrigation, spraying insecticides, fungicides or hormones, pruning and harvesting of fruits. There are many weed control measures. These can be broadly classified into four groups, viz. cultural, biological, chemical and integrated weed management.

Cultural (mechanical) control

Cultural weed control involves practices to minimize weed competition with the main crop. The most effective way to control annual weeds is to prevent their flowering and seeds setting. For this, the collective efforts of all farmers of the area are necessary. Control practices are hand weeding, mechanical weeding (mowing), tillage, growing intercrops and mulching. These methods prevent weeds from becoming established.

Once established, destruction of weeds manually by cutting with sickle or turning into the soil with spade and cultivation of the land frequently and collection of the roots, rhizomes, runners, bulbs, etc. by hand are efficient and practical methods for small orchards, but it is economical only when sufficient and cheap labour is available. Also, the manual weeding has limited effectiveness against established perennial weeds. Mowing controls the weeds by preventing them from setting seeds. The advantage of the mowing is that the soil is never exposed to erosion. This is good for the orchards established on slopes or sandy soil.

Tilling the orchard soil regularly provides a weed free environment in the orchard. Traditional tillage with the spade or hoe involves shallow cultivation and partial burying of weeds. Use of animals for ploughing reduces the labour cost but some superficial tree roots get damaged in ploughing. The use of tractor-mounted tillage implements is an improvement over traditional tillage and animal drawn plough. Tractor drawn tool can be used during initial years when the orchard trees are not fully grown. Branches hanging low, spreading in the inter-row spaces hinders the movements of the tractor even in widely spaced rows at maturity. Under such situations small power tillers can be used. Tillage controls weeds partly through burying weeds and partly through exposing underground parts of weeds to desiccation. The land around the trunk must be cleaned separately with the help of spade as it is difficult to operate a power implement in the vicinity of the trunk.

In hilly areas or where the slop of the land is high, trees are often grown under sod to protect the soil from erosion. Grasses are planted or the natural grasses are allowed to grow on the vacant space and the soil is not disturbed. The grass is mowed time to time and preferably fell on the ground. For tropical fruit orchards existing on the fairly leveled plain land, this practice is not suitable. Under 'Clean Cultivation' the objective is to keep the floor (surface soil) clean as far as possible. This is done by tilling the soil when required. Tillage not only checks the weed but also protects the surface soil from becoming hard or excessively compact for penetration of moisture or nutrients. Compact soil surface interferes with the growth and functioning of the roots owing to lack of soil air. Clean cultivation or tillage may have other beneficial effects on productivity of the orchard. For example, it can increase microbial activity by increasing gas exchange (aeration) and water penetration. Increased biological activity can promote the degradation of unwanted coarse organic material in the soil but it also can promote degradation of desirable organic matter causing decrease in the exchange capacity of the soil.

Intercropping/cover cropping for weed management

Cover cropping and intercropping are other practices followed in litchi orchards to suppress the growth of weeds, to bring additional income to the grower and to improve the health of trees until the trees begin to bear and can be continued even after that. Generally, cover crops are grown in space between the trees taking due care that no crop comes under the spread of the trees so that roots of the intercrops and the trees do not come into competition with each other for drawing nutrients from the soil. The detail of intercropping and cover cropping is presented in chapter-9.

Biological weed control

This type of weed control involves the use of natural enemies for the control of weeds. Grazing of animals like goats, cows, bovinc can be allowed to suppress weeds in the established orchards but animals may graze the litchi plants as their height is quite low and within their reach in initial stage. Other aspects of biological control of weeds involve the use of mulch or low growing crops to smother weeds and manipulation of plant density and canopy. Theoretically, biological control is the best method for eliminating weeds, but it is difficult and tedious in practice to locate and develop agents for biological control of weeds.

Chemical weed control

This refers to use of herbicides to suppress or kill weeds. A herbicide is a chemical that has phototoxic properties. It includes a wide variety of chemical compounds which are classified on the basis of certain characteristics which may include (i) chemical structure, (ii) selectively and (iii) where the herbicides act as a contact or translocated herbicides. Selective herbicides are those which kill certain kinds of plants (weeds) without significant injury to others. Selectivity depends on the reaction of both, the crop and associated weeds to the herbicide being applied. The most desirable condition exists when the crop is not injured by the herbicide but the weeds are killed.

Selectivity based on the nature of herbicides is often determined by whether it is active when applied to the foliage, to the roots or to both foliage and roots. To obtain maximum selectively to the advantage of the crops and disadvantage of the weeds such factors as rate of application, soil type, irrigation practices and cultural practices must be considered. Selectively may vary according to the weed species present, stage of growth and weather conditions. Consequently, selectively is rarely perfect and depends upon the environment, the nature of the herbicides, application rate and method of application.

Non-selective herbicides will indiscriminately kill all plants when they come in contact. An example of non-selective herbicide is glyphosate. Paraquat is another example which destroys the green tissue of all plants when it comes in contact with but perennials are able to produce new shoots from rhizomes or other such organs. Therefore, the action of paraquat is non-lasting for perennial weeds. But this is not the case with glyphosate. Its action is long lasting and thus, has been recommended widely for use in the orchards.

Some herbicides are systemic in their action. They are also referred to as translocated herbicides. They are readily absorbed by leaves, stems or roots of treated plants. Some herbicides move in the phloem of treated plants alongwith food materials. These herbicides tend to enter the plants through the leaves and stems (foliar applied). Some systemic herbicides move mainly in the xylem vessels along with water and minerals. These herbicides are generally taken up by the plant roots in the soil. Example of herbicides that move in the xylem vessels are atrazin, simazine, diuron, metachlor, alachlor and fluometuron.

Table 20 : Herbicides useful for weed control in litchi orchard

Name	Amount (a.i.) kg or l/ha	Remarks
Paraquat	0.6-1.2	Any time when weeds are young, 2-15cm high
Glyphosate	1.0-2.0	Use very carefully around trees; never use if any possibility exists for spray to touch young, green bark of tree trunk; can be used any time
Diuron	2.0-3.5	Lower rates in sandy soil
2,4-D	1.0-2.0	Any time, repeat as necessary
Dalapon	3.0-5.0	For grass control; repeat as necessary
Simazin	1.0-2.0	Good for annual weeds; can be damaging to young trees on sandy soil

Amitrol	0.6-1.8	Not recommended for trees less than 3 years old
Terbacil	1.8-3.6	Use in old orchards; use at lower rate in light sandy soils
Trifluralin	0.6-2.2	Mix with soil while tilling it

Cited from : K.S. Chauhan's Litchi :Botany, production and utilizat on , Kalayani Publishers New Delhi

Herbicides are generally safe to use if the information mentioned on the label is followed explicity. A list of different herbicides used in fruit plantation has been furnished in Table 19. However, in using herbicides, following guidelines must be strictly adhered to, otherwise crop injury or poor weed control can occur.

1. Rates other than those recommended never be used in the orchard.
2. Sprayers must be properly calibrated and nozzles are directed towards weeds and not to the tree trunks or foliage.
3. Young weeds are killed more easily than other older established ones.
4. Fine textured soils (clay) require higher rates of herbicides than coarser (sandy) soils
5. In old orchards, a relatively narrow band just covering the inter-row space is treated with the herbicides. Weeds near the trunk are controlled by tillage equipment.
6. Application must not be made when wheather condition are unfavourable (i.e. high velocity wind or rains or cloudy weather).
7. Ajuvants or melting agents may be added to cause the spray solution to spread more uniformly in order to cover the foliage thoroughly.
8. If leaves of the fruits trees are accidentally sprayed, the sprayed portion must be cut off immediately.

Integrated, weed control system suppresses weeds by combining two or more weed control methods. Environmental and socio-economic constraints make integrated weed management one of the best option for weed control measures in the tropics. Cultural weed control methods are often labourious, unattractive and ineffective. Integrated weed management has been demonstrated as the most effective in banana and other fruit tree plantations. Intercropping, tillage of the soil and use of herbicides are examples of integrated weed management followed in different fruit trees.

Mulching for weed management and moisture conservation

The process of covering plant root zone or entire basin with protective coverings is called mulching. Mulching is beneficial to all litchi orchards. It is very much useful in plants raised under rainfed conditions or in adverse soils. The covering materials may be organic or inorganic in nature. Using mulch can help a litchi orchard in many ways.

Fig. 32 : Mulching

- Mulches make unfavorable conditions for seed germination of weeds and provide a physical barrier for emerging weeds. A good mulch layer saves many hours of manual weeding. Thick layer of mulch material is very effective in preventing the number of annual weeds in the orchard, since they have difficulty to penetrate such a layer.
- Mulches maintain uniform moisture conditions in the litchi orchard. Moisture loss through evaporation is reduced and soil erosion is decreased in rainy season. Slow and steady water infiltration takes place in the soil and crusting is prevented. Mulches reduce the force of raindrops, thereby protecting the structure of the soil at the surface. Mulches may serve as vapour barriers, thus reducing evaporation of soil moisture. Water fall on bare soil surface causes splashing of soil onto the fruit. Mulches keep fruits cleaner and help in reducing the spread of disease.
- Organic mulches add nutrients and humus to the soil as they decompose, improving its tilth and moisture-holding capacity. Some mulches are tilled into the soil before planting a new crop, and therefore may have an effect upon soil fertility and soil property. In the short term, mulches may decrease nitrogen available for a given crop. A material that has a high carbon content and is very low in nitrogen and other nutrients may

actually "bind" or immobilize plant-available nitrogen temporarily. This occurs because soil microorganisms use available nitrogen to metabolize and decay the organic material. The immobilized organic nitrogen can be made available (mineralized) later as the organic matter continues to decompose.

- The increased organic matter content due to mulches increases soil aggregation, infiltration, and water-holding capacity. In poor soils, addition of organic material improves water retention and reduce water losses through deep percolation.
- Mulches shield the soil surface from solar radiation effects. Because of the increased moisture content and reduction of incoming solar radiation energy, a mulched litchi plant basin has lower difference between day and night temperatures. However inorganic mulches can act otherwise. For example, dark mulch can adsorb more solar radiation and may actually increase soil surface temperature; transparent plastic mulches may increase temperature through a"greenhouse" effect and can prevent water from entering soil thereby decreasing soil moisture and increasing water runoff.
- Mulches are very much useful for grownup as well as developing litchi plants. In bearing plant, it has direct impact on fruit drop, fruit cracking and quality development whereas in developing plants it helps in better survival, growth with less water. Natural as well as synthetic materials are used for mulching. Details of some of the mulch materials has been given as under.

Different types of mulches and their use

(i) Organic mulches

Saw wood dust :A layer of sawdust provides good weed control. Five cm thick mulch of saw dust adds nearly 450g of nitrogen, if applied 10 cubic feet around the basin. Fresh sawdust contains high carbon and low nitrogen and its breakdown requires that microorganisms take nitrogen from the soil. It is very good material for plants particularly to new plants. It is readily available from sawmills but it tends to be expensive.

Straw: A 15 cm thick layer of paddy/legume straw provides good annual weed control. These materials may decompose quickly and therefore, should be replenished as per need. They stay in place and improve the soil as they decay. Avoid straw of weeds that is full of viable seed. Legume such as cowpea, supplies nitrogen as it quickly breaks down. Wheat and paddy straw are readily available in villages.

Field grasses and weeds: The layer of mulch is made gradually using dry grass. A 5cm layer of local grass and weed cuttings provides good weed control and moisture conservation. Layer of green grass/weeds gives off excessive heat and foul smell and decompose slower than other organic material. In limited quantity, green grasses decompose rapidly and provide an extra dose of nitrogen to growing litchi plants as well as making fine humus.

Dry plants leaves: After tree pruning, all the twigs should be collected at one place where dry leaves are collected and used as mulch material. A 5-7 cm thick and compact layer of dry litchi/other plant leaves provides good weed control and moisture conservation. Leaves decompose quickly and improve the soil once decomposed. To reduce blowing of dry leaves due to wind blow, the leaves should be surrounded by a basin bund or these may be used only after partial decomposition.

Crop residue and compost: Apart from above organic mulch materials, compost, crop residues, including twigs and branches etc. can be used as organic mulch material in the litchi orchard.

(ii) Inorganic mulches

Black polythene: One layer of black polythene provides excellent weed control. It is relatively slow to decompose but broken down by sunlight and should be replaced after every two years. It increases the soil temperature by which may cause soil temperatures to rise too much in mid-summer, damaging the roots of plants. A good foliage cover can prevents direct absorption of sunlight on mulch surface. It is easy to obtain, but is fairly expensive. A new type of black polythene has recently come into the market which has a white, reflective side to prevent the overheating problems. In all cases, polythene mulching insists litchi feeder roots to confined on upper surface which increases cracking under stress.

White polythene: White polyethylene can also be used as mulch material in litchi orchard but it is very expensive. It increases the soil temperature and also not much effective in weed management hence not used commercially.

Method of mulch application

Most of the organic mulches should be applied after plants are well established. In developing litchi plants also, organic mulch can be applied but problem of termite should be taken care off. Remove all weeds before spreading the mulch. Apply mulches when there is reasonably good soil moisture and before the weather turns hot. Infiltration of rain water will be slowed somewhat by mulch, so it is best not to place the mulch on soil which is dry. Water thoroughly or wait for a good soaking rainfall before applying any mulch. Application of organic mulch in plant basin can delay the termite infestation to tree trunk for some days as termite feed on mulch material and do not move to tree. However, it aggravates the infestation in subsequent days when mulch is digested and no additional mulch is replaced. To keep these termites in check, tree trunk should be pasted with chlorpyriphos 0.02% + coal tar paste and mulch should be applied one foot apart form the tree trunk. This practice delays the termite infestation and restricts the soil mounding on tree trunk. Black polythene and similar materials should be spread on land that has been completely prepared for planting and has a high moisture level. Edge of the mulch should be buried with soil to prevent it from blowing away due to air.

❑❑❑

Chapter 9

Maximize Land Output through High Density Planting and Orchard Floor Management

Increasing population pressure and shrinking agricultural land have made it imperative to product more per unit of land area. Production per unit area can be increased through high density planting or by using orchard interspace utilization for the production of crops other than main crop, commonly known as orchard floor management.

Advancing knowledge in litchi tree architecture, plant growth physiology and use of chemical growth retardants have enabled farmers for closer planting by maintaining small sized trees. This system is popularly known as the High Density Planting (HDP). It enables profitable cropping, high regular yields and improved farm management practices, leading to higher productivity. Today new orchards of litchis are being attempted to plant in this system with a view to produce higher fruit yield and increased profitably. Use of growth retardants which restricts tree growth and encourages early flower induction, have also been found helpful for these high-density planting systems. Perhaps, presently the high density farming has been the area of most intense research in the litchi production. To meet the challenge of high productivity, one has to optimize the parameters of growth and minimize the unproductive components of plants without sacrificing the overall health of the tree and quality of the product.

High Density Planting

High density planting technique is a modern method of litchi cultivation involving planting of litchi trees densely, allowing small or dwarf trees with modified canopy for better light interception and distribution and ease of mechanised field operations. Control of pests and diseases, weeds and pruning of tree canopy are carried out by machine. Irrigation and fertigation are automatically controlled. Such systems have been reported to give precocious cropping, high and regular yields of good quality fruits and low labour requirement to meet continually rising production costs. It has been successfully adopted in the cultivation of temperate fruits especially apples and also showed its promise in mango, guava and litchi.

Reasons for HDP as against normal planting

Increasing pressure on land owing to diversion of orchard lands to other purposes like urban development and infrastructure generation as well as rising energy and land-costs, together with mounting demand for fruits have made it imperative to achieve higher productivity of litchi from limited space. In the case of relatively short duration fruit crops like banana, pineapple and papaya, the advantage being that any mistake committed in spacing of these short-term fruits can be rectified easily without much financial loss. However, in the cases of litchi, hasty attempt for closer spacing may bring negative impact in growers' fields if the complete package of high density has not been properly understood.

Most of the commercial litchi cultivars in India are location specific with long gestation period. The normal planting distance in these cultivars has been 9-10m. The orchard takes 10-15 years to provide economic returns depending upon the cultivar, planting distance and other cultural practices. Due to poor early returns and clash between the cultural requirements of the intercrop with main crop, litchi orcharding so far is done by large farmers who can afford long gestation period and or failure of intercrops particularly in the early stages of orchard life. When the tree is full of vigour, then the dominance of vegetative phase prevails over reproductive phase, which adversely affects flowering and fruiting in litchi due to high competition for photosynthates between vegetative and reproductive phases of the tree. In litchi, some times vegetative and reproductive phases do not have antagonistic effect to each other.

The tall trees of normal density at high productive stage require heavy machinery and equipment for spraying of pesticides and crop harvest because outer tree canopy is the major fruiting area extending outwardly with the enlargement of the tree. High density orcharding appears to be the most appropriate answer to overcome low productivity and long gestation period for early returns and export of litchi. In high density system, yields are improved in early years of orchard life.

Advantage of high density planting

Higher production efficiency : High density planting aims at achieving higher production efficiency which may be defined as yield per unit of production capacity. The production capacity is a function of the area of the ground covered by the tree canopy, the total area of the tree canopy, the total volume of the tree canopy and the bearing volume of the tree volume. Therefore, production efficiency can be calculated as yield per unit of ground area, canopy area, canopy volume or canopy bearing volume. In an orchard of small trees, the total volume of the trees may be considered productive or bearing while large trees have an inner non-productive portion that is significant portion of the total tree volume. This is an important factor favouring suitability of smaller trees to higher planting densities.

Prococity : In HDP,it is important for trees to early start of production and producing good fruit yield before crowding occurs. An induction of precocity by the use of rootstock is important feature of high density planting, which is very much lacking in litchi and needs to be standardized along with efficient propagation techniques. Therefore, Pruning (Summer pruning, root pruning), girdling and use of growth of regulators are few alternatives for inducing precocity where dwarfing root stocks are not available.

Early returns on investment : Though the initial investment per unit area in HDP is more than the normal planting, but high initial investment is rapidly recovered as the smaller plants in HDP have the potential for precocity and prolifcity.

Better utilization of land : The HDP permits better utilization of land to produce good quality fruits, especially in first phase of fruiting. This enables the growers to get higher returns/unit area.

Easy management : Smaller trees are easier and less expensive to manage in all agricultural operations. Thus, expenses on operations are reduced. Such trees permit the use of orchard equiments.

Flexibility in production : The HDP also offers flexibility in production since variety can be changed easily by planting dwarf trees which comes into production in 4-5 years. However, it is difficult in litchi plantation.

Reduction in cost of production : It enables the mechanization of fruit crop production and facilitates more efficient use of agricultural inputs.

Major components of HDP

For establishment of high density planting in litchi factors like site selection, site preparation and irrigation are more important. It will decide the productivity of litchi as it is propagated by air layering and use of root stock is yet to gain commercial momentum along with standardized grafting techniques. Besides, planting geometry, nutrient management and canopy management are issues of vital importance for high dencity planting in litchi.

Although the work carried out on high density planting at different places in litchi proved efficient in improving productivity. But in comparison to the other fruit crops, it needs more attention especially in Indian conditions. There are four major components of high density planting system. These are :

Planting density : Even though a small canopy with a high number of well-illuminated leaves is efficient in photosynthesis but it is very poor in light interception, which leads to low potential yield per hectare. Light interception could be improved by increasing tree density. An optimum tree density is the level of tree density that is required to produce optimum light distribution and interception which leads to high canopy photosynthesis. As a result, yield per hectare is maximized. An optimum light interception is a factor of plant form, planting density, tree arrangement and leaf response to light for photosynthesis. Optimum light interception can be defined as a level of light intercepted by an orchard system above or below which, the economic yield will be reduced. Depending on the planting system, tree density under different system of high density planting ranges from as low as 223 trees to 400 plants per hectare, however, still researches are continuing on HDP in litchi.

Planting system: Planting system is a combination of tree arrangement and plant form. Tree arrangement in HDP system must have sufficient alleyways for movement of farm machinery. The way trees are arranged also determines the light distribution pattern and light interception level. As of now, full field system is common for HDP of most of the fruit crops including litchi. This system is called full field system because in this method litchi trees are planted without alley-ways. In term of light interception, this is the most efficient system. However, the full field system is only suitable for small scale orchard due to the difficulty in field operation through machines. In litchi single hedge row and double hedge row system of planting with enough alley space is being practiced in some countries for HDP.

Pruning system to make desirable plant stature: The tree growth in low density litchi remains similar as in high density litchi upto eight years. The horticultural methods most commonly known to control tree growth are shoot pruning and use of growth regulators. The training begins when the tree is first planted and continues through out its productive life. Proper tree forms, branch angle and limb spacing in themselves aid in growth control. First training is done after one growing season. Each plant is allowed to maintain single stem (main stem) with upward growth upto 60-80cm and then four scaffold branches are allowed in four directions to make the tree frame. Thereafter, 2 shoots arising from each primary branch at a distance of 60-75 cm from main stem is allowed to form secondary and like wise the tertiary branches. After start of bearing in plants, shoots arising from secondary and tertiary branches are given 15-20cm deep pruning in soon after fruit harvest. Spray of 1% urea combined with 0.2% Blitox-50 or any other copper fungicide should be done soon after pruning.

The trees of high density orchard show slightly higher growth in terms of height from 8th year onwards than those of normal density because of more pressure from side plants and want of sun light. A reverse trend occurs in the growth of crown circumference and main stem girth and the differences in the growth of trees in different densities widenes after 10th year. Once the trees have filled their allotted spaces, crowding may occur and canopies of adjacent trees begin to overlap. This may lead to excessive shading and reduction in photosynthesis by layered leaves within the tree canopy resulting in poor yields. In fact, at this point of time litchi trees require vegetative growth control, particularly in high density orcharding. Once the tree is mature, excessive growth can be regularly removed by pruning to provide a short term or immediate benefit. In high density planting, the main objective of pruning is to develop plant form which is good in light distribution, small in size and easy to manage especially in terms of mechanization.

The tree height, circumference and main stem girth in high density can be reduced by regular back pruning or dehorning of branches after crop harvest in 11th and 15th years, besides light pruning of fruited shoots (15-20 cm top) in other years. Dehorning is a process of removal of some selected branches from their place of origin on limb for vigour control and light and air penetration. Regular pruning induces new shoot production within 45-60 days during July-August. In general, every pruned shoot produces more than one shoot. After harvest of crop, about 50% branches of the tree canopy should be dehroned in high density orchard when the plant starts intermingling of canopies and the remaining 50% branches should be dehroned in next July after crop harvest to have a proper balance. Delhroning is done to avoid overlapping of trees in the high-density orchard. In single hedge row or double hedge row system, the orchard is developed with a concept of single canopy with multiple roots and therefore only side branches are pruned to develop alley of space.

Good light distribution within tree canopy increases the number of well illuminated leaves. This condition induces a high rate of canopy photosynthesis that leads to high yield/tree. Small and compact canopy plant form is desired for high density which can be achieved by constant pruning and dehorning.

Automatic systems: Another component in high density litchi planting is the system automation which contributes to high production efficiency. One of the important farm operations that can be automated is irrigation and fertigation. In fact, irrigation and fertigation have been identified as one of the key factors for the success of high density orchards in case of litchi and many other fruit crops. Plant should not be kept under stress after pruning therefore, assured irrigation coupled with fertigation is essential after pruning and during fruit development in high density litchi orchards.

Nutrition and irrigation in HDP

Soil application of full doses of P and K should be done in July; whereas N may be given in two equal doses in March after fruit set and after crop harvest in July. In case of fertigation, the standard schedule should be followed during vegetative and reproductive growth phase and fruit development stage. Leaf litter should be allowed to decompose in the orchard soil.

High density orchard should be irrigated soon after if paclobutrazol has been applied during September-October, if there is no residual moisture in soil. Thereafter, the orchards are irrigated from panicle emergence to crop harvest during at 15-20 days interval March to June. However, no irrigation is applied from first shower of the rains in June to February-end. Tillage operations should be done in October-November to keep the orchard free from weeds and also to mix fertilizers and leaf litter into the orchard soil.

Control of pests and diseases

The pest and disease problem remains similar to normal planting except some pest such as mite. To control the pests and disease the recommended practices and chemicals should be followed. Weed cause some problems in initial stage of orchard. These should be removed and mulching may be done. In later stage the weed problems goes down in HDP.

Fruit Yield

Like normal density litchi plants under high density planting system start fruiting after 6-7 years of their planting. These orchards have more number of tertiaries per unit area, hence, orchard become economical earlier than normal planting.

Economic impact of HDP

The litchi trees under HDP system are precocious and sometimes start bearing 4-5 years after planting. Production per unit area is increased and early return to investment is possible. There is also better light interception and distribution thus improved fruit quality. The increase in yield, early and maximized return, reduced cost of production and improved fruit quality result in greater orchard profitability and provide better competitive advantage of local litchi growers. It is appropriate to adopt complete package of high density litchi production. Only planting and or pruning will not be of much use for HDP.

Utilize vacant space to compensate the long gestation period

Since litchi is slow to start bearing, it is debatable whether trees should be planted densely. If the trees are properly spaced, there is considerable land which will not be used by the permanent trees for several years (9-10 years). This space can be utilized for growing suitable inter-crops for economical return during non-bearing juvenile phase of the trees. Once an orchard is laid out and trees are established, it becomes imperative on the part of the orchardist to use the vacant space between the trees to ensure a quick return on investment. Management of ground space of orchard for additional benefits, however, should not be at the cost of main crop. It is also imperative for an orchardist to keep the orchard floor weed free. Therefore, it is desirable to grow some short-term crops in initial stage of orchard establishment when there is no income from the primary crop and should continue till the main crop plants come in full bearing. Neglect of this land means its deterioration and growth of weeds which will infest the area around the trees and also adjoining fields.

Intercropping is the growing of two or more crops simultaneously on the same piece of land or between plant rows for the purpose of utilizing land not occupied by the trees. It improves the fertility and protects the soil and orchard from factors detrimental to them, help in increasing of the yield and give additional income when orchard plants are not in bearing stage. There is another kind of cropping in the orchard, where the orchard soil is covered with some crop which in due course is turned back to the soil which supplements organic matter to the orchard soil. This is called cover crop. These two types of crops (intercrops and cover crops) are important operations in the management of a good orchard, although many growers in India do not pay much attention to it.

Benefits of space utilization

- Due to wide spacing and developing root patterns, the large unutilized interspaces can be exploited for growing inter and mixed crops successfully.
- The soil fertility can also be maintained/enhanced by careful selection of intercrops and adequate management of the orchard. However, selection of intercrops depends on agro-climatic region, marketing facilities, levels of inputs and other local considerations.
- Cover crops save the orchard soil from being eroded during the rainy season, particularly in areas where drainage is not proper. When the cover crops are turned into the soil, the organic matter of the soil is increased and this results in many advantages. For

example, if organic manure such as compost or farmyard manure is not available, the nutrient requirements of fruit trees can be met by applying only artificial fertilizers without any adverse effect on the soil. Water-holding capacity of the soil is increased and the biological complex of the soil is also improved.

- Cover crops prevent weed growth, protect the young plants from wind and even from cold injury.
- They are also a source of mulching after they have been harvested. Constant use of intercrop may reduce the soils' phosphorus and potassium content while increasing their nitrogen content. Because of this, small quantities of P and K should be incorporated to offset these losses.

Feasibility of space utilization

Though some workers have opposed the planting or sowing of any crop between the permanent trees, yet the general opinion is in favour of it. Carefully controlled intercropping with primary attention paid to permanent trees would not do any damage but on the other hand, prove beneficial to them.

Feasibility of inter-planted crop is determined by many factors, viz., valuation of land, fertility of the soil, taxes and price per acre. Where the investment is large and expenses are high, intercrop would do much help or produce profit during first few years. However, intercropping is labour intensive and is not practicable under the condition where the land values are low and labour cost is high.

Fig. 33 : Cover crop in litchi orchard

Orchardists are usually advised to grow shade loving plants in their orchards. But some farmers, without giving consideration on selection of the crop or its effect on main plant, grow any crop of their choice in the orchard. Some even neglect the main crop and pay much attention to the intercrop as it is the immediate source of their income. Such neglects prove very costly in the long run because fruiting in such orchard is delayed. Average productivity and life of the tree crop is also reduced. So intercrops must be selected with due care.

Factors determining choice of intercrops

1. For those orchards which are near to cities or towns, it is always better to grow vegetables as intercrops as it is more paying. For distant markets, crops like onion and potato can be grown which can be stored longer.
2. Only those crops are selected which are well adopted to the climatic and soil conditions of the area. Legumes are better choice for soils poor in nitrogen.
3. Water requirements of the intercrops should as far as possible coincides with the requirement of litchi plants. Such intercrops should be selected that do not exhaust the nutrients and moisture from the soil which are essential for the growth of litchi plants.
4. For young orchards, vegetables which require abundant sun-light can be selected. For old orchards, shade loving crops are preferred. It may be kept in mind that the intercrop should not affect the management of the main crop or interfere with the cultural practices. It is not desirable ordinarily to grow grain or other exhaustive crops like sugarcane, because of the exhaustion of nutrients and moisture.
5. Crops like wheat, oat, grain, oilseeds and radish (for seed) may not be grown as intercrops in litchi orchard. This is mainly because the period of seed production and subsequent development of seed in the intercrop generally coincides with the period of active fruit development and during the period the intercrop extracts maximum amount of moisture and nutrients from the soil and depletes it irrespective of the care taken to replenish it.
6. Perennial or exhaustive crops should be discouraged as an intercrop in the orchard. This may have devitalizing effect on the growing trees. For example, sugarcane, pigeonpea, maize, *jowar* should invariably be excluded from an intercropping programme in the litchi orchard.

7. The crops that may grow tall and have a tendency towards excessive growth should be discouraged. Intercrops must be compatible with the main crop. Trailing vegetables or vine crops are not desired as intercrops. Growth habit must be such that it does not affect the growth of litchi crop directly.
8. Certain pest and disease problems are enhanced by growing two or more crops together. It is also noted that intercrops favour some pests and pathogens since they serve as alternative hosts for them. Such crops may not be allowed to grow together.
9. Vegetables that have their roots within 25cm depth of the soil are considered good for intercropping. These are tomato, onion, cauliflower, beans, radish, *palak* etc. These vegetables can be taken profitably in a growing orchard. It is preferable to take leguminous crops for cover cropping so that nitrogen fixation in the orchard soil is facilitated. Another advantage of cover crops is that they will prevent the growth of weeds. During rainy season *(kharif)* greengram, blackgram, cowpea and *guar* are preferable. During winter season *(rabi)* it is desirable to take chick pea, pea, lentil, mustard, tomato, potato etc.
10. Some crops like sunhemp in light soils and *Dhaincha* in heavy soils may also be grown to protect the orchard soil from erosion and also for enriching the soil fertility. These crops are sown in beginning before rainy season and ploughed back after 45-50days before or just before flowering.
11. Colocasia, ginger, turmeric and grain crops are basically heavy feeders and extract maximum amount of nutrients from the soil for several months during which these crops stand in the field. Severe defoliation of trees is observed in the plots where ginger is grown as an intercrop, hence, such crops should be avoided. The partial shade loving crops like pineapple and amorphophallus etc. can be grown in fully grownup litchi orchards. In addition to field crops, some short duration, less exhaustive and dwarf type inter-fillers like papaya, banana and guava can also be grown till these do not interfere with the main litchi crop. These inter- fillers can be selected depending on region and other considerations.
12. Leguminous crops of like green gram, black gram, gram; etc., and vegetables crops such as cabbage, cauliflower, tomato, potato, brinjal, cucumber, pumpkin, bitter gourd, tinda, lady's finger,

etc. and spices like chillies can be successfully grown as intercrops in litchi orchard.

Crop combination

To intensity the orchard operation, a lot of work has been done on intercropping in orchard crops during initial years. For litchi orchards, crops for intercropping is selected on farmer's choice. A combination of crops for inter spaces of litchi plants have been suggested below.

(i) Vegetable based

- Cabbage (Oct-Feb.) – Cowpea (March-June)-Methi (July-Sept).
- Okra (Sept.-Dec.)-French bean (Jan-March)-Tomato (April-July).
- Potato (Nov-Feb)-Amaranth (March-May)-Cowpea (June-Aug.)-Radish (Sept.-Oct.).
- Cauliflower (Aug.-Nov.)-Potato (Dec.-March)-Dolicos (April-July).
- Cabbage (Sept.-Dec.)-Brinjal (Jan.-May)-Cucumber (June-Aug.).
- Palak (Nov.-Jan.)-Cluster bean (Feb.-May)-Brinjal (June-Oct.).

(ii) Crop based

Cow pea-potato, Green gram-Niger, Black gram-Mustard, Green gram-gram, Black gram-gram, Cowpea-Mustard, Cow pea-gram

All the crops mentioned do not perform well under dense shade, which is common phenomenon in litchi old orchard having large spreading trees. Intercropping under such condition is generally not practiced. However, such orchards should be protected from weed infestation and soil erosion, commonly seen during rainy season. For such situation, cover crops should be grown in between the tree rows or otherwise shade loving crops like elephant foot yam and turmeric can be tried. Cowpea however, has been found very suitable cover crops for growing litchi orchard. It is quick-growing, fast-spreading, broad leaved and covers the ground effectively. This crop can be turned down under the soil as green manure crop during September-October as it does not mature before October.

(iii) Filler fruit crop based or multi-tire system

Some growers do oppose to the planting of early bearing trees between permanent litchi trees feeling that the temporary gain is likely to be more than off-set by permanent damage to the trees. When primary consideration is given to the permanent trees, there is no reason why filler trees may damage the trees.

Many growers prefer some quick growing and early bearing fruit trees. In some areas, banana (vegetable type) is often grown. In a properly spaced litchi orchard, there is scope for guava trees which begin to bear in second year and produce a number of crops before it is removed. Among smaller trees, papaya, citrus or phalsa are found suitable as filler plant in litchi. In no case the roots of the filler trees occupy the soil where the roots of the permanent trees are concentrated.

The danger in using filler is that they may be allowed to remain in the orchard too long. Growers frequently fail to realize that the root system develops more rapidly than the branches and that the roots of the permanent trees come in contact with the roots of the fillers. In this way growth of permanent trees may be hindered. The safe rule is to make definite plans to remove the fillers after certain number of years. In any case, if exhaustive crops like banana are grown as fillers, the soil fertility must be maintained or improved. It is a very shortsighted policy to deplete the fertility of the soil in the years when the fruit trees are coming to bearing. Again, it must be kept in mind that the fillers must be planted away from the permanent trees and irrigated independently. When litchi orchard is developed in multi-tire mode, the third component i.e. the intercrop should also be grown as lower storey.

Pruning of filler trees : It differs somewhat from pruning of the permanent trees and care may be taken in developing strong scaffold branches. Early heavy pruning as in the case of phalsa, which delays fruiting, may be avoided. The filler trees are never allowed to crowd the permanent trees. Otherwise there would be competition for water and nutrient with the permanent trees. Therefore, as soon as they begin to crowd, fillers may be cut back. Judicious pruning to filler plants should be done and post pruning measures should be taken.

Special consideration : Pruning is made close to the point of origin of a branch and in such a way as to leave a clean smooth wound surface. Since conduction tissue runs parallel to the branch, healing of wounds is from the side. An elliptical cut heals over more than a round one. Moreover, healing of stubs or projections rarely takes place. The removal of large limbs may be done in stages. The first shallow cut is made several centimeters from the point of branch origin and on the lower side. The second cut from the top is made 2.5cm or 5.0cm further from the point of origin. The first cut serves to prevent tearing of the bark and beyond that point, while the second cut goes all the way through and thus removes bulk of the branch. A third cut is made close to the point of origin, with special care. So-called bleeding of wounds is serious only if it encourages disease. The heavy spray programme normally carried in an orchard makes the protection of small wounds necessary. Painting large wounds with blitox or bordeaux or Chaubatia paste may be considered a good practice. Disinfecting wounds and pruning tools with alcohol is important, specially if the bacterial disease is a problem.

Method of growing

The method of growing intercrop in an orchard is different in the young and the bearing orchards. In many plant almost the entire land is utilized, leaving only the basin portion of the tree. The plant is located in the middle of the bed in which the crop is grown. As the trees grow in age, the size of the basin increases and the area for intercrops gets reduced gradually. When the plant attain good bearing age, the intercrop is grown in between the tree rows in one direction, i.e. on two opposite sides of the trees, leaving the other two sides of the tree without intercrop to facilitate orchard operations. The tree rows are enclosed by a strong bund on two sides running from one end of the orchard to the other. This facilitates irrigation of trees within the bunded space independently of the intercrop. It also provides space for movement during plant protection and other operations in the orchard.

Pest management

If the cover crops or intercrops are taken regularly, the weeds will be automatically suppressed. A constant watch over the diseases and insect pests of litchi is necessary for ensuring quality crop from the trees. When the humidity builds up appreciably during the flowering time, incidence of insect / pest and fungal diseases are increased and sometimes this may result in to heavy loss in fruit set. Some times due to excessive use of pesticides, the build up of natural enemies in litchi orchard is threatened. Thus, intercropping/cover cropping in orchard may promote infestation of pests and diseases in greater intensity than under clean cultivation. This is due to increase in the humidity and vegetation around the trees due to intercrops. Hence, regular plant-protection measures against pests and diseases of the litchi crop are essential. However, no vegetable insect pests have been found to attack litchi crop and vise versa in intercropped litchi orchard. Skirting and centre opening like pruning process must be followed from time to time in litchi plant for successful intercropping.

Irrigation and nutrition

While taking intercrop, people do mistake by no application of nutrients to main crop with a view that the plant will obtain nutrients from the same fertilizer dose applied to the intercrops. This is not a good practice. The intercrop and the fruit trees must receive their nutrition separately because of their independent requirements of fertilizer and irrigation. The nutrition and moisture to intercrops must be provided in additional amount. Proper irrigation of trees, particularly during summer, is very essential to get quality crop from the bearing trees and proper growth in young plants.

Trees must be regularly observed for any setback due to any cause and steps must be taken immediately for its remedy. Round the year cultural schedule must be drawn up and operations should be done as per schedule. This will ensure production of quality crop from healthy litchi trees and good yield from intercrop.

Long-term productivity

It is generally believed that growing intercrops will adversely affect the long term productivity of the main crop. But published literature shows that the crux of the problem is the level of management. If both the main crop and the intercrop are managed well, inter-cropping is most paying exercise. This has been proved by several experiments that inter-cropping not only give additional income but also enriches the soil. On the other hand, if the additional crop is allowed to be a parasite, then the yields of both main crop and intercrop will be adversely affected.

It is true that productivity in mixed cropping mainly depends upon plant to plant interaction. Intercrop and main crop draw their nourishment from the same environmental resource pool. If selection of crop is not properly done then the extent of competition may be too high and can harm the productivity significantly. If selection is done carefully taking care of aerial growth and root distribution pattern, then considerable additional benefits can be achieved by intercropping. Plants may complement each other in sharing resource pool and thus achieve a complete utilization of the eco-climate.

❑❑❑

Chapter 10

Feed the Plant as per Need and Schedule

Litchi requires soil nutrients and water for satisfactory growth and fruiting. Yield and fruit quality are usually adequate over a wide range of leaf nutrient concentrations. Litchi roots uptake nutrient from large volume of soils. Therefore, plant is able to sustain growth even in low fertility soils. Trees take a long time to respond to fertilizer application. New growth is more dependent on reserves in the tree than on fertilizer applied just recently to the soil. When the concentration of a nutrient reaches abnormally low or high levels in litchi plant, characteristic symptoms appear in the leaves, stems, flowers or fruit. Normally, growth and yield are reduced long before visible symptoms appear. When leaf nutrient concentration falls below critical levels, tree takes several years to recover. The only way to avoid this is to monitor the concentrations of the nutrients in the plant and soil and maintain these within the acceptable range to establish healthy, high-yielding orchards. Six distinct growth stages i.e. the nursery stage (0-1 year), the field establishment and non-bearing stage (1-6 years), the junior adult bearing stage (6-10 years), young bearing period (10-25 years), adult bearing stage (25-40 years) and the ageing stage or renewal of stage (40 years onward) are found in the life cycle of a litchi plant. The nutritional requirements in each of these periods may differ. Speedy growth and proper establishment of layer are the main objectives in nursery; therefore, the nutrients available in the potting medium are the main source of nutrition at this stage. At the time of field establishment, the main objective is to ensure least mortality whereas, during the pre-bearing stage the main objective is rapid growth of the plant and the development of a sound and strong frame work which can be supplemented by regular application of manures and fertilizers near the root zone. During the full bearing period, good cropping, regular bearing and high quality of fruit are the main criteria, whereas, during post rejuvenation period, the ability to produce for longer age are the main aim in aged litchi orchard. At these stages, plant performance depends on inherent soil nutrient and supplemented nutrition.

For obtaining optimum yield from a litchi plant, the nutritional requirements of plant must be met by supplying enough nutrients in the form of organic manures and fertilizers. Other inputs e.g. water and plant protection chemicals also be supplied adequately as they are supplementary to the nutrition. In general, nutritional need of the plants in a given environment depends on its health and stage of growth, inherent nutrient status of the soil, and the climate in which it is being grown. Altogether 16 elements namely C, H, O, N, P, K, Ca, Mg, S, Fe, Mg, Zn, Cu, Mo and Cl are considered essential for normal growth and production of a litchi plant. Over 90% of a fresh plant tissue is made up of C, H and O and the share of other mineral elements is less than 10% when the fresh plant tissue is analyzed. It is interesting that the C-H-O matrix which accounts for over 90% of the dry matter is usually 20 to 50% of the fresh weight of the whole plant. The importance of the rather small content of mineral elements results from the specialized and varied nature of their interaction with the major constituents. However, plant growth is not seriously retarded by lack of non-mineral elements (C, H, O) as long as H_2O is available and CO_2 in the air is plentiful. Obviously, ability of a soil to supply the remaining nutrients in amounts, forms and proportions required for maximum plant growth is held responsible for better productivity of a litchi crop.

When the plant contains much less of an essential element, it is said to be in a state of deficiency. Deficiency results in physiological disorders in leaf and fruit which can be identified by visual symptoms but in case of acute deficiency, it leads to death of a plant. In litchi, concentrations of nutrients in fruit can be used to estimate the removal of the different nutrients by the crop (Table 21). It is therefore, considered that the rates of application of Nitrogen, Phosphorus, Potassium, Calcium and Magnesium may vary depending on crop load.

Table 21 : Nutrient composition of litchi fruit and its removal from soil

S.No.	Nutrient	Concentrations in the fruit	50 kg crop would remove(g per tree)	1000kg fresh fruit extracts from the soil (kg)
1	N	0.85%	98.00	2.20
2	P	0.19%	22.00	2.00
3	K	1.04%	120.00	6.60
4	Ca	0.10%	12.00	1.60
5	Mg	0.18%	21.00	1.10
6	Mn	29 ppm	0.30	-
7	Zn	34 ppm	0.40	-
8	Cu	36 ppm	0.40	-
9	B	15 ppm	0.20	-
10	Cl	0.01%	28.00	-

Nutrition programme in litchi should begin with soil analysis.This analysis is done prior to planting to obtain data on the nutritional characteristics of the subsoil in which the plants grow. This helps in deciding the basal dose of nutrients for pit filling and to take further corrective measures particularly where soil pH is not normal. For other initial guidelines on fertilization, the doses should be based on physiological parameters, such as stem thickness or tree age. There is no doubt that soil and climatic differences, cultural practices, the size and type of plantation and even the cultivar, have a definite role on the quantities of fertilizer required. General recommendations, nevertheless provide the grower with a good starting point of fundamental importance from any point of view. Rates can later be adjusted depending on yields obtained and results of soil and leaf analysis. Leaf analysis are particularly important when the plant enter in the bearing phase. One should always be on the alert for sign of deficiencies or excesses, which must be corrected without delay. It would, of course, be even more desirable if they could be avoided altogether.

Determining the Need

For determining the type of fertilizer elements and its quantity, litchi growers always remain in a dilemma. There are some diagnostic techniques to resolve this dilemma which should be well interpreted for judicious dose of nutrition.

Some methods (soil or leaf analysis) are adopted to know the deficiency precisely. Since only plant analysis can provide this information, assessment of the trees nutritional state is achieved by means of chemical analysis of the plant itself. The part of the plant chosen for analysis is usually the leaf, which is metabolically very active and whose composition is therefore, a good guide to changes in the plants nutritional state.

(i) Soil analysis

Soil analysis can be used to assess the nutritional status for tree crops provided that a particular site selected for sampling does not fall outside the range of fertility and the sample size is adequate for the crop and soil. Soil tests have a role for correcting or avoiding problems such as acidity, salinity and nutrient interactions and toxicities, which are not directly related to plant composition. The analysis should preferably be taken with a leaf test. Soil samples should preferably be taken at the same time as leaves collected for tissue analysis. This is normally just after panicle emergence in spring prior to the application of fertilizer. Collection of soil samples just after fertilizing is avoided due to sampling errors associated with the uneven distribution of fertilizer in the topsoil. Research has shown that the feeder roots and nutrients under the trees are concentrated in the topsoil. Samples must be taken from two zones i.e. at depths of between 0-30cm and 30-60cm. It is the layer which provides the most reliable estimate of soil nutrient reserves. Each sample must represent some 3 ha, be homogenous in colour and texture and comprises not less than 10 sub-samples. These sub-samples must be mixed well in a clean container to form a sample of approx 2 kg. Each sample should be accompanied by a leaf sample collected from the same trees. The leaf mulch should be removed before taking soil sample. Separate samples should be taken for each block, soil and cultivar. Analyzing the soil, however, only provides information on the amount of nutrients available at a given moment but not on the amount actually taken up. Recommended soil nutrient concentrations in litchi are presented in Table 22.

Table 22 : Recommended soil nutrient concentrations for litchi

Parameter	Value	Parameter	Value
pH	5.5 to 6.0	Copper	1.0 to 3.0 mg kg^{-1}
Organic carbon	1.0 to 3.0%	Zinc	2 to 15 mg kg^{-1}
Electrical conductivity	<0.20 dS m^{-1}	Mn	10 to 50 mg kg^{-1}
Chlorine	<250 mg kg^{-1}	Boron	1.0 to 2.0 mg kg^{-1}
Sodium	<1.0 meq 100 g^{-1}	Potassium	0.5 to 1.0 meq 100g^{-1}
Nitrogen	10 mg kg^{-1}	Calcium	3.0 to 5.0 meq 100g^{-1}
Phosphorus	100 to 300 mg kg^{-1}	Mg	2.0 to 4.0 meq 100g^{-1}

Source: www.fao.org/DOCREP/005/AC681E/

(ii) Leaf analysis

Leaf analysis is the only technique according to which sensible fertilisation can be applied to litchi planting. A greatest reserve of nutrients occurs in the leaves, twigs and small branches which accounts for about 75 per cent of the total reserves of the tree. The amount of nutrients in the other plant parts is usually less than 5 per cent. The high reserves in the leaves, twigs and small branches are mainly because these tissues account for a large proportion of the plant's weight. The concentration of nutrients in the leaves reflects the reserves in the rest of the plant indicating that they are a reliable index of the tree's nutrient status. These reserves are used for new leaf, flower and fruit growth, but can last a long time. For instance, nitrogen concentrations are maintained for four years after fertilizer is withdrawn. This explains why litchi can take several years to respond to changes in nutrition management.

In order to obtain the best results from this analysis, care must be taken to ensure that climatic, soil type and irrigation conditions as well as cropping practices are normal. While taking sample, following points should be kept in mind:

a) The best time for taking samples is 1-2 weeks prior to flowering. Do not take samples from obviously good or weak trees.

b) Sample must be taken from the leaves shown in Fig. 34 so that findings may always be comparable. Leaves from eight branches, uniformly distributed around the tree are selected. Sample of 4 leaves per tree should be taken.

c) Leaves should be taken from healthy trees, free of sunburn and showing no symptoms of deficiencies or damage by pests of disease.

d) To be true representative, samples should be taken from some 20 trees, well spaced an area of out over 3 ha. The trees should be homogenous in appearance and be average bearers.

Fig. 34 : Sample either of the 2 leaves coloured in the illustration

e) The results should be supported by a record of leaf colour, tree vigour and yield so that fertilizer management can be adjusted for the next crop.

f) The leaf sample should be accompanied by a soil sample from 0 to 15 cm each year. A soil analysis could probably be done at three year intervals. The findings of both these analyses must be taken into account before any general recommendations are put forward.

Interpreting a leaf analysis is not a simple matter. The laboratory findings have to be compared against established standards so that the elements in deficit or excess and the magnitude of these deficiencies or excesses may be determined. The optimum levels of a particular nutrient in leaves vary depending on climate and cultural conditions and even on the cultivar and this is why each of these aspects must be taken into account. Decision-making is further complicated by the fact that there is no strict proportional relationship between the amount of a particular element contained in a leaf and the magnitude of the corrective measures to be taken.

It is therefore not possible to establish general standards to help in the decision-making process. Table 23 is intended as a guide to nutrient content in areas where yields range from acceptable to good and where symptoms of deficiencies or toxicity are not apparent. The ultimate fertilizer programme depends on tree size; crop load, cultivar and soil type and vary considerably between different districts, orchards and years.

Table 23 : Litchi leaf nutrient standards

Nutrient	Optimum level Galan Sauce (1987)	Menzel *et al.* (1992)
N (%)	1.50-1.70	1.50-1.80
P (%)	0.15-0.30	0.14-0.22
K (%)	0.70-0.80	0.70-1.10
Ca (%)	2.00-3.00	0.60-1.00
Mg (%)	0.35-0.45	0.30-0.50
Fe (/g)	40-70	50-100
Mn (μg/g)	40-80	100-250
Zn (μg /g)	12-16	15-30
Cu (μg /g)	-	10-29
B (μg /g)	45-75	25-60
Na (μg /g)	300-500	<500
Cl (%)	0.30-0.35	<0.25

Cited from : K.S. Chauhan (2001). Litchi. Botany, Production & Utilization, Kalyani Publishers, New Delhi

(iii) Visual symptoms of deficiency

Visual nutrient deficiency symptoms can be a very powerful diagnostic tool for evaluating the nutrient status of plants. One should keep in mind, however, that a given individual visual symptom is seldom sufficient to make a definitive diagnosis of a plant's nutrient status. Many of the classic deficiency symptoms such as tip burn, chlorosis and necrosis are characteristically associated with more than one mineral deficiency and also with other stresses that by themselves are not diagnostic for any specific nutrient stress. However, their detection is extremely useful in making an evaluation of nutrient status. In addition to the actual observations of morphological and spectral symptoms, knowing the location and timing of these symptoms is a critical aspect of any nutrient status evaluation. Plants do not grow in isolation, they are part of the overall environment and as such they respond to environmental changes as that affect nutrient availability. Also, plants do influence their environment and can contribute to environmental changes, which in turn can affect the nutrient status of the plant.

Visual symptoms can be useful for identifying deficiencies. It is apparent that some symptoms are clearly defined and typical whereas others are not. Also there are instances when two or more nutrient deficiencies produce similar symptoms (compare Mn and Fe or Zn and Cu deficiencies). Thus, one may get confused and fail to draw any definite conclusion simply on the basis of deficiency symptoms.

Not all tissues of a plant show nutrient status during times of stress. Leaves on the same plant that are exposed to different environmental conditions, (such as light), or those of different ages may have considerable differences in nutrient status. Mineral nutrients are for the most part acquired by the roots and translocated throughout the plant. The distance of any part of the plant to the roots will influence nutrient availability, particularly in the case of the less mobile nutrients. In plants recovering from nutrient deficiency, the root and conductive tissues recover first. For example, in the case of recovery from Fe-deficiency, it is common to see the veins re-green, while the interveinal tissue remains chlorotic and Fe-deficient.

In order to maintain rapid optimal growth, all plant tissues must have a favourable nutrient status. Although, a plant may be marginally low in a number of nutrients, only one nutrient at a time will limit overall growth. However, if the supply of that limiting nutrient is increased even slightly, the resulting increase in growth will increase the demand for all other nutrients and another nutrient, the next lowest in availability, will become limiting.

Deficiency symptoms of major nutrients in litchi

Nitrogen

Nitrogen is the most important nutrient affecting growth and productivity litchi plant. As in the case of other plants, nitrogen plays an important part in relation to vegetative growth and has a pronounced effect on the bearing of the trees. Increase in fruit set, retention and yield with nitrogen have also been reported. A high level of fertilizers may delay the entering of trees into bearing and there are several reports also to show that the lower fertilizer doses do not affect tree vigour and fruit yield. Deficiency symptoms have been reported when leaf concentrations fall below 1.5 per cent. Because nitrogen moves from old to young leaves when concentrations are low in plants, the first sign of deficiency (yellowing) are noted in the older leaves. In cases of severe deficiency, the margins of the leaves may curl, leaves may be small or fail to develop or be shed prematurely. Plant growth is stunted, flowering and fruit set are prevented and branch development become poor in N deficient plants. When the deficiency is very severe, new leaves do not mature and leaf drop occurs at the terminals. Fruit become small with low flesh recovery and reduced eating quality. Nitrogen does not have a direct effect on floral initiation as no consistent relationship between flowering time nitrogen application and soil nitrogen concentrations have been observed.

Phosphorus

Low phosphorus concentrations are rare where mixed fertilizers are applied regularly. This is because phosphorus is not readily leached from the topsoil. Once soil Phosphorus concentrations is high, it remains sufficient for several years. The first symptoms of 'P' deficiency show as dead patches on the tip and margins (coppery brown colour) of mature leaves that progress towards the midrib. Eventually, the leaves curl, desiccate and shed. These severe symptoms have only been recorded in sand culture. Only in extreme cases some degree of necrosis occurs and leaflets become copper-coloured.

Potassium

Potassium is an important nutrient for fruit filling (size and weight) and quality. Potassium is required for the production and transport of plant sugars that increase the weight of fruit. It also reduces the detrimental effects of high plant nitrogen levels on fruit quality. Potassium also promotes more even ripening and enhances skin colour. The highest demand for potassium is just prior to harvest. Because potassium is very mobile within the plant, foliar application of potassium is very effective, especially in the later stages of growth. Soil uptake of potassium is inefficient and therefore, deficiencies often occur even in soils high in potash, especially if calcium, magnesium and nitrogen levels are also high. Many orchards have potassium concentrations below 0.80 per cent. This occurs late in the season when potassium is translocated to developing fruit after heavy nitrogen applications or after heavy rain. These problems are more likely to occur on sandy soils. The leaves start to become yellow, the leaf tips die and at last the bases of the leaves are also affected. Severe deficiency can restrict shoot and root development. Plants may come under flowering but do not set the fruits and trees may die. Plants are stunted and branch development is poor. Leaflets are small, not much dark green in colour and sun burnt at the edges. Necrosis begins at the leaf tips and spreads along the leaf edges towards the base. In severe cases, leaf drop occurs just prior to the final vegetative flush so that only terminal cluster of leaves remain.

Micronutrients deficiency

If micronutrient deficiencies that constrain high productivity are not identified, monitored and alleviated, the fertilizer use efficiency of costly chemical fertilizers and other agricultural inputs will be markedly reduced. It is essential to augment the availability of native nutrients by modifying soil environment through management practices in litchi. Some farmers have applied micronutrients in the hope of increasing crop yields even though there is little evidence to suggest a deficiency exists. There is still a widespread lack of awareness by farmers for the importance of micronutrients in litchi production. Also, a continuous research and testing is required to identify the nutrient deficient areas.

Micronutrient deficiencies have the following impacts on crops:

- Affect yields, either directly or indirectly;
- Lead to less efficient use of other essential plant nutrients;
- Can trigger nitrogen losses to the environment;
- Result in lower water use efficiency;
- Weaken crops capacity to withstand difficult conditions;
- Contain undesirable element (like cadmium) with higher risks of having non-nutritive trace elements present in edible parts.

Diagnosing a micronutrient deficiency can be a difficult and time consuming process. To identify some visual nutrient deficiency symptoms are given below:

- Ensure that poor crop growth is not the result of a macronutrient deficiency, drought, salinity, disease or insect problem, herbicide injury or some physiological problem.
- Find out if a micronutrient deficiency has been identified before in a particular crop or soil type in the area.
- Examine the affected crop for specific micronutrient deficiency symptoms.
- Take separate soil samples from both the affected and unaffected areas for complete analysis, including micronutrients.

- Send plant tissue samples from both the affected and unaffected areas for complete analysis that includes tests for micronutrient levels.
- If all indications point to a micronutrient deficiency, apply the micronutrient to a specific, clearly marked cut affected area of land to observe results in subsequent seasons.

Calcium deficiency

Calcium is the most important nutrient for fruit quality and its deficiency is a common problem in litchi crop. While moisture stress is the major cause of calcium deficiency, high available soil nitrogen or over-use of nitrogen fertilizers that promote excessive plant vigour also aggregates calcium deficiencies. Low boron levels also induce calcium deficiency as boron is required for the maintenance of transpiration (water uptake) and therefore, also calcium uptake. Calcium deficiencies also occur on soils with high levels of exchangeable calcium and this is largely due to the low mobility of soil calcium and competition with other nutrients such as, ammonium nitrogen, potassium and magnesium. For this reason, foliar sprays are needed to supplement soil calcium. Symptoms of calcium deficiency have been seen in plants grown in sand culture but are rarely in the field. Calcium-deficient leaves show necrosis around the base of the leaves. The very low mobility of calcium is a major factor determining the expression of calcium deficiency symptoms in plants. Classic symptoms of calcium deficiency include blossom-end rot of tomato (burning of the end part of tomato fruits), tip burn of lettuce, blackheart of celery and death of the growing regions in many plants. All these symptoms show soft dead necrotic tissue at rapidly growing areas, which is generally related to poor translocation of calcium to the tissue rather than a low external supply of calcium. Very slow growing plants with a deficient supply of calcium may re-translocate sufficient calcium from older leaves to maintain growth with only a marginal chlorosis of the leaves. This ultimately results in the margins of the leaves growing more slowly than the rest of the leaf, causing the leaf to cup downward. This symptom often progresses to the point where the petioles develop but the leaves do not, leaving only a dark bit of necrotic tissue at the top of each petiole. Plants under chronic calcium deficiency have a much greater tendency to wilt than non-stressed plants.

Magnesium deficiency

The concentrations of magnesium are often low, especially when trees are grown on sandy soils where it is readily leached. Deficiencies can also be induced by heavy applications of nitrogen and potassium. Magnesium is not readily transported from old to young leaves, hence, symptoms occur first on young tissues. Plants grown in sand culture without magnesium have small leaves that die between the veins and eventually drop. Flowering is suppressed when leaf concentrations fall below 0.25 per cent. The Mg-deficient leaves show advanced interveinal chlorosis, with necrosis developing in the highly chlorotic tissue. In its advanced form, magnesium deficiency may superficially resemble potassium deficiency. In the case of magnesium deficiency the symptoms generally start with mottled chlorotic areas developing in the interveinal tissue. The interveinal laminae tissue tends to expand proportionately more than the other leaf tissues, producing a raised puckered surface, with the top of the puckers progressively going from chlorotic to necrotic tissue.

Manganese deficiency

These leaves show a light interveinal chlorosis developed under a limited supply of Mn. The early stages of the chlorosis induced by manganese deficiency are somewhat similar to iron deficiency. They begin with a light chlorosis of the young leaves and netted veins of the mature leaves especially when they are viewed through transmitted light. As the stress increases, the leaves take on a gray metallic sheen and develop dark freckled and necrotic areas along the veins. A purplish luster may also develop on the upper surface of the leaves. Grains such as oats, wheat, and barley are extremely susceptible to manganese deficiency. They develop a light chlorosis along with gray specks which elongate and coalesce, and eventually the entire leaf withers and dies.

Iron deficiency

The leaves loose green colour and turn white and size of the leaf is reduced. The growth of the tree is restricted; yield and quality of the fruit is reduced. In some plants the leaves dry from tip to downwards. The symptom described is called "Bleaching". The deficiency is common in soils with high calcium content. Hence, the effect is known as "calcium induced iron chlorosis". Small plants may die due to acute iron deficiency. Orchards established on sandy soils often have low iron levels, especially after excessive super phosphate applications that interfere with iron uptake by the roots. There can also be problems in alkaline soils with a pH above 7.0, or after excessive lime applications. There is a general yellowing of the young leaves, spreading to older leaves. When the deficiency is severe, the branches may die. Concentrations below 40 ppm are considered a problem. It is best to spray iron chelate as a foliar spray because it is difficult for the tree to take up iron.

Zinc deficiency

Zinc deficiency is most likely to occur in litchi grown on alkaline soils, however deficiencies on heavily leached acid soils may be induced by lime application. Zinc is most important at different stages of plant growth and development for photosynthesis, root and shoot growth, flowering and fruit set. Deficiency symptoms are normally associated with 'little leaf disease', interveinal chlorosis of young leaves and poor fruit set and yields. Leaves become small and narrow with leaf margins bent upward or downward. Internodal length is reduced drastically and the twig with crowded leaves gives rosette appearance and intervenial areas are pale. Plant do not grow well and the yield, size and quality of the fruit is reduced. Zinc deficiency occurs on acid leached soils where native zinc is low, or on alkaline soils where zinc is not readily available to plants. These problems are often exarbated after heavy nitrogen applications. Zinc deficiency causes general bronzing or mottling of the leaves, smaller shoots and smaller fruit, small and short leaves with normal colour and occasional curving. The branches may die when leaf concentrations fall below 10 ppm.

Copper deficiency

Copper deficiency is most likely in sandy soils with an inherently low copper content. Often the young leaves roll and die. Shoots may also die when leaf concentrations fall below 6 ppm. In some soils, copper and zinc deficiencies occur together.

Boron deficiency

Boron deficiencies occur on a wide variety of soils, however boron availability reduces as pH and calcium levels increase. Litchi is sensitive to boron deficiency especially when high levels of nitrogen are applied or trees become moisture stressed. Boron is essential for flowering and fruit set and deficiencies cause flower and fruit drop as well as misshapen fruit. Boron is required for calcium uptake and movement, and calcium deficiency is significantly reduced by boron application. Boron plays a similar role to calcium in cell wall stabilization which makes it essential for quality factors such as skin strength, fruit firmness and storage life of litchi. Fruits cracking in litchi is mainly due to boron deficiency. Luster less leathery leaves with thickened veins are the other associated symptoms and brown areas in yellow fruit pulp is conspicuous. The range between deficiency and toxicity is small, so care should be taken when applying boron fertilizers.

Salt Injury

The leaves are scorched and leaves are loosing their natural colour and turn to bronze colour. Tip burning is also seen some plants due to salt injury. Raising *Diancha* (Sesbpania) as green manure crop with onset of monsoon in the inter-spaces of the litchi orchard and its incorporation in soil during September have been found effective. Farm yard manure and compost should be applied adequately every year. Keeping Gypsum filled gunny bag in flowing irrigation water reduces salt effect.

Nutrient application

Amount, quantity and type of fertilizers to be used in a litchi orchard are influenced by differences in the growing conditions including characteristics of soil, cultivar, age of trees, cultural practices, irrigation, pest and disease control and environmental conditions such as weather. It is impossible to set up a fertilization programme which would suit all litchi growing areas.

The amount of fertilizers to be used on bearing trees is determined by a careful estimation of the orchard production capacity based on tree size or age and yield records of several previous years. Soil and leaf analysis are helpful in perfecting the dose. One must also bear in mind that nutrient demands of litchi tree during the formation of extensive vegetative structure or the fertilizer rates required by growing plants or plants coming into bearing and requirement of the plants in full production would not be the same. During early years of the orchard establishment, main aim of plant nutrition is to obtain the maximum bearing surface. This is achieved by regular application of nutrients, water and appropriate pruning.

Different doses of manures and fertilizes are required to nourish the litchi plant and produce economical yield. FYM can be used as additional fertilizer at 10-20 kg, respectively per mature (10 years) tree, spread evenly in the trench around basin. Phosphorus stays in the soil a lot longer than potassium. So after the rainy season, potassium is used by the trees in large quantities during the dry season and is easily leached out, therefore, even when enough has been applied to bring the level up, it is recommended that additional applications of potassium in the dry season to maintain the level. There are indications that applying phosphorus and potassium stimulates the tree into flowering when it has not yet flowered. Muriate of potash is cheaper than potassium sulphate but muriate of potash is not good for saline soils and if a soil is very low in sulphur, potassium sulphate may be applied. For the convenience of understanding and recommendations based on research and experience gained by the orchardists, the nutritional requirement and management for litchi crop are given below.

(i) Nursery stage

In the nursery stage, the plant requires more nutrition and therefore, the nursery beds/potting mixture should be rich in humus which also improves the water-holding capacity of rooting medium and promotes root growth. Normally, 25-40 tonnes of farmyard manure is applied to a hectare where litchi layers are planted after detachment from the trees. When pots/poly bags are used, they are generally filled with a mixture of leaf mould / FYM, top-soil and sand/ clay in the ratio of 2:1:1. For improving the growth, light doses of inorganic fertilizers, viz., ammonium sulphate and calcium ammonium nitrate for nitrogen supplement and SSP for phosphorus supplement is applied once at the time of bag filling. While raising plants in nursery 3-4 foliar sprays of poly feed containing 19:19:19 NPK water soluble mixture at 15 days interval has been found effective. If the plants show nutrient deficiency symptom or weak growth, the application liquid manure comprising of cow dung + neem / karanj cake + NPK mixture should be done at monthly interval or on need basis.

(ii) Transplanting stage

The fertilizer application schedule for young plant begins with organic manures including cake. The newly planted saplings are not likely to need feeding in the very first year. If the fertilization programme starts in the first year, fertilizer should not be placed within 20cm radius of the trunk because the litchi plants are sensitive to fertilizer burn. It is generally not advisable to apply any fertilizer at the time of planting however, manures, cakes and anti termite inputs should be applied well before planting at the time of pit filling. Almost all soils suitable for planting litchi, have enough available nutrients to support growth of the plant for at least for one year. If the soils are too poor to sustain the plant even for one year, the whole pit soil should be replaced and good soils with nutrient mixture should be arranged for pit filling. In order to meet out the initial nutritional requirement, 25kg FYM + 0.75 kg bone meal/SSP + 2 kg Cake + 50g termiticide may be applied during pit filling.

(iii) Non bearing stage

Once the plant gets established in field, it requires high amount of calcium, nitrogen and phosphorus for initial growth. Potassium is also required in reasonable amounts. In soils with a high calcium content and pH (7.0 to 8.0), surface application of calcium is not necessary but in soils where pH is between 5.5 to 6.5, it may be necessary to add calcium in the form of dolomite lime. The quantity of lime depends upon the soil reaction and in certain cases (highly acidic soil) a dose of 2-3 t/ha is recommended. During the non-bearing period, nitrogen is particularly required in sufficient amount to induce vigorous growth. It would be better if some part of N is derived from organic matter so that soil texture, moisture holding capacity and ultimately the growth and development of roots improve. Phosphorus is required to cope up with high respiration rates and for translocation of carbohydrates during non-bearing stage. Potassium is required to meet out the need of photosynthesis but its need during non-bearing period is comparatively less than that of bearing period when it is required for sugar and starch synthesis in greater quantities.

(iv) Initial bearing stages

The fertilization programme for initial bearing stage may start from the third year and continues till the come in full bearing stage. In the fourth year of planting, if trees are well grown and flushing vigorously in spring, N fertilization is stopped until leaf colour falls and/or fruiting begins. The application of fertilizers at this time must be supported by soil and leaf analysis. If the approach is not adopted, trees may not begin to bear regularly for several years.

(v) Full bearing stage

In case of fully developed bearing trees, nutrition management is directed firstly to gain maximum vegetative growth immediately after fruit harvest in summer. The second requirement is to maintain the tree in state of dormancy for 3-4 months before flowering. The third requirement is to maintain nutrition at a high rate once fruits have set, to prevent any stress. To achieve this, nutrients are applied twice during the year. The first application is made 4 weeks after fruit set (March-April) to encourage fruit development and second two weeks after harvest (June-July) to encourage first growth flush. Nitrogen application during autumn and winter may encourage excessive flushing prior to the normal period of flower initiation therefore should not be applied. Application of N in early spring can promote growth flushing which may complete with the flowers and reduce shedding. Since the main aim of any fertilizer programme is to obtain maximum yields, fertilizer management should be directed towards maintaining the tree in a state of dormancy for 3 to 4 months before flowering. In order to obtain profuse flowering and good yields, it is necessary to withhold fertilizers for a year. Great care must be exercised when doing so, however, as drastic reductions of N, P and K not only leads to stunted growth, but hinder development and floral initiation. It is therefore not surprising that increasing the rates of application of these nutrients at the appropriate time has a favourable effect on flowering, fruit set and final size and weight of the fruit. It is also essential to maintain a high rate of growth immediately after harvesting so as to obtain a good vegetative flush in summer and a sufficient number of terminals potentially capable of flowering in the following spring. Fertilizers should not be withheld throughout the period extending from flowering to harvesting so as to ensure a good yield.

Fertilizer application rate

Total and relative amounts of each nutrient or fertilizer vary greatly in different litchi orchards. Since soil and climate of different places are highly diverse, there is wide variation in the response (i.e. yield) vis-α-vis the rate of fertilizer application in India The common dose of fertilizer is presented in Table 24.

- The fertilizer rates given in the table are a guide only and should be supported by the results of leaf and soil analyses. Depending on cropping patterns and soil, they can easily be modified to suit other environments. In many parts of litchi belt, most of the nutrients are supplied from organic fertilizers. The suggested applications of the different nutrients can be amended as necessary. The major nutrients are best applied to the soil.

Responses to foliar applications have been reported in some countries, but tend to be short-lived.

Table 24 : Recommended doses of fertilizers for getting optimum yield.

Year	Quantity/annum/tree						
	FYM (Kg)	Cake (Kg)	Nitrogen (g)	Phosphorus (g)	Potash (g)	Zinc (g)	Boron (g)
1	10	1.0	50	25	25	25	-
2	15	1.50	100	50	50	50	-
3	20	2.00	150	125	75	75	-
4	25	2.50	200	150	100	100	-
5	30	3.00	250	200	125	125	125
6	35	3.50	300	250	150	150	150
7	40	4.00	350	300	175	175	175
8	45	4.50	400	350	200	200	200
9	50	5.00	500	400	250	225	225
>9 year	60	5.00	600	600	250	250	250

Source : K.S. Chauhan (2001) and other reports

- For nitrogen, don't apply fertilizer if leaf concentrations are above 1.8 per cent and the trees are vigorous and have not set a crop. If the range is 1.5 to 1.8 per cent, apply the recommended amount. If the range is 1.2 to 1.5%, apply 25% more, if it is 1.1 to 1.2%, apply 50% more, and if it is less than 1.0%, apply double dose of nitrogen.
- For phosphorus, interpret the results in conjunction with soil analysis, and don't apply if the leaf test is more than 0.22% or if the soil test is above 300 ppm.
- For potassium, trees carrying a heavy crop, with less than 0.50% K in the leaf test, will require twice the amount of fertilizer listed for their size or age. If the leaf test is 0.5 to 0.6%, use another 50% more than the recommendation. If the leaf potassium is 0.70 to 1.10%, use the recommendation, but if it is above 1.10%, add nothing.

Fertilizer scheduling

During the first three years, fertilizers are used to promote rapid plant growth. Decide the nutrients application and quantity schedule that suits to local situations. In frost prone areas, do not apply fertilizer during autumn or winter and do not exceed the recommended rates. Excessive amounts of organic or inorganic fertilizers can kill plants, especially on shallow and poorly drained soils.

The following points must be considered for the strategic fertilizer scheduling of litchi orchard

- For young trees up to three or four years of age, apply a complete fertilizer beginning after the hardening of the first growth flush after transplanting.
- Subsequent applications should be made after hardening of each succeeding flush, or every two to four months.
- The fertilizer should be spread evenly around the tree at least 1 ft (30 cm)
- Stop applying fertilizer in the spring of the fourth or fifth year, when the tree is large enough to bear a crop the next year.
- Apply fertilizer to bearing trees immediately after fruit harvest.
- If rainfall is limited, apply irrigation water at that time to promote a vigorous flush.
- Bearing trees need less P than developing trees, while opposite relation is reported with K.
- Excessive amounts of available N during the winter will favor untimely vegetative flushes; thus, N application levels should be calculated so that the N is depleted before a rest period prior to flowering occurs.
- During the first four years after planting, fertilizers should be applied generously to promote canopy development.
- After fruit set, when the fruits are pea-sized, a supplemental application of NPK in the ratio 2:1:4 or 2:1:8 will provide adequate potassium for fruit development. This application should be light, because too much N at the time of fruit ripening may cause fruits to crack during rainy periods.
- Amounts of fertilizer applied to bearing trees vary depending on tree condition and location. With many fruit trees, a general rule is to apply annually 150 g NPK fertilizer for every year of tree age, reaching the maximum application level around Years 10 to 12.

- Organic soil amendments and fertilizers are useful to promote tree establishment but should be used with caution on bearing trees. Organic N is released more slowly than most chemical fertilizer N, and it may be more difficult to manage the time of availability and depletion of N from mulches and manures.

Restricting fertilizers application

The most effective method to prevent untimely vegetative growth is to maintain leaf N content at 1.75-1.85 percent during the critical 4-6 week period before flower initiation. Maintaining low leaf N levels can be achieved by applying N fertilizer only after panicle emergence and fruit set and at no other time during the growing season. This method has proven effective even in areas with heavy rainfall.

(i) Bearing trees of early-season cultivars

It is desirable to induce two vegetative flushes after harvest, one in June-July and a second in August-September. To promote these two flushes, one-half of the fertilizer allocated for the year should be applied with irrigation immediately after harvest. The balance of the year's fertilizer should be applied in two equal parts, one in spring during flower panicle elongation and the second in early summer when fruits reach pea size.

(ii) Bearing trees of late-season cultivars

A single vegetative flush is promoted after harvest. One-third of the fertilizer allocated for the year should be applied two weeks before harvest. The remaining two-thirds of the fertilizer should be applied in two equal parts, one in spring during flower panicle elongation and the second when fruits reach pea size. The first application, before harvest, ensures sufficient time for the vegetative growth to mature before winter; the second, smaller application avoids high residual leaf N levels, which may inhibit flower induction.

(iii) Bearing trees in areas with deep soil and high rainfall

The allocation of fertilizer should be applied in two equal parts, one during spring flower panicle elongation and the other when fruits reach pea size. N-containing fertilizer should not be applied after harvest.

Where leaf magnesium is low, magnesium sulphate can be applied to the soil at the rate of 40 g per m^2. Another strategy is to apply the magnesium sulphate is as a foliar spray for quick results.

For low calcium concentrations, apply gypsum at 500 g per m^2 if the pH is above 6. If the soil pH is below 6, use lime or dolomite at the rate recommended by laboratory.

As a general practice, full dose of organic manures and three-fourth of the total inorganic fertilizers are applied in June-July after harvesting. Remaining one-fourth inorganic fertilizers are applied each in January and early April.

Method of fertilizer application

Correct placing of fertilizer is necessary if the best advantage is to be obtained from their application. Young and recently planted saplings must be well established and grow vigorously before any nutrients other than base dressing are applied. Bearing in mind that the roots of air layers, (generally used in commercial plantings) are very shallow therefore, special care must be taken not to apply these compounds very close to the plant during its first six months in the field. Indeed, it is neither necessary nor useful to incorporate fertilizers until the first flush of growth appears. One alternative could be to incorporate the fertilizer to a depth of 30 cm over a square metre area one month prior to planting and to open the holes just a short time before placing the saplings in them. First of all, the weeds should be removed from basins. The mixture of recommended dose of fertilizers should be broadcast under the canopy of plant leaving about 50 cm from tree trunk in old trees.

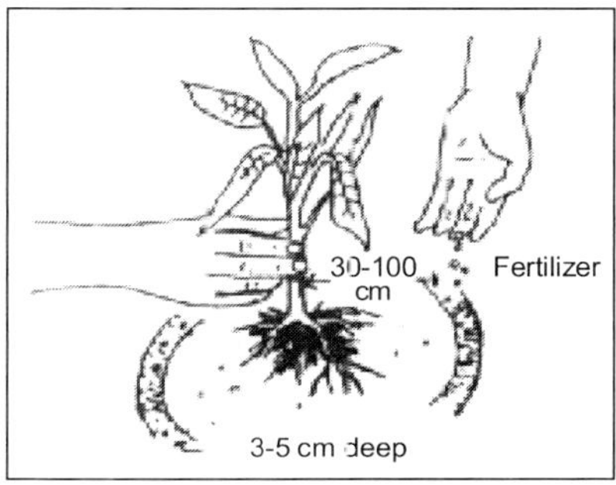

Fig. 35 : Fertilizer application method

Fertilizer application should be followed by light and controlled irrigation. The reason for this is twofold; to prevent an excessively high concentration of fertilizer around the roots and, ensure that the fertilizer is not drawn too deeply into the soil. Where irrigation is by the drip system, soluble fertilizers may be used.

In addition to time of nutrient application, the method of fertilizers placement for effectively uptake by the plant is also important in nutrient management. For efficient utilization by trees, fertilizers must be equally distributed on the root zone area. In young litchi plants, the active roots are scattered 30 cm away from the trunk in a circular fashion (drip line) and in well grown trees this may extend up to about 1.5-2.0m away from the main trunk. After broadcasting equally on the root zone area, fertilizers should be incorporated to the soil by cultivating the top soil. Care should be taken not to damage the roots too much. If thick mulch with under tree sprinklers is available for irrigation such mechanical incorporation is not required as a number of frequent irrigations may slowly incorporate the fertilizers to root zone. Special care should be taken when applying urea as the source of nitrogen. Urea must only be applied when the soils are moist. If not, trees must be irrigated immediately after urea application. Manures and fertilizers can also be applied as bands in the root zone of the tree to reduce application costs. A trench of 30cm wide and 15-20cm deep in the drip area should be prepared and the nutrient mixture should be applied in it. After through mixing of the mixture with soil of the trench, it should be covered with side soils and a raised bund is prepared on the outer periphery of the trench. The practice help improve nutrient uptake and reduces the loss particularly during rainy season.

Fertigation : application of fertilizers with irrigation

Fertigation is defined as the application of nutrients using an irrigation system by introducing the nutrients into the water flowing through the system. Fertigation has increased dramatically in the past 15 years, particularly through sprinkler and drip systems. For drip systems, the expansion is seen mostly in horticultural and high value crops. With declining water supplies and increasing irrigation, a corresponding increase in fertigation is taking place. It will continue to grow since such systems result in less water usage and better uniformity than the less water-efficient and non-uniform furrow and flood systems. Effective fertigation requires knowledge of certain plant characteristics such as optimum daily nutrient consumption rate, root distribution in the soil, nutrient characteristics such as solubility and mobility, irrigation water quality factors such as pH, mineral content and salinity and nutrient solubility.

Soil characteristics and behavior of nutrients

Cropping and fertilization history as well as basic soil chemical properties need to be considered prior to fertigation. Soil testing by the method recommended in a specific area is the best way to assess current fertility needs. Particular attention should be given to soil pH and its adjustment to an appropriate range for the litchi. Proper pH can have a great effect on the availability of residual nutrients in soil as well as those added via fertigation. Any required soil pH adjustments should be made using conventional incorporation techniques. Fertigation is not utilized to alter pH of soil. Cation exchange capacity is an important consideration in determining the quantity of cations that will be retained during fertigation. In order to maintain an acceptable concentration of cations in solution, a soil with a low CEC must have its cations replaced via fertigation more often than one with a relatively high CEC, which can hold a greater quantity of ions. Since nutrients are easily added with fertigation, it is most practical on sandy, drought affected soils with low CEC. These soils need frequent irrigation and nutrient replenishment.

Fertigated nutrients behave similarly to conventionally applied soluble nutrients followed by rainfall or irrigation. Only small quantities of nutrients applied by an overhead sprinkler system are absorbed by the foliage. There is some concern that since nitrate nitrogen (NO3-N) is highly mobile, it may leach past the root system and reach groundwater. However, leaching of NO_3-N in the sandy soil is at higher depth.

In contrast to nitrate N, phosphorus (P) fertilizer moves very little in the soil. When broadcast on the soil surface, it will not normally move more than 1 to 2 inches below the soil surface. Therefore, fertigation of P is highly inefficient when using sprinkler or flood irrigation techniques. However, P fertilizer applied in drip systems may be rather efficient. An advantage of drip irrigation is that plant nutrients can be applied directly in the root zone. Since the majority of roots in drip-irrigated crops are located within the wetting zone, drip applied P will be placed in the soil region containing the highest root density. Therefore, the P applied in this manner is generally used more efficiently by plants than if the same amount of P were surface applied.

Water quality

For successful fertigation, careful attention must be given to water quality. The quality of water required depends on several factors, but foremost is the type of fertigation. Drip fertigation requires the highest quality water, i.e. free of suspended solids and microorganisms that plug small orifices in emitters. Only solutions should be used for drip/ trickle fertigation. As the orifice size increases in sprinkler systems, suspensions can be used. Precipitation of fertilizers within the irrigation systems can be a very serious problem if solubilities exceeds. One such problem commonly encountered is with calcium concentrations exceeding about 100 ppm. As calcium concentration increases and phosphates are injected, the probability of precipitation increases. When the water contains very high levels of calcium and magnesium, an injection of ammonium phosphates will result in precipitates deposited on pipe walls as well as in orifices and sprinklers. Often this will completely plug the irrigation system. If appreciable salts are in the water used for fertigation, one needs to consider the total amount of salts applied to the crop between leaching rains rather than only the amount applied in a single fertigation. Crops vary widely in their tolerance to salts; therefore, the number of irrigations as well as the salt concentration must be considered. In some areas, high levels of boron in irrigation water is a problem. Some sensitive crops can tolerate only about 1 ppm B. Total irrigation with water containing 1 ppm B applied to sensitive crops on light textured soils should not exceed 3 acre-inches between leaching rains.

Fertilizers quality

Three qualities are necessary for a good source of fertilizer for fertigation : (1) it must contain the needed nutrient elements in a form available to plants or in a form that is readily converted (i.e., urea is readily converted to available NH4- and NO3-N forms in good soil conditions); (2) it must be uniformly distributed — either uniformly broadcast for sprinkler, furrow, or flood or uniformly distributed among emitters for drip/trickle; and (3) the chemistry of the application must be such that plants are not burned or stunted and irrigation lines, emitters, or orifices are not restricted or plugged. In order to provide for the latter two limitations listed, the materials must either be in solution or be in suspension. For drip systems, solution is always needed and only clear liquid should be used. Nitrogen is the most fertigated element due to high plant nutritional needs, to its great mobility in soil following chemical and biological transformation to NO_3-N, and because of the many soluble sources of N fertilizers available for fertigation. Many forms of N have been used in fertigation. Urea solutions are probably most commonly used, but any soluble form of dry fertilizer can be used. Anhydrous ammonia or any other N fertilizer that has free ammonia should not be applied through sprinkler systems. Phosphoric acid and ammonium phosphate solutions can be used as sources of phosphorus in drip systems. Ammonium phosphate solutions are subject to precipitation if injected into water with high calcium and/or magnesium concentrations.

Potassium fertigation is usually accomplished with potassium chloride (muriate of potash) due to its solubility and relative low cost per kg of K. For most crops, foliage burns are not a problem due to the great dilutions in irrigation systems. Relatively pure grades of potassium chloride (white, 62% K_2O equivalent) are normally used in fertigation. Potassium nitrate has been used as a source of K and N, and potassium sulfate as a source of K and S. These K sources are less soluble and cost more per unit of K than potassium chloride. Sulfur, when needed, can also be provided as ammonium thiosulfat or ammonium sulfate. It can be readily mixed with several other soluble fertilizer grades and injected. Magnesium sulfate is often used to supply Mg and S. Micronutrients such as iron, zinc, copper, and manganese may (if applied in forms such as sulfates or chlorides) react with salts in the irrigation water and cause precipitation and clogging. However, many of the micronutrients can be applied as chelates - such as iron or zinc EDTA. Chelates are generally highly water soluble and consequently, will cause little clogging or precipitation.

Some consideration

1. Distribution uniformity of the irrigation water is critical for uniform application of the fertilizer.
2. Surface irrigation systems generally produce surface runoff. Make sure that any surface runoff during fertigation is picked up and reused on the field being treated or a similar field that is also being treated. Fertigation may not be suitable on fields that are very susceptible to runoff (i.e. steep slopes and low infiltration rates).
3. Make sure the combination of fertilizers and water will not produce clogging if using a micro-irrigation system.
4. Know how much water is needed to refill the root zone during the irrigation and plan the fertilizer application accordingly. Over-irrigation during fertigation not only wastes water but could result in leaching of the fertilizer below the root zone.
5. The type of injection device is critical depending on the type of irrigation system being used. Some devices will inject at a relatively uniform rate throughout the irrigation, some will not.

Be aware of which type is being used and which type is required by the situation.

6. Be aware of the requirements for flushing the irrigation system after fertigation. This may take 10 to 15 minutes. Also run clean water through the injection meter, discharge hose and check valve.

7. Irrigation systems should be monitored much more closely during fertigation, continuously if possible.

8. Under saline conditions, salinity problems can be intensified by fertigation and improper irrigation management. The level of salinity in soil may itself be increased, even temporarily, by fertilizer application as mineral fertilizers are for the most part soluble salts.

Micronutrients application

In addition to major nutrients, other nutrients that are important include Calcium, Magnesium, Sulphur, Zinc, Boron, Iron, Manganese and Copper. Under high tree management, the requirements of these nutrients may be increased as most of these nutrients are removed from the field with harvested fruits. Under such circumstances external applications may be required.

Zinc and boron are very important for quality litchi fruit production. For some reason, zinc seems to be more effective when applied on the ground, rather than as a foliar spray. It is a good idea to apply zinc in lumps in well watered soil. This helps zinc being locked up in the soil. Boron works together with calcium so it is good to apply both at the same time. In fact boron also needs to be applied evenly. Calcium uptake relies on boron. As an added calcium boost for the trees, calcium spray may be applied during fruit set until the start of colour break stage. The smaller the fruit, the more benefit the calcium spray is to the fruit. K, Ca and Mg all compete with each other, so when K and Ca are applied, the level of Mg reduces. It is better to reduce Mg because excess levels of Mg inhibits the uptake of K. Mg does not seem to be important in litchi plants because excess Mg will result in green coloured fruit, even when it is ripe. For fruit quality, all nutrients/elements are important but the calcium is most important. Calcium improves quality of fruits and there is never any disadvantage in having more calcium in the soil, as long as the ratios are not imbalance. Deficiency of these nutrients can be met by balance fertilization through soil application and foliar feeding at appropriate time. A single spray should be used for minor deficiencies while 2 or 3 sprays should be applied where major deficiencies occur.

In recent years considerable progress has been achieved in spraying of nutrients on the trees or dripping them in the root zone with irrigation water. High levels of N can be applied to get better results but with caution as this may delay the entering of trees into bearing. If the spraying is done on foliage, it is necessary that strong solutions are not used as they cause injury to the plants. Urea can be sprayed safely on trees. Its concentration varies between 2 and 4%. Three kilograms of urea in 500 litres of water has been found good for litchi trees.

Foliar sprays of micro nutrients to supplement their soil applications have been found effective to obtain quick response and to correct deficiencies in litchi. For micronutrients, if the range is within the optimum values, use the recommended rate (Table 25), but if it is below the optimum, apply a second application. If the leaf test is above the standard value, apply nothing. Micronutrients are generally applied through foliar sprays at the time of their deficiencies. Frequency and time of spray depend on severity of the deficiency. Usually B and Zn sprays are applied yearly (i.e. once in a year). Deficiencies of Zn, B and Cu can also be overcome by single soil application during summer every 2-3 years. It is possible that water stress may interfere with the response to these nutrients. There have been many attempts to increase fruit set and fruit size with foliar applications of zinc, copper and boron. However, most of these sprays did not result in consistent increases in yield. Responses to foliar applications would only be expected if leaf nutrient concentrations are below critical values.

Table 25 : Micronutrient recommendations.

Nutrient	Product	Soil application (g per m^2)	Foliar application (g per litre)
B	Borax	2	2.0
Zn	Zinc sulphate	25	1.0
Cu	Copper sulphate	4	2.0
Fe	Ferrous sulphate	10	5.0
Mn	Manganese sulphate	5	2.5

Source: www.fao.org/DOCREP/005/AC681E/

❑❑❑

Chapter 11

Manage the Canopy and Rejuvenate the Old Trees

Besides the various biotic and abiotic factors responsible for low yield and quality in litchi, overcrowding of branches, less circulation of air and poor light inside the canopy lead to production of more unfruitful branches in a litchi tree. It is happens due to lack of knowledge about initial tree framing and canopy management. In order to make litchi competitive both in the domestic and export market, there is a need to use modern production technology i.e. high density orcharding, mechanization and canopy manipulation.

Pruning plays an important role in regulating and controlling growth, flowering and fruiting of litchi trees. It is management of both, above ground (shoot) or under ground (root) parts of a plant in order to modify and utilize plant's natural habit for maximum return to the growers. Young litchi trees are usually trained in order to obtain a framework that is strong and accessible for other cultural practices. Fully grown trees are pruned to maintain the canopy height, spread and density required for easy spraying, fruit thinning and harvest. However, the main reason of pruning of mature litchi trees is to encourage them for better fruiting by way of maintaining a balance between vegetative and reproductive growth and synchronous flowering of the two sexes for better pollination and fruit set. Pruning can also be used to limit excess number of fruits by removing some fruit bearing branches, giving a thinning effect that can improve fruit size and quality. It facilitates light penetration into the interior of the tree canopy, which improves fruit colouration.

Developing desirable plant ideotype is an important tool for litchi plantation as it allows the producer to restrain the exuberant vegetative growth of litchi to manageable sizes and forms. There are no hard and fast rules for pruning rather a matter of common sense; the main objective is to develop a good tree structure which meets desired requirements and facilitates harvesting and movement of machinery through the orchard. In senile trees with huge unproductive canopy, rejuvenation pruning forces the development of productive fruiting shoots.

Understand the importance of ideal litchi canopy

- Most of the litchi orchards are dense and over-crowded and their bearing is low. They harbour various pests and disease causing agents due to change in micro-climate inside the canopy. Pruning is usually carried out to shape trees and open up the centres, allowing free movement of air and sunlight into the tree. This facilitates the penetration of sprays through the trees making control of pests and diseases much more efficient. It reduces the cost of plantation and management by reducing the incidences of disease and pests and also expenses on their management.
- The ability of sunlight to penetrate the tree enhances the colour of the fruit and improves quality. Litchi tree bears 90 per cent fruit or one can say produce commercial yield from lower canopy in the peripheral of the tree. Therefore, developing ideal canopy in litchi is very much important from the beginning of the plantation.
- It lowers the cost of harvest by reducing the time and required labour for harvesting. This is also important to reduce production and harvest costs, synchronize vegetative growth and flowering period, extend the production cycle and sales season, increase productivity of the plants, stimulate precocious flowering, extend the productive life of the plantation; recuperate overgrowth in older plantations and increase air circulation inside the plantation, which ultimately lowers costs and losses.

- Specific activities such as irrigation, pruning and harvesting at field level and post harvest management operations like grading and packing can be synchronized if all the litchi plants have ideal canopy. Related services like pack houses, refrigerated-Van, labour force can be efficiently utilized when the plant canopy is ideal and uniform.
- Ideal canopy improves the appearance and quality of litchi fruits and helps in predicting yield of a tree which ultimately helps in planning for marketing and export of fruits.
- Pruning has direct impact on yield. Litchi has tendency to bear at the terminal ends of the shoots, however, some fruits are also produced from auxiliary panicles as well as old branches. The pruning helps to activate the quiescent fruit bearing buds owing to redistribution of the endogenous hormonal substances to favour flowering and fruiting.
- An important aspect of the periodic pruning is that by limiting the length of the branches, wind damage is reduced. Even though it might take time for the overall yield of pruned trees to approach that of unpruned trees, fruit picking and treatment are made easier by pruning, the actual amount of fruit harvested from pruned trees in the third year, or even before, might be greater than from larger, unpruned trees.
- Pruning results in, increased dry matter content of leaf and stem which reaches the maximum level by October-November, thus the C/N ratio of the shoot reaches the highest level during November-December. This becomes possible owing to increased photosynthetic efficiency of litchi plant due to improved light penetration inside the trees.
- As a result of pruning, the RNA content of leaf becomes higher. The pruning increases the ascorbic acid content in litchi leaves from September onwards which helps in more flower formation.
- Pruning also increases the total phenolics content in litchi. Phenolic acids act as inhibitors similar to abscisic acid which favours the flower bud formation. Lower IAA Oxidase activity has been observed as a result of pruning which possibly makes optimum levels of auxin available to favour flowering.
- The respiration rate of leaves during September and November-December is increased due to pruning which helps in increased flower bud formation and flowering.

- The gibberellins-like substances in the leaves of pruned trees are found to be lower than those of unpruned trees. The lower contents of GA are associated with normal flowering. Similarly, the higher level of cytokinins which favours the flowering also increases due to pruning. A higher level of cytokinins induces and increases flow in the xylem sap which activates the latent buds from their innate dormancy resulting in flowering.
- Better fruit set of pruned trees might result from increased water and nitrogen supply to the remaining wood but not from increased photosynthate resources. However, there is some evidence that pruning can stimulate photosynthetic activity of leaves. This effect is thought to be result of an enlargement of leaf and mesophyll cell size, an increase in chlorophyll content and a lengthening of daily period of stomatal opening caused by an increased leaf water content. Increased water supply to shoots and leaves of the pruned trees is the result of the change in the proportion of foliage area to root system, not the reduced transpiration per leaf area unit. Pruning reduces the transpiring area in proportion to the temporarily unaffected root system and thus increases water supply to leaves, flowers and fruits.
- It is evident from many studies that pruning changes total dry weight partitioning in such a way that more dry weight is added to new shoots than to remaining wood of frame, trunk and roots. The high production of new shoots would be expected to decrease the reserves of nutrients, particularly carbohydrates, stored in the remaining parts of the tree, which are indispensable for such important processes as fruit bud formation.
- Heavy pruning at the on-set of rainy season (i.e. in Jun-July) results in decreased foliage area at the beginning of the growing season, but due to rapid growth of shoots (particularly in juvenile litchi plants), the foliage area is restored and may be equal to that of unpruned plants. When pruning is done late in the season (September-October), leaves which are removed by pruning is not compensated and growth of young tree or fruiting of bearing trees is adversely affected. But, if the tree has excessive vegetative growth, a dense foliage influence its photosynthesis indirectly by reducing the interception of light and its distribution within the tree canopy. Although this is done to assure good photosynthesis of the fruit bearing parts, these aims are not

always realized. If pruning involves numerous heading cuts of shoots and small branches, it induces excessive sprouting of new shoots on the outer mantle of the tree canopy, which hinder light penetration to the fruit bearing parts.

- Fruit setting also seems to be influenced by tree vigour. As a rule, fruit set is better on vigourous older trees than on young ones. Vigour of the tree can be improved with the pruning of limbs or growing shoots. Also, since hormonal levels are modified by pruning, it is possible that pruning increases fruit set as a direct result of growth hormone action. The balance between the hormones supplied by the root system and the above ground part of a tree is disturbed by pruning.

- Flower formation in a severely pruned tree is retarded not by deficit of assimilates but by hormonal imbalance. Firstly, hormones supplied by the roots become unbalanced when the top part is removed. Secondly, hormones produced by the green parts of the tree cannot be utilized when the number of meristem is reduced. Quick growing shoots of pruned trees are thus overloaded with growth promoting hormones (auxins, gibberellins) that prevent flower bud initiation. Any pruning that stimulates growth of pruning trees prolongs the vegetative phase and delays reproductive processes. Because pruning of very young plants delays fruiting and lower yield, some fruit growers have replaced pruning with shoot bending in order to obtain the required tree frame and early bearing.

- The yield depressing effect of pruning operation depends on how much it stimulates new shoot growth. Thus, dormant (winter) pruning decreases yield more than summer pruning, heading back cuts more than thinning out cuts and many small cuts more than a few large cuts even when the same amount of wood is removed. Litchi tree is evergreen and never becomes dormant, but vegetative flushing is restricted during winter, thus pruning winter greatly reduces the yield. Therefore, when litchi trees come into full bearing in a commercial orchard, the grower is usually constrained to prune trees in order to control tree height, spread, and density. This may lead to imbalance in the canopy structure in the long run. All care must be taken to contain mature trees by pruning to their allotted space and all such pruning must be carried out immediately after harvesting the crop.

Proper planning for canopy management

Canopy management is a skill cum art which needs a lot of planning and knowledge about behaviour of plant after pruning. Therefore, before taking up any canopy management operation following points needs to be looked into:

(i) *Understand the litchi growth pattern and fruiting habits before pruning*: The growth pattern of all the litchi varieties is not same. Some are more vigourous whereas, others are less. Therefore, pruning depends upon the variety. All pruning is done after harvest. Litchi trees do not have to be pruned annually to bring on flowering or increase yield as is the case with deciduous fruit trees. They are terminal bearers and usually flowers are borne on mature wood. In fact, pruning at the wrong time may have a detrimental effect. Pruning tends to stimulate shoot development in litchi trees usually resulting in vigorous vegetative growth. If trees are cut back too heavily, they may not fruit for two to three years.

(ii) *Visualize the image of desired plant canopy to be developed by pruning*: The ideal canopy for litchi may be variable depending upon the needs and requirement. For high density, the small canopy will be ideal whereas, for wider plantations the canopy should be little bigger. At any cost, the canopy of two plants should not touch to each other. This can be managed by regular pruning. Regular pruning is essentially needed in high density litchi plantations which should be done just after crop harvest. The desired plant canopy can be visualized by seeing a perfectly pruned tree and gaining experience in plant canopy by visiting top gardeners and research stations etc. When one becomes able to visualize a desirable tree shape for a mature tree, it will provide significant assistance to pruning decisions on the young plants. Knowing which shoots and branches to leave on young plants to form the basic tree framework is critical, as these shoots will remain throughout the life of the tree. The ideal tree should have three and not more than four main branches, be open inside and low-set, i.e. 4-5 metres. Over this height, harvesting becomes difficult. The crop canopy refers to leaf area and its orientation, height and tree girth. The quality of the canopy is best measured as the percentage of light falling on a field that is intercepted by leaves. Normally leaves should be oriented in such a way that light passing by one is captured by the leaf below. Thus, canopy management is one of the most important steps in orchard management which is done with a goal to shape the tree for maximum light utilization and efficient cultural operations in the coming years.

(iii) *Try to keep a balance between vegetative and fruit yield*: A good balance between vegetative and fruitful growth should be maintained by pruning. Factors such as soil type, climate, variety and nutrition play critical roles, however the severity of pruning is very important. On vigorous varieties, there is need to increase fruit and slow down the vegetative growth. It has been observed that slower vegetative growth usually favours more flowers and increased fruitfulness. Selective pruning to remove vigorous, less fruitful shoots and to retain less vigorous shoots can assist in swinging the balance from vegetative to fruitful production.

(iv) *Decide proper time for pruning*: Generally, the bearing litchi plant should be pruned after fruit harvest. The developing plant, however can be pruned during dormant season (November-December). In bearing trees, delayed pruning results in weak growth of sprouts.

(v) *Decide proper method of pruning*: Marking of the trees is essential before pruning to avoid error. Cut the side branches at least 50 cm from the target. Cut the remaining branch close to the main branch and parallel to it. Make a clean cut to avoid later infection. The cut surface should be painted with Bordeaux paste or clay+cowdung, or coal tar to seal the surface and promote healing of the wound and remove small branches and dead twigs from the upper half of the canopy.

Different types of canopy management practices

(1) Framing of newly planted orchard for shape

There is a strong tendency of layered trees to form defective branch unions or to produce branches almost at ground level and this is why it is virtually important to shape the trees properly in the early years. Shape pruning begins when a branch is chosen for layering. The branch chosen for layering must have a single stem. If it does not, the other stems must be removed either at the time of layering or when the layer is transplanted into a bag or planted out. This is of course not applicable for plantations obtained from plants propagated by seed, cuttings or grafting. One definite aim in litchi shaping is to develop sufficient fruiting terminals. Some cultivars, have long branches, giving them the appearance of weeping willows and fewer points of growth than other cultivars.

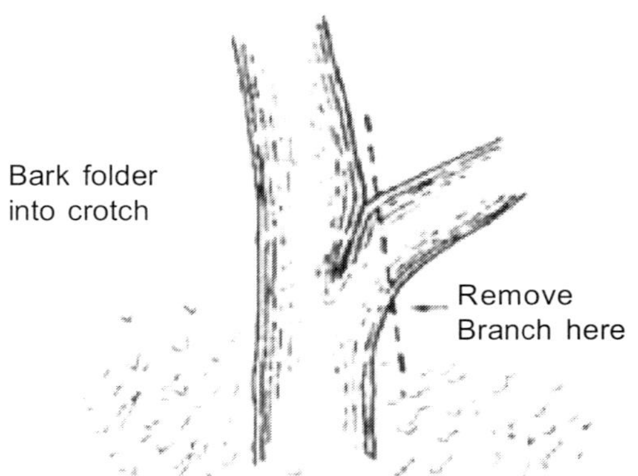

Fig. 36 : Remove weak crotch angles
Source : FAO corporate document repository Regional Office for Asia and the Pacific

During early years of orchard establishment more emphasis is given on development of framework. Pruning at this stage is a constructive operation which determines general outline of the tree. This is also referred to as "pruning for training". The first pruning cut is made on a newly planted layer to reduce the size of the top to compensate for roots that were lost during digging planting operation in the nursery and to establish the trunk height. Pruning of nearly 75% leaves of the litchi layers can be done for better survival in the nursery.

Since majority of litchi trees in an orchard may not grow naturally into a definite shape, training demands lot of drastic pruning at young stage. Litchi plants tend to branch low and thus it is necessary to select the branches to be retained as scaffolds or to be removed to get a longer trunk. Usually, litchi tree consists of four or five primary branches extending in different directions to form a well balanced top. The canopies (skirts) are raised above ground level to facilitate cultural operations beneath the top and also to minimize damage from soil-borne problems. Scaffold limbs are selected around the trunk and other weak or poorly placed shoots are removed.

The first primary scaffold is allowed within 60cm height from the ground in the second year. Second primary scaffold is also selected on the trunk in the same year. This branch is spaced 20-25cm apart vertically and as closely as possible to 120^0 laterally around the circumference of the trunk. Third and fourth primary scaffolds are selected in third year. Likewise fifth and sixth primary scaffolds are allowed in fourth year. Fifth year onward all strong laterals emerging on the main trunk are thinned out and no limb is usually allowed to be a scaffold. Two secondary scaffold limbs arise about 60-70cm above the junction of the primary branch and the trunk. Two tertiary branches fork another 60-75cm above the junction of the secondary scaffold branches and must be oriented tangentially so that the trees will finally resemble an inverted cone. In such shaped litchi trees, secondary or tertiary branches are developed to be nearly equal in size. This is achieved by suppressing the growth of the most vigorous scaffold branches through moderate pruning and by promoting the least vigorous ones through light pruning. In case of cultivars which produce only a few but excessively long branches, growing points are pruned to encourage branching.

When two equal size branches with a narrow angle between them grow for some years at the same rate, a V shaped crotch is formed at their meeting point. Eventually considerable amount of bark inclusion develops between the two branches at their junction, with little or no connecting wood. With heavy loads of fruits and strong winds, these limbs are liable to split apart. This problem can be prevented by completely removing one of the branches when they are young or heading back one of them heavily so that the other becomes the main limb and the headed branch becomes a lateral. Remove branches with weak, narrow crotch angles where the bark is folded into the crotch (Fig. 36) After such pruning, large cuts (over 2.5cm in diameter) are covered with a fungicidal paste to prevent drying or decaying. This type of pruning may be done at any time of year whenever shaping is necessary.

Trees should be inspected regularly during the first four years and the required action is taken where necessary. Severe pruning should, however, be avoided as it hinders plant development and significantly increase time to first harvest. This process of allowing new branch sprouts to grow and then to cut off their growing tips after the leaves have turned dark green is repeated for two years. After this the trees should start branching by themselves. It is therefore, recommended that the shoots on terminal branches be 30-35cm from the apex just prior to a new vegetative flush. In this way, for each terminally pruned branch on an average 2.5 or 3 new terminals will be produced and top-spread will be denser. Downward and inward growing branches or branches that cross over each other should also be removed. Young trees can be pruned at any time of the year. The aim should be to develop a spreading tree rather than a tall tree.

(2) Tip pruning

Tip-pruning is done in the cultivars which produce long branches or dominant leader branches. Remove approximately 15 cm in spring of 2nd year (Fig. 37). This increases the number side branches. Tip-pruning in growing points of long branches provides more flowers and fruit. It also reduces the risk of limbs breaking. In spite of pruning, some cultivars still produce weak crotches that can split. To minimize such risk, growers can adopt the device of strapping and bracing system using heavy gauge wire to link the main branches may be used. The last tip pruning should be done five months before the farmer wants the tree to flower and nine months before the farmer wants the fruit ready for harvest.

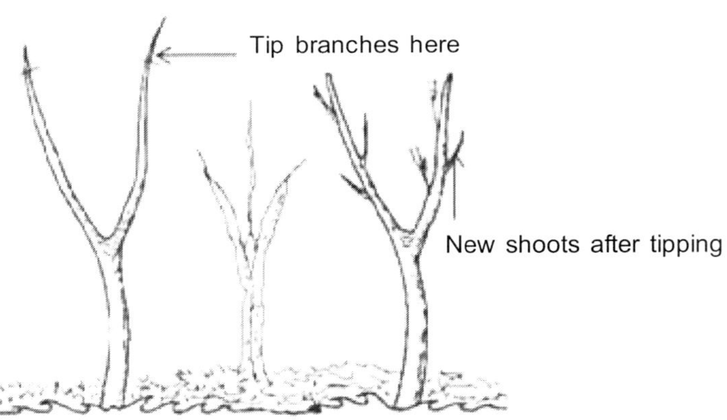

Fig. 37 : Tip pruning in long branches

(3) Annual pruning

Young litchi plants are pruned to provide a strong structure, minimize wind damage and increase fruit bearing area. Wind damage is an important issue for some cultivars. Cultivars with long branches are susceptible to branch splitting, while others with short dense crowns can break off at the ground. When the bunches of fruit are cut, branch lengths of 20-30cm are lopped off with them. The litchi flowers are borne mostly on new shoots. The old branches rarely produce flowers. Snipping of old branches to produce fresh growth is considered desirable. In India, the fruit is harvested in bunches along with shoots and this serves the purpose of pruning. No additional pruning is usually done. In a pruning trial conducted at Malda, pruning panicle at harvest caused maximum panicle emergence and yield in subsequent year. Quality of fruits improved with respect to T.S.S., sugar, acidity and ascorbic acid.

The severity of pruning varies with tree age. Old litchi trees in low vigour are pruned more heavily. On long term basis, however, the orchard will be more productive than if left unpruned. In this practice, up to 50cm of the branch is removed and is believed to help improve subsequent fruiting since it encourages terminal shoot production. However, severe pruning of terminals is not usually as effective as other practices such as cincturing or fertilization. In many areas where litchi is cultivated, usually after the harvest, removal of dead wood, small internal branches, as well as all branches preventing the sun's rays from penetrating the tree is recommended. At many places, one to two pairs of leaflets are removed together with the floral stem at harvesting .

Flowering and fruiting of older branches must be encouraged, at the same time growth of additional new shoot system is also required. In a practical sense, no single pruning technique is adequate to achieve the different goals. Managing litchi trees through this transition requires co-ordination of pruning with other practices such as limb positioning and fruit thinning so that the equilibrium between shoot growth and fruiting is maintained. At this point the interactions among all forms of growth, including growth of the crop that constitute vigour must be thoroughly considered before the appropriate technique is selected.

Old litchi tree does not produce fruit uniformly. Only certain branches of the tree produce during a given season. Often, trees have only a few fruit on some branches whereas, most of the tree has no fruit. This means that the tree is not producing up to its potential. A well-managed tree produces fruit on all of its branches. If the tree does not produce its maximum, this will not only reduce the return per tree, but will also increase the cost of harvesting as pickers must visit more trees to harvest a given amount of fruit. Pruning all branches of a tree ensures that all branches will produce at the same time. Doing this to all trees in a plantation can bring all the trees into harvest during a relatively short time and can significantly improve the efficiency of harvesting and marketing activities. Maximization of production in litchi is achieved by pruning, which stimulates all the branches to flower and produce fruit together.

(4) Girdling or cincturing

Girdling or cincturing can be used in the same way as the auxins or ethephon to improve flowering. Girdling presumably redirects assimilates that normally support stem and root growth. However, the long-term effects of these treatments on tree health are unknown. In this process a ring of bark is removed around the major lower branches of a tree to a depth just below the soft tissue under the bark or cambium layer. The operation may be carried out on the main stem or on primary branches also, provided the diameter at the point of cincturing is not less than 1.5cm. Often only half the branches are cinctured while the others are treated in alternate years. The ring is cut with a hacksaw, 3 mm wide and 3mm deep. Healing time depends on tree vigour, the cultivar and the size of the incision. Cuts must be painted with a pruning paint. The cut is covered with a saturated paste of copper oxychloride to prevent disease infection.

Cincturing can induce vegetative dormancy in autumn and winter giving rise to better flowering, fruiting and yield. This is a technique used in shy-bearing trees to induce flowering. This prevents new shoot growth for about three months, so that the next activity of bud growths occurs when conditions are favourable for flowering.

Cincturing stops the interchange of sugars and other growth promoting materials between the leaves and the roots. After cincturing, sugars and growth inhibitors tend to build up in the leaves and assist flower development. With time, the tissue on each side of the cincture cut grows across and rejoins the conducting vessels, so that the plant resumes normal development. The length of time that this healing process takes is vital to the health, performance and survival of the trees.

Cincturing is not always satisfactory and may differ significantly depending on the cultivar and on climatic conditions. Spectacular results have been observed in warm, moist conditions which do not promote flowering. Responses to cincturing in some areas have been much less consistent probably because low temperatures and low humidity. Cincturing is not recommended as a general practice, since continuous use of the technique often leads to retarded growth, small fruit, leaf scorching, branch die-back and loss of trees. It may also develop the tendency toward alternate bearing. Cincturing a few months after pruning has been done to overcome the alternate bearing, but this practice has not been effective even though it improves flowering the following season. Main factors which influence the response of litchi trees to cincturing are : seasonal conditions, the cultivar, tree vigour, stage of flush development and the timing of cincturing. Therefore, cincturing or even girdling should be practiced on case-to-case basis. Growers intending to try cincturing must firstly be aware of the factors which control its success. They must also be aware of the likely damage it can cause to the health of the trees.

Seasonal conditions : Differences in seasonal conditions in different localities bring the variable response to cincturing of litchi. Good natural flowering normally occurs in years in which low temperatures and low soil moisture conditions extend from November to February. In litchi, flowering may occur because of the natural vegetative dormancy, which would mask the effects of cincturing.

Cultivar : Both low vigour and high-vigour cultivars can respond favourably to cincturing. The response to cincturing is therefore, not influenced by the cultivar itself, but rather by the degree of winter flushing that the cultivar can develop. This varies from orchard to orchard, locality to locality and year to year.

Tree vigour : Sometimes, cincturing is a means of correcting litchi's erratic bearing pattern The tear made in the phloem cells induces dormancy and favours floral initiation. Trees must be fairly vigorous to respond to cincturing without damage to the tree and reduced fruit size, so it is essential that they are well nourished. Only healthy, vigourous trees are cinctured. Trees that have not flushed well after harvest, respond poorly to cincturing. Considerable yield reduction, injury or death can occur if trees are cinctured when they have water or nutrient stress.

Trees to be cinctured are adequately irrigated and fertilized immediately after harvest, especially if they have set a heavy crop. Repeated annual cincturing of litchi is not recommended unless the trees are vigourous and routinely irrigated and fertilized.

Stage of flush development : The best response to cincturing is achieved when branches are dormant and before the emergence of the next vegetative flush. Cincturing at early, mid or late flush may reduce flowering when compared to uncinctured trees. The response at different stages of flush development is considered to be a reflection of the varying proportion of young to mature leaves at the time of cincturing and its influence on the relative accumulation of flower inhibiting and promoting substances above the cincture.

Timing of cincturing : Girdling or cincturing is normally carried out when the post-harvest flush has matured and just before an autumn flush begins. Flowering is not promoted when trees are cinctured at the wrong time. The best response is obtained when litchi plants are cinctured three months prior to the commencement of flower development, which occurs in October-November in India. Trees cinctured late in the season do not develop sufficient dormancy prior to flowering. In addition, the cincture does not have sufficient time to heal before fruit set and fruit formation is adversely affected. Cincturing trees too early in the season allows the wound to heal early. As a result, there may be a premature vegetative flush which will have an adverse effect on flowering. For ensuring a successful response to cincturing, a grower must be aware of the flushing times of the different litchi cultivars in his orchard and must gain experience in managing water and fertilizer application to manipulate the time of flushing.

Shock management : Heavy or continued cincturing or cincturing of unhealthy trees often leads to problems of leaf chlorosis, leaf scorching, defoliation, stem swelling, retarded growth, a tendency towards alternate bearing, poor fruit set, small fruit, branch die-back (especially in dry years) and loss of trees. Tree decline due to cincturing has been attributed to a number of factors which includes poor regrowth of tissue in the cincture cut, prolonged interference with water transport to the leaves and impaired root development.

Several techniques have been recommended for minimizing the damage. These include the use of small cincture cuts, cincturing only half of the major limbs of a tree each year, cincturing each tree in alternate years only, application of growth promoting chemicals to the cincture and covering the cincture with plastic tape to promote growth by preventing desiccation of the cut tissue.

(5) Root pruning

Root pruning is usually practiced when the tree is making too much vegetative growth particularly in heavy soil. The objective is to restrict the vegetative growth and encourage the tree to bear a good crop. Like cincturing, the responses of root pruning also have been somewhat disconcerting. It involves ploughing the land to a depth of 5cm accompanied by adequate fertilization which promotes floral initiation. Some times a wide trench is dug around the canopy area to prune the root. However, in some cases, root pruning in winter to a depth of 40cm has no discernable effect on flowering. This difference may be due to a combination of other factors which inhibit or promote floral initiation. Because of the severe damage done to litchi's shallow root system, this practice should be done carefully.

Root pruning may be a last option for the old trees which are not bearing. However, in no case removal of roots be done too extensive or drastic because a reduced root area cannot sustain a large canopy required for good yield. Sufficient amount of carbohydrates stored in the roots are removed from the plants through root pruning. To replace the removed roots, the plants regenerate new roots. Regeneration consists of elongation of existing roots and initiation of new laterals with subsequent elongation. Root generation is stimulated by increased levels of growth promoting hormones or reduced levels of growth-inhibiting hormones in the rhizosphere. The new roots require carbohydrates and other assimilates which are translocated from the top discouraging excessive vegetative growth. This finally produces a dwarfing or growth restricting effect. As the root system regenerates, uptake of nutrients and water may increase accordingly.

Fig. 38 : Root pruning

When the litchi saplings are lifted from the nursery and planted in the orchard, root pruning is unavoidable and carried out automatically. If too much roots are cut at this stage, survival of the sapling in the field will be poor. Thus, care must be taken to keep root pruning as little as possible between lifting and planting.

In more recent years, with the advent of mechanical tree planters, it is possible to plant large trees (3 to 10 years of age). Such trees are compulsorily transplanted with a relatively large soil ball. When the trees of this size are transplanted, special precautions, such as application of anti-transpirants or severe pruning of canopy are needed to avoid severe moisture stress.

(6) Pruning to control growth

Under high density planting system litchi trees are pruned regularly to control size and shape. Hedging or topping may be required by the fifth or sixth year after planting, depending on location and initial tree spacing. This is in addition to the normal pruning done during initial years. Hedging is typically accomplished using large saws. It is most effective if done in June-July or in August-September under North Indian conditions. Hedging during this period allows improved light penetration into the canopy during vegetative flushing, reduces canopy volume and promotes growth of lateral buds. Hedging at this time will not remove flower and thus does not cause any substantial decrease in yields. Usually hedging and topping of litchi trees decrease yields in the following year compared with unpruned trees. Nevertheless, average yields over a period (7-8 years) are similar and hedging and topping improve fruit quality and increase fruit numbers. Severe pruning decreases yields compared with unpruned trees in the pruning year, but yields are almost the same as unpruned trees in the next fruiting year when no pruning occurs.

Different patterns of hedging or pruning is used. Some growers cuts only one side of the tree in a particular year, while others cut both sides along the row in the same year. In the first case hedging cost are less on a yearly basis, but fewer advantages of hedging viz. improvement of regrowth and increased light penetration in the canopy are realized. In the second instance, tree which is controlled in both direction further improves light penetration into the canopy. Most effective hedging programmes will remove only the outer 15-20cm of the canopy, thereby cutting limbs on greater than 0.5 to 1.0cm in diameter. A routine maintenance programme avoids cutting of large limbs and provides opportunity for growth renewal each year and reasonable for fruiting on all sides. In contrast, trees allowed to grow together between rows over a number of years will become unproductive and require more severe pruning into older woods. Generally, more severe the cuts, the more vigorous and vegetative the regrowth will be, often requiring 2 years or more to return to normal crop level on that surface. By then, the surface will require severe cuts again.

Topping is also done with mechanical saws mounted on adjustable arms. Topping improves light penetration into the upper canopy and decreases tree height to reduce the cost of harvesting and spraying operations. It also improves spray coverage. Under high density planting, tree height should not be greater than twice the width between the outer perimeter of the tree canopies between rows for optimum interception of light. Limiting tree heights becomes increasingly important in high density plantings with very narrow between row spacings.

Trees are topped in different patterns. Some growers prune straight across the top of the tree in a flat top. Others set the cutting angle of the saw between 10 and 25^0 creating a root top effect. These patterns also remove varying amounts of wood from the sides of the trees. It is preferable to reestablish required tree height by making small cuts over several seasons rather than cutting into large woods. With hedging, removing large limbs produces vigorous, vegetative shoots from new growing points, thus defeating the purpose of topping. Therefore, a much greater concentration on balancing the competition between the two operations is required.

In orchards where tree spacing is narrow, hand or mechanical pruning must be done to prevent overcrowding and to make branches most exposed to sunlight. Because litchi trees are slow growing, this can be done only once every five or six years. The best time for this type of pruning is immediately after harvesting in a very productive year to allow good vegetative flushing the following year. It could be postponed until late winter / but no later, because the trees must be well covered by early summer to prevent scorching. If the trees are pruned later than this, the stem and branches would have to be painted with agricultural lime or some similar substance. Although it would be advisable to apply a healing substance to the wounds caused by this type of pruning, this is in fact not done because it is a complicated and costly operation.

Similar work is being done in other areas on very erect trees. The trees are cut back to a height of three metres from the ground to allow easier harvesting, which is one of the aims of the pruning.

(7) Skirting

Pruning of lower branches of the tree is done for weed control, surface mulch applications, mini-sprinkler irrigation distribution and to prevent fruit of the lower branches from touching the ground. This is the removal of low hanging branches which could hinder orchard operations such as fertilizer application and under tree spraying for weed control. The best time to do this is at early flowering during cool weather, when trees are not in a strong vegetative-growth phase. Repeated skirting by selective hand pruning can eventually result in strong, robust branches able to support fruit without touching the ground.

Skirting is done from the third year onwards by removing all branches and shoots to a height of 50 cm leaving a clean single trunk (Fig. 39). Some cultivars, which have compact foliage are easily twisted out of the ground in strong cyclonic or stormy winds. This can be reduced by removing 20% of major branches within the canopy and by skirting. This allows wind through the trees and stops the twisting effect. There are indications that less compact cultivars may also benefit from such pruning.

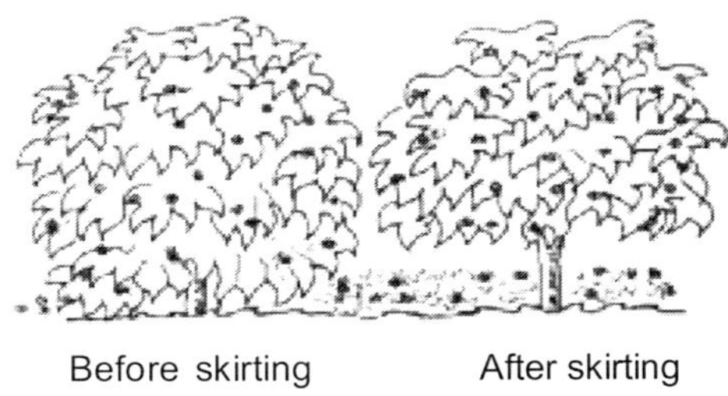

Fig. 39 : Tree skirting

The skirt of the mature tree in most of the cases is at least 1-2 metres above the ground and major branches are removed after harvest in every alternate or 2-3 years. Ant and scale control is made easier and fewer fruit are damaged by insects and rots if skirting is done.

(8) Centre opening up

Litchi plant is a multi-branched evergreen tree having a compact canopy. As a result of number of growth flushes (3-4) during a year, overcrowding of the foliage in the canopy often takes place. This leads to poor penetration of sunshine inside the canopy, resulting in long unfruitful branches and higher incidence of insect pests like bark eating caterpillar and trunk borer. It has been observed that in litchi, the lower 2/3rd of the canopy contributes for the maximum economic yield. The fruiting at the top of the canopy generally becomes uneconomical due to more fruit cracking and difficulty in harvesting. Thus, the branches inside the tree which cross over or cluster up the centre of the tree restricting the penetration of sun light and sprays may be pruned. The litchi trees are often trained to open-centre system because this helps in the development of low spreading tree-top. Also, the open-centre trees affords maximum light penetration, which gives a more uniform distribution of fruit on the tree. Among the three systems of training, the plants under open vase system record maximum height and girth of the stem besides well spread canopy. The main disadvantage of this type of training is that the arrangement and equality in size of the scaffold system may produce weak, crowded crotches. All possible care is taken to minimize development of such crowing crotches.

Fig. 40 : Centre opening

(9) Hygiene

This involves the removal of any diseased or dead branches from the tree, which could be a source of infection.

(10) Thinning

Thinning out of very dense cultivars is necessary. Removing approximately 10 to 20 per cent of the branches within the canopy in the third year is recommended (Fig. 41). One should be able to see broken sunlight on the ground under the canopy when thinning work has been finished. This practice allows wind to move through the canopy and reduces the risk of the tree twisting out during heavy winds. Check the trees and repeat each year if necessary. An alternative to periodic pruning in densely stocked orchards is the removal of culls. Since litchis are usually propagated by layering, there is little difference between plants of the same cultivar. It is therefore, desirable to follow the initial planting layout to avoid crowding. Thinning may begin gradually, but must be done as soon as the branches of neighbouring trees begin to touch. Plants that are removed may be used as resource of propagation material. The decline in yield which immediately follows thinning is overcome within two or three years.

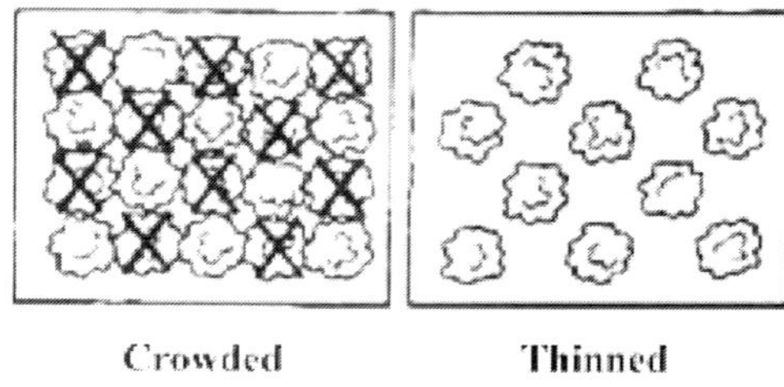

Fig. 41 : Thinning of crowded orchard

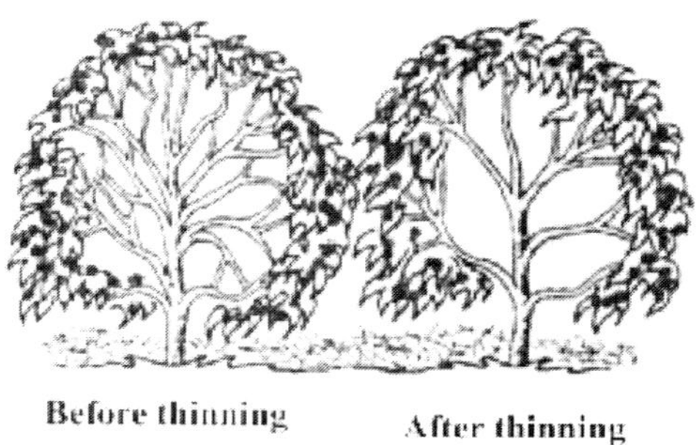

Fig. 42 : Tree thinning

(11) Regeneration pruning

Frost or cold sometimes become severe enough to kill the trees which occasionally occurs in some subtropical regions. Where damage is not so severe but the branches or part of the trunk has been affected by cold, pruning is necessary to enable the tree to recover. Dead branches should not be cut off immediately after a freeze since there is likely to be an additional dying back of branches due to latent cold damage or sun-burn and cutting will have to be repeated, which may increase the cost of the operation. Pruning should therefore be delayed until the full extent of cold damage can be determined. Where frosts are relatively frequent, it is better to wait until the following spring when new shoots have grown before cutting back the dead wood. In this way, the freeze-damaged branches, whose leaves have dried but not fallen can protect the undamaged part of the tree from sun-burn. Furthermore, new growth on unpruned trees is generally as good as on pruned trees. Another advantage of delaying pruning is that the dead wood and leaves provide additional cold protection for the rest of the tree.

Fig. 43 : Regeneration pruning

When several shoots on a stump reaches a height of about 1.5m, it is desirable to remove weaker shoots, repaint the cut surface of the stump to prevent wood rot. Remaining shoots are tied to wooden or metal posts near the stump to increase wind resistance. If the main stem has been affected, great care must be taken when choosing the shoots which are to form the new crown.

(12) Rejuvenation of old tree

There are number of causes which makes the orchard uneconomical. Ageing and poor management is most prominent which can be rectified by rejuvenation pruning. Some of the such factors have been mentioned.

i. Growth of wild shrubs and grasses

Wild shrubs and grasses continue to grow without any check under the tree and in between the rows. At a stage they develop into bushes here and there in the entire plantation. It has been realized that vegetation removes quite a larger quantity of moisture from the soil and also ties up nitrogen so that it is not available to the trees at the time, though it may be returned later when the weeds die and decompose in the orchard soil. Weeds harbour insects, pests and diseases too. When they become large enough they interfere in the orchard operations and some of them climb up the tree and by shading the leaves or due to their parasitic nature cause great damage to the main crop.

ii. Overcrowding and unsystematic planting

In some of the orchards no regular or systematic planting is followed. The trees are also found to be planted so close that they form a dense overhead canopy which does not allow sufficient light to reach at different parts of the fruit trees uniformly. The lower branches remain under perpetual shade. As a result of it the bigger branches remain devoid of small twigs or branchlets and ultimately this abnormality reduces the fruiting surface. In such cases, trees for want of space grow tall and bear fruits almost on periphery only. These fruits in hotter localities are affected adversely by sun burn. This way the growers unfortunately suffer a double loss because on one hand the quality of the fruit and on the other the total produce both are considerably reduced. Free circulation of air is also checked and this results in development of pests and diseases.

iii. Inferior varieties

The poor selection of planting material at the time of planting will cause a great loss throughout the life of the orchard. Some of the orchards trees have been raised from seeds and of inferior varieties. Seedlings of poor varieties are potentially poor and their performance is subjected to a wide range of variation so that even if they grow and function normally, they do not produce a crop that can fetch good price in the market.

iv. Damage due to adverse weather conditions, rodents and other enemies

Trees with damaged trunks, broken limbs and big hollow in the trunks are not uncommon in such orchards. Some trees may be found standing with only half their base. Many trees live for hundred years or more and grow to huge dimensions. Old trees with the advanced age have a tendency to develop hollows in their trunks and main branches starting from a crotch or from the point of bifurcation. The branches which cross one another rub together when fast wind blow and injure their bark. Rain water is deposited in the hollows produced and cause the bark and wood inside to rot.

v. Infestation of pests, diseases and parasites

The orchards which have been established in neglected localities are seriously attacked by number of pests, diseases and plant parasites. These enemies are so rampant that even in orchards otherwise in good condition, the production is very poor. These organisms devitalize the plants and destroy flowers and fruits in all stages of development.

Rejuvenation need and process

When the trees become too old and produce fruits of small size, heavy rejuvenation pruning is required to promote new vegetative growth and improve yield. One must bear in mind that rejuvenation pruning is used only when trees have grown into a densely foliated canopy orchards and yields have become severely limited, but tree major limbs to promote the production of vigourous new growth from latent buds. Rejuvenation eliminates production for about 2 years after pruning, after which yield will generally increase.

Rejuvenation as name implies is restoration of youthful vigour. It helps in restricting the deterioration of the health of plants, thereby extending their life and aims at improving yielding capacity of quality produced of the existing plant in minimum period of time as compared to establishment of new orchard which attains the commercial bearing stage after 10-12 years.

Rejuvenation of old, dense and unproductive orchards is a medium term strategy for better productivity and quality. After rejuvenation of old senile orchard of litchi can regain and restore new productive life for another 12-15 years. The quantum increase in percent level of production through horizontal expansion i.e. through new plantation will be a long term venture and the existing orchards particularly old plantations becoming to unproductive / uneconomical, can be brought back to productive stage by restricting further deterioration through rejuvenation techniques.

Rejuvenation of old orchard is vital under orchard management, which needs careful, visionary and scientific approach be followed in skillful manner. Heavy pruning a tree should be considered carefully from financial point of view. The expected increase in harvest value after pruning may be less than offset the lost harvest value. There may be a significant drop in production with the radical pruning to reduce the height of the tree. Therefore cutting back large limbs to reduce tree size is always risky as one may lose two or more years' production, depending on the amount cut back. Some of the value of this loss is compensated through the sale of the wood cut from the tree. In later stage, pruned tree produce several times more than the un-pruned tree because it has more branches capable of producing fruits. Additionally, some lost harvest value may be gained through pruning practices manipulation of harvest times to capture higher priced markets.

The following strategies and steps are involved in the rejuvenation of old senile orchards/trees of litchi.

(i) To be done only where needed

In general, litchi plant starts bearing after 5-6 years of planting and attains commercial bearing stage at least after 10 years and commercially viable life (period) lies between 11 to 40 years. At this commercial juncture of bearing stage, the plant needs skillful management by proper pruning and training for status and canopy management apart from proper nutrition, irrigation and disease-pest management for quality production. It has been observed in general that orchards after attaining the age of 30-40 years even spaced at 10m x 10m, turns dense with compact top canopy covering most of the branches at the bottom and bearing fruits only on the high-tops. These plants pose problems in proper management such as pest control and, harvesting etc.

Under such circumstances, it is rather difficult to assess longevity of trees. On the other side, with rising cost of management, it may not be economical to maintain these old senile orchards of above 40 years of age. Such orchards need to be rejuvenated for further higher production of quality produce. Prior to selection of orchard or tree, it is important to look for the health, location and value of the litchi trees, which must be judged/determined before going for rejuvenation. Trunks must be healthy, wood portion/part unexposed, intact with bark and live, such trees only be selected for rejuvenation or reiterative pruning which will regain vitality and vigour. If above situations do not prevail, reiterative pruning may not result anything. These are important factors to be considered with the trees to be rejuvenated.

(ii) Awareness to the growers for its importance

Considering the various pros and cons of rejuvenation, it will only be programmatic to undertake it after giving proper training for acquiring knowledge to the orchardist to develop the expert system and apply the methodology in scientific manner. The orchardist should have equipped himself with technological development and ideas that litchi tree has strong renewable ability and the breakeven point is only two years, where as establishment of new orchards will take 10-12 years to come into commercial bearing. The harvest of enhanced quality production like young commercial bearing orchards can further be obtained for atleast another 12-15 years.

(iii) Repairing of wounds

Any wound on the tree if allowed to remain exposed may attract the organisms of diseases from the surrounding atmosphere. So they should be properly treated to encourage healing. If wound is small, simply painting with colour or any other disinfectant may suffice the purpose. Big hollows may be strengthened by scrapping off the inside diseased or rotten parts smearing the exposed portion with coal tar and filling them with bricks and kankar. These are finally plastered with cement.

(iv) Reiterative pruning

In case of rejuvenation, heavy reiterative pruning of litchi tree is done at the height of 2 meters to 3 meters depending upon the girth and type of main trunk. About 3 to 5 main branches with outward growth from the base are marked for pruning at required height, with a plan of developing umbrella like or semi circular frame work of tree canopy. Pruning can be done either with manual saw or power operated saw. Care should be taken to avoid bark splitting or debarking at the cut end due to falling of heavy branches at the time of pruning. It is always advised to go for reiterative pruning for rejuvenation in phase manner starting from the top light pruning to reduce weight upto the final point. After care is very important for successful rejuvenation. To avoid any external infection at the cut portion, it should be pasted with Bordeaux mixture or Copper-oxy-chloride (Blitox) immediately after pruning.

(v) Time of reiterative pruning

By adopting and following all the required scientific procedures, the best time to go for reiterative pruning for rejuvenation is August, September i.e. mainly after the rainy season. Rejuvenation is labour intensive operation and it is extremely necessary to complete the operations in time (August-September) for achieving the desired / results.

(vi) Nutrition management

After the reiterative pruning, the tree should be fed with optimum fertilizer doses, followed by irrigation near the root zone, just like the commercial orchards. The manures and fertilizers should be applied through ring method. The dose per tree as an adult bearing stage i.e. 75-80kg well rotten FYM, 2 kg Neem / Castor cake, 1.00kg Urea, 1.50kg Single Super Phosphate and 500g Muriate of Potash should be applied preferably in two split doses one during August-September and another in February-March. The micronutrients like Zinc ($ZnSO_4$, @ 2-3 g/l) and Boron (Borax @ 2g/l) may be applied through foliar application from one year after the reiterative pruning.

(vii) Thinning and canopy development

It has been observed that the pruned trunk / branches during the month of August, starts putting forth vegetative sprouts just after 25-30 days in case of cv. Shahi and 40-45 days in case of cv. China. The enormous numbers of branches come out and cover the entire body of the pruned plant. At this juncture, careful thinning operation is required to be done with the ultimate aim of developing solid, semi-circular or open umbrella type canopy. Shoot removal should not be done just after their emergence, rather, these should be removed sequencely after 6-8 months of rejuvenation. In this attempt, it is required to remove the sucker growth around the bottom of the trunk and thin out the excessive branches as well, keeping only 3-4 numbers nearly top-side portion on each pruned limbs of the trunk projecting all the directions and further training for acquiring desired shape. The air and light must have access to all the parts of the tree. The horizontally developing limbs should be cut back to the point where they will not hang under the weight of fruit into the row or to the ground on lower limbs.

(viii) Irrigation

Rejuvenated trees require immediate irrigation if any dry spell occurs, to avoid drying out of the trees. Irrigation is must after the manure and fertilizer application. Irrigation at regular interval of 8-10 days during hot summer season and 15-16 days during winter season helps faster vegetative growth and good canopy development. Irrigation applied to the intercrops also gives the added advantage to the rejuvenated trees, apart from maintaining / improving soil physical condition (soil health) and complete check over weed growth. Mulching during the months of April-May and September-October have been found beneficial in conserving soil moisture beneath the tree canopy, reducing the frequency of irrigation and enhancing growth.

(ix) Space utilization for yield compensation

Just after the completion of reiterative pruning for rejuvenation, the open interspaces between the plants/trees is created in the orchard like newly planted orchard. Intercrops during summer season like Black gram, Mung, Maize, Cucurbitaceous vegetable, fodder crops etc. and during rabi season, the crops like Cowpea, French bean, Pea, Potato, Mustard etc have been found most suitable. Flowering plants and annual fruit crops like papaya and banana have also been found to give good income for 2-3 years. Apart from the significant income obtained by the intercrops in rejuvenated orchards, the added advantages like improvement in the soil tilth (health), almost complete check over weed growth as well as less incidence of pests diseases are also obtained.

(x) Plant health management

Plant protection measures are equally important for healthy growth. Intensive care to mange the infestation of important pests like stem/shoot borer, bark eating caterpillar, mite, leaf roller, leaf miner and leaf cutting weevil etc. and the diseases like microbial infestations are required. Control methods may be biological, mechanical or chemical or the combination depending upon the growers' preference and specific circumstances. Leaf roller, leaf miner and leaf cutting weevil start damaging right from the beginning of the rejuvenated trees, hence care is required from that moment itself, spraying with Endosulfan / Monocrotophos @ 2ml/l or Cypermethrin @ 0.5ml/l 2-3 times during new flush emergence will save the damage from these insects. The mite can be controlled by applying miticide like dicofol / omite @ 3ml/l atleast 2 to 3 times during the months of September-October and February-March. The serious damage is being observed by the attack of bark eating caterpillar, which can be controlled by the spray of insecticide at initial stage but at tunneling stage, it is require to clean the fresh frass and webbing, finding the hole, notching the hole with and pointed spoke/poker, plugging with DDVP / Nuvan (0.1%)/Kerosene oil soaked cotton and ultimately sealing the hole with mud. Spraying of copper fungicide during the month of October and March takes care of most leaf spot and other algal / fungal diseases. Brushing/pasting of Bordeaux paste to the main trunks / branches up to one meter from the bottom keep the plant clean and save the crop from many pests infestation.

(xi) Fruit yield and quality

It has been found that the yield obtained from the old trees (non rejuvenated) is high but fetching very less price in the market due to inferior quality particularly with respect to size and wastage due to attack of many physiological disorders and attack of pests-diseases. Fruit yield and physiochemical characters of mature fruits were found to be better in fruits obtained from rejuvenated trees. Maturity period is found to be slightly delayed in rejuvenated plants.

(xii) Support to prevent wind damage

Whenever possible, the branches of the trees, particularly when heavily loaded with fruit, should be adequately propped up to prevent them from cracking. Where trees are small, this can be done using the grape arbour system, whereby the branches of one part of the tree are laid along a horizontal pole each extremity of which rests on two poles arranged in a V-shape. The propping up of older trees is more complicated, as a pole structure must be erected around the tree. This structure also helps to protect the branches from wind damage. The cost of this operation makes it impracticable in many plantations. To minimize wind damage, growers can adopt the device of strapping and bracing system using heavy gauge wire to link the main branches may be used (Fig. 44).

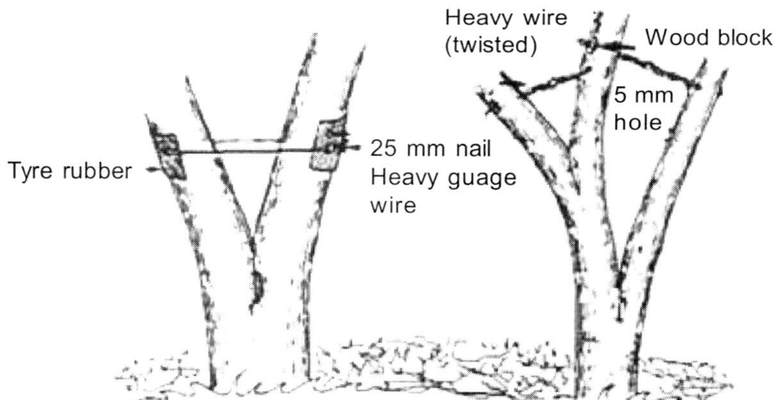

Fig. 44 : Braces to support split branches. Modified : FAO

Basic pruning tools

Secateurs : Secateurs are handy for tip pruning and cutting out shoots and suckers. They should be kept sharp at all times. Regular cleaning and a drop of oil prevents the blades sticking. A good pair may cost a few Rupees more, but are usually more robust and stay sharp longer. They often have the added advantage of having replaceable blades.

Lopping shears : These are useful for removing quite large branches, up to 50 mm in diameter. There are two types, one has a cross over cutting action where the blade is curved and passes across a cutting plate. This type is prone to splaying if used on very big branches. The other has an anvil type cutting action, the blade is straight and cuts down onto a cutting plate/anvil. This type is more robust and can cut through larger branches by taking two or more cuts.

Pruning saw : This tool is capable of cutting through large branches and trunks up to 150 mm in diameter. These have a fairly narrow, slightly curved blade, which enables them to be used in quite restrictive situations such as narrow forks of trees. Some brands are also designed to fit onto a long handle to enable the user to reach higher branches in the tree.

❑❑❑

Chapter 12

Ensure Regular Production in Bearing Orchard

To sustain productivity and quality of fruits in bearing litchi orchard, appropriate horticultural practices at right time are very important. Management systems for bearing trees are quite different from that of adopted for non bearing trees, especially with regard to nutrition and irrigation management. The main objective in managing bearing trees is to promote the tree reserve level for subsequent flowering and fruiting. Depending upon the age of tree, 2-3 stages have been delineated in the bearing litchi tree. The plants with 5-10 years age should be nourished and maintained in such manner so that a balance between vegetative growth and fruiting can be achieved. At this period, trees should be allowed to grow and maintain sufficient vigour for canopy building and fruiting terminals for producing moderate yield. After 10 years of age, litchi plants should be maintained for moderate to low vigour or in other words, constant vigour and higher food reserves. The constant vigour can be maintained by annual pruning, centre opening and promoting plants to develop shoots which can bear panicles. The tree reserve levels in litchi plant can be increased by two ways i.e. by optimizing conditions required for high rate of photosynthesis and by controlling excessive vegetative growth that consume stored reserves unnecessarily. Photosynthesis rate and excessive vegetative growth can be manipulated by controlled nitrogen and water supply to the trees. During bearing stage, the grower has to think its management for long term productivity of trees and to maintain the sustainability of tree production capacity for long period. To ensure this certain steps are needed to be taken in bearing orchards.

Adopt proper nutrient management practice

Application of plant nutrients in right amount at correct time and through right method helps plant to take it up efficiently with minimum losses which covers the whole issue of nutrient management. Proper nutrient management has significant effects on tree productivity. The main purpose of nutrient management is to keep nutrient levels in a desired range in the tree and to have balance between growth and fruiting of trees as desired by the grower.

The most important aspect of nutrient management in bearing litchi plant is its diagnosis and timing of nutrient application in relation to tree phenology or growth cycle. The diagnosis can be got done by analyzing leaf and soil samples of orchard. However, some recommendations have been made for different age group plants which may vary from one location to other. In bearing plants, application of nutrient should be based on the quantity of nutriment removed by harvesting fruits and falling leaves. Excessive or less use of nutrient affects the crop performance. As per principle, the organic manures (FYM, Cakes etc.) should be applied in abundance whereas, the macro and micro nutrients (N, P, K, Ca, Mg, S, B, Mn, Cn, Zn etc.) should be applied as per need. It is advisable to use agriculture grade micro-nutrients for application through soil and fine/laboratory grade micro nutrient for foliar feeding.

Apply full dose of organic manures and half dose of the major and full dose of micro nutrients in June after crop harvest and pruning with the onset of monsoon. The additional dose of micro nutrient if required, should be applied through foliar feeding before panicle emergence and after fruit set. The half of nitrogenous fertilizer should be applied during fruit growth and development.

With the exception of nitrogen, the essential nutrients for growth can be applied to the soil at any time without forcing competitive and non productive vegetative flushing. This does not mean to say that some rationalization of fertilizer application is not needed. However, looking at tree nutrient requirements, application timing can be adjusted in relation to growth cycle. Now a days, nitrogen management in litchi orchard is one of the most important and powerful management tool. Knowledge and learning to use the nitrogen effectively is very essential to improve orchard productivity because the application of nitrogen stimulates vegetative flushing and withholding temporarily fruitfulness however, one has to be careful as overall productivity declines if trees become nitrogen deficient. The timing in nitrogen application is also critical in balancing the tree between vegetative and reproductive growth. Therefore, it is better to apply required nitrogen just after harvesting to induce growth flush immediately.

Under normal condition, phosphorus may be added during rainy season preferably along with nitrogen after pruning, whereas, under rainfed conditions it should be essentially applied in rainy season. Phosphorus is a slow release fertilizer which is very much required in plants to initiate flowering and fruiting process. Litchi plant requires potassium during the period from flowering to fruit maturity when the trees are very active in translocating food reserves to maturing fruits. Trees need potassium during flush development also. Therefore, potassium may be applied after harvesting and again during flowering to fruit maturity.

Now for a farmer, it is imperative to understand that trees are not over fertilized with any of the major or minor nutrient as excess amounts may cause problems like excessive vegetative growth, low yields, inferior fruit quality and nutrient toxicity.

Lime application may be necessary

Lime application is an integral component in litchi nutrient management. Litchi prefer a soil pH of around 6.5. At this pH most soil nutrient elements become available to plants. If the soil is acidic, finely ground lime or dolomite may be applied to adjust the soil pH. These liming materials may be applied as bands along the tree row or just around the trees on the root zone. The application of lime can be combined with fertilizer application in acidic soil. The exact lime requirement at a particular place can be estimated by using soil analysis results however, for moderately acidic soils (5.5-6.5pH) prevailing in high rainfall zones of Eastern India, 5 kg lime per tree should be applied once in 3 years.

Irrigate the plant as per need

Like nitrogen, irrigation is another effective and powerful management aspect in litchi to improve the tree productivity. Irrigation is also important in enhancing the fertilizer use efficiency of trees.

While talking about irrigation management, it is necessary at the beginning to understand the critical period of irrigation requirements and required quantities of water in line with the tree growth cycles. In an annual growth cycle, litchi has two critical periods of high water demand i.e. one is after harvesting for flushing which continues up to flush maturity and the second during flowering and fruiting which continues up to fruit maturity. Irrigation in the second stage of plant cycle is very important and essential as it is required for the developing fruits which govern the outcome of the orchard. During these periods, the quantity of water demands are very high and are determined by the variety, tree age and tree size, soil type and climatic conditions such as rainfall temperature, ET and wind velocity prevailing on during these periods. As a general practice, the irrigation must be stopped after the flush mature up to flowering. This is because litchi need 2-4 month dormant period before flowering as a pre condition for the development of floral primodia. If excess water is applied before the onset of flower primodia, unnecessary growth flushes may come which affects the subsequent flowering and fruit production. Therefore, during the period from flush maturity to flowering, irrigation may be completely cut off. Under extreme drought conditions if trees show signs of water deficiency, light irrigation may be given to keep the plants alive and active in photosynthesis. In most parts of the country, the period of flushing and flush maturity is coincides with monsoon and therefore, there is no need of irrigation except in case of drought. The period of flowering and fruiting and fruit development commence during dry periods of the year and therefore, irrigation during fruiting is very essential. As a principle 25mm water per weak is required in root zone of litchi plant but the growers can decide the frequency of irrigation and quantity of water depending upon the water availability. Quality of water also plays crucial role on productivity of litchi orchard. For good yield, the irrigation water should contain less than 1000 ppm total salts. The critical time for irrigation is during fruit development and stress at this time can lead to fruit cracking and reduced yield. It is essential to provide adequate water from the fruit first set to fruit up to harvesting.

Usually, in areas where suitable climatic conditions are found for commercial litchi production, the period from flushing to flowering must have a dry period with very little rains. Main irrigation schedule should start from panicle initiation. The plant require constant and sufficient moisture during fruit development and prematurity stage. Therefore ensured irrigation should be there. Final irrigation may be given about 10 days prior to the expected date of harvesting. The time taken from flowering to maturity also may slightly vary according to the climatic conditions and locations. Growers must record observations on flowering and harvesting dates which are very valuable in deciding irrigation timing.

Make attempts to increase pollination

Litchi flowers are visited by a number of insects pre-dominately by honey bees who are seeking the nectar during full bloom. In the event of nectar collection, they transfer the pollen grain from one flower to other and the process of pollination takes place. Many of the female flowers are shed or fail to set fruit, or the set fruits are dropped at very young stage due to lack of pollination. Therefore, pollination process in litchi orchard should be ensured. For this, keep bee hives at the rate of 10-15 boxes / ha. When litchi trees are at the flowering stage, sunflower plants should be cultivated on the sides of bunds. Honeybees, which are attracted by the sunflower, also rest on the litchi flowers. This facilitates cross-pollination. Heavy rains wash off pollen and thus prevent fruit setting. Some cultivars tend to produce a high percentage of small fruits without a fully developed seed because of unfavorable weather during the fruit-setting period. Setting of honey bee boxes in flowering litchi orchard has been recommended as standard practice.

Fostering strong and healthy bearing branches

This is a composite but important mechanical and cultural practice for bumper litchi production. Timely fostering of strong and healthy shoots gives healthy inflorescence which is key for higher production. Restrictions in winter flushing through mechanical means helps reproductive bud formation. The practice also increases the ratio of female to male flowers in cluster and fruit set rate. The package of fostering includes:

- Timely application of quick-acting fertilizers before fruit harvest for recovery of tree vigour after harvest.
- Timely and adequate pruning after fruit harvest.

- Shoot control and flowering enhancement in last autumn branches through binding, ringing and spiral barking of main trunk or main branches.
- Artificial or chemical killing of winter sprouts by manual removal or chemical sprays.
- Use of plant growth regulators (paclobutrazol) for maturation of shoots and flower promotion.

Perform pruning after harvest

Litchi trees must be pruned after harvesting. The objectives of pruning are to remove dead or diseased wood, to remove additional growth flushes, to allow more light penetration into the leaf canopy, to enable the fruiting terminals to produce more sprouts which can bear panicles in the next season and also to control tree height to facilitate cultural management practices. Erect branches are less fruitful, compared with spreading branches. Crop maintenance and harvesting from upright growing branches is also difficult and beyond the reach of grower and therefore, these must be removed. At the same time lower branches are pruned to about 1 meter from ground level to facilitate cultivation practices such as weed control, fertilizer application and irrigation.

Thin out excess fruits

A tree can sustain a certain number of fruit depending on its size. If more fruit are set than this, excess fruit may drop naturally. In certain instances, however, even after natural fruit drop, a large number of fruit may remain on the tree and due to this only a few fruit will set in the following year. For a commercial cultivation, this is not a good thing to happen. Remove all the excess fruit. This is important to minimize yield variation over the years. Also it is helpful to obtain a higher proportion of well grown high quality fruit every year.

Manage the tree basin

Root zone area of the trees must be kept weed free all the time. Manual cultivation is not recommended as it will disturb active roots. Weed killers with heavy mulch may be quite effective in controlling weeds. Mulching materials may be continuously added every year. Mulching along with chemical weed control and under tree sprinkler irrigation has the added advantage of easy tree manipulation for inducing water stress required as a precondition for excellent flowering from flush maturity to flowering. Under these circumstances, the active root zone may be confined to the top 15-25 cm of soil depth and tree manipulation therefore, become easy with controlled irrigation.

Banking

This technique, which involves pilling earth around the tree-trunk up to a certain height to protect it from cold, has been recommended for trees upto 5 years old in any location where temperatures may drop to 1^0C for an hour or more. Banks usually reach a height of 45cm. Sandy and loamy soils are especially suitable for banking, but mucky, rocky and peaty soils may also be used.

Use growth regulators

Growth regulators have been used on litchi in order to obtain a number of effects, especially bigger and better-quality yields. Some of the positive results have been obtained with growth regulators, however it worthwhile to understand the certain basic concepts regarding the use of growth regulators.

a) Applications which have given success in certain regions or with certain cultivars do not always give the same responses in other regions and with other cultivars.

b) Response variations are due to the fact that these substances act at extremely low rates and within a restricted range of concentrations. The major problem is ensuring that an adequate quantity penetrates the plant within a acceptable time frame and at the required point.

c) Internal and external factors influence penetration of the growth regulators. For a given species, internal factors are cultivar, thickness of cuticle, age and physiological state of the plant which, may change from region to region and rate of absorption and assimilation. External factors include the volume and concentration used, time of application, type of water and equipment used, mode of application, quantity of the substance deposited on the leaves and climatic conditions.

d) In many cases, the use of plant regulators tends to meet the requirements of a certain type of market, but this may not be justified in other circumstances particularly if the reasons for using the substances are economic.

e) In some circumstances, response variations may be to the extent of crop-phytotoxicity.

It is therefore, recommend to proceed with caution while using growth regulators and it is always useful to test the substance on small scale before proceeding with larger-scale use.

(i) For Increased fruit set and retention

In litchi heavy fruit drop have been observed at different stages through fruitlet abscission. Although the fruit drop continues up to maturity but most of the flowers and fruitlets abscise during the first month after pollination. The main cause for flower and fruit abscission in litchi is lack of nutrients and hormones needed for embryo and fruit development, lack of pollination/fertilization and environmental factors in certain cases. Therefore, fruit retention in litchi cannot be achieved by merely one or two simple measure but a comprehensive set of practices to be attended step by step throughout the season. High fruit retention in litchi can be ensured by controlling winter shooting, flowering period and flower quality. Timely treatment of plants for strong and healthy shoot production delays flowering phase and increases the female flower ratio and finally the fruit setting rate. Honey bees are the main pollinating insect in litchi. Increasing the bee population in orchard ensures high pollination and fertilization which increases the fruit retention. If necessary, artificial hand pollination can also be done for high fruit retention. In vigorous litchi cultivars, particularly young bearing trees produce more vegetative flush and less flowers. Under such situation, girdling of flowering shoot after 10-15 days of pollination is very effective practice for higher fruit retention.

1. 20 p.p.m. NAA and 10 p.p.m. 2,4-D applied prior to anthesis, 2 and 6 weeks after fruit set and 2 weeks prior to harvesting increases fruit set and retension.
2. Spraying with NAA (10, 15 or 20 ppm) and 2, 4, 5 - T (25 or 50 p.p.m.) alone or together with micronutrients 3-4 times during flowering and fruit set reduces fruit drop.
3. 100 p.p.m. GA; 10 p.p.m. 2, 4, 5 –T and 250 p.p.m. chlormequat applied when the fruits are the size of a pea, reduces fruit drop.
4. 20 ppm IAA applied when 50-100 percent of flowers are open, significantly increases fruit set. GA_3 at 50 mg/l, applied on the same day, has a beneficial effect on fruit retention.

(ii) For increased fruit size and quality

1. Spraying with 25 or 50 ppm. GA four weeks after fruit set increases fruit size. At higher rates, soluble solid content also

increases. 10 ppm. NAA improves fruit weight, but reduces quality (loss of sugar content). Application of 5 ppm 2, 4-D significantly improves quality (i.e., increased total soluble solids/ acidity ratio and reduced seed diameter). Application of 10 p.p.m. 2, 4, 5 – T also has a positive of effect on quality.

2. Application of 250 ppm chlormequat when the fruits are the size of a pea has a positive effect on fruit colour and quality and somewhat accelerated maturation. Maturation is faster with 20 ppm, NAA and 10 p.p.m. 2, 4, 5-T.

Adopt pest and disease control practices

Pest and diseases must be effectively controlled to get a good harvest from bearing trees. Pest and disease activity must be closely monitored from flowering to fruit harvest in order to prevent major pest or disease out breaks which might destroy most of the crop if left unnoticed. Major pests include, litchi mite, litchi leaf minor, litchi leaf rotting, leaf cutting weevil, litchi bark eating caterpillar, litchi stem borer and litchi fruit and seed borer. No serious diseases problem is noticed in litchi crop however, timely action is necessary to prevent economic losses from these pest and disease problems.

Suggested management protocol

The following practices to ensure fruit production work well. Results may vary in locations with different soil types and micro climates. This could be a general guide for bearing trees with May-June harvest which can be customized based on location.

- Prune back 20 cm from all branches during or immediately after harvest.
- Apply one-half of the years allocated fertilizer and irrigate to synchronize and promot shoot growth. The first round of growth should mature around early July.
- Irrigate if needed in Aug-Sep to promote vegetative growth. This second vegetative growth fresh matures around late October to early November.
- Do not fertilize or irrigate after the second flush which matures in late October.
- In some years, cold temperature does not occur until December or January and a new flush of vegetative growth may emerge in late November. If this occurs, pinch off each new shoots at 2.5-5 cm above the base as the first leaf begins to expand.

- The complete removal of a shoot without leaving a stub generally results in immediate vegetative growth from axillary buds. By using the recommended pinching method, the growth of the axillaries can be delayed for approximately 6 weeks to allow buds to be exposed to the cool temperature necessary for flower initiation.
- In January, when flower panicles reach 10-13 cm, apply one fourth of the years fertilizer and irrigate.
- Apply fertilizers only to plants with well formed and fully elongated flower ponicles, otherwise, emerging shoots may revert to vegetative growth.
- Apply the remaining one-fourth of the years fertilizer and irrigate when fruit reach pea size.
- Irrigation should be stopped 2-3 weeks before harvest.
- Spray the trees with Zinc Sulphate 0.2 per cent (2 gm per litre of water) at least 15-20 days before flowering. Application of Zinc Sulphate promote flowering in the crop and balance the ratio of male and female flowers.
- During flowering, keep 5-6 bee hives boxes in the orchard for proper pollination and fruit setting. Pesticides should not be applied during flowering as it adversely affects the activity of honey bees.
- About 10-15 days after fruit setting, apply micro nutrients like boron 0.2 per cent (2 gm per litre of water) or biozyme 0.2 per cent (2 gm per litre of water) to control fruit dropping, skin cracking and for proper fruit development. Three application of micronutrient should be done at the interval of 15 days commencing 15 days after fruit setting.
- During second and third application of micronutrients, apply insecticides like indoxacarb (0.3 ml per litre of water) or (spinosad 0.3 ml per litre of water) or (endosulfan 2 ml per litre of water) to control the incidence of fruit borer. Growers should not repeat the same insecticide to avoid resistance problem and get good results.
- If incidence of fruit rot and brown spot is prevalent in the area, spraying thiophenate methyl 1 gm per litre of water or carbendazim 1 gm per litre of water may be done. It should be applied at least 20 days before harvesting of the fruits.

❑❑❑

Chapter 13

Adopt a Schedule to Keep Orchard Healthy and Productive

It has been observed that except few, many litchi growers are active in their orchards only when flowers or fruits are present on the trees. They have low concern about their orchard for rest of the period. This practice is not good and leads to many problems and poor productivity of the orchard. To maintain vigour, growth and productivity, it is necessary to perform proper operations throughout the year. As such, proper care and management is needed in litchi throughout its life span. Manuring, fertilizer application based on test report and timely orchard management practices provide continuous fruiting. The calendar of operation for field work in litchi orchard as described may be followed. The calendar of operations can be divided into two, i) annual calendar for establishing and non fruiting orchards and ii) annual calendar for fruiting orchard.

Monthly calendar for establishing and non-fruiting orchard

January

- Light irrigation at weekly interval may be done.
- Remove the mite, leaf miner and weevil infested shoots. At bud stage, *Trichogramma* card may be placed @50,000/ha.

- Termite kills small and young litchi plants. Apply suitable control measure as described in this book.

February

- The process of light irrigation and termite management should be continued.
- Leaf miner problem starts at this stage. Control them by using methods as described in this book.

March

- In case, thatches were erected to prevent plants from cold waves, these may be removed and plants may be shaped by training.
- Soil may be turned in basins and 20kg FYM and 500g pongamia /neem cake may be mixed thoroughly followed by needful irrigation.

April

- Soil testing may be done.
- Leaf analysis may be done for nutrient status.
- It case, new orchard is to be established, layout and pit digging may be started. Open pits may be left for 15 days and then filled with soil + FYM mixture.
- Irrigation may be done.
- Remaining plants may be properly trained, irrigation may be given and protective covering may be given to protect plants from hot air (loo).
- Apply control measures for litchi bug and fruit borers as descried.

May

- Leaf testing may be done.
- Soil Testing may be done.
- Keep plant basin wet by light irrigation till wet season.
- Pending work of layout and pit digging / filling may be completed immediately in this month.
- Regular irrigation may be given and plant may be protected from loo
- Orchard palying may be done.

June

- Orchard ploughing may be done.
- If rain has started, apply fertilizer and mix it properly.
- In case, there is no rain, irrigate the plants.
- Plants trained in the month of May give rise to water shoot. Remove these unwanted water shoots.

July

- Planting work may be done after the commencement of rains.
- If manure and fertilizer have not been applied in the month of June, these may be applied in the first week of July.
- Litchi bug and mite may be controlled as described.
- Leaf cutting weevil attack starts in this month. DDVP 1.5 m/l may be sprayed as and when damage is recorded.
- Plant basins may be racked and weeds may be removed.

August

- Weed control operations and basin cleaning may be done.
- In case litchi mite infestation is seen. Sulphur 2g/l may be sprayed as per need. Bark eating caterpillar may attack plants , try to control them.
- Make provision to drain excess water due to rain.

September

- Remove the mite affected minor twigs and spray sulphur 2 g/l.
- For appropriate tree ideotype, training and pruning work may be done.
- Stem borer / bark eating caterpillar may damage shoot. Control them using method as described. Bordeaux paste lime: copper sulphate: water (5:5:50) may be pasted to a height of 4 feet on trunks.
- Manage the leaf roller on new flushes by spraying chemicals as described.

October

- Orchard plouhing may be done for weed management
- In case copper deficiency is seen, copper sulphate may be sprayed @2g/l twice at 15 days interval.
- Basins cleaning work may be taken up and mulching may be done.
- Litchi mite control may be taken up, if needed.

November

- Cypermethnin 1 ml/l or Cartep hydrochloride 2g/l may be sprayed to control the shoot borer. It will be more useful if Bt @ 1g/l is sprayed against this pest.
- Termite management may be taken up, if needed.
- In order to protect young plants from frost, thatching and light irrigation may be done.
- Water shoot may be removed.
- Orchard ploughing may be done.

December

- Light irrigation may be done.
- Thatching may be done invariably in frost prone areas

Table 26 : Visual Monthly calendar for establishing orchard

Operation	Months											
	J	F	M	A	M	J	J	A	S	O	N	D
Soil testing				*	*							
Leaf testing				*	*							
Ploughing					*	*				*	*	
Pre plantning preparation				*	*							
Irrigation	*	*	*	*	*	*					*	*
Fertilizer application			*				*					
Planting							*					
Weeding			*					*		*		
Tree protection				*							*	*
Pest control	*	*	*				*	*	*	*	*	
Pruning									*			

Monthly calendar of operation for fruiting orchards

Fruiting orchards need special attention and management. Timely fertilizer application, pruning, pest management and stress management leads to higher production of quality fruits. Monthly work schedule for fruiting orchard is presented hereunder.

January

- Orchard ploughing may be done but irrigation work must be stopped.
- First spray of zinc sulphate @ 2 g/l may be done.
- Pruning of mite affected twigs may be done.
- Remove the mite, leaf miner and weevil infested shoots. At bud stage, *Trichogramma* card may be placed @50,000/ha.

February

- Leaf analysis may be done for nutrient status.
- Second spray of zinc sulphate may be done after 15 days of first spray.
- Honey bee boxes @ 10/ha may be introduced at the time of flowering in the orchard.

March

- Nymphs of litchi bug may be seen at this stage. Metasystox 1 ml/l or Dimethoate 1 ml/l may be sprayed at 10-15 days interval. Timing of spray may be scheduled carefully to avoid bee toxicity.
- After one week of fruit setting, apply planofix 2 ml / 5lit of water or NAA 20mg/l once.
- After 15 days of fruiting, spray borax 4 gm / l once.
- When fruit setting is over, apply 1/3 amount of nitrogen and irrigate the basins at 5-7 days interval.
- Irrigation may be done.

April

- After 15 days of first boron sprays, apply second spray and repeat it at 30 days.
- Manage litchi bug as and when needed.
- Soil and leaf malysis may be done.

- Apply irrigation water as per need. Sprinkler/drip irrigation is more useful at this stage.
- In order to control fruit and seed borer, apply cypermethrin 1 ml/l before 45 days of fruit maturity and repeat it at 10-15 days interval.

May

- Irrigate plants at regular interval until fruit maturity.
- Harvest fruits when colour is deep red and TSS comes nearby 17-19%. Morning hour is the best time for litchi harvesting. Before intense sun and heat, fruit must reach the grading and packaging houses.
- Fruit may be harvested with twigs of 15-20cm.
- Unwanted and insect infected twigs may be removed. Manage the canopy to a desired level. Plant height should not exceed 4-5mt. Pruning may be done to allow sunlight and aeration to entire canopy and plant parts.
- Soil and leaf analysis may be done for nutrient management.

June

- Pruning work, if was not done, must be completed in this month.
- Medium and late varieties may be harvested.
- Ploughing may be done to incorporate mulch materials in the soil.
- Make trench under canopy 150 to 200cm apart from trunk. Leave it for 5-6 days, afterwards fill it with a mixture of FYM + Soil + Fertilizer. This process may be repeated at 3 years interval. During rest of the years, FYM and fertilizer may be mixed in plant basins.

July

- Late varieties may be harvested.
- If fertilizer and FYM application work is pending, complete it in first week of July.
- Manage leaf cutting weevil as suggested.

August

- In case litchi mite is seen, apply Kelethane 1.25 ml/l

September

- ❐ Apply second spray for mite control.
- ❐ Plough the orchard.
- ❐ Leaf roller management during new flush is required

October

- ❐ Ploughing may be done if it has not been done. Mulching may be done in basins.
- ❐ Mite control may be continued.

November

- ❐ For shoot borer carbaryl 2g/l or Cypermethrin 1 ml/l may be sprayed twice at 2 days interval.
- ❐ Don't apply fertilizer or irrigation water in the orchard.

December

- ❐ Remove mite infested twigs.
- ❐ Don't apply fertilizer or irrigation water in the orchard.

Table 27 : Visual Monthly calendar for fruiting orchards

Operation	Months											
	J	F	M	A	M	J	J	A	S	O	N	D
Soil testing				*	*							
Leaf testing				*	*							
Ploughing					*	*			*	*		
Irrigation			*	*	*							
Fertilizer application						*	*					
Micronutrient	*	*	*	*	*							
Hormone application												
Weeding and ploughing	*								*	*		
Tree protection				*							*	*
Pest control	*		*	*	*		*		*	*		
Harvesting					*	*	*					
Pruning						*	*					

❐❐❐

Chapter 14

Learn the Role of Pest Management in Litchi Orchard

Pests, diseases and various disorders cause enormous loss to litchi through direct and in direct invasion on various plant parts. Use of chemicals has been one of the conventional methods to reduce these losses, however, due to various unwarranted side effects now-a-days, pest management is relied upon many other options along with pesticides. The integration of all these options is called IPM (Integrated Pest Management). Integrated Pest Management is a strategy to manage pests and diseases on the basis of a systems approach that looks at the whole orchard ecosystem. This includes understanding how the pests and diseases interact with their plant hosts, with the general climatic conditions, plant health and nutrition and with each other. When implementing an IPM system, growers should select ways to reduce overall pest levels in their orchard and ensure that the management of pests and diseases is compatible with their other crop management strategies. It is important that growers realize that IPM system is updated from time to time in response to biological changes that occur in their orchard and new techniques or technologies are introduced as soon as additional relevant information becomes available.

Key aspects and concept of IPM

Although the concept of IPM is broad, there are certain aspects that need to be incorporated into a practical program. Litchi growers should select those aspects which they feel could be implemented on their farm and keep in mind others which may become feasible at some later time. The key aspects of IPM are:

- Accept that IPM programs are not a tool, rather it utilizes a range or combination of strategies and tools. Farmers should select those that best suit their situation, including the use of chemicals. These tools are discussed and suggested in forth coming chapters against different pests and diseases.
- There are many pests and diseases that are present in the orchard. All of them do not cause economic damage. A few of them (key pest / disease) may cause damage to that extent where application of control action becomes essential. Identify them and do not apply pesticide just by seeing a few insect in the orchard. Identify other climatic and cultural factors that may reduce pest numbers or damage. There are many natural enemies of insect pests; these are called predators and parasite. Recognize these beneficial insects, which are found in unsprayed orchards and reduce pest and damage levels on cultivated trees.
- Be innovative and have some basic idea about biology and ecology of the pests, their life cycles, the damage they do, their natural enemies and the effects both of the pests on the crop and the crop on pest numbers.
- Visit your orchard even if there is no fruit, exclusively for pest record. Monitor the incidence of pests and beneficial insects to determine the abundance, life cycles and levels of damage.
- The pests and diseases vary with age and stage of the tree. Therefore, collect information on pest damage in relation to the phenology (growth patterns) of the trees. Read books, take help of experts or decide yourself for the optimum time to apply control methods.
- If possible, keep a pest record register. Record and summarize the monitoring information so that relationships and seasonal trends of pests' damage and beneficial insects can be recognized within the orchard.

- Accept that some pests may be present on tree crops. Pest residue provides food for natural enemies. Low levels of damage should be tolerated and control is initiated only when pest numbers increase (or likely to increase) and raise the level of damage.
- There is a list of pest control options. Consider the control options available and select from the list, which is the best suited to your current circumstances and minimize the use of chemicals.
- There are chemicals which have less effect on natural enemies. Therefore, select chemicals which minimize adverse effects on beneficial insects.
- Natural enemies may survive on many other plant/pests species which have no relation with litchi. Learn to search for associations such as those between pests and diseases or those between beneficial insects and weeds or flowers. For example, some parasites may require a nectar source, which is available from flowering weeds but not from the litchi trees.
- Keep your orchard vigourous. Make attempts to optimise the nutrition of orchard because pest attacks are usually reduced in vigorously growing healthy trees.
- Realise that centre opening, skirting and hygiene like pruning operations can reduce some pest problems (e.g. scale / mite) by altering the density of the tree, but may increase others (e.g. shoot borer) by encouraging flush growth.
- Acknowledge that other factors such as cultivars, tree architecture, soil type, water logging, and orientation to the sun and localised aspects (slope of hill or on flat land) may also be important in pest and disease management.

Methods of IPM

Look for insects, diseases, plant growth habits and other visible signs in litchi trees by inspecting them regularly. This will enable to know crops and the effects of pests and diseases on them. Knowledge and experience is needed to make the right decisions on pest and disease management strategies. An IPM program can incorporate some or all of the following factors that may affect pest and disease incidence in litchi crop.

Cultural practices

Diversity: Orchard that grows with several tree crops in close proximity often has fewer pest and disease problems than those where monoculture dominates. The pest and disease complexes on the various crops may be different and frequently the growth stages of the different trees are not synchronized. Natural enemies may build up and survive on crops that are not treated and invade where high populations of pests occur. Intercrop/ cover crops provide good crop diversity.

Nutrition: Balanced nutrition is very important. As part of this strategy, growers should aim to keep the nutrition of trees at an optimum level. Pest attack is usually less in vigorously growing healthy trees. Many insects are attracted to high nitrogen content in leaves whereas, high potassium levels have a repellent effect or are possibly toxic to some pests. Application of need based micronutrient helps in maintaining of good orchard health.

Pruning: Opening the tree to improve air movement, light penetration and plant protection chemical to access and remove sources of infestation. Skirting helps in reducing the crawling up mealy bugs and hygiene pruning helps in reduction of diseases inoculums.

Tree shape and size: The shape and density of the tree could affect both the level of disease incidence or pest attack and the efficiency of control measures. Tall, dense and compact tree do not receive sunlight at inner side causing development of many diseases and pests. Such trees do not permit proper application of plant protection measures.

Cultivars: Varieties and cultivars may vary in their susceptibility to pest and disease infestation. Varieties known for susceptibility of particular diseases or insect pest, therefore, should be avoided.

Availability of alternate hosts: Many fruit trees and forest plants serve as host for mealy bugs and scales. Thus, weeds and other plants may allow pests and diseases to increase within, or in close proximity to the orchard. However, they may also provide food and shelter for natural enemies.

Location of the orchard: Soil type, water logging, orientation to the sun and localised aspects (e.g. slope of hill) may also be utilised to reduce pest and disease incidence on trees within a block.

Pest free planting material: Bring the planting materials from stock which is free from pest and diseases. Scales, mite and many other problems can be avoided by using disease free planting materials.

Planting density: High density planting is more susceptible to mite in comparison to low density planting. Spraying of chemicals in un pruned dense HDP orchard is very difficult which helps pests to escape the pesticides.

Field practices: Ploughing to expose hibernating pests, regulation of moisture through irrigation, application of alkathene band to prevent mealy bugs are some of the practices that prevent pest incidence.

Biological factors

Maximization of natural biological control: A wide range of natural enemies attack pests of litchi. These include predators, parasites and insect diseases. Growers should learn to recognise these organisms and attempt to attract them into their orchards and give them the best chance of survival there.

Application of bio control agents: If necessary, biological agents can be introduced when pest populations are low and released to coincide with susceptible life cycle stages of these pests. Some of these may be purchased from commercial sources or collected and introduced from neighboring areas.

Spot application and judicious use of pesticide: It is vitally important that orchard managers should use pesticides judiciously to reduce the effect on the natural enemies of the main pest species.

Chemical factors

Pesticides: The use of chemicals is still very much a component of the IPM control strategies and growers should regard pesticides as another tool in the IPM arsenal. Equally, it is important that it is not the only method considered or used to manage pests and diseases in litchi crop.

Selective chemicals and bio-pesticides: Growers should choose pesticides that have the least effect on non- target insects. For example, neem products, pongamia oil, virus based pesticide, fungal insecticide, *Bacillus thuringiensis,* selective chemicals and the "softer" insecticides like potassium soaps should be used whenever possible. Home made preparations such as, garlic extract and bio-dynamic products may also be useful under certain situations.

Timing: The timing of sprays can affect the efficacy of the pesticide applications and the presence of pests. Considerations may include the time of day when treatments are applied, the prevailing weather conditions and the phenology of the trees (e.g. whether they are flushing).

Application techniques: Whenever possible, the use of spot spraying rather than total block coverage should be considered since it will have an effect on both the cost and on the natural enemies within the block. In some situations, only a few trees around the boundary may be infested and require treatment while in others, only parts of the tree (e.g. the trunk) may need to be sprayed. High volume applications and higher pump pressures to achieve penetration of the chemicals within the canopy may also be important.

Others: Use of pheromone and early harvesting has been found successful in management of litchi fruit borer.

Proper care and handling of pesticides

In integrated pest management, pesticides have a major share, though the use of other control methods like biological control agents and bio-pesticides is on the increase. Synthetic pesticides are popular because of the ease of application, effectiveness and high economic returns. Most pesticide accidents result from careless practices or lack of knowledge about safe handling of pesticides. Being poisonous in nature and harmful to human beings and other creatures, pesticides should be used with care after purchase onwards.

At the time of purchase

At the time of purchase, select the pesticide according to the recommendations of competent authorities. According to the Insecticides Act, 1968, certain information is mandatory to be given on the label of the container. These include technical and trade names of the pesticide, composition (percentage of active ingredient and other auxiliary materials), manufacturer's address, registered uses, designation of the toxicity class, net contents, date of manufacturing and expiry, etc. Purchase the insecticide by trade name of high standard. Do not purchase pesticides after their expiry date.

Before use

After the purchase, read the label information carefully. The toxicity label is in a diamond shape, divided into two triangles. The lower triangle is red/ yellow/ blue/ green; these colours depict the toxicity class, red being extremely toxic, followed by yellow, blue and green. The handling of the red-label pesticides needs utmost care. It is necessary to ensure that employees are adequately informed about the potential physical and health hazards of working with pesticides and the adverse effect on the environment if used improperly. This is applicable to all personnel who apply, transport or store pesticides. Give proper instructions to pesticide applicators enable them to learn safe pesticide management procedures. Taking time to use pesticides to properly is an investment in the health and safety of the employee, his or her family and others. To alert pesticide users to the toxicity of a pesticide, a signal word will appear on the label. Signal words will tell the user whether the chemical is highly toxic, moderately toxic, or slightly toxic. There are three different signal words used on pesticide labels. Look for the signal word, prominently placed on the front panel of the label. The signal word will indicate what degree of hazard can be expected when handling the product.

- CAUTION means the product is relatively low in toxicity to the handler or applicator.
- WARNING means the product is moderately toxic to the handler or the applicator.
- DANGER, usually appears with the word POISON and a pictogram of the skull and crossbones, means the product is highly toxic to the handler or applicator.

During use

While opening the packing, take care not to blow or splash the contents. Use only the recommended dose as a higher dose, besides being costly, can also cause phytotoxicity, residues and other health hazards. On the other hand, a lower dose will not be effective. Mix the pesticide in a ventilated place to avoid inhalation. Do not mix in containers meant for general use. For mixing, use rod and never bare hands. Do not use mouth to siphon pesticide out of a container. Those handling pesticides should not be sick or have any bruise or cut. Wear protective clothing while using pesticides. Do not eat, smoke or chew during the spraying operation and never rub the mouth or eyes with your hands while spraying. One should not work alone while handling the extremely toxic pesticides. Do not spray against the wind, as it will cause drift, leading to harmful effect to non-target organisms, and the wastage of pesticide. Never apply granules with bare hands as these formulations are highly toxic and are absorbed through the skin. Generally, with packs containing granules, gloves are provided; if not, then wear synthetic gloves and use a granule applicator. The risk of skin absorption is increased in warm weather when sweating. Also, people tend to avoid protective clothing in such weather. Do not use mouth to blow open clogged nozzles, instead use a fine wire. When there is bee activity in the field, avoid pesticide application. But if necessary, then do it late in the evening, when there is no bee activity. Most ready-to-use formulations are inflammable; do not bring these near a flame.

After use

After use, clean the spray equipment thoroughly with water. The unused pesticide should be kept in a labeled container. Do not use empty pesticide containers for any purpose, particularly for storing water or feed for animals. The containers should be destroyed and buried. Wash hands thoroughly with soap, take a bath and change your clothes. After application of the pesticide, observe the waiting periods recommended for different pesticides before harvest.

Storage and disposal

Pesticides are to be stored in a location which is dry, adequately ventilated and is free from fire hazards. The pesticides are to be stored in their original containers that are labeled according to content and the containers are to be free from leaks. The storage site must have an immediate supply of clean water available in case of an emergency. Pesticides are never to be disposed of in dumpsters, sinks or drains of any kind. Store the pesticides in labeled containers out of the reach of children.

Routes of entry of pesticide

Pesticides can enter the body in three major ways: 1) through the mouth (orally); 2) through the skin and eyes (dermally) and 3) through the lungs (by inhalation). People may be poisoned without realizing the seriousness of the exposure especially if pesticides enter through the skin and lungs. Acute exposure is a single incident of exposure to a pesticide. Usually the symptoms of a poisoning begin quickly and leave little doubt about the cause of the illness.

Chronic exposure is repeated exposure to pesticides over a period of time. Chronic exposure may go unnoticed since some pesticides may persist in the body for a long time without any obvious signs or symptoms of poisoning. If a person using the pesticide continues to be exposed to these pesticides, residues in their body may increase. An additional risk is that even low-level chronic exposure may lead to serious illness.

Personal protective equipment

Most of the pesticide applicators in India are illiterate and unfortunately, they are not given proper training and protective equipments by their literate owners. To prevent pesticides from entering in the body, the person applying pesticides are to wear personal protective equipment. The lack of any statement or the mention of only one piece of equipment does not rule out the need for additional protection. No safety recommendations can cover all situations, therefore knowledge of pesticide toxicity will allow employees to assess potential hazards and guide them in selecting the kind of protection needed. As with all types of PPE, it is important to inspect the equipment before and after each use.

Body covering: Any time pesticides are handled, the user is to wear at least a long-sleeved shirt and long-legged trousers, or a coverall-type garment. Coverall-type garments are to be made of woven or laminated fabric, manufactured specifically for protecting skin exposures to chemicals. When handling pesticide concentrates or very toxic materials, a liquid-proof apron is also to be worn.

Gloves: When handling concentrated or highly toxic pesticides, wear gloves. For liquid formulations, liquid-proof neoprene gloves are best. Gloves are to be long enough to protect the wrists. When working with hands and arms overhead, put the gloves outside of the sleeves.

Hat: Wear something to protect the head. A wide-brimmed hat will help keep pesticides off the neck, eyes, mouth and face.

Shoes and boots: While sturdy shoes and socks are sufficient for some pesticide applications, neoprene or rubber boots are a wise precaution. Many canvas, cloth, and leather shoes can readily absorb pesticides. Wear unlined boots with trouser legs outside the boots so the pesticide will not run down the leg and collect in the boot.

Goggles or face shield: Wear liquid/chemical splash goggles or a face shield when there is any chance of getting pesticide in the eyes. Eyes readily absorb pesticides and may cause temporary blindness. Users may wear goggles alone or with a respirator.

❑❑❑

Chapter 15

Identify the Pests and Manage them to Avoid Direct Loss

Litchi is relatively free of any serious pests or diseases, regardless of where it is grown, however, some groups of insects attack the flowers, fruit, new flush, leaves and branches. Fruit borers, bugs, leaf rollers and mites are the most important pests affecting litchi production. Other important species include various leaf and flower eating caterpillars and beetles, bark borers, scales, nut borer, fruit-piercing moths and leaf minors. Various workers have reported nearly 42 insect species and mite pests to attack litchi trees and fruits in different stages of growth (Table 28).

Table 28 : Insect and mite pests of litchi

S.No.	Name of the pest	Common name / nature of damage
A.	**Lepidoptera**	
1.	*Conopomorpha* sp.	Leaf miner
2.	*C. cramerella*	Fruit borer
3.	*Indarbela tetraonis* Moore.	Bark eating
4.	*Selagena* sp.	Bark eating

Contd...

5.	*Argyroploce leucaspis* Meyr.	Leaf roller
6.	*A. illipida* Butl.	Fruit borer
7.	*Platypephus aprobola* Meyr.	Leaf roller
8.	*Cryptophleba carpophaga* Wism.	Nut borer
9.	*Rapala sp.*	Fruit damage
10.	*Cocoecia epicyrta* Meyr.	Leaf damage
11.	*Statherotis discama*	Leaf damage
12.	*Cricula trigenestrata* H.	Leaf damage
13.	*Deudoris epijabas* Moore.	Pomegranate butter fly / Fruit damage
14.	*Thalassodes veraria* Guen.	Leaf damage
15.	*T. quadraria* Guen.	Leaf damage
16.	*Chlumetia transversa* W.	Shoot tip borer / shoot damage
17.	*Selepta celtis* Mo.	Leaf damage
18.	*Lymantria mathura* Moore.	Leaf damage
19.	*Ephestia cautella*	Fruit borer
B.	**Coleoptera**	
20.	*Myllocerus delicatulus* Boh.	Leaf cutting weevil
21.	*M. discolor Boh.*	Leaf cutting weevil
22.	*M. undecimpusturlatus* Faust	Leaf cutting weevil
23.	*Anoplophora macularia*	Trunk borer
24.	*Cryptocephalum insubidus* Suffia	Leaf damage
25.	*Carpophilus ebsoletus*	Fruit borer
26.	*C. mutilatus*	Fruit borer
27.	*Apoderus blandus* Faust	Leaf damage
28.	*Xylleborus fornicatus* El.	Leaf damage / shoot white borer
29.	*Depaurus marginatus* Pascal	Leaf damage / weevil
30.	*Potosia angustata* Gernar	Fruit damage
31.	*Ptochus* sp.	Leaf damage
C.	**Hymenoptera**	
32.	*Oecophylla smaragdina F.*	Red ant/ Annoyance to orchardists
D.	**Thysanoptera**	
33.	*Megaleurothrips eusitatus* (Bagn.)	Thrips /Flower damage
34.	*Dolichothrips indicus* Hood	Thrips/ Leaf damage
E.	**Hemiptera**	
35.	*Drosicha mangiferae* Green	Mealy bug / Sap feeding
36.	*D. dalbargiae* Green	Mealy bug / Sap feeding
37.	*Parlatoria pseudopyri* Kuwana	Sap feeding
38.	*Chloropulvinaria psidii* Maskell	Mealy bug/Sap feeding

Contd...

39.	*Dialeurodes elongate* Doz.	White fly / Sap feeding
40.	*Aleurocanthus husaini* Carb.	White fly / Sap feeding
41.	*Toxoptera aurantii* (BdF.)	Aphid/ Sap feeding
42.	*Saissettia nigra* Nietn.	Sap feeding
43.	*Chrysocoris stolii* (Wolff.)	Litchi bug / Sap feeding
44.	*Tessaratoma quadrata* Dist.	Litchi bug / Sap feeding
45.	*T. javanica* (Thunb.)	Litchi bug / Sap feeding
46.	*T. papillosa*	Litchi bug/ Sap feeding
47.	*Halys sulcata* (Thunb.)	Sap feeding
48.	*Ceroplastes pseudoceriferus*	Scale insect/ Sap feeding
49.	*Kerria lacca* (Kerr.)	Lac insect/ Sap feeding
50.	*Metachordia sp.*	Sap feeding
E.	**Diptera**	
51.	*Bactrocera dorsalis*	Fruit fly / internal feeding
G.	**Acarina**	
52.	*Acerya litchi* Keifer	Leaf curling / mite
53.	*Oliganychus mangigerus* Rah. and Sapra	Leaf damage / mite

Source: The Litchi : M. Rai, Vishal Nath and P. Dey (2003)

1. Erinose mite (*Aceria litchi* Keifer)

Erinose mite, *Aceria litchii* (Keifer), also known as hairy mite, hairy spider or dog ear mite occurs throughout India i.e. Bihar, U.P., Assam, Tripura, West Bengal, Orissa and Punjab where litchi is grown. As a major litchi pest, the incidence of litchi mite is seen during March which remains active up to June-July. Severe infestation has been observed in Bihar during March-September but its population declines from November-February. However, it has also been reported that the mites remain active throughout the year in one stage or the other. It is active on litchi trees from January to October and hibernate in adult stage in the hairy and velvety growth (erineum) from November to December under the surface of the leaf. There is a direct and positive correlation between the growth of new leaves and the mite infestation, and negative correlation between the population density of the mite and temperature and humidity.

The adults are 0.15-0.2 mm long, while the young nymphs are microscopic. Both nymphs and adults are vermiform.

The adults start multiplying from the end of Feb/March and the peak activity is noticed around July. The female adults lay eggs singly at the base of the hair on the lower surface of the leaves. The eggs hatch within 2-3 days and newly emerged nymphs feed on soft leaves. The nymphs are smaller and have less number of lateral setae. The nymphs and adults are similar in appearance, whitish and four-legged. The life-cycle is completed within 8 to 12 days with two nymphal stages before becoming adults. The longevity of the adults varies between 2 to 3 days. The litchi mite completes its life cycle in 15 to 20 days and there are 10 to 12 generations.

Dispersal of the erineum mite from one orchard to the other usually takes place through old infested litchi leaves or by winds. The current of the wind carries the erineum along with the mite from one orchard to the other and initiates infestation in un -infested orchards. Planting material obtained as layers may also be source of infestation if they have been taken from trees with the mites. Later infestations occur when the mites are moved around the orchard by direct contact between trees or carried around by orchard workers, wind and bees.

Litchi erineum mite is the most destructive and endemic pest causing curling of leaves. Both nymphs and adults damage the leaves, inflorescence and young developing fruits. They puncture and lacerate the tissues of the leaves with their stout rostrum and suck the cell sap. As a result of infestation, undersurfaces of the infested leaves show abnormal growth of epidermal cells in the form of hair like velvety growth of chocolate brown colour. In some cases, the mites cause galls or wart-like swellings or depressions on the upper surface of the infested leaves. Chocolate-brown velvety growth on the ventral surface of leaves indicates the presence of this pest. The attacked leaves become thick, curl, wither and ultimately fall off. The infestation generally begins from the lower portion of the trees and gradually extends upwards. In addition to leaves, the mite also infests the newly formed leaf bud; inflorescence and epicarp of the newly formed fruit. The attacked leaf buds fail to bear flower or fruit. The growth of green patch on the epicarp indicates the mite infestation on fruits. Young plants and plantlets in nurseries are very susceptible and could even be killed if leaf drop is excessive.

Management

Control measures must be preventive, for once the mite is established, it is almost impossible to eradicate. It is therefore recommended that :

- Layers be prepared only from non-infested plants.
- Layer saplings may be sprayed with 0.05 per cent dimethoate when they leave the nursery. Prior to planting out, the spraying should be repeated twice at 10-14 day intervals.
- The leaves should be checked regularly for symptoms over summer and autumn. All trees in an orchard are not be flushing or infested at the same time. Therefore, branches infested with the mite should be cut off and burnt.
- After harvesting in June, infested branches must be removed.
- In September-October, trees must be treated just prior to vegetative flushing with Hexythiazox 0.5 ml/l or 0.05 per cent dimethoate either alone or in combination with 0.12 per cent dicofol. Spraying should be repeated two weeks later and monthly thereafter until new growth is free of all symptoms of infestation. Infested leaves should be gathered and burnt or buried deeply into the ground.
- In December-January, just before flush/flower buds, the affected shoots must be removed and spraying of 0 15% kelethane may be done. In the month of February, two sprays, one each before and after flowering have been found useful.
- Numerous species of predatory mites have been recorded with *A. litchii.*

2. Bark eating caterpillar (*Indarbela* spp.)

Litchi is damaged considerably by the bark-eating caterpillars, which attack trees of all ages, particularly the older ones, lowering tree vitality. There are two species of bark-eating caterpillars, *Indarbela tetraonis* (Moore) and *I. quadrinotata* (Walker) which bore into trunk, main stems and thick branches of litchi trees in India. Both are polyphagous with a wide range of host plants including aonla, ber, citrus, phalsa, guava, jack-fruit, jamun, loquat, mango, mulberry and pomegranate. The old, shady and neglected orchards are more prone to attack by this pest. When severely infested, the entire branch or tree may die.

The moths are stout, pale brown with brown spots and steaks on forewings with whitish hind wings. Wing expanse of 35 to 38mm in male and 46 to 50mm in females. The full grown caterpillars are 46 to

52mm long with dark brown head and dirty brown body. The female moth lays 250-300 eggs in cuts and crevices in the bark in cluster in early June. Each egg hatches in 8 to 10 days and newly emerged caterpillars come out. The full grown caterpillars develop after 4 to 5 moultings within 10 to 11 months. Pupation takes place within the tunnel in March-April. The pupal period lasts for 15-25 days. There is only one generation in a year

The newly emerged caterpillars start nibbling at the bark. The attack by this pest is characterized by the presence of long-winding, thick, blackish or brownish ribbon-like masses composed of small chips of wood and excreta, both of which intermix with the help of adhesive material secreted by the caterpillar. After 2-3 days, larvae bore into the trunk or main branches usually at the forking place and make tunnel downwards. There is only one larva in each hole, and there may be 2-16 holes in each tree, depending upon the intensity of infestation and age of the tree. By continuously devouring the tissues, it tunnels through the stem and branches. The caterpillars remain within the bored holes during day and come out at night to feed upon the bark. The attacked trees show the presence of windings and silken galleries full of frass and faecal matter. As a result of feeding in the trunk or main branches, the translocation of sap is disrupted and in case of severe infestation, the growth of tree is arrested. Fruiting capacity of tree is adversely affected. Severe injury weakens the stem, resulting in drying of the branches and finally of the tree itself.

Management

- The caterpillars can be killed by inserting an iron spoke into the tunnels.
- This insect has also been successfully controlled by injecting ethylene glycol and kerosene oil in the ratio of 1:3 into the tunnel by means of a syringe and then sealing the opening of the tunnel with mud.
- Another method of control is dipping a small piece of cotton in any of the fumigants, like carbon bisulphide, chlorosal or even petrol and introducing it into the tunnel and sealing the opening with clay or mud.
- A mixture of equal parts of creosote and chloroform or petrol or cheap kerosene oil may be injected into shelter tunnels to kill the active boring caterpillars.

- Remove the webs from tree trunks and put emulsion of DDVP (0.05%) in each hole and plug them with mud. Mix chlorpyrphos 2 ml per litre of water and apply on the bark eating caterpillar infested area with a brush at 15 days interval.
- As a preventive measure, spraying of the attacked trunk and branches with 0.05%DDVP may be done.

3. Litchi Leaf roller (*Platypephus aprobola* Meyer)

Pletypephus aprobola (Meyer) was recorded for the first time in Bihar causing severe damage to litchi foliage at and around Muzaffarpur. Its infestation was also observed on the leaves of *Cassia flora* and *Polyalthia longifolia*. It is a cosmopolitan pest and has been reported as a pest of various fruits.

The incidence of leaf roller is found on litchi trees during July to February. The number of larvae is the highest during December to February, preceding flowering season of litchi. The breeding season of the leaf roller on litchi leaves is from August to February when new leaf flush is available and restricted breeding takes place during off-season (March to April) on alternate hosts such as kath-jamun (*Eugenia jambolana*) and *chhota amaltas* (*Cassia tora*) growing around litchi orchards from March to July. It may attack flower also. Adults are 24-29 mm long. The head, antennae and thorax are brown and face is dark fuscous.

Mating and oviposition of adult moth generally takes place during night. The female moth lay 150 to 200 eggs under the surface of newly emerged tender leaves. Eggs are creamy white which hatch within 2-8 days. The newly emerged larvae undergo five moults before pupation. The larval period varies from 7 to 14 days and pupal period from 7 to 24 days. There are 5 larval instars in the life history of the moth. The last instar larvae pupate in larval clip, a small portion of the leaf on the margin, both anteriorly and posteriorly and conceal themselves by bending and sealing the clipped piece of the leaf. The adults emerge within 7 to 10 days after pupation. The life cycle is completed within 21.6 days in August and 46.5 days in February-March.

The symptoms of leaf injury by the larvae are manifested through rolling of tender leaves and feeding inside. As a result of larval injuries, the infested twigs distort and wither. Litchi trees whose foliage is attacked by the larvae, very poor flowering or in case of younger tree, there is no flowering at all in the flowering season. Thus, the crop yield gets reduced considerably. Besides, the general growth of the younger trees infested by this pest is also distinctly affected. Leaf injured by the leaf roller varies between 16.70 to 71.60% while trees infestation varies 12.88 to 53.54% during August to February.

Management

- The damage caused by leaf rollers is tolerated as long as it is restricted to the foliage and unlikely to affect flower initiation.
- The rolled leaves that contain larvae may be removed manually during light infestations.
- If necessary, carbaryl 2g/l can be applied when 20 percent of leaf flushes are infested to minimize damage to young trees or at critical periods of leaf growth in older trees.

4. Litchi fruit borer (*Conopomorpha sp.)*

This pest is known as litchi fruit borer or litchi stem end borer or litchi seed borer. Bradley (1986) clarified the identity of the cocao pod borer, (*Conopomorpha cramerella)* and described three previously unrecognized co-generic species (*C. oceanica, C. sinensis* and *C. litchiella)* from South East Asia. However, historically the host records of this group are confused. According to recent literature, the primary host plant of *C. cramerella* is cocoa (Bradley 1986). Subsequent literature and tentative sorting of earlier records suggest that of the four species treated by Bradley, only *litchiella* and *sinensis* have been associated with litchi fruit. Other species of litchi fruit borer are *Carpophilus obsoletus* Er., *Carpophilus mutilates* Er., *P. illepida* Meyer. and *Ephesia candela* Wlk.

Of all the species in the group, *C. sinensis* is apparently found infesting litchi most often, and elicits the greatest economic effects. The moths are active fliers, reddish-brown, having characteristic long antennae. There are irregular zia-zag white patches on the fore-wings and hind wings fringed with long hairs. Dorsally, the abdomen shows black cross bands, while the ventral surface is brown. The wing span of the male ranges from 6.0 to 7.5mm and that of female, 8.0 to 9.5mm.

Adults are nocturnal in habit and mating usually takes place during dusk. Eggs are laid singly soon after mating, on the under surface of the leaf or near the calyx of litchi fruits. During winter months, the leaf buds are preferred for oviposition. The total number of eggs laid by a single female varies from 30 to 49. A single female usually lays

1-3 eggs per day. Freshly laid eggs are grayish-white, spherical, 0.42±0.09mm long and 0.28±0.02 mm wide and spherical. The incubation period lasts for 4-5±1 days. There are five larval instars. The larval period varies from 5 to 8.48 days. The fourth and fifth instars feed faster than the early three instars.

Newly hatched larvae are milky white, slender with distinct light brown head and minute setae all over the body, 3.28 ± 0.05mm long and 0.31 ± 0.05mm wide. This instar lasts for 1 to 1.4 days. The second instar larva is also cylindrical and its head is dark brown. Thorax and abdomen are greenish with white tinge on dorsal side and pale yellow on the ventral side. Body length, width and width of capsule are 5.26 ± 0.06mm, 0.41±0.10mm and 0.39±0.022mm, respectively. The third stage larvae are green with dark brown head, 6.4 ± 0.04mm long and 0.61 ± 0.1 mm wide. The body of fourth instar is green with dark brown head. Small irregular white patches are formed on dorsal side of the body. Body length is 7.3 ± 0.04mm and width is 0.74 ± 0.03mm. Fifth instar larva is greenish yellow and measured 8.0 ± 0.1mm long and 0.8 ± 0.01mm wide. The full grown larva starts spinning cocoon, which is usually formed on the old litchi leaf.

The pupa is greenish on dorsal side and greenish yellow on the ventral side. The pupa measures 5.9 ± 0.19 mm long and 1.6 ± 0.18mm wide. The duration of pupal stage is 5 to 9 days. The longevity of the adults varies from 3.12 to 6.84 days.

The newly emerged larvae start boring into the fruits and feed on its pulp. The infested fruits do not attain normal size and can be identified by the formation of black spot near pedicel. Larvae do not enter much deeper into the pulp but feed below the calyx about 15 mm deep. When fully grown, they come out of the fruit and pupate on the leaf surface. In this way, the larvae cause direct damage to litchi fruits.

Females of *C. sinensis* clearly prefer fruits over shoots for oviposition. If no fruits are available, they are constrained to lay their eggs on shoots. The survival rate is higher on shoots than in fruit. Males emerge earlier than females and live slightly shorter. The larvae mine the young leaves, shoots and also bore the litchi fruits. The pest has now established itself as one of the major pests of litchi in India particularly in Bihar and Uttar Pradesh.

The fruit loss has been estimated from 24 to 32 per cent in Muzaffarpur district, Bihar. During July, they cause indirect damage by making mines in young shoots. The young larvae make mine in the lamina and bore into mid-rib of young leaves and tunnel through it, as a result branches wither and drop. Leaf infestation varies between 7.1 to 72.5%, while tree infestation ranges between 47.70 to 88.9% during August to February.

The highest density of the pest is observed in September, while the lowest in December. Extent of damage is positively correlated with the amount of precipitation. Its infestation increases considerably with the onset of rains. The maximum fruits are damaged by this pest during May and its population is very high in August. From October-March its population is not present in the orchards but reappears in the month of April.

5. Leaf miner

C. litchiella is a much less frequently recorded pest on litchi than its congener *C. sinensis*. All stages of the leaf-miner, *C. litchiella*, are similar to those of the fruit borer *C. sinensis*. The female lays its eggs on new shoots and the small, light-yellow eggs hatch three to five days later. The newly hatched larva is creamy white, and bores into shoots and leaf blades, usually into the mid-rib and veins. Affected shoots often wilt. The larvae pupate on the leaves and give rise to new leaf mining and shoot boring generations. The moths are attracted to leaf flushes that emerge during the rainy season. The pest is active on litchi trees as a leaf miner during August to February, while during off-season (March-April) it restricts itself mainly on alternate hosts viz., *kath jamun* (*Zyzigium jambolana*) and *chhota amaltas* (*Cassia tora*) growing wildly in litchi orchards. When litchi fruits become mature in May, the moths migrate to litchi orchards and lay their eggs on the calyx. All members of the species complex pupate underneath leaves in a cocoon before emerging as adults. The life cycle takes about a month.

Management

- Moths can be excluded by enclosing the fruit panicles in nylon mesh bags, but is uneconomic in areas with high labour costs.
- Parasitoids *Phanerotoma* sp. and *Apanteles* sp. are important parasites. NRC on Litchi, Muzaffarpur has recommended the use of *Trichogramma chilonis* @50000eggs/ha.
- NRC on Litchi, Muzaffarpur has recommended the use of pheromone trap, however, this pheromone is not giving consistence results in all the situations.

- Fallen fruit should be removed to reduce the build-up of moths and ploughing may be done after fruit harvesting.
- Bearing trees should be inspected during early flush development and sprayed if necessary. The leaf flush before flower initiation is very important as it supplies the carbohydrates needed for fruit development. Young, non-bearing trees do not need to be sprayed. This also allows the parasitoids to build up in the orchard.
- Fruits may be inspected weekly from fruit set to detect eggs of borers, which are very small and almost invisible to the naked eye. Infested fruit should be picked and destroyed at infestation levels of 1 to 2 per cent. When the pest becomes more active, first spraying of 0.05 per cent fenitrothion or dicholorvos or carbaryl 2g/l may be done.
- Permethrin, cypermethrin, deltamethrin or fenthion during early fruit set is recommended to prevent damage later in the season.
- Second spray may be done with neem based insecticide (0.4 per cent) or Bt based biopesticide (Halt) 1 gm/l. In case damage in high, indoxacarb (0.3 ml/l) or spinosad (0.3 ml/l) may be sprayed.
- Affected shoots may be removed and all agronomic efforts such as ploughing and nutrient management should be done so that flushing takes place before the month of September.
- Neem based products may be applied at the time if new shoot emergence to avoid heavy population of the pest.
- Monocrotophos 0.05 per cent may be applied in the case of severe shoot damage.

6. Shoot borer (*Chlumetia transversa*)

The shoot borer has been observed causing serious damage to new flush of litchi throughout the country. This pest is usually active from August to October. The adult moths are shining grey in colour and measure about 17.5 mm with expanded wings. Hind wings are light in colour. Female moths lay eggs on tender leaves. Eggs are oval in shape and pale yellow in colour. Full grown larva is dark pink with dirty spots and measures 24-24 mm in length. After hatching, the larvae first bore into the veins of young leaves and later into the soft stems of shoots situated close to the apex, working their way downwards. New shoots are damaged by the caterpillars by tunnelling from growing tip downward. As a result, the shoots droop and finally dry and wither. Larvae also bore into the inflorescence stalk. When full grown , these caterpillars come out and enter into the slits and cracks in the bark of the tree, dried inflorescence or cracks or crevices in the soil for pupation. Upper and lower parts of the plants are more infested than middle parts. The affected plants become stunted. Young trees up to the age of 8-10 years are more susceptible. There are four overlapping generations of the pest in a year and it overwinters in pupal stage.

Management

- The attacked shoots may be clipped off and destroyed.
- Spraying of Carbaryl (0.2%) or Quinalphos (0.05%) or fenitrothion 0.047% or 0.05% malathion at fortnightly intervals from the commencement of new flush gives effective control of the pest. A total of 2-3 sprays may be done depending on the intensity of infestation.

7. Fruit sucking moths (*Othries* sp., *Achaea* sp.)

Fruit-piercing moth attacks many fruit crops such as apples, apricots, bananas, breadfruit, coffee, figs, grapefruit, guava, kiwifruit, litchi, longan, mandarins, mangoes, nectarines, oranges, papaya, passion fruit, peaches, persimmons, pineapple, plums, and star fruit. Fruit-piercing adult moth is the damaging stage and the larvae are essentially not harmful. The mouth parts of the moth are about an inch (2.5 cm) long and strong enough to penetrate through tough-skinned fruit. Once the moth has punctured the skin of the fruit, in a few seconds and feeds upon the juices of the fruit. Feeding occurs at night. Fruit flesh damaged by this moth becomes soft and mushy.

Damage caused by this pest is not only a result of the direct feeding but also by the fungal and bacterial infections that develop at the wound site. When moths are abundant green fruit is attacked, causing premature ripening and dropping of fruits. Incidence of damage by this moth is normally low, however when outbreaks occur, most of the crop is affected.

The duration of the life cycle from egg to egg-laying adult female is 35 to 49 days. The small eggs are about 1 mm in diameter and are yellowish green. Eggs are generally laid on the underside of leaves in forest areas. Eggs hatch in 3 to 4 days. This moth has 5 larval stages, or instars, separated by 4 molts. Caterpillars are cylindrical in shape and are 1/5 to 1/3 inch long during the first larval stage. They reach up to two inches in length when fully grown. They are either dark green to black or pale green to yellow

Mature larvae pupate within a cocoon spun between leaves and woven together with silk. The leaves containing the cocoon may remain on the host plant or dry and fall to the ground. Pupae are very dark brown with a purplish cast and about 28 mm long. Pupation lasts for 14 to 21 days. If pupation occurs under very dry conditions, the adult may not be able to emerge successfully. The adult moth is large and robust. It has a wingspan of almost 10 cm and a stout body, about 5 cm long that does not extend, or slightly extends, beyond the hind wings. The eyes are large. The thorax is pale to purple-brown and the abdomen is pale brown at the base brightening to yellow-orange at the tip. After emerging from the pupa, females have a pre oviposition period of 4 to 8 days before she begins to lay the eggs. Each female may lay up to 750 eggs during her lifetime.

Caterpillars mostly feed between 5:00 PM and 10:00 AM, but may feed at any time. They are usually located beneath or on the edges of leaves. Young larvae drop to the ground at any sign of danger, while the older larvae take an aggressive attitude by hanging on to the food plant with their hind legs and swaying the rest of their body from side to side. The adult moths are very strong fliers and can travel great distances from their breeding grounds in search of food. Adults fly mainly between the hours of dusk and 11:00 PM. They are not usually attracted to light. When disturbed, the moth flares its forewings, exposing its conspicuous hind wings

Management

i) *Capture and destruction of moths* : Since this moth is not easily disturbed from fruit once it has begun to feed, netting and killing of moths is possible. This is best accomplished an hour after sunset when there is sufficient darkness with the aid of torches or a strong flashlight. This method is most feasible when fruit are easily accessible and population of this moth is small. This method is not very effective once a large population of this moth exists and is usually recommended only if no other control means is possible.

ii) *Lanterns and lights* : This moth prefers darkness and avoids light, therefore the illumination of orchards has been tested as a possible means of deterring moth attack. Although damage is less severe under illuminated conditions, this method is determined to be mean of reducing damage based on a small scale trial.

iii) *Bagging or screening fruits* : Mechanical protection of fruit against moth attack is achieved by covering the fruit with brown paper or transparent oil paper bags. Brown paper bags last for about a month in the field and the transparent oil paper bags may be used for up to two seasons if the weather is not very wet. This method is most practical when fruits are easily accessible and compact bunches. Although this method is labour intensive, it is especially good if regular attention cannot be given to the crop or it is desired to have the fruits fully ripen on the trees. This method is not practical when fruits are small, scattered, and not within easy reach.

iv) *Smoking of the orchard* : This method involves the obscuring of the odour of mature and ripening fruit that attracts the moth with the odour of the smoke. Containers full of inflammable material, oil, tar and some green plant trimmings to enhance the smoke are placed within the orchard at a rate of 2 to 4 per acre. The smoking process is started a half an hour before dusk and continued for 2 to 3 hours after nightfall. This period represents the time in which the moths are seeking their night time feeding grounds. If the smell of the orchard is masked, the moths choose various wild hosts and remain on them throughout the night to feed.

v) *Orchard sanitation* : This method involves the regular collection and proper disposal of all attacked and spoiled fruit. Both fallen fruit and attacked fruit on the tree should be collected, buried deeply or boiled in water for 10 minutes then broken up for compost. These procedures dissipate the odour emanating from the spoiled fruit so they cannot serve as an attractant for the moths

vi) *Forced harvesting* : When severe infestations are anticipated, damage may be avoided by harvesting the entire crop as soon as signs of fruit ripening is observed. The grower must determine if the revenue loss from premature harvesting is less than that from possible damage, litchi, however, cannot be harvested premature.

vii) *Chemical control* :The effectiveness of chemical control is variable and remains ineffective in most of the cases.

8. Litchi bugs (*Tessaratoma javanica* thunberg)

Among those reported in India are a number of pentatomide such as *Halya dentatus* Fabricius, *Tessaratoma javanica* Thunberg, *T. quadrata* Distant, *T. papillosa* Drur. and *Chrysocoris stolii* Wolff. The *T. juvanica* is specific to litchi. Another bug, *Lygaeus* (Syn. *Spilostethus pandurus* Scopoli), a polyphagous pest which attacks numerous fruit crops, is also harmful to litchi.

The bug, *Tessaratoma javanica* lays globular and off pink eggs, mostly in bunch of fourteen on lower surface of leaves. The insect passes through five instars, which are sub rectangular and dark brick red except first instar, which is nearly sub-rectangular. Newly emerged adult is dirty white and soft bodied insect but colour changes to yellow red after few days. The average duration of male and female adults is 43.2±7.7 and 47.2± 9.5 days, respectively. Both adults and nymph feed mostly on tender plant parts such as growing buds, leaf petioles, fruit stalks and tender branches of litchi tree. Excessive feeding causes drying of growing buds, tender shoots and ultimately fruit drop.

Other species which attack litchi include the field bug *Nezara viridula* L., which pierces the fruit on the tree which allow micro-organisms to rot the fruits. The bug can cause serious leaf-shed in nursery plants. Leaf-shed may also be caused by one of the coreids, *Eutochta galeator* Fabricius. The bugs feed on the developing seed and this causes the fruit to fall a couple of days later. The puncture mark is invisible on the fruit surface and the only way to distinguish damage from natural drop is to dissect the fruit. Fruit that has been attacked, typically has a tan lesion on the seed testa.

Management

- This pest is combated by shaking the trees in winter, collecting the bugs and dropping them into kerosene.
- The eggs of *T. javanica* are in group and visible which can be removed and destroyed.
- There are natural enemies such as egg parasitoids, *Encyrtus* (*Ooencyrtus*) sp., *Anastatus* sp. *Microphanurus* sp. and *Blastophaga* sp. which parasitize 70 to 90 per cent of eggs laid late in the

season. The adults are attacked by several fungi, birds and red ants may also be used as biological means of control.

- If chemicals are used, the timing of sprays is critical because the bugs vary in their susceptibility to insecticide at different times of the year, depending on body fat content and its nature. Many of these bugs may be controlled with endosulfan, dimethoate and fenthion.

9. Grey weevil *(Myllocerus* species)

They are polyphagus pests. Adults are poor fliers but are very active feeders on the leaves of litchi. *M.discolor* has long snout with grey color. It attacks leaves, shoot and flower. In winter they are not active. *M. laetivirens* feed on many plants in addition to litchi leaves. Adult weevils congregate on the tender leaves and nibble irregular holes on the leaves and some times consumes the entire leaf leaving the midrib only. Grubs feed on roots.

Management

- Hand picking of the adult weevils and spraying of 0.07% endosulfan will solve the problem.

10. Litchi nut borer (*Cryptophlebia ombrodelta* Lower)

The insect originally referred to as *Argyroploce illepida* Butler (Syn. *Cryptophlebia carpophaga* Walsingham) in India is actually *Cryptophlebia ombrodelta* Lower. Commonly it is called as nut borer. It is a polyphagous pest and infests 33 food crops including *Acacia* spp. Carambola, *Cassia* spp., soyabean, bean (*Lablab purpureus),* lima bean, yard long bean (*Phaseolus vulgaris),* tamarind, cowpea and avocado.

The moth is reddish-brown and measures 23-25 mm across the extended wings. The male is smaller than the female, which has a large dark triangular blotch two thirds of the way along the hind margin of each forewing. The scale-like eggs, which are laid singly on the fruits are oval, approximately 1.0 x 0.8 mm in size, and vary in colour from ivory-white when first laid to pink then black just prior to hatching. During summer they hatch in 4 to 6 days. The larval stage lasts 3 to 4 weeks. Fully-grown larvae are up to 20 mm long, and pinkish with discrete, dark green spots. Moths emerge after a pupal period of 8 to10 days. Three or four days before pupation they construct tightly woven silken cocoons that are sealed with an unobtrusive flap that provides and exit for the emerging moth. Pupation occurs in damaged fruit and sometimes in sheltered sites in other parts of the tree. When the moth emerges, the pupal case is left protruding from the exit hole. The life cycle, from egg-laying to moth emergence takes about 5 weeks.

The newly hatched larva feeds on the fruit skin and then tunnels towards the seed. When this occurs in green fruit, the fruit drops, although larva is most likely to be able to develop to maturity in the fallen fruit, if ground predators don't discover it. Ripening fruit generally does not fall, and the larva often drowns in the juice if the skin is penetrated in the equatorial region where the flesh is thickest.

The rind tissue around the entry hole may appear to be scalded and such damage is sometimes wrongly attributed to fruit fly which rarely attacks litchi. Entry on the shoulder or near the peduncle is more likely to ensure survival of the larva enabling it to reach the seed. Mature fruit damaged by nut borer may stain other fruits in a cluster or those hanging below. One larva can cause perhaps 10% more damage through this secondary staining effect. In immature fruit, the young larva bores directly into the seed which is completely eaten. A single larva may damage two or three fruit, if the fruit are small. However, they prefer mature colouring fruit with larger seeds.

Management

- Examine 5 fruit panicles on 20 trees widely spaced throughout the crop, commencing when green fruit are 20mm long. Spray if more than 5 out of 100 panicles are infested with live, unhatched and unparasitised eggs. Check developing fruit weekly for larval entry holes and or frass. Infestation levels increase as the fruit mature due to immigration of moths from alternative hosts. Oozing juice from maturing fruit may also be an indication of nut borer infestations.
- The nut borer are attacked by various species of egg, larval and pupal parasitoids; however, these do not always keep borers below economic thresholds. Egg parasitoids such as *Trichogrammatoidea fulva* Nagaraja and *T. cryptophlebiae* offer the best prospects for biological control. Numerous native larval parasites that attack the borer include the braconid wasps, *Apanteles briareus* Nixon, and *Bracon* sp., the ichneumonid wasp, *Gotra bimaculatus* Cheesman, and an unidentified tachinid fly parasite.

- Spray with carbaryl 2 g/l when 5% of panicles have fruit with fresh, unparasitised eggs. These will be white or pink, as opposed to parasitised eggs, which are black. Full cover spray 40 days before harvest, or two sprays at fortnight apart commencing when the fruit are 10 mm in diameter are recommended.
- Alternatively, the panicles can be covered with paper bags. The bags also improve fruit colour and overall quality.

11. Thrips

Three species of thrips have been reported to attack litchi in India. Two of them, *Dolicothrips indicus* Hood and *Magalurothrips* (Syn. *Taemothrips*) *usitatus* Begnall cause damage to flowers, while the other, *M. details* Karny attacks the leaves. They are small, thin, fragile insects, which in adulthood have two pairs of broadly fringed wings. Nymphs and adults scratch the epidermis and suck the sap oozing from the wound. Other thrips found on litchi are the red ribbon thrip (*Salenothrips rubrocinctus* Giard), the greenhouse thrip (*Heliothrips haemorroidalis* Bouche) and the Florida flower thrip (*Frankliniella cephalica* Crawford). Those which attach the leaves are usually found on the leaf-tips. Attack by these insects causes the drying of leaves and flowers. So far, thrips have not been reported to cause serious damage in litchi in India.

In addition to above, foliage defoliating semilooper in Ranchi and a tender leaf damaging weevil (matching with *Apoderus sp)* in Muzaffarpur have been reported.

❑❑❑

Chapter 16

Prevent the Diseases to Avoid Direct Loss

Litchi plants, as compared to many fruit bearing species, are least affected by diseases. Diseases are however, more important after harvest, although many of the fruit are infected before picking. There are a few organisms that infect the leaves, flowers and fruit, and a few others are associated with tree decline and tree deaths. A few leaf spot diseases have come to light which are caused by fungal pathogens. No bacterial or viral infections have been reported so far. A few reports of algal infections are also available. A commercialized account of pathogens as reported by various works has been listed in Table 29.

Detailed information and method of control of some of the disease of litchi has been enumerated.

1. Brown blight

It is known as blossom blight or downy mildew and caused by *Peronophythora litchi* which is a facultative necrotroph (feeds on dead host tissue) and produces colourless, aseptate mycelium 4-6 µm wide, irregularly branched at right or acute angles. Asexual reproduction is

Table 29 : Common diseases reported on litchi

Common name	Scientific name	Place of incidence
A. Mukerji and Bhasin		
Fruit rot	*Alternaria alternate*	Allahabad
Fruit rot	*Aspergillus sp.*	Muzaffarpur
Spot on leaf, fruit and flower	*Botryodiplodia thebromae*	Muzaffarpur
Leaf spot	*Cephaleuros virenscens*	Dholi (Bihar)
Leaf spot	*Cladosporium herbarum*	Muzaffarpur
Fruit rot	*Colletrotrichum gleosporoides*	Muzaffarpur
Fruit rot	*Cylindrocarpon tonkinensis*	Muzaffarpur
Fruit rot	*Gleotricium candidum*	Allahabad
Leaf decay	*Memnoniella echinata*	Lucknow
Spot on leaf and flower	*Nigrospora oxyzae*	Muzaffarpur and Meerut
Spot on leaf and fruit	*Pestatolia pauciseta*	Muzaffarpur
Fruit rot	*Rhizopus sp.*	Muzaffarpur
B. Bilgrami ***et al***		
Fruit rot	*Absidia blakesleeana*	Calcutta
Spot on fruits and young flowersPusa, Bihar	*Clanydomyces sp.*	Allahabad
Found associated with mite on leaf	*Ceuthospora litchi*	Pusa, Bihar
Found associated with fruits	*Coniothyrium sp.*	Bangalore
-	*Gyrothrix circinata*	Bangalore
Leaf spot	*Microdiplodia litchi*	Udaipur
Associated with fruits	*Symcephalastrum recemosum*	Calcutta
C. Om Prakash and Mishra		
Leaf spot	*Arthrinum phaeospermum*	-
Leaf spot	*Drechslera hawaiiensis*	Pusa, Bihar
Leaf spot	*Alternaria tenuis*	Allahabad
Leaf spot	*Nigrospora sphaeriea*	Allahabad
Twig blight	*Curvularia ovoidae*	-
Twig blight	*C. verruculosa*	-
Twig blight	*Nigrospora sphaerica*	-
Twig blight	*Dictyoarthrinium sp.*	-
Flower blight	*Chatomium globosum*	-
Flower blight	*Fusarium sp.*	-
D. Diseases from Abroad		
Downy mildew	*Peronophythora litchi*	China
Sour rot	*Geotrichum candidum*	Taiwan

Source : M.Rai,, V. Nath and P.Day (editors),CHES-IIHR, ICAR, Plandu, Ranchi initiated by sporangia developing from the mycelial hyphae in the presence of water movement. Production of zoospores occurs at temperatures of 8-22°C, by hyphae at 26-30°C, and by both sporangia and zoospores at 24°C. The kidney shaped zoospores are highly motile and have two flagella one long and smooth, the other shorter and bearing a row of hairs. Zoospores swim around and may swarm in response to a suitable plant exudate. They infect host tissues directly when dispersed to surface water films on aerial plant parts. Mycelial growth inside the host tissues follows, on flower, fruit and leaf tissues, repeating the asexual phase. A sexual stage follows this development where gametes are produced, but sexual reproduction has not been observed on fruit.

The optimum temperature for lesion formation and enlargement is 25°C. Incubation takes less than one day at this temperature and 2-3 days at 18°C, while at 11°C incubation is prolonged to 7 days. Only a few sporangia are produced at 30°C. Continuous rain and re-infection are the most important factors leading to the wide distribution of this disease.

It attacks both young and ripe fruit, pedicels and leaves of litchi. The infection reduces production and shelf life. Flower panicles are particularly susceptible. Browning of flowers and dessication of panicles attacking young and ripe fruits causing brown lesions and white downy growth or premature fruit drop are the symptoms. Immature fruit turn brown, while those infected before harvest have a white mildew growing on the skin. The pathogen probably persists as oospores or dormant mycelium in the soil or in plant debris. Higher temperatures during the day are suitable for sporulation, germination and infection by the pathogen, and lower temperatures and high humidity at night facilitate zoospore release and distribution.

Management

- Cleaning up the orchard by removing shaded, infected and dead branches after harvest.
- Copper oxychloride spray during winter and copper sulphate in spring also help to reduce inoculum levels.
- Metalaxyl spray during flowering and fruit development, reduces the disease occurrence

2. Anthracnose

(i) Anthracnose caused by *B. theobromae* : The spots usually start from the tip or the margin of the lamina. These spots are deep chocolate in colour. The limiting margins of the spots with irregular outline are vandyke brown. Black pycnidia appear on both the surfaces of the leaves but more often on the upper one. The pycnidia bear one celled (two celled at maturity), oval to ellipsoidal chestnut brown conidia (20.96-27.51 x 10.48-14μ) which possess a thick wall of two layers and a granular matter inside. The hyphae are chestnut brown in colour, well branched and are provided with a septa at very short intervals. At one side of the septum the hyphal end is swollen and on the other side it is tapering. The hyhae are 2.62-5.24 μ broad.

(ii) Anthracnose caused by *Collectrotrichum* : The spots usually start from the tips of the lamina and extend towards the base. These spots are irregular in outline and are brick brown in colour with prominent marsh brown margins encircling them. Mummy brown waxy subepidermal acervuli appear on both the surfaces of the infected leaves especially on the upper one. The acervuli are deep brown with multicellular setae having a swollen basal cell. Anthracnose, *Colletotrichum gloeosporoides*, attacks leaves and branches, along with flowers and flower stalks and fruit. Lesions on the leaves may appear as small round light grey areas or irregular brown marks at the tips. In contrast, infections are much more obvious on the flowers and fruit. Outbreaks are common after warm wet weather. The infection remains restricted to lower surface. Under no circumstances upper surface of healthy leaves get infection. During the rainy season, the rain water trickling through the diseased leaves is richly laden with the spores of the pathogens to cause further infections. The fungus may not always cause immediate disease, which sometimes only becomes apparent after harvest. Fungicides are used during an initial outbreak, but are not always effective. Prevailing climatic conditions (high temperature and high relative humidity) are highly conducive for the development and spread of Anthracnose on litchi fruits caused by *Colletotrichum* sp. The higher latent infection rate of anthracnose of the fruits causes more serious the postharvest decay and browning of the fruits. The rate of latent infection of the fruits could be evidently controlled by integrated management of the disease in the growing season and the postharvest decay and browning of the fruits are effectively reduced. The effect of storage can be improved when the measure is applied to control the latent infection of anthracnose on litchi fruit before harvest.

Management

1. Avoidance of over crowding of trees and branches in orchard.
2. Pruning of affected plants and burning has been suggested to minimize the chances of fresh infections.
3. Copper oxychloride spray has been found effective but calendar sprays of copper are costly and could lead to unacceptable residues if used close to harvest. Spray of 3:3:50 Bordeaux mixture in February, April and September-October or application of Caption 50 WP @ 0.2% is effective.

3. Tree decline and root rot

Slow decline and sudden tree death have been recorded in China, Viet Nam and Australia, especially in poorly drained soils. It can affect the whole tree or just one or two branches. The symptoms include a sudden branch wilt is followed by the decline of new growth on the affected branch over a period. In other situations, the tips die without wilting. The tree or branch may recover temporarily but subsequently dies. Parts of the tree flush and grow while other sections die. A number of organisms including *Phytophthora, Pythium* and *Fusarium* have been isolated from the roots of trees but it is not known where they cause the disease. Growers are advised not to plant on waterlogged soils. The fungus may survive in the soil or on stumps and roots of various trees for many years. The planting sites need to be fumigated before establishing new trees in the orchard.

In some parts of litchi belt, trees are killed by the root rot. Host plants include citrus, guava, banana and grapes. Some isolated cases of root collar rot caused by *Botryodiploidia theobromae* Patr. have been reported. This can quickly kill the trees and no satisfactory control method has been established. A species of *Fusarium* has been isolated which is associated with the sudden death of the plant. One side of the tree's crown may be perfectly sound and the other totally necrosed. Leaf shed never occurs (it does in the case of a nematode attack) and the internal parts of the roots are characteristically red in colour. As with the previously mentioned cases, no method has been found to save the tree once it has become infected. Growers are recommended to remove all roots from other host trees before planting litchis.

4. Red rust

A parasitic algae, *Cephaleuros virenscens*, occasionally attacks trees causing loss of vigour. The alga starts its appearance as small dark isolated patches which spread very fast and ultimately develop into a velvet reddish brown to orange coloured cushion like growth. Algal filaments associated with both asexual and sexual reproductive organs, grown on both the surfaces of leaves and penetrate deep between cuticle and epidermis and sometimes extend between adjacent epidermal cells into the layer of parenchyma below epidermis. The growth of this alga initiates the development of cork tissue in few upper layers of leaves, thus causing their death. Severely infected leaves exhibit curling inward towards dorsal side. The disease first appears on the young unfolded tender branches. On the infected young leaves, small lesions of velvety white growth appear on the lower surface. On the upper surface, just the opposite site of the lesion, chlorotic patches occur. As the leaves unfold and increase in size, the velvety growth becomes more prominent and dense. Later, larger areas of leaves are affected with this growth. Old and thick leaves show various types of malformation. The velvety growth turns light brown to brick-red.

Control

- Foliar application of copper oxychloride @0.3 per cent should be done in the month of July and Oct.
- The spray of 5:5:50 Bordeaux mixture during autumn (September-October) and Spring (February-March) at 15 days interval depending upon degree of infestion.
- Spray of 0.25% ziram also reduces the disease.

5. Fruit rot

Fruit rot of litchi has been a serious problem. Litchi is host to a range of post-harvest pathogens, often with quite different modes of infection. Several fungi have been found to be associated as reported by several workers. Usually large water soaked lesions appear on the surface of fruits. Initially the disease symptoms are perceptible on injured portion of the fruits. With the advance of the disease the decayed areas get depressed. The rot gradually penetrates deep into the pulp. Ultimately rind of infected fruits cracks off exposing the pulp which subsequently is covered with thick cottony mycelium. Such affected fruits emit an odour of fermentation.

Management

- Low temperature storage is the most successful means of slowing rot development. For instance, fruit stored at 22°C rotted three times more quickly than fruit stored at 5°C.
- Fungicides are also effective. Hot benomyl dips at 48° to 52°C slow rot development compared with undipped fruit. Rots still affect the dipped fruit, although the fungicides slows the spread of the diseases. This technology has not been used due to health concerns surrounding fungicides.
- A great range of other possibilities exist for controlling rots, such as the 3 to 5 per cent O_2 and 3 to 5 per cent CO_2 mixture. However, these need to be studied in a commercial environment.

❑❑❑

Chapter 17

Manage the Physiological Disorders

Litchi cultivation and its potential yield is affected by many physiological disorders like fruit cracking and splitting, flower and fruit drop, sunburn, retarded fruit development, irregular bearing and black spot etc. Therefore, it is necessary to know the causes and symptoms properly to manage these disorders for increased quality production. The incidence and severity of these physiological disorders vary with locality, season, cultivar and orchard management practices.

1. Poor fruiting / non-fruiting / irregular fruiting

Poor fruiting/non fruiting/irregular fruiting are the problems in litchi growing area of the country. It has been observed that capacity of litchi shoots to bear is varietal in nature. While some bearing terminals of current year are more productive (70-95%) in the next year others are not. Non adoption of good management practices lead to irregular bearing, even in regular bearing cultivars. Improper management practices include late application of manures and fertilizers (particularly nitrogenous fertilizers) and high frequency and heavy irrigation just before the panicle emergence and flowering period. Cultivar China is generally known as irregular bearer, can bear fruits every year with adoption of management practices accurately. Irrigation, nutrition and environmental factors are also responsible for irregular bearing in litchi. Type of soil coupled with management practices also found to be associated with irregular bearing or alternate bearing. Late harvesting and severe pruning and training operations may lead to non bearing in that particular year of operation.

Usually, heavy crop is followed by a light crop in some varieties, while in some varieties production is quite consistent every year. In some cases at the time of bearing panicle/ inflorescence, trees start giving vegetative flush and that particular year becomes off year, while in some case only part of the tree bear long panicle with leafy growth at the base, resulting only few flowering terminals and very poor fruit set. Sometimes, inspite of heavy bearing of flowers and enough fruit set, harvest of mature fruits is very poor due to heavy flower and fruit drop at various phases during flowering to fruit maturity.

Management

- Varieties chosen for planting must be regular in bearing, apart from other desirable characters.
- The cultivars Shahi, Rose Scented, Dehradun are regular bearer, while China, Seedless and Late Bedana are irregular/shy bearers.
- On the basis of age of the trees, the commercial orchards should be properly fed with balance nutrition immediately after fruit harvest and irrigation at right time through right method, particularly at aril development stage.
- The plant protection measures to control major pests and diseases should be followed.
- Proper pruning and training at the time of harvesting and just after harvesting is needed. While harvesting of the fruit bunches, the branches having 8-10 inches length should be plucked to give rise strong and healthy flush to bear fruits in next year.
- Late pruning and training should be avoided. The tree should be trained to give canopy shape in a semicircular manner or an open umbrella shape.

2. Flower and fruit drop

Sometimes even after profuse flowering and fruit set, poor yield has been observed as because litchi trees suffer a heavy flower and fruit drop between flowering and fruit maturity. Only a small proportion of it (2-18%) is carried to maturity in different cultivars.

The flower and fruit drop is thought to be due to failure of fertilization, embryo abortion, nutrition and hormonal imbalance and external factors like high temperature, low humidity and strong westerly winds as well as due to fruit borer and heavy mite attack. The young bearing tree suffers fewer drops than the older trees. Studies revealed maximum fruit drop during the first fortnight after fruit set and at harvest the retention varies between 3.0-39.6% among the cultivars. The specific relationship between auxin production in developing fruit and abscission in litchi has also been taken into consideration.

In litchi, male functional flowers drop after pollen formation. The heavy fruit drop have been observed at different stages due to the formation of abscission layer. Although fruit drop continue up to maturity but most of the flowers and fruit drop occur during the first month after the pollination. The initial fruit set in litchi is very high but a very small proportion finally mature. The premature fruit drop commences soon after fruit set and continues till fruit maturity, with most fruit abscising in the first 2-4 weeks.

Management

- High fruit retention in litchi can be ensured by controlling winter shooting, flowering period and flower quality.
- Timely treatment of plants for strong and healthy shoot production delays flowering phase and increases the female flower ratio and finally the fruit setting rate.
- Honey bees are the main pollinating agent in litchi. Increasing bee population in orchards ensures better pollination and fertilization which increases the fruit set and retention.
- It is advised not to do any type of spray at the blooming stage of the orchard.
- Spraying of $ZnSO_4$ @ 0.2% at 30 days before panicle emergence induces healthy inflorescence leading to more fruit set.
- Foliar application of 0.2% Boron, 2-3 times during the period of fruit growth and development enhance fruit retention, minimize cracking, improve fruit colour, sweetness and enhance maturity leading to increased percentage of quality production.

- Two foliar application of planofix @ 4ml/5 litre water may be done at an interval of 15 days from peanut size fruits.
- Treatment with growth regulators like NAA at 20-30 ppm, GA, at 20-25ppm, 2,4-D at 10-20 ppm are effective in minimizing fruit drop when sprayed on panicles, before the flower opening.
- The fruitlet drop can be reduced by the sprays of $ZnSO_4$ and synthetic auxins like 3,5,6-trichloro-2pyridyl-oxyacetic acid (3,5,6-TPA) on fruits increasing yield.

3. Fruit cracking

It is the most important disorder in litchi found occurring in almost all the important litchi growing tracts of the country or even in the world (mainly throughout Asia and the Pacific). It is a major problem in litchi in India which causes loss as high as 5-70%. This may occur due to varietal characters, orchard soil management, inappropriate levels of water at maturity stage, light, mechanical injuries, temperature and micro-nutrient deficiency. All cracked fruits loose their value for fresh market and they are used for processing only (especially for fruit juice) if they are not effected by fungus. Cracked fruits are susceptible to storage disease, have shorter storage as well as shelf-life.

Fig. 45 : Fruit cracking

In India most of the cultivars are found susceptible to fruit cracking. It occurs when trees are droughted soon after fruit set and if the drought is severe enough, fruit development is affected, particularly the development of the fruit skin, resultantly the cell division is reduced and the fruit skin becomes inelastic and often splits when the aril grows rapidly before harvest. This can occur after irrigation or heavy rain or just an increase in relative humidity. The poor managed orchards receiving heavy irrigation at once or heavy rains abruptly after a long gap creating imbalance between temperature and moisture in the environment, cracking becomes more pronounced. Deficiencies in nutrient affect fruit development causing favour to fruit cracking. Role of Calcium is also found associated with fruit cracking in boron deficient soils. Injury to skin by pest or mechanically during pericarp development phase may also lead to fruit cracking. The relationship between fruit cracking and endogenous growth substances have been also established the fact that the higher concentrations of abscisic acid (ABA) and lower level of Gibberellins found in the fruit pericarp, seed and aril of cracked fruits. Insects, hail, and the sun can damage the skin during cell expansion and induce cracking towards harvest. These damaged areas cannot expand with the rest of the fruit, creating a weakness in the skin that splits.

Field observations showed that the litchi cracking occurs mainly after the colour break in fruits, coinciding with the peak phase of aril growth. The cultivars with thick skins are less susceptible to cracking than those with thin skin and that of early ripening cultivars observing poor management practices are more susceptible to this disorder.

The cracking injury in litchi may vary from short cracks, generally skin deep to larger ruptures extending to most of the length of the fruit. As the underlying aril gets exposed to dry air, it withers rapidly. The cracks on the skin may be longitudinal or horizontal. The cracked fruits deteriorate rapidly and thereafter suffer secondary infestation by disease causing organisms and insect-pests and become non-marketable.

Factors influencing the fruit cracking

(i) Environmental factors

Influence of temperature: Temperature plays a very important role in the ratio of fruit cracking. Arid & semi-arid zone where temperature is more and humidity or rainfall is very low favours cracking. In general, there is a linear increase in cracking with temperature increase. Temperature also affects many other factors such as permeability of the cell walls and bio-chemical processes of the cells etc.

Wind: In north India, occurrence of loo/hot wind during summer months is a common phenomenon. When such hot wind passes through the litchi fruit surface, due to water loss, it becomes hard and inelastic. Sudden fluctuation in moisture level in fruit skin leads to cracking in litchi fruit.

(ii) Fruit Characteristics

Maturity: Over maturity leads to cracking of fruit epidermis.

Fruit size: It is generally supposed that large fruits are more prone to cracking than the smaller one.

Fruit firmness: It is found that firm fleshed fruit such as litchi is more susceptible to fruit cracking than soft fleshed ones. Cracking of fruits is caused by excess uptake of water resulting in bursting of skin.

(iii) Lack of orchard management

Moisture stress: Moisture imbalance and heavy rainfall or irrigation after a prolong dry spell, sudden and high fluctuation in the water supply to plants may cause cracking of the fruits.

Nutrient: The deficiency of boron and calcium is responsible for cracking in cherry, pomegranate and litchi.

Insect-pest and disease: In litchi, due to sun burn, there is the appearance of small dark water-shaped spots, which is finally assuming the shape of raised spots. These areas on the fruit, develop longitudinal cracks and starts oozing out from the splits.

Bagging: It is also a remedial operation to escape the sunlight from the plant surface because water is loosed by transpiration though stomata.

Early picking: Early picking of fruits is also a remedial measure to overcome cracking. This does not allow to over maturity or over ripening which cause cracking of fruit, however it is not practicable in litchi.

Management

- Selection of site is a most effective way to minimize rain damage of fruit orchard. The appropriate site should be with little or no rain incident at near harvesting time. Cultivars which show some resistance to the rain induced fruit crack may be selected. Physical protection of fruit by water resistance covers from heavy rainfall may be done.
- To save the cracking problem at maturity stage, fruit plants may be covered only on the top and permit free airflow into the sides.

- Application of calcium @ 2m/l liquid formulations and Gibberellins @ 20 ppm, reduces the activity of cellulose and thereby reduced cracking.
- Sprayings of 2,4-D and NAA at concentrations 20 ppm or 20 mg/litre reduces cracking.
- Constant moisture and appropriate humidity are needed at the time of fruit maturity. Irrigation at 30-40% depletion of available soil moisture is quite helpful in reducing cracking of fruits. Installation of drip as well as micro sprinkler below the canopy area has been reported to be effective in reducing the fruit cracking.
- Mulching can play a big role in stabilizing the temperature and moisture level in root zone. Planting of maize or sugarcane around the litchi orchard and daily irrigation in such border crop creates congenial microclimate which reduces fruit cracking.
- Planting wind break around the orchard provides protection from desiccating hot winds. Suitable wind break should be planted around the orchard boundary and it should be at a right angle to the direction of prevailing wind. A row of tall growing trees, such as seedling mango and jamun are suitable wind breaks
- Boron sprays in the form of Borax or Boric acid @ 2g/l at the initial stage of aril development with enough soil moisture in the root zone checks fruit cracking significantly.
- There is also need to develop or find out cracking resistant cultivars as only one cultivar i.e. Swarn Roopa developed at Ranchi, Jharkhand (India) is reported to be tolerant to this disorder. Litchi cultivars which have relatively thin skin few tubercles per unit area and rounded to flat in shape are less prone to cracking. Early cultivars are more susceptible than late cultivars.

4. Sunburn

Sunburn also known as lesion browning or pericarp necrosis is a serious problem in litchi producing areas. Climatic factors and cultivars in particular growing areas are determinants for incidence and severity of sunburn disorder. The damage caused due to sun burning occurs up to 0.9-19.13% in different varieties.

Apart from environmental factors, varietal, hormonal, nutritional and soil moisture factors are associated with this disorder. This disorder is physiologically related with PPO (Poly-phenol Oxidase) activities and it also varies with cultivars. Sunburn is pronounced in ill managed orchards having sandy or sandy loam soils or light soils receiving/ exposed to high temperature (>40^{0}C) and very less RH (<50%). It is a type of direct thermal injury and in case of higher temperature, the tissue coming in contact/exposure gets sunburnt/ sunscalded. Sunburn problem is also seen more in early ripening cultivars. Fruits on shaded branches suffer less damage than those more exposed to sun. Lower translocation of Calcium in the pericarp region also found to favour sunburn disorder.

In case of sunburn, light brown blotches appear on the portion of the fruit skin facing direct sun rays. In severe cases more than half of the surface area becomes discoloured, blotchy light brown. The blotches become intense in few days and the blotchy area dries up blocking the the aril growth and finally destroying the market value of the fruit. The symptoms appear more on the south west side than the north-east side as on the latter side fruits remain almost in shade except during early hours of the day, which is not harmful. Under similar climatic conditions, the bearing of the fruit may suffer with this disorder due to the position of the fruit on the tree.

Management

- Irrigation at regular interval during the fruit growth and ripening stage reduces the sunburn.
- Planting wind break around the orchard provides protection from desiccating hot winds, thereby low sunburn.
- Irrigation through overhead sprinkler system during hot hours increases humidity, cooles the orchard atmosphere thus decreases the incidence of sunburn.
- In light and sandy soil, only light irrigation with increased frequency (4-5 days interval) is found beneficial.
- Feeding trees with sufficient quantities of organic manures particularly compost, FYM, cakes, green manure, vermi-compost along with irrigation at regular interval during fruit development and ripening stage have been found useful.

- Planting of maize or sugarcane around the litchi orchard and daily irrigation in such border crop creates congenial microclimate which reduces sun burn.

5. Retarded / under developed fruits

This disorder is not well documented but somewhat causing considerable loss due to poor quality fruits or ill developed fruits. The fruit size remains smaller than the normal and juice content or aril development is also very poor. This type of fruits remain in glossy pale green colour for a longer period and then only dull reddish colour appears. In fruit growth of litchi, seeds develop initially at a high rate, followed by membranous mesocarp and aril, which grow very fast towards the later stage of fruit development. During these phases, sometimes it happens that there is no aril development or aril development remains incomplete with portion of testa uncovered also give rise to the development of retarded/underdeveloped fruits. Lack of proper fertilization by less visits of pollinators may lead to improper fruit set and retarded fruit growth. Insect-pests attack and physical jerk also cause retarded or underdeveloped fruits. Unavailability of moisture during development stage of fruit is also one of the cause for retarded / underdeveloped fruit. It is more prevalent in poor managed orchards having poor pollination.

Management

- Proper nutritional management of orchard needs to be done.
- Putting enough number of honey bee colonies during flowering to fruit set stage increases fruit set with better fertilization leading to normal and healthy fruit growth.
- Spraying plain water in early morning hours of the day during the advanced stage of growth and development have been found highly effective in better growth of fruit and quality.
- Arrangements of sprinkler system of irrigation has also been found to be effective in reducing this disorder.

6. Blackening on fruit surface

The occurrence of blackening is not wide spread in litchi but it has been found infesting and damaging the quality harvest of litchi. In Indian condition, this disorder has been observed in existing plantation mainly in vicinity of urban areas. There are a few reports in litchi about its occurrence. It may be due to deleterious effect of smoke fumes which contain sulphur dioxide, acetylene and carbon dioxide as found in case of mango.

The symptoms of blackening become apparent when the fruit attain certain size and remain in developmental phase i.e. from March end to mid May. The first symptom appear as the development of a small etiolated area at the distal end of the fruit against the normal green colour of the fruit pericarp, which gradually spreads, turns nearly black and covers the distal end completely. The infested fruit show discolouration and overall slow pace of development resulting very poor development of aril, seed, size and colour.

Management

- The disorder can be reduced by applying proper nutrition and irrigation particularly during fruit growth and aril development stage.
- Spraying with micronutrients like zinc ($ZnSO_4$ @ 0.2%) a month prior to flower panicle initiation and boron (borax @0.2%) during fruit growth stage can reduce the problem.

7. Pericarp browning

Pericarp browning is the first visual sign of fruit decline. Browning that occurs during the first few days after harvest is usually caused by dehydration of the pericarp. This problem has been reported in several litchi producing areas and involves the appearance of necrotic spots which may expand and affect a significant portion of the fruit. Fruits start to brown once they lose a few percent of the harvested pericarp fresh weight. Below 50 percent of its initial fresh weight, the pericarp is entirely brown. Browning of the pericarp occurs at ambient temperatures of 20-30°C within 24 hours of harvest. Water loss (desiccation) of litchi results in brown spots on the bright-red shell (pericarp). Other factors also cause the fruit to brown, including mechanical stresses of various sorts (tugging the pedicel at harvest, sliding the fruit down a rough picking bag, dropping fruit from short heights); microbial and insect attack; and extremes of temperatures. In short, anything likely to accelerate cell breakdown is likely to increase fruit browning. This occurs as a result of damage (due to causes unknown) to the parenchymatic cells of the thin wall of the mesocarp. This damage speeds up a characteristic hypersensitive reaction, leading to necrosis of these cells, which then spreads to the epicarp and endocarp. Skin-browning is associated with an increase in polyphenol oxydase and perioxydase enzyme activity and with ascorbic acid oxidation. Losses may be considerable, although the aril is not affected; the fruit is not suitable for marketing. In extreme cases, the pericarp may crack. Through the use of fungicides and refrigeration, litchi fruit have a storage life of about 30 days. However, when they are removed from storage, their shelf life at ambient temperature is very short due to pericarp browning and fruit rotting.

Management

- Packing fruit into moisture-proof (plastic) bags and punnets can substantially reduce water loss and slow the rate of browning. More permeable barriers such as paper, wicker baskets and cardboard, offer less protection.
- Cool temperature storage also slows browning. Low temperatures slow evaporation as well as respiration and probably slow tissue senescence. It has been reported that fruit treated with polyamines (suspected anti-senescence agents) then wrapped and stored at 5°C, have lower membrane permeability and less browning than controls. This implicates senescence as a significant co-determinant of the life of well-packed, cool-stored litchi.
- A controlled atmosphere of 3 to 5 per cent O_2 and 3 to 5 per cent CO_2 has shown to slow water loss. Fruit stored under such an atmosphere for 30 days at 1°C lost only a quarter of the water lost by the controls. However, the mechanism of the response is not clear. Such an environment may affect the metabolism of the fruit as well as that of the pathogens.
- Sulphur dioxide fumigation has been used extensively in South Africa and Israel. There have also been many experiments in China and Thailand. However, sometimes the fruit are tainted. There are also concerns about high sulphite residues in relation to sulphur-sensitive individuals. Sulphur and acids, or combinations, can be used to stabilize the red colour of the pericarp but sulphur results in the formation of a colourless anthocyanin-sulphite complex. Consequently, sulphur-treated fruit are somewhat bleached relative to controls. Sulphur is also anti-fungal if applied correctly. Sulfur dioxide fumigation effectively reduces pericarp browning, but some countries do not

permit due to concerns over sulfur residues in fumigated fruit. Thus, sulfur-free postharvest treatments that maintain fruit skin colour are increasingly important. Alternatives to SO_2 fumigation for control of pericarp browning and fruit rotting are pre-storage pathogen management.

- Israel has recently promoted a hot water brush/acid/prochloraz treatment, but acid treatments sometimes give an artificial and persistent red colour to the fruit that masks poor eating quality.

❑❑❑

Chapter 18

Specific Problems and their Remedy

This section of the book has been specially included in order to enable the farmers who are not able to take appropriate action against a problem owing to confusion. Quite often, farmers encounter these problems and due to lack of proper guidance, they fail to rectify the situation even after spending the money. The diagnosis given hereunder is indicative, more detail information can be had from the respective book chapters.

1. Poor rooting in air layering

- While making layers appropriate shoot has not been selected. Thin or too old branches may not give proper rooting.
- Time of layering is too early or too late. Layering at adverse time such as dry weather may be harmful due to change in temperature and atmospheric humidity.
- Proper medium of air layering has not been selected.
- Layering has not been done properly.

2. Layers are healthy but there is heavy mortality in potted layers

- Generally most of the leaves fall after the shifting of layers from plant to pots, however, this does not lead to layer mortality. The roots of layers are brittle in nature. High temperature and low humidity or low temperature during night increases brittleness of root. Delay in potting and mishandling while removing wrap materials may lead to root damage.
- Excess vegetative growth of layers causes imbalance between roots and shoot system.
- After planting, the layers have been put under stress condition such as sun and proper irrigation has not been done.
- There may be incidence of some diseases.
- Delay in potting after detaching from mother plant leads to mortality.

3. Poor growth/non flush in litchi layers under nursery condition

- Shoot for layering was selected from recently flushed wood which leads poor growth or non-flush.
- Delayed planting of layers has been done in the nursery.
- Irrigation has not been done properly leading to moisture stress condition.
- Proper potting mixture was not prepared and poor management of nursery has been done.

4. Poor or no flush or wilting of field planted layers

- Weak and poorly rooted planting material has been used.
- Planting of immature layers have been done. Planting should be done after two continuous growth flushes of plants to avoid the problem.
- Poor size pit, poor soil or site of the field has been selected.
- Excessive soil moisture/water logging condition of the field.
- Faulty planting method, excessive root damage while planting, poor initial care or poor management.
- Incidence of pest/diseases.
- Moisture stress condition or frost injury may be the reason.

5. One directional flowering in plant

- Overlapping of canopy/plant receive unidirectional light.
- Very dense planting and poor canopy management.
- Plants facing road side get raised temperature because of auto mobile smoke.
- Related with maturity of shoots, June-July flush matures and flower in February-March but other flushes do not flower.

6. Continuous vegetative flushing

- High temperature during winter season because litchi plants require low temperature stress during winter.
- Winter rain/irrigation produces vegetative flush.
- Luxury consumption of nitrogenous fertilizer, especially during winter.

7. Poor growth and development of plants in field

- Poor root system of air layered plants (brittle root system) which take time to develop fibrous root system.
- Small pits have been made in problematic soil.
- Abnormal soil condition or michorhiza is lacking in the soil.
- Poor management practices such fertilizer and watering or heavy weed incidence.
- Incidence of pest and diseases.

8. Non flowering in litchi even after 7-8 years

- Air layered plants produced from thin branches, produces smaller plants, which takes longer time to produce flower and fruit.
- Plants of seedling origin take longer time to produce flower and fruits.
- Heavy soil with poor drainage system inhibits growth of plants and delayed flowering and fruiting.
- Poor orchard management practices leads to non-flowering in litchi.
- Orchard has been established in unsuitable climate/area.

9. Good flowering was recorded but there is heavy flower drop

- High temperature during flowering causes poor pollination, leads poor fruit set.
- Very low temperature during flowering checks activity of pollinating agents, which leads to poor pollination and fertilization, causes poor fruit set.
- High wind velocity and moisture stress may cause flower drop.
- Poor nutrient status of soil and leaf and incidence of pest and diseases also cause flower drop.

10. Heavy fruit drop

- Failure of pollination/fertilization causes fruit drop.
- Embryo abortion.
- Poor nutrition of orchard.
- Hormonal imbalance.
- High temperature, low humidity and poor irrigation management.
- Incidence of insects such as bugs and fruit borer.

11. Heavy fruit cracking

- High temperature and low humidity during fruit development may lead to fruit cracking.
- Moisture stress condition.
- Lack of wind break in hot desiccating wind areas.
- Isolated orchard condition.
- Nutritional imbalance such as boron.
- Irrigation at irregular interval leads fruit cracking.
- Poor management of orchard.
- Varietal differences and soil type.

12. Poor fruit color development/sourness in fruit

- Over lapping of canopy & poor light penetration leads to poor colour development.
- Poor management of orchard.

- Green variety-In this variety even after maturity fruits remain green.
- Harvesting before proper maturity.

13. New shoots and flush looks like burnt

- Attack of leaf minor give appearance of shoot burning.

14. Breakage of branches

- Initial framing not done properly and V shaped branches are formed.
- Heavy vegetative growth and crop load may put pressure for breakage.
- Heavy wind velocity and lack of wind break.
- Excessive use of fertilizers (leads extra vegetative growth).

15. Irregular flowering

- Genetic character of variety for example China variety gives alternate year fruiting.
- In other variety may be because of regular vegetative flush and heavy crop load during previous year.
- Excessive use of fertilizers specially nitrogenous fertilizers may lead to irregular flowering.
- Excessive irrigation before flowering and poor management of orchards may cause irregular flowering.

16. Leaves turn like leather

- This happens due to mite attack.

17. Spliting of bark in grown up plants

- Great difference in temperature between surface and inner layer of tissues especially when branches are exposed due to breakage or pruning.
- Very low temperature may cause splitting.
- Exposure of stem to sunlight after heavy pruning may cause bark splitting.

18. Plant grows excessively but no fruiting.

- Very fertile soil leads to profuse regulative growth.
- Excessive use of fertilizers specially nitrogenous fertilizers may cause vegetative growth.
- Hormonal imbalance and irrigation before flowering may start vegetative growth.

19. Long unfruitful branches inside the canopy

- In general litchi bears fruit outer periphery of canopy, therefore, inside branches are unfruitful and becoming liability to the plant. It should be removed at initial stage, with out growing for longer period.

20. Slow growth of new flush after fruit harvest

- Over exhaustion of plant because of heavy fruiting.
- Poor status of nutrient in the soil and leaf because of higher consumption during fruiting period.
- Moisture stress condition and high temperature.

❒❒❒

Chapter 19

Reduce Harvest, Handling and Marketing Losses

Today litchi production is characterized by many contract growing schemes between producers and processors/exporters. The post harvest losses in litchi have been estimated in the range of 25-40% from harvesting to consumption stage. If proper methods of harvesting, transportation and storage are adopted, such losses can be minimized. The earliest fruit in the market usually commands premium prices (depending on quality). To this end, it is advantageous to pick fruit soon after it reaches maturity. This also ensures that fruit has the maximum shelf life. Fruit left too long on the tree, until it has already started to soften, does not travel well and lacks shelf life. Immature harvested fruit shrivels and softens but does not develop flavour is either rejected at the market place or receives very low prices. Flowering and fruit-set normally occur over a three to four week period resulting in the fruit on any one tree being at various stages of maturity. Depending on the flowering, the crop can usually be harvested in about three picks (for any one variety). Harvesting usually begins with a light first pick of early fruit, about 10-20% of the crop. The second pick is done a couple of weeks later which is generally the heaviest, about 40-70% of fruit being mature. A final pick is done about a week later which comes from the late set fruits.

Ensure that harvesting is done at right stage of the crop

Marketing immature fruit damages future sales potential. Litchi harvest criteria can vary with local consumption patterns and distance to the market. Time from flowering combined with fullness of colour are the commonest criteria. Producers must decide whether to harvest as soon as the market price ensures a reasonable return, or to leave the crop in the field to obtain maximum profit. However, litchi crop does not permit longer waiting period as it may drastically shorten the marketable life of the produce and lower the sale price. This balance is a critical factor in determining the growers income from the crop. In practice the total harvest period is very short and the grower has very little time to make the correct decision. Litchi takes about 3 months from the time of fruit set to mature and ripen. Some varieties may take 3½ months. More precisely, a heat unit submission of 1500-1650 GDD has been found sufficient for litchi maturity. If picked immature, fruits develop lower amounts of Brix, acid ratio, taste and flavour, whereas over-mature fruits lose their storage life. Such fruits present numerous problems during handling. It is therefore, desirable to pick the fruits at the correct stage of maturity to facilitate ripening, distant transportation and maximum storage life to increase their quality and market value.

Maturity indices

Fruit harvesting is always a delicate task, but it is even more so where litchi is concerned, since the fruit is not climacteric and maturation virtually stops as soon as it is harvested. Thus, if picked too early, colour and sugar/acidity ratio, which are vital factors far fruit marketing and flavour will suffer. The later the fruit is harvested the larger the edible part and the higher the total mass. However, in too late harvesting, the level of acidity will drop significantly and the fruit may become rather tasteless. Following criteria may be followed to ascertain the litchi maturity.

- Skin colour is usually a determinant of ripeness. The fruit is said to be ready for harvesting when the pericarp is a uniform red, (just when the orange colour has disappeared completely or almost completely, except in the case of cultivars which produce yellow or green fruit) and the protuberances have become smoother. In varieties which produce red fruit, these should be an even deeper red at harvest time.

- A useful test of ripeness is to check the colour of the inner surface of the skin, which should also be red. Class of surface redness is prime method to judge the fruit maturity. The youngest fruit (<33 per cent red) and the oldest fruit (100 per cent red) have lower aril concentrations of total soluble solids and higher pericarp superoxide dismutase activity than intermediate fruit. The youngest fruit also have higher rates of water loss than older fruit.
- Sugar testing is another way of determining whether the fruits are ready for picking. As litchi fruit mature, the concentrations of sugars, principally those of sucrose, glucose and fructose increase, while the concentrations of organic acids, predominantly malic acid decrease. Other criteria used to determine the correct harvest time are minimum fruit weight and the sugar/acidity ratio. These two criteria vary depending on the cultivar and the country or climatic area in which it is grown.
- In Israel, for the some cultivar, the sugar/acidity ratio is 20 : 1 at maturity and not more than of 29 : 1 when the fruit is over ripe. This contrasts sharply with a ratio of 80 : 1 obtained in subtropical Africa with fruit just prior to rotting.
- The most reliable guide to maturity is titratable acidity (TA) or the ratio of total soluble solids (TSS, degree Brix) to titratable acidity. Recommendations vary, but a TSS:TA of 40 or greater is recommended for commercial fruit. In practice, however, most orchards of litchi are harvested on the basis of taste and general appearance.
- The flattening of the fruit segments on the skin is a good way of telling when the fruit are mature. Over-ripe fruit are sweet, but bland. Indices used to judge maturity include fruit size, skin colour or texture, the aril sugar: acid ratio(18-21^0 Brix), and flavour. There is little information on the effects of fruit maturity on storage life. Overall, there seems to be a large range of fruit maturities, with similar physiological characteristics and storage potential.
- The numbers of days from full bloom to harvest are considered to be the best maturity index for litchi fruits. Most of the varieties arrive at harvest maturity between 55-80 days after full bloom. The optimum time for harvesting Bombai litchi comes between 105-110 days after anthesis.

Table 30 : Maturity indices for different litchi cultivars

Name of Cultivar	Optimum harvest Period	TSS (%)	Physical Indicators
Calcutta	53-55 days after fruit set	18-20	Development of bright pinkish red colour
Muzaffarpur	73-78 days after anthesis	18-19	Tubercles become slightly flattened and pulp becomes smooth at maturity
Shahi	58-65 days after fruit set	19-20	Tubercles become slightly flattened and pulp becomes smooth at maturity
Dehradun	55-60 days after fruit set	17-19	Turn deep red when fully ripe

Do proper planning for harvesting and disposal

The objective of harvesting is to get the crop without damage and to send it to the market in the best possible condition. Thus, the objective is to move the crop from field to buyer with the minimum number of handling operations compatible with the quality requirements of the buyer; and to minimize exposure of the crop to stresses such as extremes of temperature, or of compression pressures caused by over-loading.

Although the scale of production, availability of labour and type of produce may vary. Certain basic factors must be taken into account in the planning of litchi harvest operation. Equipment must be obtained, labour organized, marketable produce identified and provision of collection and removal from the field is made. Each of these tasks must be planned, managed and implemented efficiently if the value of the crop is to be fully realized. To achieve this following things are needed.

- *Good production planning*: It ensures that crop maturity coincides with market demand. This can be achieved by proper care while selecting the variety and proper orchard management.
- *Continuous communication with buyers*: It is necessary to identify exact requirements as harvest time approaches, and to let the buyers know the best time of harvest and expected quality.
- *Forward planning*: It is necessary to coordinate equipment, labour and transportation facilities.
- *Field supervision*: The efficiency of the harvesting operation itself depends upon the use of experienced or trained staff, and the

adoption of methods which will meet the buyers' requirements. For this, proper field supervision is needed. Good management of the harvest operations is usually reflected in the speed at which produce moves from field to market place, packing station or storage centre, not at the expense of careful handling and subsequent quality downgrading.

Give proper training to workers

Training and supervision of labour are critical to a successful harvesting operation. Constant supervision is necessary to maintain quality and reduce subsequent spoilage of produce. Training is required in both general principles and crop specific techniques relating to maturity selection, detachment method, maintenance of equipment, field hygiene and division of labour. Some of the more important areas are:

- Teams of workers must work systematically through a plot or field, experienced staff removing the crop and others carrying it to collection points. If crops are relatively inaccessible, as with older litchi, great care must be taken by pickers climbing on the trees if fruit is to be harvested free of damage. For this, planting densities and pruning techniques should be followed which minimize tree size.
- It is essential that crops should be harvested at the correct size and maturity for the market. The workers must be given strict specifications before entering the field and their performance supervised carefully.
- During long harvesting sessions, some individuals develop habits of slapping, pressing and rubbing produce. Others become tired and start to throw or drop produce into field containers. Such practices can cause damage to the produce. They should be controlled by checking performance, by restricting the lengths of work shifts and by promoting comfortable working conditions.
- Sharp edges on rings and bracelets, and long fingernails, are significant cause of post harvest abrasions and should be removed before harvesting starts.
- Unmarketable produce must not be left in the field to rot and contaminate healthy standing crops. Routine collection of waste is an important part of the harvesting operation and all workers can contribute to this. Cleaning, sterilization or replacement of

picking containers must be carried out regularly to prevent the build-up of infection.

- Each individual should be issued once only with necessary equipment and given clear instruction and training in its maintenance. It should then become that individual's responsibility to keep knives and clippers both clean and sharp and to keep other equipment such as boxes, nets and bags in a good state of repair.

Decide proper time of the day for harvesting

The time of day at which to harvest the litchi depends on the availability of transport and other facilities, weather and environmental conditions and human factors, as well as market demands. The factor which assumes greatest importance depends upon the crop and the local situation.

- Most favourable condition for harvesting is early in the morning. In some areas, where markets need night transportation, it may be advisable to avoid harvest during the middle of the day. Produce harvested must be balanced against the possibility of local heavy rains which could have detrimental effects. Packing of wet produce often leads to greater post-harvest spoilage and fully turgid tissue can split more easily.
- Since harvested produce standing in the field usually starts to deteriorate rapidly unless specialized facilities are available, it is unwise to start harvesting until transportation is assured.
- If the crop is to be transported to a relatively distant market, storage centre, packing station or processing facility and harvesting should be timed to allow for delivery at a convenient time.
- Harvesting can only take place when sufficient labour of the required skill and strength is available. Thus the distance which workers must travel, their domestic, religious and sometimes social arrangements may have to be considered. The water content of fruit on the tree fluctuates throughout the day. Harvesting early in the morning or late in the afternoon maximizes fruit water content and reduces the risk of desiccation.

Adopt proper picking techniques

- To obtain the best quality fruit, trees with proper ripening must be selected. In peak of the season, most fruit can be picked from a tree in one harvest and all trees of one cultivar in a orchard may be harvested within 8-10 days. If fruits are allowed to reach a good state of ripeness or if bags have been placed around the clusters, each panicle can be cut from the tree as one.

- Harvest in the morning when fruit is cool and keep harvested fruit out of the sun.
- Harvesting may be carried out by removing whole panicles using secateurs.
- Fruits are best harvested using clippers, and placed in harvesting bags carried by the harvester.
- Furthermore, the fruits should never be harvested when they are wet and should be kept out of the sun after harvesting to prevent the skin from turning brown.
- In some countries a length of branch is cut off some 20-50cm above the panicle. Picking platforms and ladders are used to reach higher fruit, and where trees are propped up, the props could be used as supports to help workers engaged in harvesting. Litchis should always be handled with care, although the skin is fairly rough, the slightest blemish could mean a loss of commercial value.
- Pulling out of bunches during harvesting has to be avoided at all costs because broken skin at the point of attachment of the branches is particularly susceptible to a decay condition. Injury caused by rough handling contributes to fast rotting and fungus infection. No matter what harvesting method is used, care must be taken to avoid injury to produce.
- If harvested clean, the produce should be kept clean and not stacked on the soil, even momentarily. Clean the crates and equipments regularly in detergent.
- After rain, postpone harvesting for at least three hours of sunshine to dry out fruit as skin is susceptible to handling marks after rain.

Use field crates and containers

- Harvested produce may be transferred from the harvest containers into field containers, for transport to a packing house, or in some cases for transport direct to local markets.
- Wooden crates are made of boards or slats, with a capacity of up to 25 or 50 kg for manual handling. If the wood used in such

crates is rough then sharp edges or splinters may cause damage to the contents, especially if the full crates are carelessly handled. Wooden crates are usually returnable or re-usable for local marketing operations. However, rough handling methods usually limit their usage to very few occasions, sometimes as little as three to five return journeys.

- Plastic crates are now in common uses. Most are made from high density polyethylene which contains ultra-violent inhibitor (UVI), which extends the life of the container when it is exposed to sunlight. Plastic returnable crates have a longer life than wooden ones because they are more durable. They are usually made to pack inside each other when empty to save space during transport and storage.
- Sacks, plastic bags and net bags made from natural or synthetic fibres are also used where produce are transported directly from grower to market. They are relatively cheap and require little storage space and usually provide good ventilation to their contents. However, they give little protection to their contents and when made from synthetic fibres tend to disintegrate when exposed to sunlight.
- Packing bin or basket heights of 30 cm or less are recommended. The bins should be clean since soil and debris can increase rots during storage.

Know the main harvesting season of litchi in the country

Litchi distribution in India has cascading production pattern commencing from the east, and spreading to the west. Litchi production commences during the first week of May in Tripura, while litchi in Bihar matures by the third week of May and continues up to the second week of June. Litchi in Uttar Pradesh and Punjab is ready for harvest during the 2-3rd week of June, while in Himachal, it comes to harvest by the last week of June, thereby extending the total season of its availability to over two months in a year.

Take proper care of produce in the field

Once harvested, exposure to the sun and air can increase water loss by a factor of ten. Therefore harvested fruits can be kept at cooler place in the orchard or in temporary thatched house for cooling. A tarpaulin can be used to protect the fruit, but must be kept clean to prevent the build-up of pathogens. Spraying the fruit with water may help to maintain fruit quality in hot, dry weather. The transfer of fruit to the pack house soon after harvest minimizes the opportunity for water loss in the field. Transporting the fruit dry and fairly tightly packed reduces the risk of vibration damage.

Much of the required preparation and sorting of produce can be done during the harvest operation itself. Scarred, cracked, discoloured or otherwise affected fruits can be placed in separate harvest containers or discarded completely and removed from the field later. More rigorous preparation such as critical size and colour selection may be better performed in a pack house. There is a growing tendency towards the practice of packing produce directly into final marketing containers in the field, reducing the number of times the produce is handled with a corresponding reduction in the amount of injury caused. This practice however, requires that appropriate selection and packaging facilities are set up for the purpose, together with adequate supervision.

Collect the produce promptly and safely from the field

Aggregation and bringing harvested produce together for the purpose of further process is necessary. Field assembly should be planned bearing in mind the best location and the provision of basic facilities. Containers filled with produce should not be left exposed to sunlight. As with all aspects of post-harvest, the objective of assembly should be to develop a system which minimizes stress on the produce and keeps handling operations to a minimum but with maximum care and to keep the time between harvest and first destination of the produce as brief as possible. Clearly, reaching such an objective does not happen by accident, but only by carefully planning the assembly process as carefully as all other operations are planned.

To get the produce from big orchard to a collection point by the roadside may involve passing over one or more kilometers of farm roads. Growers rarely give sufficient attention to the logistics of this operation and then wonder why they get problems. In the certain areas this problem is compounded by the topography. In many cases 'heading-out' or mule are the only alternatives and crates may not be as suitable as sacks. If vehicle is used for the purpose, it should not be overloaded and it should have good suspension and with preferably larger wheels and lower tyre pressures to take out some of the jolts.

Final sorting

Product quality is maintained by removing damaged and inferior fruit during sorting. Close attention to all the fruits under good lighting is required at this stage. The entire surface of each fruit must be observed to ensure that damaged specimens are not packed. Damage extending to the aril rapidly leads to rots, which may spread to sound fruit within the package. For this reason, fruit with pulled stems, splits, cracks and insect damage should be rejected at this stage. Immature fruit and fruit showing any signs of rot are also removed during sorting. Some markets have low tolerance for cosmetic defects, such as scale infestation, small fruit, infection or superficial browning. Fruit showing these defects are generally downgraded and not sent to the central markets, but can be processed or sold at roadside stalls.

Final grading

If the fruits are graded according to their size, weight, colour and maturity, both the producer and consumer are benefited. Grading separates fruit into different grades to suit different markets. Most producers have at least two grades of fruit. Grading is normally carried out during or after sorting. Grading systems depend on market requirements, but are normally based on fruit size and colour, and the area of blemish. Export markets usually have higher standards than domestic markets, requiring uniform, unblemished fruit. There can also be differences within different sections of the domestic market.

As per fruits and vegetables grading and marking rules, 2004 in schedule, the grade designation and quality of litchi for export will be:

(i) Litchi shall be:
   (a) whole, sound, fresh in appearance.
   (b) clean, free of any visible foreign matter.
   (c) free of pests affecting the general appearance of the produce.
   (d) free of damage caused by pests.
   (e) free of abnormal external moisture excluding condensation following removal from cold storage.
   (f) free of any foreign smell and/or taste.
   (g) free of damage and abrasion.
   (h) free of brown markings.

(ii) Litchis shall have minimum equatorial diameter of 23 mm.

(iii) Litchis shall comply with the residue levels of heavy metals, pesticides and other food safety parameters as laid down by the Codex Alimentarius Commission for exports.

Table 31 : Criteria for grade designation:

Grade designation	Grade Requirements	Grade tolerances
Extra class (Equatorial diameter 33 mm)	Litchis must be of superior quality. They must have the shape, development and colouring that are typical of the variety and/or varietal type. They must be free of defects, with the exception of very slight superficial defects, provided these do not affect the general appearance of the produce, the quality, the keeping quality and presentation in the package.	5% by number or weight of litchi not satisfying the requirements for the grade, but meeting those of Class I grade or exceptionally coming within the tolerances of that grade.
Class I (Equatorial diameter 28 mm)	Litchis must be of good quality. They must be characteristic of the variety and/or commercial type. The following slight defects however may be allowed provided these do not affect the general appearance of the produce, the quality, the keeping quality and presentation in the package: - slight defects in shape, - slight defects in colouring, - slight skin defects, Provided these do not exceed a total area of 0.25 sq.cm.	10% by number or weight of litchis not satisfying the requirements of the grade, but meeting those of Class II grade or, exceptionally coming within the tolerances of that grade.
Class II (Equatorial diameter 23 mm)	This grade includes litchis which do not quality for inclusion in the higher grades, but satisfy the minimum requirements specified in general characteristics. The following defects may be allowed, provided, the litchis retain their essential characteristics as regards the quality, the keeping quality and presentation: -defects in shape. -defects in colouring. -skin blemishes provided these do not exceed a total area of 0.5 sq. cm.	10% by number or weight of Litchis not satisfying the requirements of the grade, but meeting the minimum requirements.

Litchis shall be presented under one of the following forms:

(a) *Individually*: In this case the pedicel must be cut at the first knot and the maximum length of the stalk must not extend more than 2 mm. beyond the top of the fruit. Litchi in special grade must be presented individually.

(b) *In bunches*: In this case the bunch must include more than three attached and well formed litchis. The branch must not exceed 15 cm. in length.

Pre-cooling

To remove the field heat, litchi fruit should be hydro-cooled or air-cooled. Pre-cooling requirement of a fruit is largely determined by its physiology in relation to harvest maturity and ambient temperature at harvest time. Highly perishable litchi must be pre-cooled before storage, as the ambient temperature at the time of harvest in India varies from 25-30°C with 40-45% RH. This is highly detrimental for storage of litchi fruits causing rapid desiccation. There are two methods for pre-cooling of fruits after harvest i.e. hydro-cooling and air-cooling. Cooling speed is generally greater in hydro-cooling system without dehydration than air-cooling. When hydro-cooled, the internal temperature of the fruit is reduced to 5°C within 18min, whereas air-cooling takes longer time. Hydro-cooled fruits packed in polyethylene-lines boxes remain fresh and give the best appearance after 6 days of storage at 5°C. The pre-cooling treatment, however, has also been found advantageous even for storage at relatively higher temperature. Pre-cooled litchi fruits, when treated with 0.5% lectithin and 2.5% sodium bicarbonate and packed in 0.04mm polythylene bags, retain their fresh colour upto 8 days of storage at room temperature (25°C). Hydro-cooling of litchi at 0-2°C and maintenance of a cold chain throughout its post harvest life i.e. from picking to retailers shop has been found useful. Fruit which are to be sold in local markets should not be refrigerated, since on subsequent exposure to heat or humid air rotting will quickly set in.

Packing and transport

Sorting, grading and packing are often carried out in a pack house or shed to protect workers and fruit from the heat. Where shelter is not available, operations are best located in a cool, shady area. This is more common in China and Thailand and other parts of Asia. Good hygiene in the pack house is required to avoid the spread of diseases during handling. Pathogens can build-up on packing surfaces and fruit crates. These surfaces should be washed with sanitizing agents such as chlorine every day. Water and fungicide dips also require frequent replacement or sanitizing. Waste fruit need to be regularly removed from the packing area to reduce the spread of spores. Proper packing, labeling and safe transport are to be done for good revenue.

(i) Packing

After harvesting, fruit should be packed as quickly as possible, as their quality deteriorates markedly, if they are exposed to sun even for a few hours. Litchi can be packed in panicles or as individual fruit. Fruit are often sold on panicles in Asia, whereas loose fruit are more common in Australia, Europe and North America. De-stalking is required when fruit harvested on panicles and packed individually. The first step in the packing process consists of cutting but allowing a portion of the pedicel to which fruit. Next, all the unripe, cracked or damaged fruits must be removed. Unripe fruits are low in quality and cracked or damaged fruit ferment within 1 or 2 days due to bacterial activity, thus creating favourable conditions for a fungus attack, which could then spread to healthy fruits. In Australia dipping for 2 minutes in active Benomyl (0.5 g/l) heated to 52^0C is recommended. This provides adequate protection without leaving excessive amounts of residue in the fruit. After dipping, the fruits must be air dried and cooled.

Some times darkening of fruits take place. This darkening occurs within 24 hours at ambient temperatures of 20-30^0C and is due to enzyme activity, probably an aerobic system of oxidative poly phenolase which is accelerated by drying. Two conditions, humidity and low temperatures are vital to prevent darkening and to preserve the taste qualities of the fruit. This was a well-known fact in ancient China where it was customary to spray the fruits with a solution of cold salt water and then pack them in bamboo containers sealed with clay. It is worth noting that while low temperatures help to prevent rot from developing, high humidity has a negative effect and care should be taken to ensure that humidity is not excessive.

Selection of packaging depends on market preferences and availability. The ideal package protects fruit from damage and minimizes water loss and condensation. For domestic markets litchi is usually packed in small bamboo baskets or wooden crates. Much of the fruit marketed in Asia is transported in bamboo baskets. Square baskets less than 30 cm high give good protection against injury. Circular baskets can be improved by tying strings in two or three directions across the top, with padding underneath. The outer layers of fruit in these baskets are prone to rapid water loss. This can be alleviated by lining the baskets or by covering them with a tarpaulin. Many of the larger commercial operations now consign their fruit in plastic trays or cardboard boxes, which provide better control of water loss. Some growers in Asia dip their fruit in cold water or cover their fruit with ice. However, very low temperatures can injure the fruit. Free water around the fruit as the ice melts can also increase the risk of diseases.

Proper packing of fruits is important in maintaining freshness and quality and preventing fruit decay during transit for marketing to distant places. These are lined with litchi leaves or other soft packing material as paper shavings, wood-wool, etc. *Sissum* leaves are supposed to be the best material for packing. A good box for packing fruits should be light in weight, shallow and rigid enough to protect the fruits. It should have few holes for ventilation and rope handles on either side for lifting the box. Fruits are packed in clusters along with few leaves. Plastic packaging combined with cool storage or fungicides has also been recommended. However, without good temperature control, plastic covers result in condensation and an increased risk of rots.

(ii) Labeling

Growers packing litchi for sale on the domestic markets must mark on the end panel of the box the following information:

- The name of the grower or packing establishment.
- Street/road and area, not the postal address.
- The litchi variety, class and count if packed in litchi cartons.
- Date (or date code) on which fruit was packed.
- Fruit in bulk packs must have the weight shown instead of count number.

(iii) Transport

Poor transport conditions are a major problem in India. The main limitations, including rough roads, lack of refrigeration and poor truck suspension which are out of the control of growers. However, with the advancement of transport facilities and road connectivity, now-a-days, litchi is transported by tractor or mini trucks within a radius of about 8 km of the market. In almost all the litchi growing states, the huge quantities of fruit are transported from distant producing centers mainly by road and some times by rail. For local market the harvested fruits are packed and transported by trucks. While transporting litchi, following care should be taken.

- Packing between vehicles and baskets with soft straw or leaves can help cushion the blow but it should be clean. Shelves can be used in the truck to avoid stacking of baskets for reducing damage to fruit. Padding and strapping the baskets can restrict movement during transport. Vehicles should be driven slowly. Do not ride on top of the produce.
- Fruit are often damaged when baskets are overfilled, dropped, stacked or packed on their sides. Vehicle should be shaded from both sun and rain.
- Fruit can be protected by covering the baskets with a clean tarpaulin or similar material. However, it should not prevent ventilation of the produce.
- Often the above points are overlooked because the operators argue that the produce does not have far to travel and the produce is not expected to be on the vehicle for long. Unfortunately, drivers do have the habit of making unscheduled stops during their journeys. They may stop due to road blockage or due to personal work and often the Van is parked in the full sun.
- Exposure to warm air can dry out the fruit very quickly, so transport during the warmer part of the day is best avoided, if possible.
- For better quality fruits of litchi cool chain should be maintained.

  The major concern to traders and growers is the exorbitantly high rates charged by truck operators for transporting litchi. Taking advantage of the very short shelf life of litchi, truck operators are demanding double the normal freight charge. The railways also provide facilities for transporting litchi by attaching bogies on a daily basis. However, the desired impact of this initiative is yet to be felt at the field level.

Storage and marketing

(i) Storage

Storage is essential for extending the consumption period of fruits, regulating their supply to the market and also for transportation to long distances. Factors affecting storage life are, pest and disease infection, pressure damage caused by over filling packages, package failure due to over stacking (collapse of bottom layers) and moist cartons losing compression strength when stacked. Litchi fruit cannot be kept for more than a few days after harvest at room temperature. If marketing is delayed, fruits should be kept in cold storage to avoid rapid loss of colour and quality. Pre-cooling of fruits should be done before cold storage because without pre-cooling, browning could be more rapid compared to the storage of pre-cooled fruits at 2-5^{0}C. It brings not only appreciable reduction in the rate of aeging and water loss in the fruits but also makes litchi hardy on prolonged cold storage.

The cold storage without pre-cooling and chemical treatment has less impact on colour maintenance. Pre-cooling as well as chemical treatment has recorded shelf life upto 27 days at 5^{0}C. Fruits treated with 2% sodium hypochlorite can be stored satisfactorily in perforated polythene bags at 0-3^{0}C for 25 days. For short-term storage less than two weeks, a temperature of 7°C is satisfactory. A relative humidity should be kept at 90-95% throughout storage and transport. Controlled atmosphere storage (3-5% O_2 and 3-5% CO_2) reduces skin browning and slows down the losses of ascorbic acid, acidity, and soluble solids. Exposure to oxygen levels below 1% and/or carbon dioxide levels above 15% may induce off-flavors and dull gray appearance of the pulp.

(ii) Marketing

Pre harvest contract in the most prevailed method of litchi marketing as most of the of growers prefer it; however, some of them undertake self-marketing. Growers, who do not intend to market their litchi harvest, give litchi orchards to Pre-Harvest Contract (PHC) during the month of January. PHC sells the orchard to the wholesale or commission agent, who undertakes the harvesting and packing, in addition to transporting the produce to the market. On the other hand, self-marketing in the distant domestic market is more profitable with higher net gain but is also costly due to high cost of marketing risk. Absence of price information and poor leadership are the constraints in self-marketing. Big orchards, high marketing expenditure, long distance to main market centres, remote production areas, absence of market regulation and exploitative marketing practices by market intermediaries are other important factors impeding self-marketing. The major marketing difficulty with litchi involves preserving its attractive red colour.

The potential of litchi in India is unexploited so far. High price disparities exist between the returns that the producers get and the consumer pay. The Pre-Harvest Contractor or the commission agent makes the maximum margin in litchi marketing, as he only performs a transfer function without involving any other cost. The stockiest in litchi sale adopt the undercover system and realize higher margins.

Many litchi growers sell the standing crop to pre-harvest contractors at flowering stage for a year or more. In certain cases, the crop is leased out to pre harvest contractors for 2-3 years. The crop contractors negotiate and settle the price to be paid and the terms and conditions of payment directly with the growers. Most of the produce is sold through this mode. The pre-harvest contractors have a clear picture of the yield potential of the orchard as a whole and performance level of individual trees in the orchard. The price offered on a per tree basis varies according to the age category. The farmers receive 50 per cent of the settled price in advance (at the point of firming up deal) and the rest at the time of harvest. Harvesting, sorting, packaging are done in the farms by the contractor. Loading the truck for transport to distant cities is done at the farm gate itself.

The pre-harvest contract system prevailing in the country has an impact on the health and life of the litchi orchards. In cases where the contract is for 2 to 3 years, he works with a profit motive in mind and does not care much for the upkeep of the orchard. Contractors are not keen on investing in the orchard as they are not sure of continuing the contract during the next term. Most of such orchards, belong to absentee land lords and suffer in the long run. Further, if the orchard owner is not present, harvesters may damage the twigs and branches while harvesting causing harm in the long run.

Pre-harvest contractors market maximum of litchi. Most of the pre-harvest contractors are commission agents of wholesale merchants in cities like Mumbai, Delhi, etc. Some of them are financed by wholesale fruit merchants and are obliged to sell the fruits to their financiers. The pre-harvest contractors directly supply the commodity to the local processing units and export houses located in the region. Some of the established processing units and export houses in the region directly purchase the orchards from growers by making advance payments. The commission agents and wholesale merchants play a greater role in the distribution of the produce after purchasing it from the pre-harvest contractors. They forward the produce to important cities in the country. At the terminal market, agents who operate on commission basis arrange to sell the produce. The daily arrivals at terminal markets are sold either by auction or through negotiations.

Only the best quality fruits are selected, packed and forwarded for marketing from the interstate trade while the slightly inferior quality produce is sold in markets within the state. The fruits are transported and handled through out the marketing chain at normal temperature conditions i.e. without pre cooling. The market reach is limited to places/cities which could be covered within 36 hours of harvest since the fruits loose their attractive red colour after 48 hours, rendering the fruits unmarketable. Fruits are graded and packed in the farm itself. Two types of packs are used. Traditionally 16 to 18 Kg wooden boxes were used. However, drying up of sisam, the raw material for wooden boxes and the prohibition on cutting trees under forest law has forced the traders to go for 15 kg cardboard cartons. The major chunk of the produce is transported through trucks. The standard practice is to load 500 boxes per truck irrespective of the weight of individual boxes. Thus the weight of the fruits transported per truck varies from 7.5 T (cardboard carton) and 8 to 9 T (wooden box). The fruits are loaded to trucks at the farm gate or at points near to the farms. Delhi, Mumbai, cities of Punjab, Mumbai and UP cities like Kanpur and Lucknow, Kharakpur are the major destinations for litchi. 80% of the litchi marketed and sold through interstate trade.

Presently, pre cooling and cool chain transportation are not adopted for want of cool chain at the terminal markets. Though efforts have been made for cool chain linked interstate trade of litchi, the volume has remained very small.

(iii) Price structure

Like many other fruit crops, minimum support price for litchi is not fixed therefore; there are no official records or publication showing the prices of litchi. Growers do not maintain any record of their produce and price and the same happens with the contractors. The commission agents and wholesale merchants do keep accounts of their business and their records provide some information regarding the quality and the variety procured and sold by them. Litchi prices vary a great deal from year to year, depending upon each year's total production and various other factors like demand, transport and marketing facilities. Daily arrivals have a direct impact on the prices. Thus, there is a seasonal and annual variation in prices. Ordinarily, the prices are high at the commencement of the season, declining gradually as the supplies increase. Later on, when the arrivals decrease, a high level price is again seen before the close of the season. There is no uniform pattern for price quotations. The unit of sale, both in wholesale and retail trade, varies from place to place. However, to get good price one should make good quality of litchi because quality is not a chance but it comes through year round efforts.

❑❑❑

Chapter 20

Reduce Post Harvest Losses and Convert into Value Added Products

Litchi is non climacteric fruit which is harvested at a mature stage. Post harvest practices are devised to reduce the rate of losses associated with post harvest life. Litchi is delicate and highly perishable, with a short storage life, hence, fruit should be sent to consumers on the day of harvest to avoid physiological disorders for maximum fruit quality. Insects, diseases, disorders, nutrition deficiencies, injuries due to transit and storage environment may shorten the post harvest life of litchi. All these have to be controlled by proper pre and post harvest treatment and management to give litchi long shelf life without any loss of nutrients, shape, colour, size and other things. Post-harvest diseases occur due to heat produced due to accelerated rate of respiration, high temperature in transport containers and storage place. Post harvest technologies involve cooling and heating the fruit, use of various packages and packaging materials and the application of fungicides and other chemicals. The objective of post harvest management should also include the preservation of litchi in the form of various products. The litchi post harvest life depend on post harvest practices as well as on the protocol followed in the field, hence the quality of litchi is affected by:

1. Pre-harvest practices followed by growers to ensure healthy harvest.
2. Handling conditions from field to pack house or transport life which determines the storage life (this phase occupy 1-2 days).
3. Methods adopted to increase the shelf life or period in which fruit remain in an edible condition (this phase occupy 2-3 days).
4. Methods adopted to convert litchi into processed products for long shelf life.

(1) Pre-harvest practices followed by growers to ensure healthy harvest life

(i) Healthy production protocol

The quality of the fruit at harvest determines subsequent shelf life and market performance. Blemishes at harvest are only magnified during handling and marketing. Trees require a good supply of nutrients and water to produce sound fruit. Fruit splitting, for example, has been correlated with low concentrations of calcium in the pericarp and uneven watering during fruit development. Insect control is also important as fruit damaged prior to harvest deteriorate rapidly.

Sensible orchard hygiene needs to be practiced to reduce the risk of rots during storage. Trees pruned to open canopies are better ventilated than non-pruned trees and provide a less favourable environment for rots, while skirting lowers the risk of infection from the soil. Removing dead wood from the orchard eliminates a source of pathogens. Field sprays with fungicides can also reduce post-harvest diseases in some circumstances. A range of farm machinery can be used for general orchard management, including fertigation, irrigation, spraying and pruning. Trees can also be netted to protect the fruit from birds, bats and the large insects. There is, as yet, no accepted mechanized means of harvesting fruit. Cherry pickers and other elevated picking platforms, along with ladders are generally used.

Some times due to over-ripening on trees, they have increased susceptibility to develop diseases. Similarly internal breakdown in ripe fruit results in un marketable fruits. Presence of insect inside the fruit reduces the reliability and profit of litchi growers. Pathogenic decay of fruits and processed products and pest infestation are the important factors affecting the post-harvest quality of litchi.

(ii) Covering litchi fruit clusters

It is a common practice in some areas where litchi is grown for export. This is particularly necessary in areas with a high incidence of insect attack, particularly fruit sucking moth and litchi borer moths. Since chemical control is not very effective, the grower is obliged to pick the fruit early to protect them from damage by these pests. Since, fruit development virtually ceases after harvest, the quality of early-picked fruit impaired by an unsatisfactory sugar / acidity ratio and a somewhat dull colour. Covering the fruit prevents damage from litchi moths, as well as sunburn and cracking. The bags even keep fruit bats and birds away. Another advantage is that the fruit is handled less during harvesting and this limits damage caused by rough handling. Covering the fruit also extends the harvest period because fruit can be left on the tree for a longer period. The fruit develops a very attractive red colour inside the paper bags.

Paper bags are at present the most suitable and also the cheapest covering material. Other bags are usually made either of raw cotton or moisture-proof paper. They measure 30cm x 45cm and can hold about 25 fruits. Once the bags a put in place, they are not opened until the harvest.

The best time for covering the clusters varies from locality to locality but it should be done approximately 1 ½ to 2 months prior to harvesting. Before the bags are put in place, most of the leaves around the clusters have to be removed, otherwise they will fall into the bags and rot. The paper bag at present used is open at both ends. The fruit must, however, be covered in the correct way and during the right time of the day (not early in the morning or immediately after rain), because of the risk of decay when covered while still wet. Spray against litchi moth may be done just before covering. The bags must be fixed to hang to an angle to allow water to flow out Both sides of the bag must be closed and, if necessary, a small opening must be left for rainwater to run out. The open ends can easily and quickly be stapled together. When removed carefully, the bags can be used for another season.

The paper, which is cheaper, does not withstand more than one harvest. It is not convenient, for economic reasons, to cover all the fruits on a tree with bags, particularly when they are dispersed. Despite the fact that this practice presents certain drawbacks, such as the cost of labour, which is needed to put the bags in place, and material, the advantages outweigh the disadvantages. Fruit quality is higher, colour and size are better and less labour is required for harvesting. Where several bunches are placed in each bag, picking takes about one-fifth of the normal time. This practice also protects the fruit from hail damage.

(iii) Preventing sunburn and fruit cracking

Sun burn and fruit cracking in litchi are the two disorders, which can be serious, resulting in a great loss to litchi growers. Therefore, these needs to be manages as discussed in this book.

(2) Handling conditions from field to pack house or transport life

As explained in previous chapter, the proper care and appropriate methods should be adopted right from harvesting to storage to ensure better shelf life of litchi. This includes harvesting at right stage with a proper planning of disposal. The workers must be properly trained to decide proper time of the day for harvesting with proper picking techniques. The use of field crates and containers to collect the produce promptly and safely from the field to sorting, grading and packing house are useful in post harvest life of litchi. The transport technology is another aspect which affects the litchi life.

(3) Methods adopted to increase the edible life of fruit

(i) *Reducing the fruit browning*: Pericarp browning is the first visual sign of fruit decline. Browning that occurs during the first few days after harvest is usually caused by dehydration of the pericarp. This problem has been reported in several producing areas and it can be managed as mentioned elsewhere in this book.

(ii) *Retaining attraction colour*: Fruit colour is considered to be a reliable index of freshness and quality of the litchi fruits. It plays a vital role in trade, as there is assured market for red fruits. The red colour of the litchi fruits is due to the presence of cyaniding-3-glucoside, cyaniding-3-glactoside, pelargonidin-3-glucoside, plargonidin-3-5-diglucoside and some other flavonoids. The browning of the pericarp is due to degradation of anthocyanin by polyphenol oxidases (PPO) and peroxidase. The PPOs comprise a large group of enzymes, all of which are characterized by their ability to utilize molecular oxygen during oxidation of phenolic substrates. Both the levels of PPO substrates contributing to browning. To slow down the browning process some post harvest treatment is required immediately after grading of the harvested fruits for this sulphuring may be done.

Fumigation of litchi fruits with 1.2% SO_2 has been reported to inhibit PPO activity. For the process, fruits are placed in a closed chamber where 50-150g sulphur / cubic meter of air is burnt for 20-30 minutes. To fumigate one tone of fruits for 20 min., 625g of sulphur is required. There is increase in the sulphur content of the pericarp from 0.046% to 0.076%, when SO_2 fumigation is done for 30 min. but there is no change in the sulphur content of the aril and its eating quality. The tolerance limit of residue in the pulp at the time of marketing has been fixed at 10 ppm. If SO_2 concentration exceeds 500 ppm, it gives disagreeable flavor. The problem of disagreeable flavour has been resolved by placing two slow releasing paper sheets impregnated with sodium metabisulphite on the outer edge of the polythene bags used for packing. Use of SO_2, bleaches the natural red colour and reduces the pH of the rind. Experiments have revealed that treatment of SO_2 fumigation followed by acid dip is effective in retaining the red colour of the pericarp. Dipping litchi fruits in a "zero" pH solution after blanching in boiling water at 98^0C for 30 sec is most effective in retaining the red colour of the pericarp up to 5 weeks, when stored at 1^0C however, internal browning of the flesh is observed. Using steam instead of boiling water have been reported to prevent internal browning. The fruits are treated with steam for 2 sec. (95^0C), followed by immersion for 4 min. in a "zero" pH solution and then dipped in 1% vapourguard solution. The fruits remain firm with no internal browning and are in acceptable condition after 28 days of storage at 1^0C. However, these practices needs to standardized in the context of situation of Indian farmers.

iii) *Controlling the fruit rot*: Fruit rot is another post harvest problem which reduces the fruit shelf life. This should be managed as mentioned elsewhere in this book.

iv) *Pest risk reduction in harvested fruits*: Insect disinfestation has become increasingly important for the expansion of export markets because of quarantine issues associated with some insect species. Thus, effective disinfestation protocols need to be developed. Heat treatment has shown promise as a quarantine technology, but it injures pericarp tissue and results in skin browning. Some export markets require disinfestation of fruit for insect pests. For example, marketing of litchi from Australia to Japan and the USA is limited because these countries consider litchi to be a host of fruit flies in Australia. There are several methods available to kill the insects, but fruit are often damaged in the process. There are also health concerns regarding treatments such as ethylene dibromide and gamma irradiation. Further research is required to develop a safe and effective strategy for these pests.

(4) Methods adopted to convert litchi into processed products for long shelf life

Very negligible quantity of the litchi fruits at present is being utilized for preservation and processing to get the different products out of the season. Due to short harvesting season, tenderness and perishable nature of fruits, large quantities go waste. Indians are fond of table enrichers, which are regularly used along with the main course as well as snacks. Apart from individual households, restaurants, eateries, roadside dhabas, clubs, hostels, caterers etc. are the bulk consumers. There are some branded litchi products. The products like canned litchi and litchi squash are available in the Indian markets but they are costly. Some Indian households make these items during the season. But this practice is gradually disappearing due to changing lifestyles, hassles of making these items and their availability throughout the year from market. There are many variants of these products and with certain change in the ingredients and taste. Litchi processing or preservation is a process in which decay or spoilage of fruits is prevented allowing it to be stored in a fit condition for future use. Every region has its own taste or liking and a care has to be taken to understand it and accordingly the recipe has to be finalized. Compliance with FPO and PFA Act is necessary, if processing is done for the market.

Basic inputs for small processing of litchi

There is no need to undertake construction by purchasing land. Instead, a readymade premise of around 100 sqm with 2 or 3 rooms would be sufficient for production, storage and packing. Majority of the operations are manual and therefore, production capacity is primarily determined by market. Processing would require facilities such as, gas furnace with burners, cutting and peeling equipments, plastic sealing machine, cap sealer, weighing scales etc., stainless steel utensils, plastic jars and tubs and other tools, bottle washing machine, mixer grinder, miscellaneous assets: furniture and fixtures, storage racks, packing tables. The products like canned litchi, squash, cordial, syrup, R.T.S. (ready to serve), jam, jellies, juice and dried or dehydrated products (nuts) can be manufactured.

Value added products

(i) Freezing of litchi

Freezing of whole fruits is the best and most easy method of preserving litchi with its natural flavour and quality for a long time. The fruits can be frozen in syrup with or without stone after peeling. Litchi fruits remain in excellent condition for 12 months when rapidly cooled and kept at -25°C.

(ii) Litchi canning

The canned litchi is of excellent quality and in great demand both in India and abroad. Fully mature ripe fruits of Shahi, Early Large Red, Early Seedless, Rose Scented Purbi or Bedana are selected. After thorough washing, the peels are removed and pulp (aril) destoned with the help of knives. Requisite quantity of pulp (aril) is filled in empty cans to which sugar syrup of 30-35^0 Brix is added. Sugar syrup is mixed with 0.2% citric acid and flavoured with rose or vanilla essence. Exhausting of filled can is completed when the temperature in the centre of the can reaches to 85^{0}C and this temperature is maintained for five minutes. The upper lids are then closed with the help of double seamer. These sealed cans are sterilized or processed in boiling water for 30 minutes. After sterilization, the cans are immediately cooled in running water. After cooling the cans are wiped out with dry cloth and stored in dry and cool place. Pink discoloration after 3 months of storage is the main problem which increases gradually and diminish the consumers appeal. This is possibly associated with formation of a tin anthocyanin complex.

(iii) Juice preservation

Litchi juice extracted with the help of basket press is preserved by 700 ppm SO_2. Juice is first pasteurized at 85^{0}C for 15-20 minutes. During pasteurization 0.2% citric acid dissolved in a little quantity of water may be added. After cooling 0.12 per cent potassium meta-bi-sulphate dissolved in little quantity of water is thoroughly mixed. Juice is then filled in sterilized bottles and after capping stored in cool and dry place. Such preserved juice may be used in preparation of various beverages i.e. squash, nectar, R.T.S., cordial and syrup etc.

Juice as such added with sugar is also a very refreshing drink. In its preparation juice may be mixed with 15% sugar and heated to 85^0C for 15-20 minutes. The hot juice is immediately filled in the bottles, crown corked and again filled bottles are sterilized at 85^0C for 20-30 minutes by keeping them in water. A layer of cloth is placed at the bottom of the bottle to prevent breakage during heating.

Another method of litchi pulp preservation is by heating at 85^0C and then adding 500 ml/l SO_2 and 1% citric acid and filling in glass bottles which are sealed with corks and made air tight by dipping in molten paraffin wax. The pulp remains acceptable organolepitcally for 6 months at room temperature (25^0-35^0C) and upto 12 months at low temperature (4-5°C). Brownish discolouration is commonly observed when juice is preserved at room temperature which can be prevented by keeping bottles of juice at low (4-5°C) temperature.

(iv) Squash

Squash is prepared by blending the juice with sugar, citric acid and water and preserved chemically with 350 ppm SO_2. As per F.P.O. specification the squash must contain 25 per cent juice and at least 40^0 Brix (T.S.S.). The recipes contains juice-1 kg, sugar-1.5 kg, citric acid-8gm, water-0.750 litre mol, potassium meta-bi-sulphite-0.06%. First of all sugar syrup is prepared by dissolving sugar in water by heat application. During heating, citric acid dissolved in little quantity of water is mixed with syrup. The syrup is then allowed to cool and after straining, the sugar syrup is mixed with juice. Potassium meta-bi-sulphate dissolved in little quantity of water is mixed with prepared squash. The squash is served after diluting three times with water. Another recipe for the litchi squash is juice - 1litre, sugar -1.25 kg, water -0.75 litre, citric acid-30g and potassium meta-bi-sulphate-11.8g.

Mixed squash can be prepared from the fruit juices of orange, lemon, mango, litchi, aonla, carrot, raspberry, pineapple etc. The recipe for mixed squash contains mango juice- 0.5 litre, litchi juice-0.5 litre, lemon juice-0.5 litre, sugar - 2.5 kg, water-2 litre, citric acid - 45g and potassium meta-bi-sulphate-3.25g.

Indian Agricultural Research Institute, New Delhi, has provided recipe for litchi squash is as preserved juice-1 kg, sugar-1 kg, water -400ml, citric acid-8g, potassium meta-bi-sulphate-1 g and rose essence-20 drops (0.8 ml).

Calcuttia and Desi cultivars have been found superior in comparison to other cultivars and maintained its organoleptic quality for a longer period as squash. The browning of litchi squash which is usually observed at room temperature storage (25-30^{0}C) after 3 months can be reduced by low temperatures storage (4-5^{0}C) and addition of 100 mg / litre ascorbic acid in the squash.

(v) Litchi syrup

A fruit syrup is the sweetened juice of a fruit usually with a high concentration of sugar and containing a small quantity of fine pulp of the fruit usually with a low acid content. Syrup does not spoil due to high sugar content. Litchi juice can also be utilized in preparation of fruit syrup. There are many recipes for litchi fruit syrup. One recipe includes juice - 1 litre, sugar - 1.5 kg, citric acid- 10.4g and essence-½ ml and another consist of juice - 1 litre, sugar - 7 kg, water-2litre, citric acid - 28 to 56g and essence-3-12 ml.

Juice sugar and citric acid are made into syrup by applying heat. The scum, if any, is removed and syrup is strained and cooled. In second recipe, sugar is first dissolved in water by applying heat. Scum is removed, strained and cooled. Juice is then mixed with sugar syrup. After adding required quantity of vanilla or rose essence, the syrup is filled in sterilized bottles and capped. After proper dilution with water, it is served as soft drink.

Cordial

Cordial is a clear, sparkling, sweetened juice from which the pulp and other suspended materials are completely eliminated. It is mostly used after duration with certain alcoholic drink. For preparation of litchi cordial, first of all juice is extracted and filtered. Filtered juice is stored in large sized glass container. This is preserved by adding 1.2g potassium metabisulphite per litre of juice after dissolving in a little water. The preserved juice is allowed to stand for at least 10 to 15 days. Within the period, suspended matter will settle down. The supernatant clear juice is siphoned out and used in cordial making. There are three recipes which may be adopted in the preparation of cordial. The first contains juice - 1 kg, sugar - 1.25 kg, water - 1 litre, potassium meta-bi-sulphate-2 g and vanilla or rose essence -1/2ml. The second recipe has juice-1 kg, sugar-2.25 kg, water -200ml, potassium meta-bi-sulphate-2.5 g and vanilla or rose essence-1.25ml. The third recipe contains 1 litre litchi juice, 1.5 kg sugar, 600 ml water, 2.2 g potassium metabisulphite, 30 gms citric acid. It is prepared by dissolving sugar in requisite quantity of water. All other ingredients are added according to the procedures followed in squash making. The prepared cordial is strained through a fine muslin cloth and then bottled.

Jam

Litchi jam is prepared from the fruit pulp. The pulp is taken out and mixed with $1/4^{th}$ quantity of water and then heated to make it soft. It is then cooked with $3/4^{th}$ quantity of sugar. During cooking when temperature reaches to 102^0C, citric acid @ 2g per kg dissolved in little quantity of water is added. The cooking is stopped when the temperatures reaches 106^0C. Before filling in the wide mouth glass jar, sodium benzoate @ 0.02 per cent dissolved in little quantity of water is mixed with jam. A layer of molten paraffin is spread on the top of the jam filled in the jar.

Litchi serbet

Litchi lends a refreshing flavor to this simple sherbet. The ingredients are peeled and seeded 24 litchi, $1^{1/4}$ packet unflavored gelatin -1/4 cup, cold water -2/3 cup, milk -1/2 cup, sugar -1 cup, and 1 teaspoon lemon juice. Litchi fruit are placed in a blender or food processor and pulsed until finely chopped. Two layers of cheese cloth over a bowl are placed. Chopped litchi is poured into cheese cloth. Litchi juice is squeeze out to fill 1 cup. Solids are discarded. In a medium bowl, unflavored gelatin is sprinkled over the cold water. It is left for 5 minutes to bloom. 1/3 cup of the milk is poured into a small saucepan. Heating is done until small bubbles begin to form around the edge of the pan. The pan is removed from heat and scalded milk is added to the gelatin water. It is stirred vigorously until gelatin is dissolved. Then sugar is added and stirred until no granules remain, then remaining 1/3 cup milk, half-and-half, litchi juice, and lemon juice are added. It is mixed well with the help of manual or electric ice cream maker.

Litchi drying for nut

Litchis are also marketed as the traditionally known "litchi nuts", and canned, in syrup. In these forms, they can be distributed without refrigeration. To obtain a quality product, the drying operation must begin with "washing" the fruit well in boiling water or steam at atmospheric pressure for 15 minutes. This is followed by a sulphur dioxide treatment similar to that applied to fresh fruit to obtain a good colour. If for any reason the colour has not returned after treatment, rapid dipping in a weak solution of hydrochloric acid is recommended. The fruit are then washed in clean water. Actual drying should begin in the sun, which will help convey a more natural colour to the final product. Care must be taken to turn the litchis from time to time so that the colour will be uniform over the entire surface. Drying may continue in the sun or in an oven at 26-27^{0}C until the pulp contains only approximately 30 percent of its initial water content. To prevent the skin and pulp from shrinking, care should be taken to avoid rapid drying in the first 6-12 hours. It is also advisable to begin drying in the shade for the first two hours after treatment with SO_2. This reduces the risk of skin-cracking.

Fig. 46 : Litchi Nuts

Jelly

Jelly can also be prepared from the mature litchi fruits. The juice is first extracted and the quantity of pectin is tested by alcohol test. As per pectin strength the required quantity of sugar is added and cooked. Citric acid @ 7g per kg of sugar is mixed with little quanity of water and added while cooking. The end point reaches at 105^0 to 107^{0}C. Then cooking is stopped and after removing the scum, product is filled in wide mouth glass jar. Excessive sweetness and tartness, singly or combined masks natural flavours. Sweetness is increased by heating sucrose in acid solution or by higher sugar concentration. Good jelly formation requires a high ratio of juice to sugar. Litchi jelly made with juice of 18% soluble solids and combined with sugar in a 55.45 ratio exhibits a strong natural flavor and high quality. However, because of proportionately high content of litchi juice, the expense of manufacturing is high. A good litchi jelly may be prepared with the juice of 16% or more soluble solids and combined by weight with sugar in a 50: 50 ratio. The juice pH should be adjusted with phosphoric acid to pH 3.7 and subsequently to pH 3.2 with citric acid. The addition of 1% commercial slow set pectin to litchi jelly was found sufficient for good gel formation.

Litchi salad recipe

Litchi salad is prepared by tossing litchis with potatoes, peas, cucumber, macaroni, carrots and cheese with lemon. The ingradients are: 1 cup peeled litchis, 1 cup boiled and quartered potatoes, 1 cup boiled peas, 1 cup quartered cucumbers,1 cup boiled macaroni, 1 cup boiled and quartered carrots, 2 tblsp chopped fresh green coriander,1/2 cup diced cheese, 1 cup quartered tomato, capsicum 1/2 cup, juice of three lemons, salt and pepper. All ingredients are mixed together and tossed in the lemon dressing. It is chilled before adding the dressing.

Litchi souffle

The ingredients of souffle include 100 ml milk, 5 tblsp sugar, 3 egg yolks, 3 egg whites, 2 tsp gelatin, 3 tsp water, 4 tblsp cream, litchi pulp 1/2 cup and juice of 1 lemon. Milk is heated and then removed from fire. Sugar is added and beaten egg yolks in the lukewarm milk and cooked on slow fire till it reaches custard like consistency and the cooled. Gelatin is dissolved in warm water and added to cold custard. Egg whites is beaten till stiff and folded into the cold custard. Beaten cream and litchi pulp, lemon juice mixture are mixed and kept in a pre-cooled mould and chilled in refrigerator till set.

Litchi based alcoholic drinks such as wine are also prepared.

❑❑❑

Chapter 21

References

Anonymous, (2000). Taiwan Agricultural Yearbook. Department of Agriculture and Forestry, Taiwan Provincial Council.

Batten, D.J. (1986). Towards an understanding of reproductive failure in lychee (*Litchi chinensis* Sonn.). *Acta Horticulturae.* **175:** 79-83.

Batten, D.J. and McConchie, C.A. (1995). Floral induction in growing buds of lychee (*Litchi chinensis*) and mango (*Mangifera indica*). *Australian Journal of Plant Physiology.* **22:** 783-91.

Batten, D.J., McConchie, C.A. and Lloyd, J. (1994). Effects of soil water deficit on gas exchange characteristics and water relations of orchard lychee (*Litchi chinensis* Sonn.) trees. *Tree Physiology.* **14:** 1177-89.

Bradley, J.D. (1953). Some important species of the genus *Cryptophlebia* Walsingham,1899, with descriptions of three new species (Lepidoptera: Olethreutidae). *Bulletin of Entomological Research.* **43:** 679-89.

Bradley, J.D. (1986). Identity of the South-East Asian cocoa moth, *Conopomorpha cramerella* (Snellen) (Lepidoptera:Gracillariidae), with descriptions of three allied new species. *Bulletin of Entomological Research* **76:** 41-51.

Butani, D.K. (1977). Pests of litchi in India and their control. *Fruits.* **32:** 269-73.

Chadha, K.L. (1968). Litchi cultivation in India. *Indian Hort.,* **12:** 13-16.

Chaikiattiyos, S., Menzel, C.M. and Rasmussen, T.S. (1994). Floral induction in tropical fruit trees: effects of temperature and water supply. *Journal of Horticultural Science.* **69:** 397-415.

Chauhan, K.S. (2001).Litchi. Botany, production and utilization. Kalyani publishers, New Delhi, pp 227.

Chen, H. and H. Huang. (2000). China litchi industry : Development, achievements and problems. Abst. 1st Int. Symposium on litchi and longan. Guangzhou, China, p. 18.

Dass, C.S. and K.R. Choudhary. (1958). Floral biology of Litchi (*Litchi chinesis Sonn.*). South Indian Hort., **6:** 17-22.

De Villiers, E.A. (1992). Fruit fly. In *The Cultivation of Litchis.* Bulletin of the Agricultural Research Council of South Africa. **425:** 56-8.

Drew, H. (1999). Pepper spot - a new disease affecting lychees in Australia. *Proceedings of the Fifth National Lychee Conference,* Twin Waters pp. 21-3.

Ghosh, B., Mitra, S.K. and D. Sanyal. (2000). Litchi cultivars of West Bengal, India. Abst. 1st Int. Symposium on litchi and longan. Guangzhou, China, p. 25.

Ghosh, S.P. (2000). World trade in litchi : Past, present and future. Abst. 1st Int. Symposium on litchi and longan. Guangzhou, China, p. 16.

Ha Minh Trung, (2000). Lychee Production in Viet Nam - prospects and problems. Proc. 5th National Lychee Conf., Twin Waters, Australia, pp. 83-87.

Huang, H.B. (2001). Towards a better insight into the development of the arillate fruit of litchi and longan. *Acta Horticulturae.* **558:** 185-92.

Huang, X.M., Li, J.G., Wang, H.C., Huang, H.B. and Gao, F.F. (2001). The relationship between fruit cracking and calcium in litchi pericarp. *Acta Horticulturae.* **558:** 209-15.

Kotur, S.C. and H.P. Singh, (1993). Leaf Sampling technique in Litchi (*Litchi chinesis Sonn) Indian J. Agril. Sci.,* **63:** 632-638.

Kotur, S.C. and H.P. Singh, (1994). Varietal differences in leaf nutrient composition of litchi (*Litchi chinesis Sonn*). *Indian J. Hort.,* **51:** 59-62.

Kumar, K.K. and Kumar, A. Integrated management of pest complex in litchi. Extension folder. NRC for litchi, Muzaffarpur, Bihar.

Kumar, K.K. Kumar, A., Kumar,R. and Pandey, S.D. (2005). Litchi- Samekit Keet Prabandh. Technical Bulletin, NRC for litchi, Muzaffarpur, Bihar.

Kunwar, R. and R. Singh, (1993). Note on nutritional survey of litchi orchards in Doon Valley of Garhwal Hills of Uttar Pradesh. *Prog. Hort.,* **25:** 164-5.

Li, D. and Wu, X. (2001). Toxicity of four fungicides for controlling *Peronophthora litchii. Acta Horticulturae.* **558:** 435-7.

Li, J.G., Huang, H.B., Gao, F.F., Huang, X.M. and Wang, H.C. (2001). An overview of litchi fruit cracking. *Acta Horticulturae.* **558:** 205-7.

Li, J., Huang, H., Gao, F., Huang, X. and H. Wang, (2000). An overview of the research in litchi fruit cracking. Abst. 1st Int. Symposium on litchi and longan. Guangzhou, China, p. 46.

Liu, X.D. and Lai, C.Q. (1998). Experiment on control of litchi stink bug by using *Anastatus japonicus* Ashmead. *South China Fruits* **27:** 31.

M.Rai.V.Nath,P.Dey and B.Das, (2002). Litchi Ki Vaigayanik Kheti. Extension Bulletin. HARP, Ranchi, Jharkhand. NRC for litchi, Muzaffarpur, Bihar.

M.Rai, V.Nath. S. Kumar, P. Dey, B. Das, R. Kumar and A. Kumar, (2002). Litchi Baag Prabandh. Technical Bulletin. NRC for litchi, Muzaffarpur, Bihar.

Mallik, P.C. and D.L. Singh, (1965). Hunger sign in litchi. *Indian Agriculturist;* **91:** 127-138.

McAlpine, J.R., Keig, G. and Falls, R. (1983). *Climate of Papua New Guinea.* Commonwealth Scientific Industrial Research Organization, Canberra pp. 89-101.

Menzel, C. (2000). The physiology of growth and cropping in lychee. Abst. 1st Int. Symposium on litchi and longan. Guangzhou, China, p. 37.

Menzel, C. and D.R. Simpson, (1994). Lychee. In The Handbook of Environmental Physiology of Fruit Crops. Vol. II. Subtropical and Tropical. (B. Schaffer and P.C. Anderson, Eds.). CRC Press, Boca Raton, Florida, USA, pp. 123-141.

Menzel, C. and D.R. Simpson, (1989). The lychee nutrition story. Proc. Second Nat. Lychee Seminar, Cairns, Australia, pp. 49-58.

Menzel, C.M. (1983). The control of floral initiation in lychee: a review. *Scientia Horticulturae.* **21:** 201-15.

Menzel, C.M. (1984). The pattern and control of reproductive development in lychee: a review. *Scientia Horticulturae.* **22:** 333-45.

Menzel, C.M. (1991). Litchi. In *Plant Resources of South-East Asia Vol. 2. Edible Fruit and Nuts* (E. W. M. Verheij and R. E. Coronel, Editors). Pudoc, Wageningen, The Netherlands pp. 191-5.

Menzel, C.M. and Simpson, D.R. (1987). Lychee nutrition: a review. *Scientia Horticulturae.* **31:** 195-224.

Menzel, C.M. and Simpson, D.R. (1988). Effect of temperature on growth and flowering of litchi (*Litchi chinensis* Sonn.) cultivars. *Journal of Horticultural Science.* **63:** 347-58.

Menzel, C.M. and Simpson, D.R. (1992). Flowering and fruit set in lychee (*Litchi chinensis* Sonn.) in sub-tropical Queensland. *Australian Journal of Experimental Agriculture.* **32:** 105-111.

Menzel, C.M. and Simpson, D.R. (1994). Lychee. In *The Handbook of Environmental Physiology of Fruit Crops Vol. II. Sub-tropical and Tropical.* (B. Schaffer and P. C. Andersen, Editors). CRC Press, Boca Raton, Florida USA pp. 123-41.

Menzel, C.M. and Simpson, D.R. (1995). Temperatures above 20°C reduce flowering in lychee (*Litchi chinensis* Sonn.). *Journal of Horticultural Science.* **70:** 981-987.

Menzel, C.M. Olesen, T. and McConchie, C.A. (2000). Lychee, Longan and Rambutan. Optimising Canopy Management. *Final Report to the Rural Industries Research and Development Corporation,* Canberra 92 pp.

Menzel, C.M., Oosthuizen, J.H., Roe, D.J. and Doogan, V.J. (1995). Water deficits at anthesis reduce CO_2 assimilation and yield of lychee (*Litchi chinensis* Sonn.) trees. *Tree Physiology.* **15:** 611-617.

Menzel, C.M., Rasmussen, T.S. and Simpson, D.R. (1989). Effects of temperature and leaf water stress on growth and flowering of litchi (*Litchi chinensis* Sonn.). *Journal of Horticultural Science.* **64:** 739-52.

Menzel, C., Olesen, T. and C.A. McConchie. (2000). Marketing profit from lychees in Australia. Proc. 5th National Lychee Conf., Twin Waters, Australia, pp. 5-15.

Mitra, S.K. and B. Ghosh, (1991). Description and performance of some lychee cultivars in West Bengal, India. Australian Lychee Yearbook, **1:** 64-70.

Mitra, S.K. and D. Sanyal, (2000). Effect of cincturing and some chemicals on flowering of litchi cv. Bombai. Abst. 1st Int. Symposium on litchi and longan. Guangzhou, China, p. 40.

Nanta, P. (1992). *Biological Control of Insect Pests.* Biological Control Branch, Entomology and Zoology Division, Department of Agriculture, Bangkok, Thailand 206 pp.

Olesen, T., Menzel, C., Wiltshire, N. and C.A. McConchie, (2000). Manipulating flushing cycles and flowering in lychee. Proc. 5th National Lychee Conf., Twin Waters, Australia, pp. 47-52.

Pandey, R.M. and H.C. Sharma, (1989). The Litchi. ICAR Publication, New Delhi.

Pandey, S. and A.P. Misra, (1975). Mycorrhiza in relation to growth and fruiting of *Litchi chinesis Sonn.,* J. Indian Bot. Soc., **54:** 280-293.

Rajesh Kumar, (2008). Technical bulletin. Rejuvenation of old senile litchi orchard. National Research Centre for Litchi Muzaffarpur, Bihar, p. 8.

Rajesh Kumar, (2009). Extension bulletin. Managing physiological disorders in litchi. National Research Centre for Litchi Muzaffarpur, Bihar.

Rao, D.P., S.K. Mukherjee and R.N. Ray. (1985). Nutritional Studies in litchi orchards in different districts of West Bengal. *Indian Hort.,* **42:** 1-7.

Ray, P.K. S.B. Sharma and K.A. Mishra, (1984). Important litchi cultivars of Bihar. *Indian Hort.*, **30(1):** 9-13.

Ren, S. and M. Tiang, (2000). Litchi pests and their natural enemies. Abst. 1st Int. Symposium on litchi and longan. Guangzhou, China, p. 74.

Roe, D.J., Menzel, C.M., Oosthuizen, J.H. and Doogan, V.J. (1997). Effects of current CO_2 assimilation and stored reserves on lychee fruit growth. *Journal of Horticultural Science & Biotechnology.* **72:** 397-405.

Roy. R.N., Rao, D.P. and S.K. Mukharjee, (1984). Orchard efficiency analysis of litchi. *Indian J. Hort.*, **41:** 16-21.

Shukla, R.K. and Bajpai, P.N. (1974). Blossom bud differentiation and ontogeny in litchi (*Litchi chinensis* Sonn.). *Indian Journal of Horticulture.* **31:** 224-228.

Singh H.P. (1992). Physico-chemical evaluation of litchi cultivars in South Bihar. In National Seminar on Recent Developments in Litchi Production, 30-31 May, RAU, Pusa.

Singh H.P. and I.S. Yadav, (1988). Physico-chemical changes during fruit development in litchi cultivar. *India J. Hort.*, **45:** 212-18.

Singh H.P. and I.S. Yadav, (1992). CHES-1, a promising selection of litchi. In National Seminar on Recent Development in Litchi Production.30-31 May, 1992, RAU, Pusa.

Singh H.P. and I.S. Yadav, (1992). Studies on seed germination in litchi cultivars affected by physiological maturity. In National Seminar on Recent Developments in Litchi Production 30-31 May, RAU, Pusa; 16p.

Singh, H.P. (1986). Cracking of fruits a problem in litchi growing. *Chona Hort;* **3(2):** 8-9.

Singh, H.P. (1998). Genetic Diversity, Breeding and Utilisation of the Genepool of Litchi. Tropical Fruits in Asia (Arora and Rao Eds) IPGRI. 171-184.

Singh, H.P. and K.K. Kumar, (1988). Litchi growing in India. *Chona Hort.*, **5(1):** 22-32.

Singh L.B. and O.P. Singh, (1954). The Litchi. Superintendent Printing Press, Allahabad. 110p.

Stern, R., Adato, I. and Gazit, S. (1990). Autumn water stress as a means of increasing flowering and improving fertility of young litchi trees. *Alon Hanotea.* **44:** 391-394.

Stern, R.A. and Gazit, S. (1997). Effect of 3, 5, 6-trichloro-2-pyridyl-oxyacetic acid on fruitlet abscission and yield of 'Mauritius' litchi (*Litchi chinensis* Sonn.). *Journal of Horticultural Science & Biotechnology.* **72:** 659-63.

Subhadrabandhu, S. and A. Koo-Duang, (1987). Effect of ethephon on flowering of two lychee (*Litchi chinensis* Sonn.) cultivars. *Acta Horticulturae,* **201 :** 181-86.

Verma, S.K., Jain. B.P. and S.R. Das, (1980). Preliminary Studies on the evaluation of the effect of growth substances with minor elements in controlling fruit drop in litchi (*Litchi chinesis Sonn.*). *Haryana J. Hort.*, **10:** 4-10.

Waite, G.K. (1990). *Amblypelta* spp. and green fruit drop in lychees. *Tropical Pest Management.* **36:** 353-5.

Waite, G.K. and Gerson, U. (1994). The predator guild associated with *Aceria litchii* (Acari:Eriophyidae) in Australia and China. *Entomophaga.* **39:** 275-280.

Waite, G.K. and Huwer, R.K. (1998). Host plants and their role in the ecology of the fruit spotting bugs *Amblypelta nitida* Stål and *Amblypelta lutescens lutescens* (Distant) (Hemiptera:Coreidae). *Australian Journal of Entomology* **37:** 340-349.

Waite, G.K. and McAlpine, J.D. (1992). Honey bees as carriers of lychee erinose mite *Eriophyes litchii* (Acari:Eriophyiidae). *Experimental and Applied Acarology* **15:** 299-302.

Wu, S.X. (1998). Encyclopaedia of China Fruits: Litchi. China Forestry Press, Beijing.

Wu, W.N., Lan, W.M. and Liu, Y.H. (1991). Phytoseiid mites on litchis in China and their application. *Natural Enemies of Insects* **13:** 82-91.

Yadav, I.S. and H.P. Singh, (1988). Studies on germination of seeds in litchi as influenced by physiological maturity of fruits. In National Convention of Strategies of Horticultural Development for Tribal Region held at Ranchi, March 1988.

Yadav, I.S. and H.P. Singh, (1992). Genetic Resources of Litchi in India. In National Seminar on Recent Developments in Litchi Production. 30-31 May 1992, RAU, Pusa.

Yapwattanaphun, C. and S. Subhadrabandhu, (2000). Litchi cultivars in Thailand. Abst. 1st Int. Symposium on litchi and longan. Guangzhou, China, p. 24.

Yuan, R.C. and Huang, H.B. (1988). Litchi fruit abscission: its pattern, effect of shading and relation to endogenous abscisic acid. *Scientia Horticulturae* **36:** 281-92.

Yuan, R.C. and H.B. Huang, (1993). Regulation of root and shoot growth and fruit drop of young litchi trees by trunk girdling in view of source-sink relationship. *J. Fruit Sci.*, **10:** 195-98.

Zhang, Z.W., Yuan, P.Y., Wang, B.Q. and Qui, Y.P. (1997). *Litchi Pictorial Narration of Cultivation.* Pomology Research Institute, Guangdong Academy of Agricultural Science (no page numbers).

Zhou, X.J., Wu, D.Y., Huang, H.B., Lai, J. and D.F. Huang, (1996). Manipulating growth and fruiting of young litchi trees by spiral girdling. *Acta Hortic. Sinica*, **23:** 13-18.

❑❑❑

Chapter 22

Colour Photographs

1. Rejuvenation of old orchard

1. Old senile orchard

2. Light limb removal

3. Light limb removal

4. Second level limb removal

5. Pasting of cut portion

6. Sprouting in cut plants

7. Boudreaux pasting

8. Irrigation in rejuvenated plant

9. Intercropping in rejuvenated orchard

10. Fully developed canopy

11. Flowering in rejuvenated plant

12. Fruiting in rejuvenated plant

(Courtesy: NRC for Litchi, Muzaffarpur, Bihar)

2. Pest damage photos

Fig.1:Shoot borer infestation

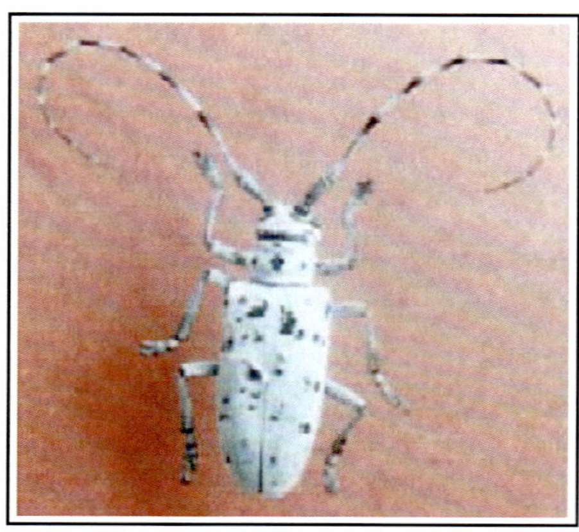

Fig.2: Stem borer

Fig.3: Nut borer infestation

Fig.4: Litchi fruit borer

Fig.5:Bark eating caterpillar infestation

Fig.6: Litchi mite infestation

Courtesy : HARP, Ranchi

7. Litchi Bug

Courtesy : NRC on Litchi, Mazaffurpur

8. Unknown Weevil

Mastering PostgreSQL for Data Engineering and Cloud Deployment

Design, Optimize, and Manage PostgreSQL Databases for Data Engineering, High Availability, and AWS Cloud Integration

Jaques Chiankem Fozao

www.orangeava.com

First Published: August 2025

Published By: Orange Education Pvt Ltd, AVA®

Address: 9, Daryaganj, Delhi, 110002, India

275 New North Road Islington Suite 1314 London,
N1 7AA, United Kingdom

ISBN (PBK): 978-93-49888-38-8
ISBN (E-BOOK): 978-93-49888-19-7

Scan the QR code to explore our entire catalogue

www.orangeava.com

Dedicated To

In loving memory of my late parents, Mama Sophina Lepefor and HM Fon Fozao Nzota 1st, whose unwavering guidance instilled in me a deep sense of responsibility, vision, and innovation. Their wisdom and values continue to shape my journey, inspiring me to pursue knowledge, lead with purpose, and create with passion.

About the Author

Jaques Chiankem Fozao is a Senior PostgreSQL Database Administrator and AWS Cloud Specialist, dedicated to helping professionals and organizations optimize database performance, security, and scalability. With over eight years of expertise, he has worked extensively in backup and restore, upgrades, replication, high availability, maintenance, and cloud integration.

He holds a Master's degree in Mathematics Education from Birmingham City University (UK) and certifications in Database Design and Programming (Oracle Academy) and Linux System Administration (Linux Professional Institute). His deep understanding of PostgreSQL and cloud technologies has enabled him to teach and mentor database professionals, simplifying complex concepts for learners at all levels.

Beyond his technical expertise, Jaques is also a passionate educator and author, committed to knowledge-sharing and leadership development. He is the author of *The Last Gold of Africa*, a book on effective leadership and resource preservation. His dedication to PostgreSQL, and database management drives his mission to empower professionals with the skills needed to build high-performance, scalable database systems.

Originally from Africa, Jaques now lives and works in the United Kingdom and Canada. He enjoys spending holidays traveling across Africa, Europe, and North America with his wife and two children. His experiences across different regions and industries have enriched his understanding of global database infrastructure challenges and best practices, ensuring his work remains relevant, practical, and impactful for professionals, students and learners of all ages.

About the Technical Reviewer

Yamini Panchal holds a **Bachelor of Engineering degree in Computer Science from Gujarat Technological University**, and boasts **over 8 years of extensive experience in the IT industry.** Throughout her career, she has contributed her expertise to diverse domain-based applications for web development, including IoT, Telecommunication, Healthcare, Cloud Services and others.

She currently serves as a Technical Lead, Senior Software Engineer at Azilen Technologies, specializing in open source development. With strong expertise in Node.js, she excels in building server-side applications and APIs. Yamini has a solid grasp of database systems, encompassing SQL as well as NoSQL databases.

Additionally, she is author of the book, *Building Scalable Web Apps with Node.js and Express*, which guides developers through designing and developing scalable web applications.

Yamini has contributed to different projects, such as robotics base applications, showcasing her programming skills in Node.js, Nest.js, Typescript, React.js, and AWS Services to enable real-time communication between robot and application. She has also developed a WebRTC-based application supporting audio and video conferences for multiple users.

She has demonstrated her proficiency in building robust and scalable backend services using Node.js for a myriad of client projects. Additionally, she is highly skilled in integrating AWS services with Node.js, within microservice architectures. She delights in embracing new technologies, indulging in reading and writing, and leveraging her skills to foster innovation.

"One can reach out to her via LinkedIn: https://www.linkedin.com/in/yamini-panchal-12939ab3/"

Acknowledgements

I extend my heartfelt gratitude to my lovely wife, Nadesh Fozao, and my two incredible sons, Divine-Favour Nzota and Lemuel Ntsindieh, for their unwavering support, patience, and encouragement throughout this journey. Their belief in me has been a constant source of inspiration.

Special thanks to HM Dr. Fozao Afou, District Medical Officer in the North West Region of Cameroon and leader of the M'Muock clan, for his extraordinary guidance, and continuous support. His encouragement has been invaluable in shaping the completion of this book.

I am deeply grateful to Professor Kennedy Folepei, Head of the Department of Petroleum Engineering at the University of Bamenda, and Dr. Dennis Mofor, Director of Construction at the Ministry of Public Works, for their insightful contributions and depth of knowledge which have enriched my understanding and strengthened this work.

A special recognition goes to Mr. Alain Fietsop for his meticulous technical editing, ensuring the accuracy and precision of PostgreSQL concepts in this book. His expertise and attention to detail have been invaluable.

I sincerely appreciate Ms. Priyanka Arora, whose persistence and dedication kept me focused on completing every chapter. Her unwavering commitment has been truly motivating.

My deepest appreciation to Mr. Rafiu Ajiwura, Team Lead of the Database Department at S&P Global, for continuously pushing me to pursue deeper knowledge in databases, and providing invaluable resources that have helped shape this book. His mentorship and support have been instrumental in refining my expertise.

Finally, my deepest gratitude to the team at Orange Education Pvt Ltd for their tireless efforts in reviewing the grammatical and syntax errors, refining content, and ensuring the finalization of this book. Their contributions have been instrumental in bringing this project to life.

Preface

Databases form the backbone of modern applications, businesses, and data-driven decision-making. PostgreSQL, an advanced open-source relational database system, has earned its place as a powerful, scalable, and secure choice for organizations, worldwide. From small startups to large enterprises, PostgreSQL enables efficient data management, high availability, and robust performance tuning.

This book takes a practical, structured approach to mastering PostgreSQL, covering installation, configuration, backup and restore, replication, security, performance tuning, and cloud integration with AWS. Thus, whether you are a beginner seeking foundational knowledge, or a seasoned professional aiming to optimize database operations, this guide provides step-by-step explanations and real-world insights.

With over eight years of expertise in database administration, I have encountered various challenges, performance bottlenecks, and industry best practices that shaped the depth of this book. My goal is to empower readers with hands-on techniques, making PostgreSQL accessible and practical for managing complex database environments.

This book is divided into **thirteen chapters**, each carefully crafted to build knowledge progressively–from understanding PostgreSQL architecture to implementing high availability, performance tuning, and database migration strategies. Additionally, cloud deployments with AWS have been explicitly explored in this book, ensuring readers are thoroughly equipped to handle the modern database infrastructure.

Chapter Summaries

This book begins with **Chapter 1**, introducing databases and PostgreSQL's history, architecture, and benefits. It explores how organizations leverage PostgreSQL for structured data management and scalable solutions.

Chapter 2 guides readers through installation and initialization, covering VMware Workstation setup, Linux OS installation, PostgreSQL deployment in Ubuntu, and the essential Linux commands.

In **Chapter 3**, readers gain insight into SQL subsets, including DDL, DML, DCL, and TCL, foundational for database operations.

Chapter 4 delves into PostgreSQL's database structure, covering schema creation, table management, data manipulation, and retrieval techniques.

Security and configuration take the centre stage in **Chapter 5** which explains Linux commands, file permissions, PostgreSQL authentication, and configuration settings to ensure a secure and optimized database environment.

Chapter 6 covers the backup and restore strategies, including logical (pg_dump, pg_dumpall) and physical (pg_basebackup) backups, along with Point-in-Time Recovery (PITR).

For ensuring database reliability, **Chapter 7** discusses replication and high availability, covering streaming replication (physical replication), logical replication, and failover management using pg_promote.

Chapter 8 details upgrade strategies, focusing on version compatibility, PostgreSQL upgrades, and using pg_upgrade for seamless transitions.

To enhance performance, **Chapter 9** provides optimization techniques, addressing vacuuming, indexing, memory tuning, OS-level tuning, logging, and handling performance bottlenecks.

Chapter 10 introduces PostgreSQL functions, such as aggregate functions, PL/pgSQL procedures, trigger functions, and string manipulation techniques.

Relational database design is explored in **Chapter 11**, covering table joins (inner, outer, natural), subqueries, Common Table Expressions (CTEs), and views, including materialized views, and interview questions for practical application.

Chapter 12 introduces cloud computing with AWS, covering EC2 instance setup, PostgreSQL installation on AWS, Amazon RDS, and migration techniques.

Finally, **Chapter 13** focuses on IAM users and AWS migration, explaining access control, AWS Database Migration Service (DMS), replication strategies, and secure database migration techniques for seamless transitions into cloud environments.

By the end of this book, you will have mastered all the essential PostgreSQL skills, gained insights into database security, scalability, and optimization, and acquired practical knowledge to apply PostgreSQL effectively in various domains.

So, let's embark on this interesting journey to unleash the full potential of PostgreSQL!

Get a Free eBook

We hope you are enjoying your recently purchased book! Your feedback is incredibly valuable to us, and to all other readers looking for great books.

If you found this book helpful or enjoyable, we would truly appreciate it, if you could take a moment to leave a short review with a 5 star rating on Amazon. It helps us grow, and lets other readers discover our books.

As a thank you, we would love to send you a free digital copy of this book, and a **30%** discount code on your next cart value on our official websites:

www.orangeava.com

www.orangeava.in (For Indian Subcontinent)

☑ **Here's how:**

Leave a review for the book on Amazon.

Take a screenshot of your review, and send an email to **info@orangeava.com** (it can be just the confirmation screen).

Once, we receive your screenshot, we will send you the digital file, within 24 hours.

Thank you so much for your support – it means a lot to us!

Downloading the code bundles and colored images

Please follow the links or scan the QR codes to download the
Code Bundles and Images of the book:

https://github.com/ava-orange-education/Mastering-PostgreSQL-for-Data-Engineering-and-Cloud-Deployment

The code bundles and images of the book are also hosted on
https://rebrand.ly/9fd5fd

In case there's an update to the code, it will be updated on the existing GitHub repository.

Errata

We take immense pride in our work at **Orange Education Pvt Ltd,** and follow best practices to ensure the accuracy of our content to provide an indulging reading experience to our subscribers. Our readers are our mirrors, and we use their inputs to reflect and improve upon human errors, if any, that may have occurred during the publishing processes involved. To let us maintain the quality and help us reach out to any readers who might be having difficulties due to any unforeseen errors, please write to us at :

errata@orangeava.com

Your support, suggestions, and feedback are highly appreciated.

DID YOU KNOW

Did you know that Orange Education Pvt Ltd offers eBook versions of every book published, with PDF and ePub files available? You can upgrade to the eBook version at **www.orangeava.com** and as a print book customer, you are entitled to a discount on the eBook copy. Get in touch with us at: **info@orangeava.com** for more details.

At **www.orangeava.com**, you can also read a collection of free technical articles, sign up for a range of free newsletters, and receive exclusive discounts and offers on Print Books and eBooks.

PIRACY

If you come across any illegal copies of our works in any form on the internet, we would be grateful if you would provide us with the location address or website name. Please contact us at **info@orangeava.com** with a link to the material.

ARE YOU INTERESTED IN AUTHORING WITH US?

If there is a topic that you have expertise in, and you are interested in either writing or contributing to a book, please write to us at **business@orangeava.com**. We are on a journey to help developers and tech professionals to gain insights on the present technological advancements and innovations happening across the globe and build a community that believes Knowledge is best acquired by sharing and learning with others. Please reach out to us to learn what our audience demands and how you can be part of this educational reform. We also welcome ideas from tech experts and help them build learning and development content for their domains.

REVIEWS

Please leave a review. Once you have read and used this book, why not leave a review on the site that you purchased it from? Potential readers can then see and use your unbiased opinion to make purchase decisions. We at Orange Education would love to know what you think about our products, and our authors can learn from your feedback. Thank you!

For more information about Orange Education, please visit **www.orangeava.com**.

Table of Contents

CHAPTER 1
Introduction to PostgreSQL

Introduction

This chapter will cover what PostgreSQL is all about and why organizations are moving their applications to PostgreSQL. We will look at what databases are and the different types of databases. The memory structure of PostgreSQL and a deep dive into its architecture will also be learned here. Configuration parameters will be mentioned. We will also review its background knowledge, which will explain how and why PostgreSQL was put to birth over 35 years ago. This chapter will mostly be about reading and understanding the basics of PostgreSQL.

Structure

In this chapter, the following topics will be covered:

- Definition of a Database
- The Explanation of PostgreSQL
- Background of PostgreSQL
- Benefits of PostgreSQL to Organizations
- PostgreSQL Architecture
- PostgreSQL Database Structure
- PostgreSQL Configuration Parameters
- Task

Definition of a Database

A database is a structured method for storing data (information) so it can be retrieved, manipulated, transformed, updated, and migrated to other databases. To understand databases, consider the functionality of Excel sheets within an Excel workbook. This workbook stores information in a tabular format, allowing for efficient data organization and access. Similarly, relational databases are structured with tables consisting of rows and columns. Thus, a collection of spreadsheets (or tables) systematically connected can be viewed as a simplified version of a database.

Databases typically store collections of records or files containing varied information, such as student details, employee details, product information, sales records, orders, customer details, and financial records. They serve to store, maintain, and access a wide array of data, consolidating information on various entities such as things, places, or people in one place for easy retrieval, manipulation, and analysis. Therefore, databases can be aptly defined as organized collections of information.

Uses of Databases

Businesses and organizations rely heavily on data stored in their databases to make informed decisions. Here are some common uses of databases:

- **Increase Business Revenue**: Companies collect data on sales, orders, products, and customer service, analyzing this data to enhance processes and increase profits. By identifying trends and patterns, businesses can develop strategies to optimize their operations and drive growth.
- **Maintain Customer Records**: Businesses use databases to store and manage customer and employee information, such as names, email addresses, and department details. This centralized information repository allows for efficient customer relationship management and streamlined internal communication.
- **Store Personal Data**: Individuals can utilize databases to store personal information, such as photos and important documents. Databases offer a secure and organized way to manage personal data, ensuring easy access and retrieval when needed.

Different Types of Databases

In computing, databases are classified based on organizational use cases and approaches. Here are some major types:

- **Relational**: Relational databases consist of tables organized into rows and columns. This structure facilitates easy information retrieval using SQL. Relational Database Management Systems (RDBMS), such as PostgreSQL, adhere to ACID (atomicity, consistency, isolation, durability) principles, ensuring data consistency and transaction integrity. Relational databases are ideal for structured data where relationships between different data entities are well-defined.

- **NoSQL**: Non-relational databases address big data performance issues better than relational databases and do not rely on tables. They are well-suited for unstructured data where the data type is not fixed or known in advance and may change over time. NoSQL databases are perfect for real-time data from sensors in IoT applications, allowing for flexible data modeling. Examples include Tokyo Cabinet, Hadoop, and MongoDB.
- **Distributed**: Distributed databases spread data across multiple physical locations, making it available in different network parts through replication. These systems can be heterogeneous, where each physical location has different underlying hardware and software, or homogeneous, where all locations use the same hardware and operating system. Distributed databases enhance data availability and reliability.
- **Cloud**: Cloud databases are created in hybrid, private, or public clouds. Users are charged based on resource usage (CPU, bandwidth, storage). Cloud databases offer on-demand fault tolerance and scalability, making them suitable for applications requiring high availability and performance. They can work with applications deployed as software as a service (SaaS), providing flexibility and cost-efficiency.

Components of a Database

According to IBM, the schema, data structure, and data types differentiate databases, but they all comprise the following basic elements:

- **Hardware**: The physical machines running the database software, including computers, hard drives, and servers. These machines provide the necessary infrastructure to support database operations.
- **Software**: Database software or applications that control the database, managed by a Database Management System (DBMS). The DBMS handles tasks such as data storage, retrieval, and manipulation, providing users with tools to interact with the database.
- **Data**: Raw information stored in databases, organized by database administrators to make it meaningful. Data can be structured, semi-structured, or unstructured, depending on the database type and use case.
- **Data Access Language**: The programming language controlling databases, with SQL being the most common. This language must work with the DBMS to execute queries and commands for data manipulation.
- **Procedures**: The methods determining how the database handles and operates data. Procedures include backup and recovery processes, security measures, and performance optimization techniques.

Database Challenges

Setting up, operating, and maintaining databases can present various challenges:

- **Data Security**: Protecting data from unauthorized access and use is crucial, as data is a valuable business asset. Implementing robust security measures, such as encryption and access controls, is essential to safeguard data.
- **Data Integrity**: Ensuring data integrity involves restricting database access to qualified individuals, such as database administrators and trustworthy data analysts. Maintaining data accuracy and consistency is challenging, especially when multiple users need access.
- **Database Performance**: Ensuring proper support and maintenance is vital to prevent functionality decline as technology evolves. Regular updates and maintenance are required to keep the database performing optimally and to address any performance bottlenecks.

Database Management System (DBMS)

A DBMS is a software application that interacts with users, applications, and the database to capture and analyze data. It allows users to create, manage, and modify databases and assign read and write permissions. There are two types of DBMS:

- **Object-Relational Database Management System (RDBMS)**: Uses tables to store data in relations, such as PostgreSQL and MySQL. RDBMSs provide support for complex queries and transactions, ensuring data integrity and consistency.
- **Non-Relational Management System**: Does not use concepts of relations, tuples, and attributes, such as MongoDB. Non-relational systems offer flexibility in data modeling, making them suitable for handling large volumes of unstructured data.

The Explanation of PostgreSQL

PostgreSQL is an open-source RDBMS. It is widely used by large web applications. Its popularity as a stable and reliable open-source relational database system has made it very powerful. It is commonly pronounced as Postgres. It has a good reputation for its support of major operating systems, its extensibility of SQL language, nested transactions, reliability, flexibility, and its MVCC (multi-version concurrency control) architecture. This MVCC architecture enables the owner and user of the database to extend the system without any problem. It also supports flexible API, database validation, and a vast amount of data types which makes it the most compliant, stable, scalable, and available among relational databases.

PostgreSQL is designed for environments with large amounts of data, it is multi-platform and supports variable workloads and concurrent users.

The PostgreSQL development program was started in 1986 by a team of computer scientists led by *Michael Stonebraker*. Its purpose was to aid developers in the development of enterprise-level applications by ensuring data integrity and fault tolerance in systems. It has a large international developer community that has constantly contributed to its development. It has achieved a significant appeal amongst developers because of its fault-tolerant characteristics. It continues to evolve year-by-year, as it is maintained by a worldwide team of contributors that is passionate about regularly improving this free and open-source database project.

PostgreSQL runs on Linux, macOS, Windows, BSD and Solaris operating systems. This shows that it runs on all major operating systems. In 2001, PostgreSQL was approved ACID (atomicity, consistency, isolation, and durability)-compliant. It has a powerful add-on such as the popular PostGIS geospatial database extender. All the features mentioned earlier have contributed to PostgreSQL becoming the open-source relational database of choice for many people and organizations.

Advantages of PostgreSQL

According to the PostgreSQL.org website, PostgreSQL comes with many features to help with developing applications by developers, protecting data integrity by administrators and building high-availability environments. It helps you manage your data no matter how big or small the dataset. Aside from the free and open-source nature of PostgreSQL, it is highly extensible and enables you to create custom data types, build out custom functions, and even write code from different programming languages in your database.

PostgreSQL is expandable and its versatile and approachable design supports a variety of specialized use cases. It has a powerful extension ecosystem, which makes it applicable in all major operating systems. Thus, maintaining dynamic database systems in today's digital landscape is made possible and easy. Organizations looking for cost-effective and efficient ways to improve their databases and data integrity are turning to migrate to PostgreSQL.

PostgreSQL is the best-of-breed relational database management system and is completely free from licensing restrictions. Expert developers who understand its unlimited use cases compared to the traditional database systems heavily support PostgreSQL.

Due to PostgreSQL flexibility, compatibility, and support of multiple programming languages such as Python, JavaScript, C/C++ and Ruby, letting developers perform database tasks in whichever language they are proficient in without generating system conflicts, it has become the database of choice.

Many of the features required by the SQL standard are supported by PostgreSQL, though sometimes with slight differences in syntax or function. Out of the 179 mandatory Core conformance features of SQL:2016, PostgreSQL accords to 170 of them. As of this writing, no relational database meets full conformance with this standard (PostgreSQL.org).

Some features of PostgreSQL 15 are shown here. Some of the features mentioned as follows have been touched on and the majority of them will be learned during the practical sections:

- **Data Types:** The data types stated as follows are supported by the current PostgreSQL version 15:
    - **Character types (Varying character and character):** Used for storing strings of text with varying character allowing for variable-length strings and character for fixed-length strings.
    - **Numeric types:** Used for storing numerical data, including integers, floating-point numbers, and decimals.
    - **Boolean:** Represents a binary value of true or false, used for logical operations.
    - **UUID:** Stands for Universal Unique Identifier, a 128-bit identifier used to uniquely identify information.
    - **Geometric primitives:** Used for storing geometric data such as points, lines, and polygons for spatial databases.
    - **XML:** A markup language used for storing and transporting data, allowing for hierarchical and nested data structures.
    - **Date/Time:** Used for storing dates, times, and timestamps, supporting various formats and time zones.
    - **Arrays:** Used for storing collections of elements of the same data type, allowing for indexing and manipulation.
    - **Json/jsonb:** Used for storing JSON (JavaScript Object Notation) data, with json storing text and jsonb storing binary representation for efficient querying and indexing.
- **Data Integrity:** The following constraints are used to specify a limit on some of the above-mentioned data types to reinforce data integrity.
    - **Check :** Ensures that all values in a column satisfy a specific condition.
    - **Not null:** Ensures that a column cannot have a NULL value.
    - **Default:** Sets a default value for a column when no value is specified.
    - **Unique:** Uniquely identifies a single row in a table. Ensures that all values in a column or a combination of columns are distinct across rows.

  - **Primary Key:** Uniquely identifies each row in a table and enforces a unique constraint, meaning no two rows can have the same primary key value.
  - **Foreign Keys:** Maintains referential integrity by enforcing a link between data in two tables (parent and child table).
  - **Exclusion Constraints:** Ensures that if any two rows are compared on specified columns or expressions, at least one of those comparisons will evaluate to false, preventing certain combinations of values.
- **Concurrency and Performance:**
  - **Multi-Version Concurrency Control (MVCC)**: PostgreSQL is renowned for its performance and speed, making it one of the best RDBMS options available. It supports data authentication, various performance optimizations, and features typically found in proprietary database technologies such as geospatial support and unrestricted concurrency. As a result, PostgreSQL excels in running data analysis across multiple data types efficiently.

    One of the key features contributing to PostgreSQL's efficiency is its MVCC architecture. MVCC allows PostgreSQL to manage concurrency effectively, ensuring that read operations do not block write operations and vice versa. This means multiple transactions can access the database simultaneously without interfering with each other, which is not always possible in database systems lacking MVCC architecture. Those systems often lock records to prevent read/write conflicts when multiple users try to access the data concurrently.

    MVCC provides transactional isolation without blocking concurrent transactions, allowing multiple transactions to read data simultaneously without interfering with each other. This significantly enhances database performance and ensures data consistency even in environments with high concurrency, making PostgreSQL an ideal choice for demanding applications requiring robust data handling and analysis capabilities.
- **Query Planner/Optimizer:** The query planner/optimizer determines the most efficient way to execute a database query by analyzing various execution plans. It chooses the best plan to minimize the query's execution time and resource consumption, enhancing overall database performance.
- **Indexing (B-tree, GiST, GIN, and BRIN):**
  - **B-tree**: B-tree indexes are widely used for their balanced structure, making them ideal for general-purpose indexing of sorted data, improving query performance.

  - **GiST**: Generalized Search Tree (GiST) indexes support a wide range of query types and complex data types, enhancing the flexibility of querying.
  - **GIN**: Generalized Inverted Index (GIN) is optimized for indexing composite values such as arrays and full-text search, allowing efficient querying of such data.
  - **BRIN**: Block Range INdexes (BRIN) are efficient for large tables with naturally ordered data as they store summary information for block ranges rather than individual rows, reducing storage space.
- **Table Partitioning:** Table partitioning involves dividing a large table into smaller, more manageable pieces called partitions. This technique improves query performance and simplifies maintenance by allowing operations to target specific partitions rather than the entire table.
- **Transaction Isolation Levels**: There are four transaction isolation levels, each providing different degrees of isolation:
  - **Read Committed**: Ensures that any data read during a transaction is committed at the moment it is read, preventing dirty reads.
  - **Read Uncommitted**: Allows reading uncommitted changes from other transactions, which can lead to dirty reads, useful for scenarios requiring minimal isolation.
  - **Repeatable Read**: Ensures that if a transaction reads the same row multiple times, it sees the same data each time, preventing non-repeatable reads, suitable for operations requiring high consistency.
  - **Serializable**: Provides the highest level of isolation by ensuring transactions occur in a completely isolated manner, preventing all concurrency issues and ensuring maximum data consistency.

These features and concepts play crucial roles in maintaining the integrity, performance, and scalability of a PostgreSQL database.

Disaster Recovery and Reliability

- **Write-ahead Logging (WAL):** WAL ensures data integrity by logging changes before they are written to the database. This technique allows for recovery after a crash, ensuring that all transactions can be reconstructed from the logs. Use case: Recovering data after a system failure to maintain database consistency.
- **Replication**
  - **Asynchronous Replication**: Copies data to standby servers with a slight delay, improving performance by not waiting for the backup to complete. Use case: Creating backups without impacting the performance of the primary database.

  - **Synchronous Replication**: Ensures data is copied to standby servers in real time, providing high availability. Use case: Achieving zero data loss in mission-critical applications.
  - **Logical Replication**: Replicates specific changes rather than entire databases, allowing for more granular control. Use case: Real-time data synchronization across different systems for analytics.
- **Point-in-Time Recovery (PITR):** PITR allows for restoring the database to a specific point in time, useful for recovering from data corruption or accidental deletions. Use case: Reverting the database to a state before a destructive operation or data corruption incident.
- **Active Standbys:** Active standbys are secondary servers that are continually updated with changes from the primary server, ready to take over in case of a failure. Use case: Ensuring high availability and quick failover in case of primary server failure.
- **Tablespaces:** Tablespaces allow for the distribution of data across multiple storage locations, optimizing storage and performance. Use case: Managing large databases by organizing data storage efficiently and improving I/O operations.
- **Extensibility**
  - Stored functions and procedures
  - Procedural Languages (for example, PL/pgSQL, Perl, and Python)
  - SQL
  - Foreign data wrappers (FDW: connect to other databases or streams with a standard SQL interface.

PostgreSQL continues to be one of the most well-known, popularly used, and supported amongst all relational databases available after more than three decades.

PostgreSQL offers the following benefits to developers looking to create highly scalable computing environments.

Background of PostgreSQL

The PostgreSQL project traces its origins back to 1986, emerging from the Postgres project at the University of California, Berkeley, under the guidance of Professor Michael Stonebraker. This initiative was conceived as a successor to an earlier project known as Ingres, which stands for Interactive Graphics and Retrieval System. Ingres was a pioneering database system project running from 1977 to 1985 and was subsequently acquired by Computer Associates. The name Postgres is a portmanteau of "*post-Ingres*," signifying its development after Ingres.

Postgres was initiated to extend the capabilities of Ingres, incorporating advanced database concepts such as object-relational technologies. The development of Postgres spanned from 1986 to 1994, during which it introduced groundbreaking features in the database field. Among these features were Multi-Version Concurrency Control (MVCC), which provides transactional isolation without blocking concurrent transactions, and advanced recovery mechanisms such as Point-In-Time Recovery (PITR).

PostgreSQL, pronounced "Post-GRES-Q-L" to highlight its support for the SQL standard, has evolved into one of the world's most advanced open-source relational database systems. It offers robust features such as asynchronous replication, write-ahead logging (WAL), and a sophisticated query planner or optimizer, making it highly efficient and reliable.

PostgreSQL runs on all major operating systems, including Linux, Windows, and macOS. Its open-source nature has cultivated a dedicated and diverse community of contributors. The PostgreSQL Global Development Group oversees its development, ensuring regular major and minor releases that enhance its functionality and performance. This ongoing development effort ensures that PostgreSQL remains a cutting-edge database system suitable for a wide range of applications.

Benefits of PostgreSQL to Organizations

Organizations need their databases to be up and running at all times. They require sustainable solutions to ensure that production databases are always available to both developers and customers. PostgreSQL offers numerous benefits to meet these demands:

- **Cost-Effectiveness**: PostgreSQL is an open-source technology, which means it is free to use and does not require costly licenses. This offers organizations better cost savings and flexibility compared to proprietary databases. Additionally, the open-source nature allows for community support and a wealth of extensions to enhance functionality.
- **Robustness**: PostgreSQL ensures high availability of services through its asynchronous or synchronous replication methods across multiple servers. With replicas set up that can be easily promoted to the primary server, PostgreSQL ensures continuity of service even when disasters strike. PostgreSQL maintains WAL files, allowing developers to use PITR to restore databases to a specific moment in time, ensuring data recovery and integrity. The WAL capability logs every database change in a transaction log before the actual changes occur, providing a robust recovery mechanism.
- **Integration**: PostgreSQL integrates well with other applications and

technologies, making it easy to incorporate into existing systems. Developers can create custom stored procedures due to PostgreSQL's support for multiple procedural languages. These procedures can be created and called on a given database. With the use of extensions, procedural languages are also used for development in many other programming languages, including Perl, Python, JavaScript, and Ruby.

- **Scalability**: PostgreSQL can handle large datasets and heavy workloads, making it suitable for organizations with growing data demands. Its SQL compliance and query optimizer make it useful for general-purpose analytics on organizational data. PostgreSQL is built on a strong foundation of skills, expertise, and knowledge, making it one of the world's most advanced databases. It is rich with many extensions, which make the system robust and efficient in managing Online Transaction Processing (OLTP) protocols.

PostgreSQL Architecture

PostgreSQL has a physical structure that comprises some data files, background processes, and shared memory.

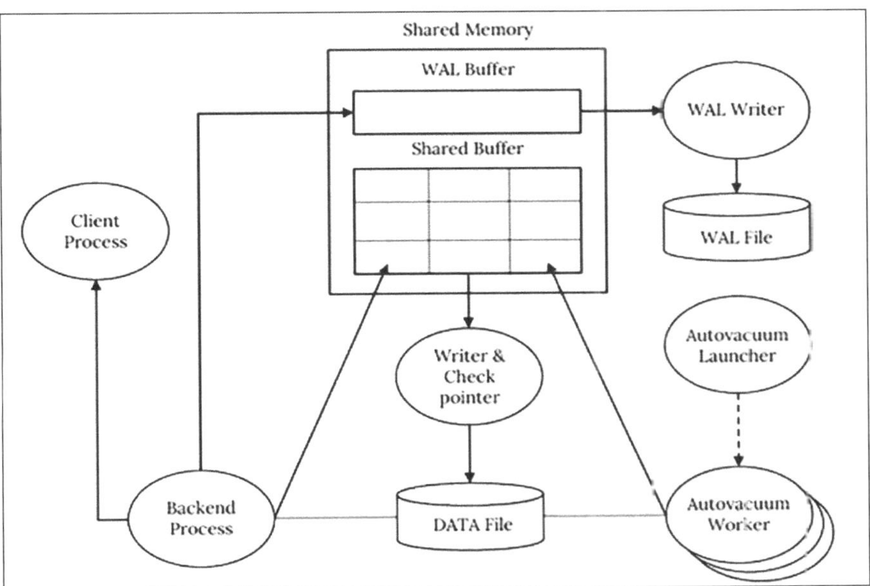

***Figure 1.1**: PostgreSQL structure adopted from Serveralnines*

Shared Memory

This is the amount of memory needed for caching (database caching and transaction log caching). WAL buffer and Shared buffer are the most important components of shared memory.

Shared Buffer

Shared buffer is the portion of shared memory used for shared memory buffers by the database server. The default size of the shared buffer is 128 MB. Nevertheless, to get good performance, the settings of this value should be more than the minimum required size. The said parameter can only be altered or set at the server start. If the RAM of the dedicated server is equal to or more than 1GB, 25% or more of it should be allocated for a shared buffer. It is improbable that allocating a value of more than 40% of the RAM to shared buffers will make it any better than a value less than that. Though there are workloads that will benefit from a larger value of shared buffers, this value should not be more than 40% of RAM as PostgreSQL also relies on the operating system cache.

Larger setting for a shared buffer usually involves increasing the amount of **max_wal_size**. Over a long period, the process of writing large volumes of new or changed data is spread out by **max_wal_size**. One of the most important goals of the shared buffer is to minimize DISK IO. It enables very large buffers to be accessed quickly. Contention between users is reduced when many users access it at the same time. Constantly accessed blocks are kept in the buffer for as long as possible.

WAL Buffer

Write ahead logs or WAL buffers are also called transaction log buffers. This is the chunk of memory granted for storing WAL data.

Database changes are temporarily stored by WAL buffers. At a predestined point in time, the WAL buffer's content is flushed to the WAL file.

WAL buffers and WAL files are very important from the perspective of backup and recovery. Actual data can be reconstructed or restored to its original state in the database following a disaster using the WAL files.

WAL buffers memory, allocated from the operating system's RAM, is controlled by WAL buffers' parameters. This memory area is not part of the shared buffers though it is accessible by all the background servers and user processes. The WAL buffers are very small compared to shared buffers and are external to the shared buffers. WAL buffers are written to WAL segments on disk after the WAL data has been modified.

PostgreSQL Processes

There are four process types in PostgreSQL:

- Postmaster Process or Postmaster Daemon
- Background Process

- Backend Process
- Client Process

1. **Postmaster Process:** When PostgreSQL is started, the first process started is the postmaster process.

The postmaster process then runs other background processes, performs recovery, and initialize shared memory. Whenever there is a connection request from the client process, the postmaster process will create a backend process. (Serveralnines).

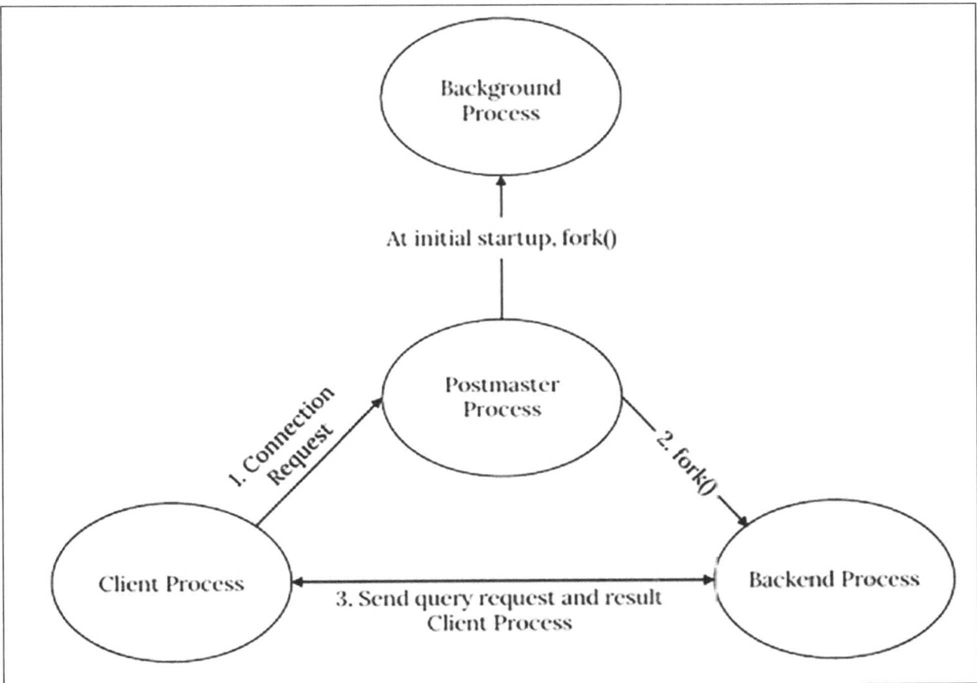

Figure 1.2: *Process relationship diagram adopted from Serveralnines*

Postmaster process is the parent process of all processes.

2. **Background Process:**
   a. Error messages are written to the log file by a Logger process.
   b. Whenever a checkpoint takes place, dirty buffers are written to the file by the `checkpointer` process.
   c. Dirty buffers are periodically written to a file by a writer process.
   d. WAL buffers are written to the WAL files by the WAL writer process.

e. Whenever **autovacuum** is enabled, **autovacuum** launcher forks **autovacuum** worker which is responsible for carrying out the vacuum operations on dirty tables.

f. When in **archive.log** mode, WAL files are copied to the specified directories by archiver processes

g. Table usage statistics in **pg_stat_all_tables** and session execution statistics in **pg_stat_all_tables** to be used by PostgreSQL are collected by the stats collector.

3. **Backend Process:** The user process' query requests are performed by the backend process and the result is spread out. Local memory is required for query execution. The **max_connections** parameter defines the maximum number of backend processes. The default value for the number of connections is set to 100.

Main parameters associated with local memory:

- **Work_mem**: Query operations such as sorting, bitmap operations, hash joins, and merge joins use the **work_mem** memory. **Work_mem**'s default value is 4 MB, and when this amount is exhausted, data is written to temporary disk files.

  Several sort or hash operations can be run in parallel when there is a complex query. Each of these operations is allowed to use as much memory as this value of **work_mem** specifies before writing data into temporary files. Similarly, several running sessions could be doing such operations concurrently. Consequently, the amount of memory used by these operations could be many times higher than the value of **work_mem**. Before altering the value of **work_mem**, it is very important to bear this fact in mind.

  Order by, distinct, and merge joins are sort operations. Hash tables are used in hash joins, hash-based aggregation, and hash-based processing of subqueries. Hash tables available memory is calculated by multiplying the **hash_mem_multiplier** by the value of **work_mem**. Sort-based operations are generally less sensitive to memory availability than the hash-based operations. Because of this, hash-based operations can use an amount of memory that exceeds the usual **work_mem** base value.

- **Maintenance_work_mem:** Vacuum, create index and add foreign keys are maintenance operations, and **maintenance_work_mem** is the amount of memory for these operations. The default setting is 64 MB. Because a database session can only process one of these operations at a time, its value can be set larger than **work_mem**. Vacuuming performance and restoration of database dumps can be improved by having higher settings set. It is advisable to say do not set the value too high because the **autovacuum_max_workers** times this memory may be assigned when an **autovacuum** runs.

- **Temp_buffers:** The space used for temporary tables is called **temp_buffers**. Each database session requires a certain amount of memory for temporary buffers. The default setting of this value is 8 MB.

  For this setting to be changed, it must be done before the first use of the temporary tables within the session and can only be changed within individual sessions. There will be no effect, if there are any subsequent changes of this value during the session. Temporary buffers are allocated in the session as needed up to the limit given by **temp_buffers**.

4. **Client Process:** For every user connection established, there is a client Process known as the background process assigned. Each user connection is served by a child process that has been forked by the postmaster process.

PostgreSQL Database Structure

PostgreSQL is initialized using **initdb**. Once initialized, there are three default databases created namely: Template0, Template1, and Postgres. These are template databases for user database creation. They also contain the system catalog tables. All these template databases and the once created by a user are collectively called a database cluster.

New databases are created by cloning the template1 database when the user issue the **createdb** or the create database command. This is because the template1 database can create objects needed by the user. Both template0 and template1 databases contain the same list of tables immediately after initialization using **initdb**.

Items related to the tablespace:

- There are two default tablespaces namely **pg_default** and **pg_global** tablespaces which are created immediately after PostgreSQL initialization.
- All tables created without assigned tablespaces at the time of creation are stored in the **pg_default** tablespace.
- The **pg_global** tablespace stores all the tables that are managed at the database cluster level.
- **$PGDATAbase** is the physical location of the **pg_default** tablespace and **$PGDATAglobal** is that of **pg_global** tablespace.
- Multiple databases can use a single tablespace.
- A symbolic link to the user tablespace is created linking the user tablespace in the **$PGDATAtblspc** directory when a user tablespace is created when a user tablespace is created. This also creates a symbolic link to Items related to the table. Each table contains the following three files:
  - A file for sorting table data called **OID** file.

- A file for managing table free space called **OID_fsm**.
- The visibility of the table block is managed by the **OID_vm** file.

OID is the name of the file at the time of creation of index and table creation.

Other Associated Memories to PostgreSQL

Aside temp files and **work_mem**, **maintenance_work_mem**, shown here are other associated memories to PostgreSQL:

- **Autovacuum_work_mem:** **Autovacuum_work_mem** is the chunk of memory reserved for each autovacuum worker process. The default value of **autovacuum_work_mem** is set to -1. This setting means that the **maintenance_work_mem** should be used instead of **autovacuum_work_mem**.
- **logical_decoding_work_mem:** For decoded changes to be written to a local disk, **logical_decoding_work_mem** is the chunk of memory to be used by logical decoding before this happens. This thus limits the amount of memory used by logical streaming replication connections. The default value is 64 MB. This value can be set higher than **work_mem** as each replication connection only uses a single buffer of this size. This thus reduces the number of decoded changes written to disk.

Each installation normally does not have many such connections concurrently.

PostgreSQL Configuration Parameters

Most of the parameters mentioned here will be explained and elaborated in later chapters. The PostgreSQL parameters are in a file called **PostgreSQL.conf.** This is a very important and fragile file for PostgreSQL. Before making any changes in this file, always take a backup of the original file. This file is a text file, and thus can be edited using any text editor. **PostgreSQL.conf** is in the data directory.

The configuration parameters of PostgreSQL are classified under the following categories: memory, database Connection, logging, WAL, and query tuning parameters.

These configuration parameters play an important role in the optimization and enhancement of database performance. Default parameters are usually not enough in an environment, thus an understanding of these parameters and how to configure them is paramount. A better configuration will greatly influence database performance.

Reducing server downtime is critical, thus understanding the set parameters mentioned above is necessary for all PostgreSQL Database professionals. A good understanding of these parameters will help you to adjust the right parameters at the right time when you want the changes to be implemented.

`Postgresql.conf` is the file where the changes to these parameters can be done respectively either manually using a vi editor or using alter system commands.

Task

Following are some questions to consolidate the knowledge of *Chapter* 1. There are also possible interview and exam questions.

1. What are constraints? What is data integrity?
2. What is the meaning of DBMS and what are the different types?
3. What is the difference between varying character and character data types?
4. What do you understand by the word PostgreSQL?
5. What does the abbreviation ACID stand for?
6. Explain the use of `work_mem`.
7. What is `maintenance_work_mem`?
8. Give a brief explanation of shared buffer and why it is important to allocate it.
9. Explain what you understand by PostgreSQL architecture to your best friend or partner.
10. Shed some light on the word WAL buffers.
11. What does MVCC stands for?

Conclusion

In this chapter, we learned and understood all about PostgreSQL, and comprehended the definition of a database and its objects. We learned about PostgreSQL Architecture and its background. Learners understood PostgreSQL configuration Parameters, benefits of PostgreSQL to Organizations.

In the next chapter, we will learn how to install the Linux operating system and PostgreSQL.

CHAPTER 2
Installing and Initializing PostgreSQL

Introduction

In this chapter, we will cover PostgreSQL installation and initialization. PostgreSQL is an application that sits on top of an Operating System (OS). We will look at installing PostgreSQL on Linux OS namely Ubuntu. The Linux OS also sits on either a virtual machine called VM Box or VMWare workstation player. We will be using VMWare Workstation as our virtual environment.

Structure

In this chapter, the following topics will be covered:

- VMware Workstation Installation
- Linux Operating System (OS) Installation
- Installing PostgreSQL

VMware Workstation Installation

The first part will be to prepare our virtual environment by downloading and installing the VMware Workstation player.

VMware Workstation Player

VMware Workstation Player is a tool for running a virtual machine on Windows or Linux. There is a free version designed for learning and training purposes and a commercial version required for Organizations to use by purchasing a commercial license. Students and educators use the free version for learning and training. The free version is non-commercial, and thus is available for personal use.

To download and install VMware Workstation Player, perform the following steps:

Step 1: Open your favorite search engine such as **Google Chrome or any browser of choice** and type the following link.

https://www.vmware.com/ca/products/workstation-player/workstation-player-evaluation.html

Alternatively, type on your search engine:

Download **`VMware workstation player >`**

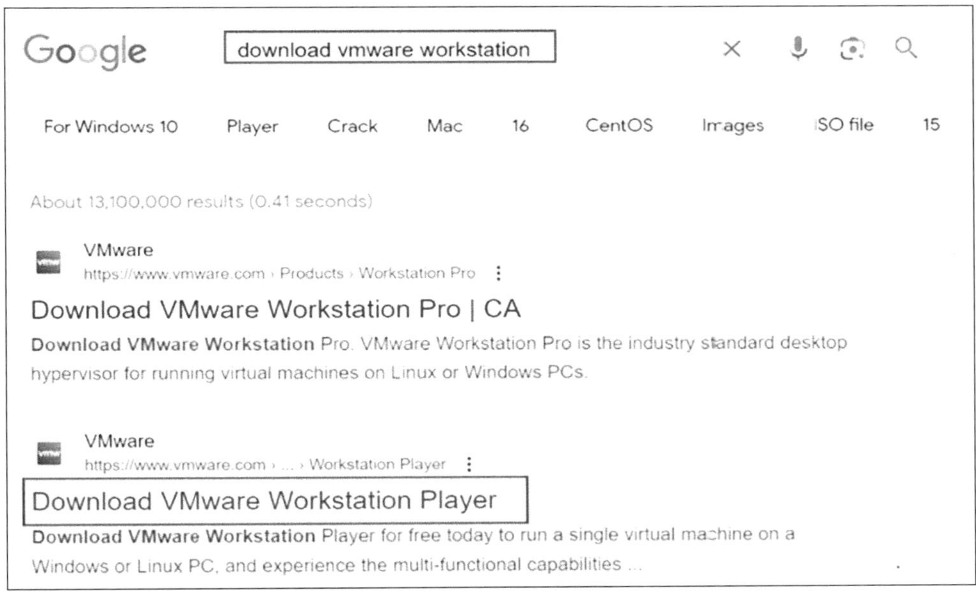

Figure 2.1: *Ways to download VMware Workstation Player.*

Click the link as shown above to be directed to the VMware official website, as illustrated in the following figure:

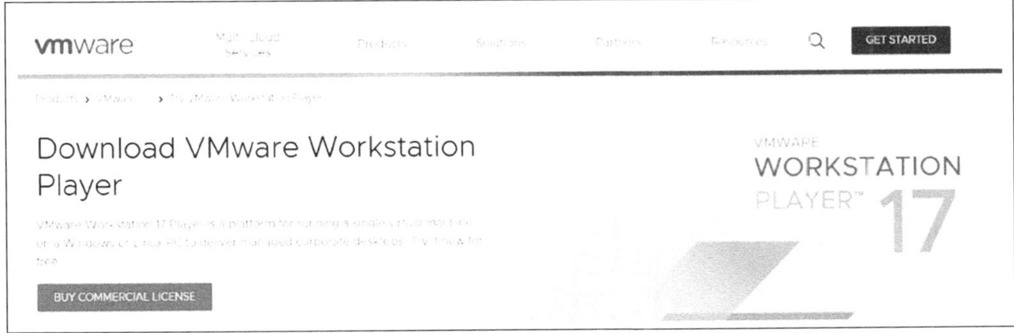

Figure 2.2: *VMware site*

Scroll down to the section that says, "**try Workstation 17 Player for Windows**" or "**try Workstation 17 for Linux**". As of the time of writing this book, the latest version was **17**. Click **DOWNLOAD NOW** as shown in the screenshot:

Figure 2.3: *Download button*

Once the download is completed, navigate to your computer **Downloads** and double-click the **VMware-player** as shown:

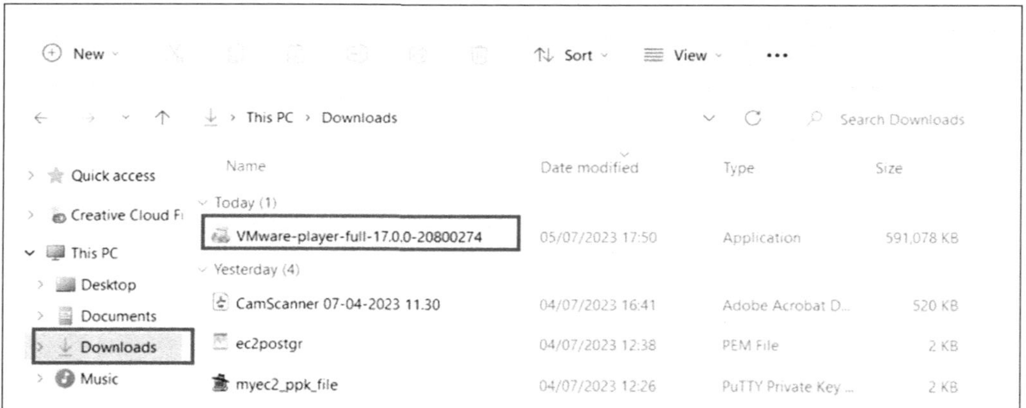

Figure 2.4: *VMware Download steps continue*

It will start the installation process for you. Use the default selection and continue by clicking **Next** till you finish the installation.

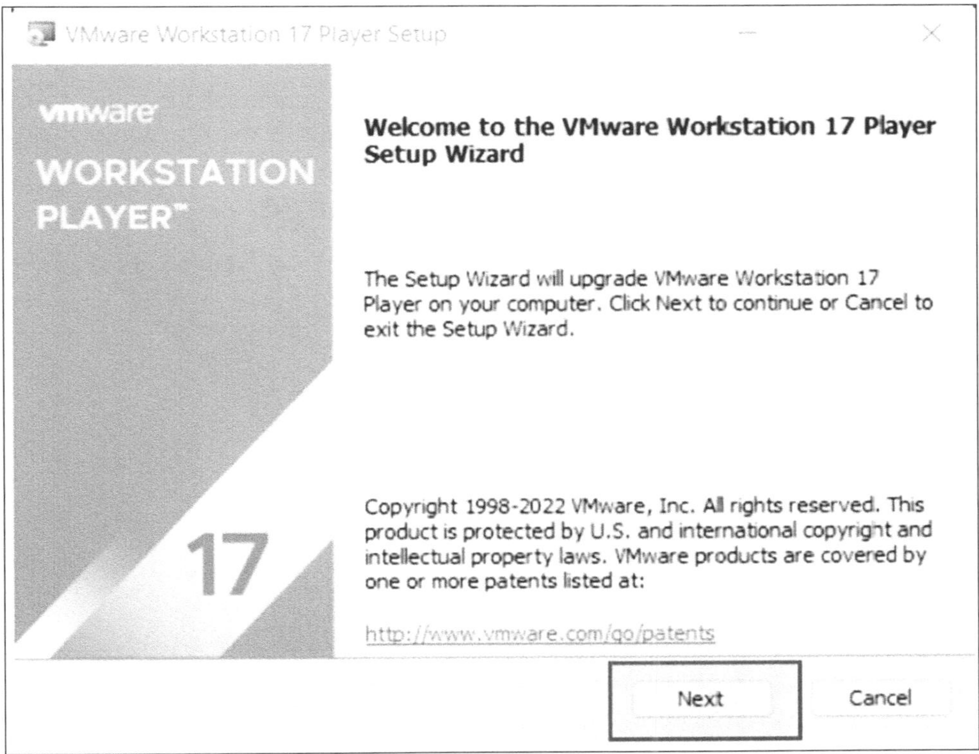

Figure 2.5: *VMware Workstation Download continue*

Once the installation is completed, you will see a shortcut on your desktop as shown here:

Figure 2.6: *VMware Workstation logo on desktop*

We are going to proceed with other installations and will revert to this later.

Linux Operating System (OS) Installation

We are going to install Ubuntu 22.04, which is the latest version at the time of writing. We will be downloading the ISO image of the specified Linux distribution.

Installing Ubuntu Using an iso Image

Open your favorite search engine again and type "**`Download ubuntu iso`**" as shown. Choose "**`Download Ubuntu desktop`**".

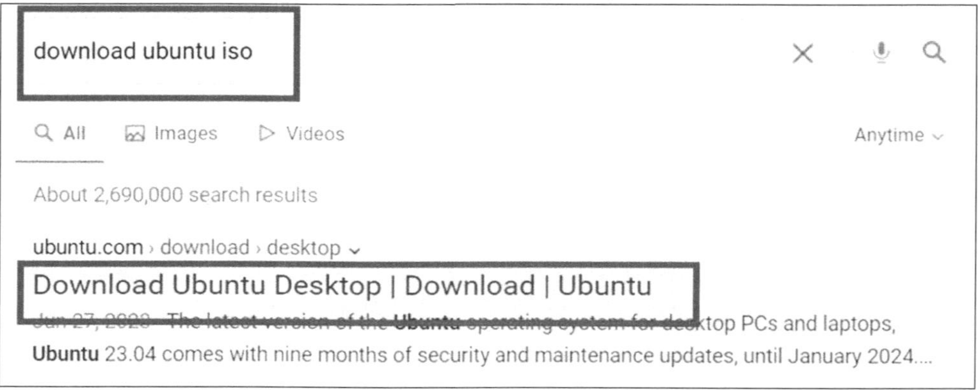

Figure 2.7: *Downloading Ubuntu ISO Image*

Perform the steps highlighted in the following figure:

Figure 2.8: *Downloading Ubuntu ISO Image (continued)*

Scroll down to "**`Download 22.04.2`**". Note any version prior to this or post this version can be downloaded.

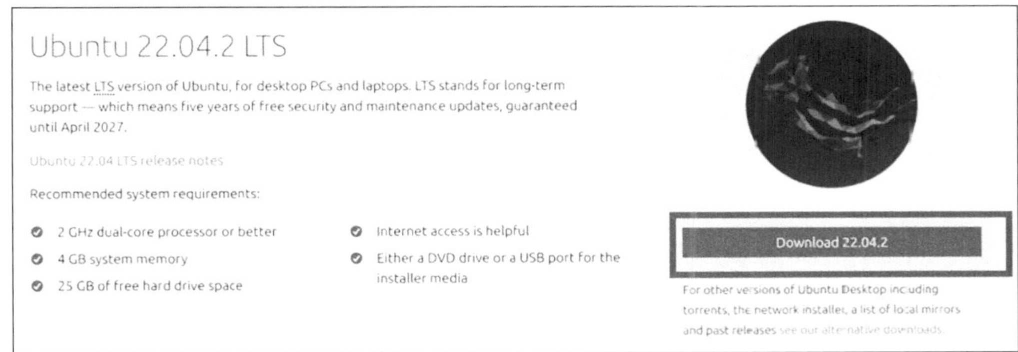

Figure 2.9: *Downloading Ubuntu ISO Image (continued)*

You should now find Ubuntu 22.04 or the version you downloaded in your Windows' download.

Figure 2.10: *Ubuntu Iso Image available in computer Downloads*

Now go back to your VMWare workstation and open it by double clicking it:

Figure 2.11: *Back to VMware Workstation Player logo*

Once it is opened, it will look like the following figure. Click **`create a new virtual machine`**.

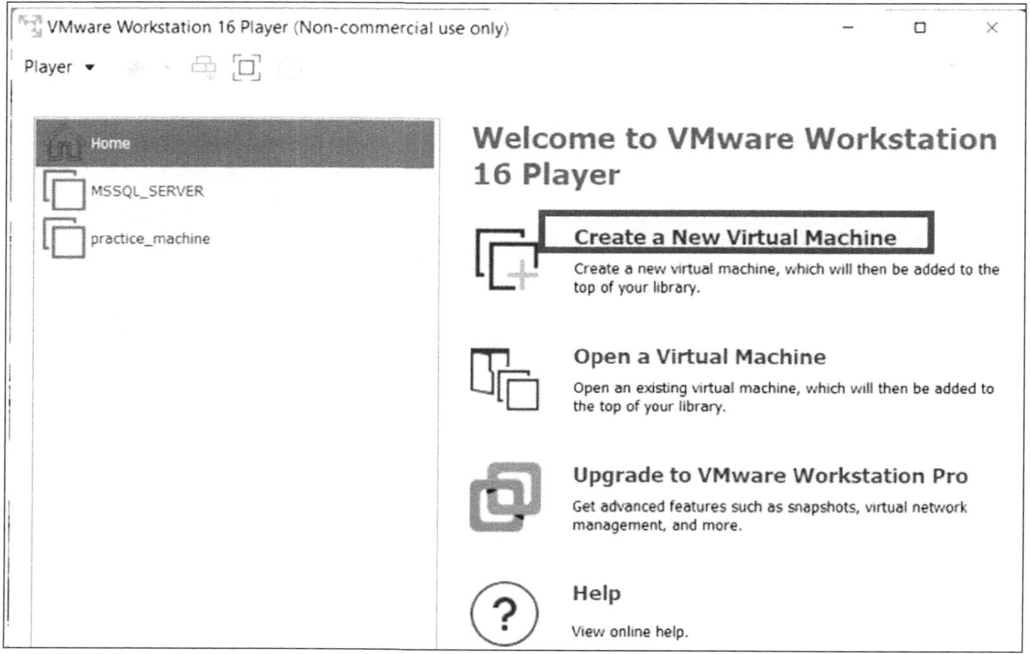

***Figure 2.12**: Creating a new virtual machine*

Perform the steps as shown highlighted in the following figure:

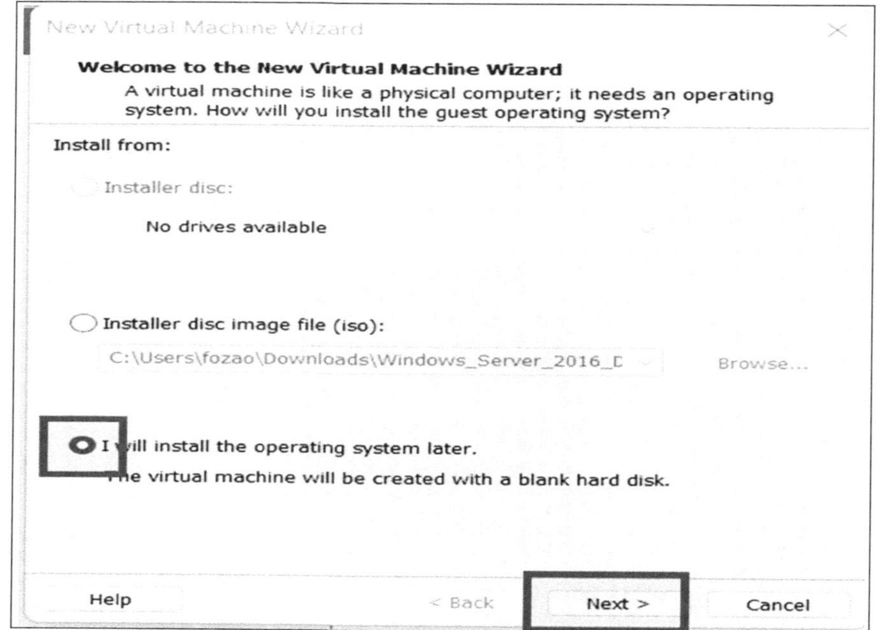

***Figure 2.13**: Creating a new virtual machine (continued)*

In the next figure, you will select the `Linux` operating system.

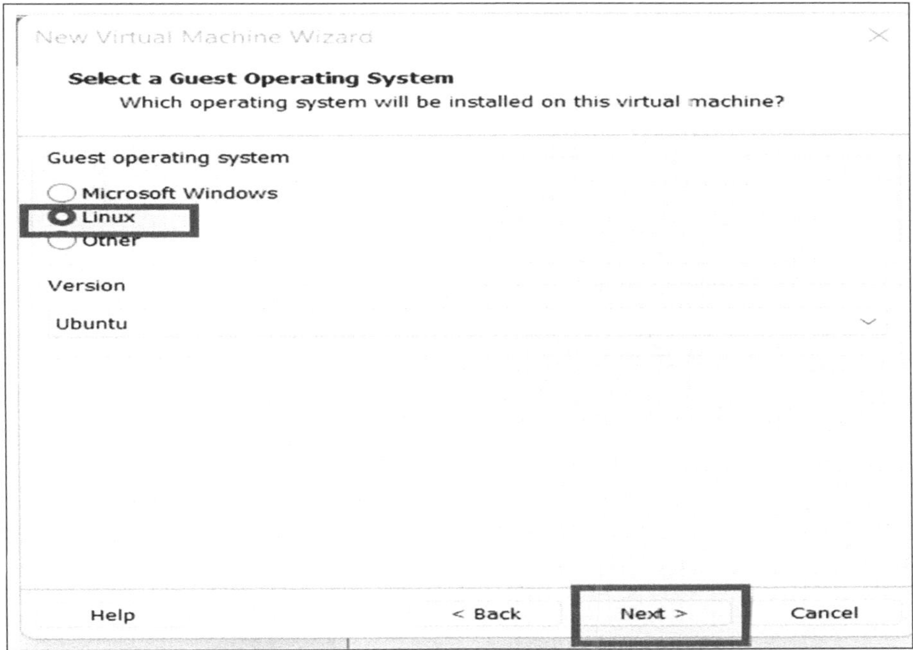

Figure 2.14: *Creating a new virtual machine (continued)*

Give your virtual machine a name of your choice as shown in the following figure:

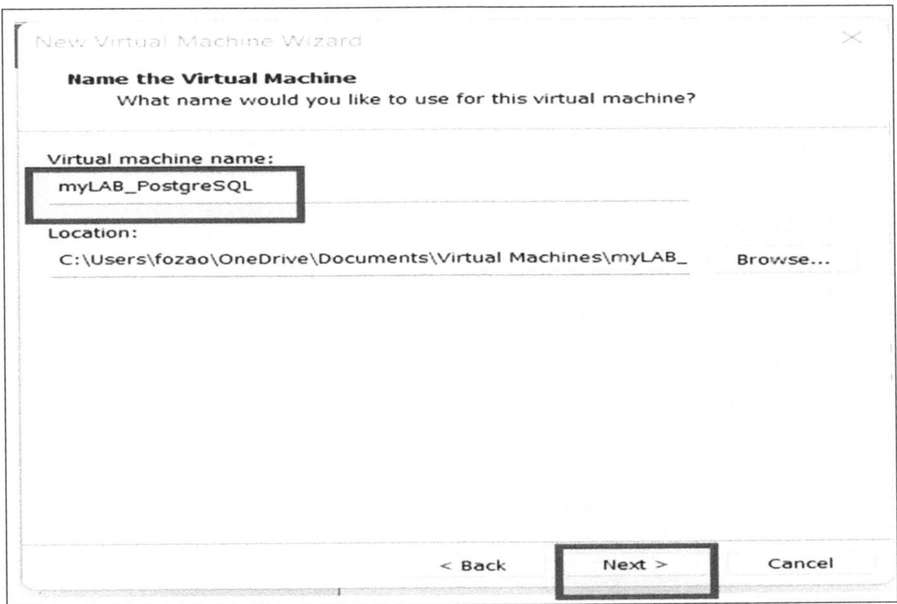

Figure 2.15: *Naming the new virtual machine*

You can adjust the disk size by clicking the up or down arrows beside the value. Since this is a learning session, we will go with the default size of 20 GB.

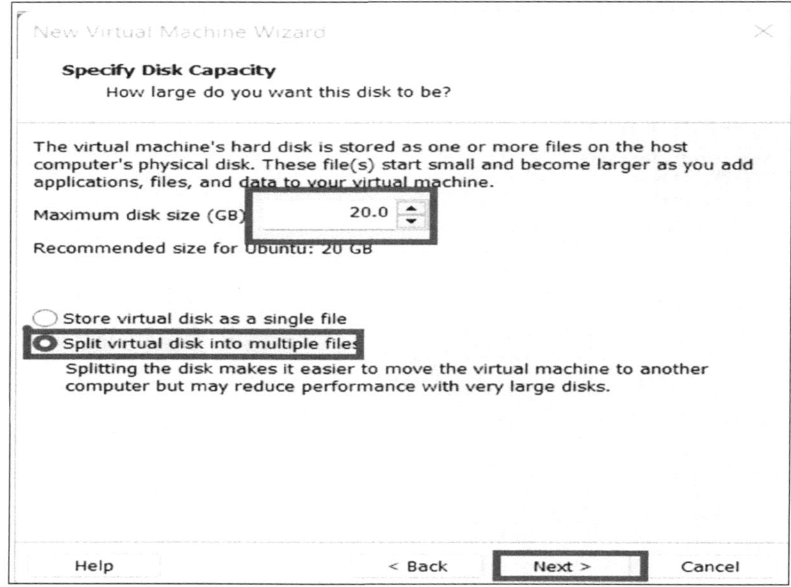

Figure 2.16: *Allocating virtual machine storage*

The next screen will show you an overview of your new virtual machine. Click **`Finish`** to create your virtual machine.

New Virtual Machine Wizard

Ready to Create Virtual Machine

Click Finish to create the virtual machine. Then you can install Ubuntu.

The virtual machine will be created with the following settings:

Name:	myLAB_PostgreSQL
Location:	C:\Users\fozao\OneDrive\Documents\Virtual Machines\...
Version:	Workstation 16.2.x
Operating System:	Ubuntu
Hard Disk:	20 GB, Split
Memory:	2048 MB
Network Adapter:	NAT
Other Devices:	2 CPU cores, CD/DVD, USB Controller, Printer, Sound C...

Customize Hardware...

< Back | Finish | Cancel

Figure 2.17: *An overview of Virtual Machine details*

Once your virtual machine has finished creating, it will look like the following figure:

Click **`Edit virtual machine settings`**

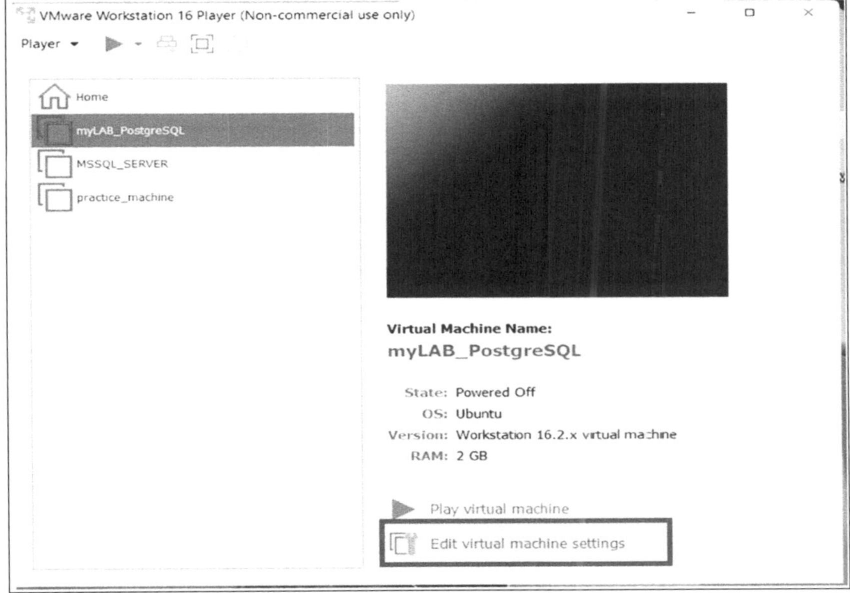

***Figure 2.18**: Edit Virtual Machine settings to add iso image*

The following figure shows the virtual machine setting overview. You can increase the memory for this machine by sliding up or down the big blue slider.

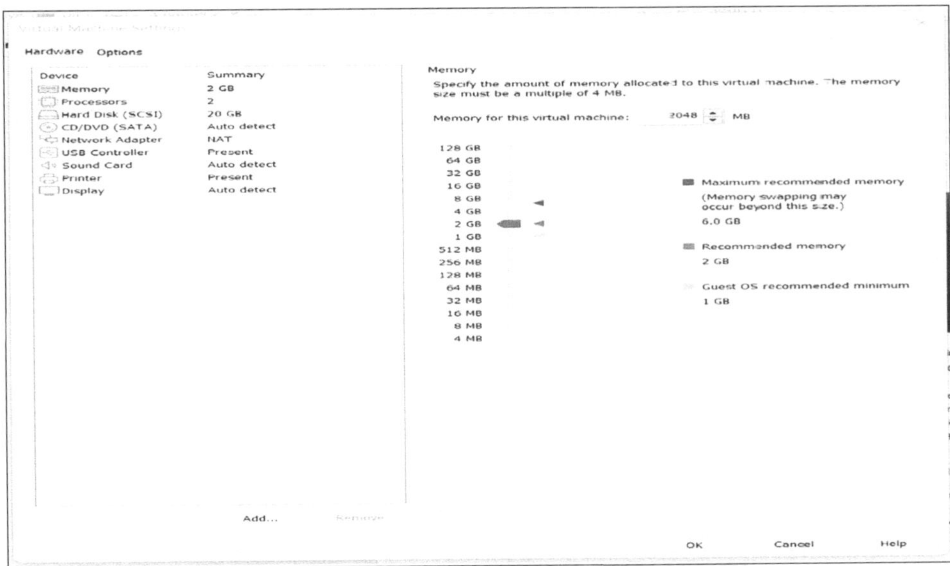

***Figure 2.19**: Overview of Virtual Machine settings*

You need to pay attention here to install your **ubuntu iso image** you downloaded earlier.

Click **"CD/DVD"** and then select "**use ISO image file**". Click **"browse"** to take you to your Windows folders, click **"Download and select the Ubuntu ISO image"** you downloaded in the previous section.

Only click **OK**, once all the preceding steps are completed.

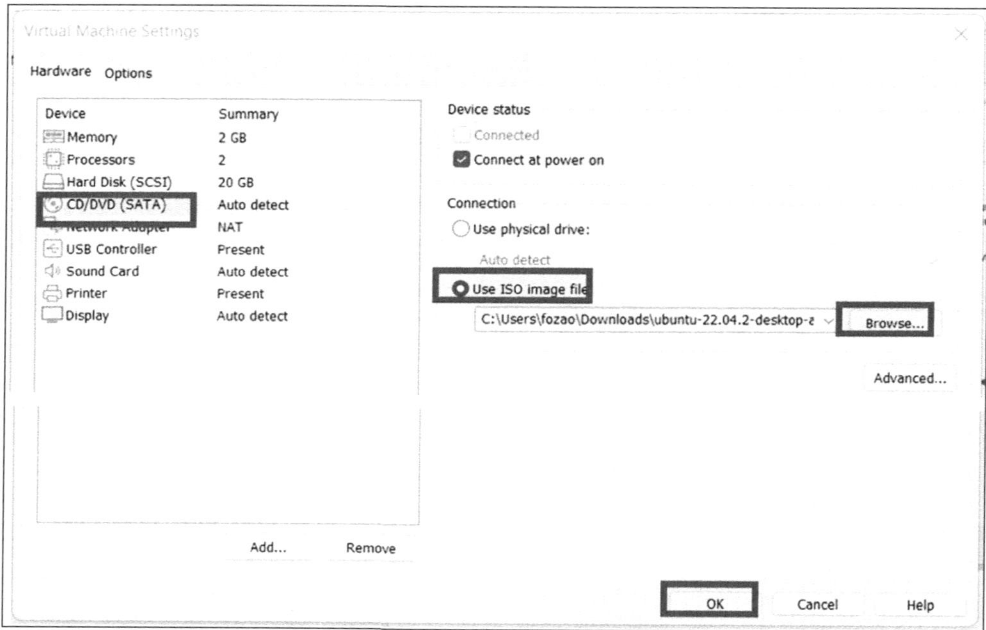

***Figure 2.20**: Installing the ISO image on the Virtual Machine settings*

On the next screen, select your newly created virtual machine and click the play as shown in the following figure:

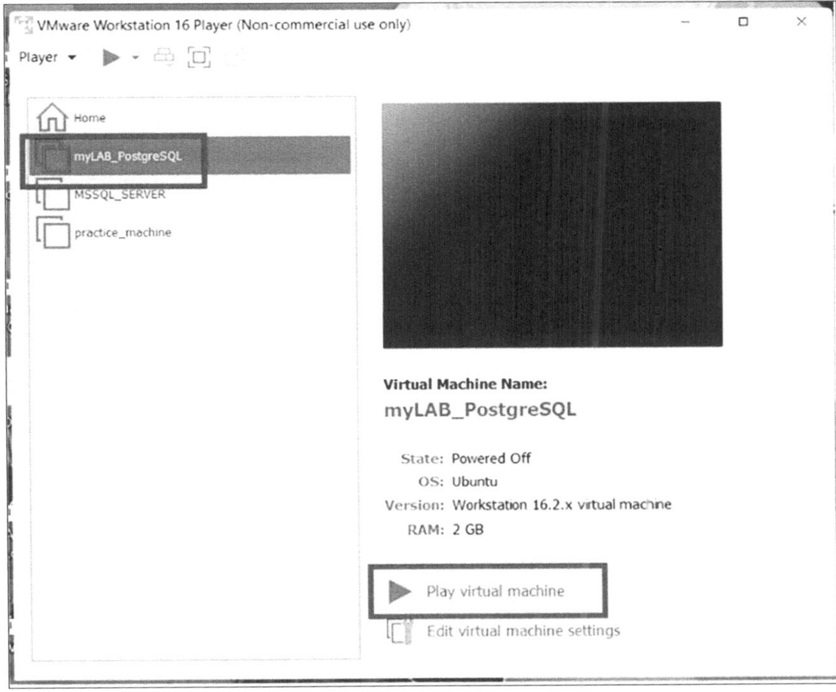

Figure 2.21: *To power up the new Virtual Machine*

A screen like the one shown here will run. Be patient and give it a few seconds to run.

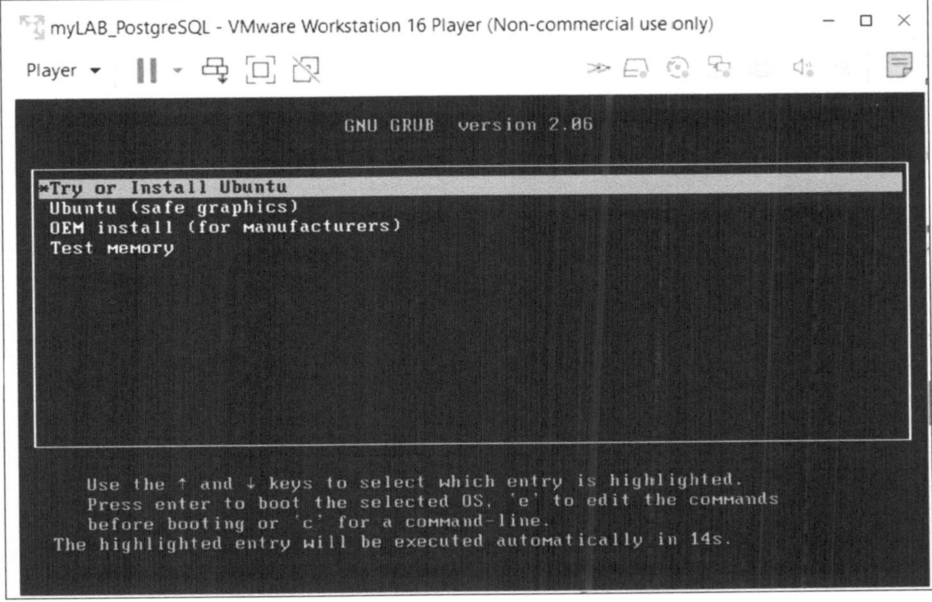

Figure 2.22: *Virtual Machine is powering up*

On the following screen, click **Install Ubuntu**.

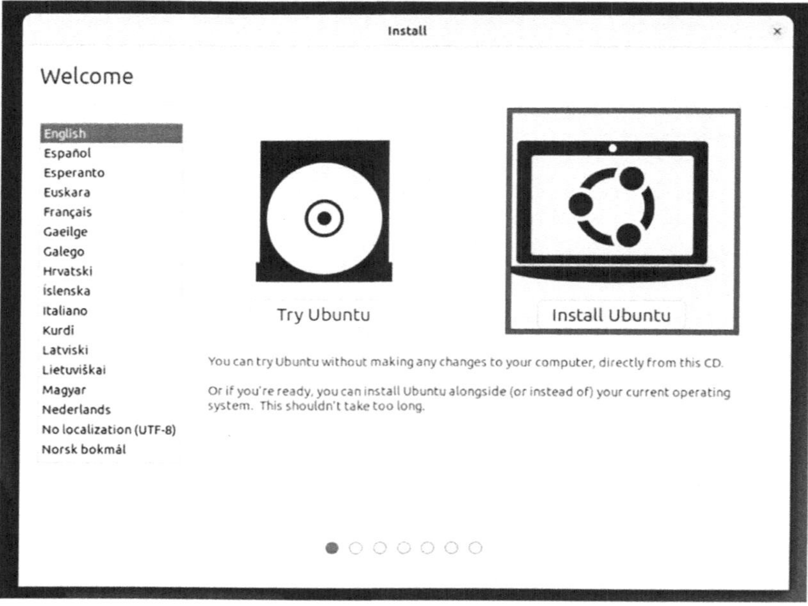

***Figure 2.23**: Installing Ubuntu*

Choose your language and click continue. You can see from the screenshot that we chose **English (UK)**.

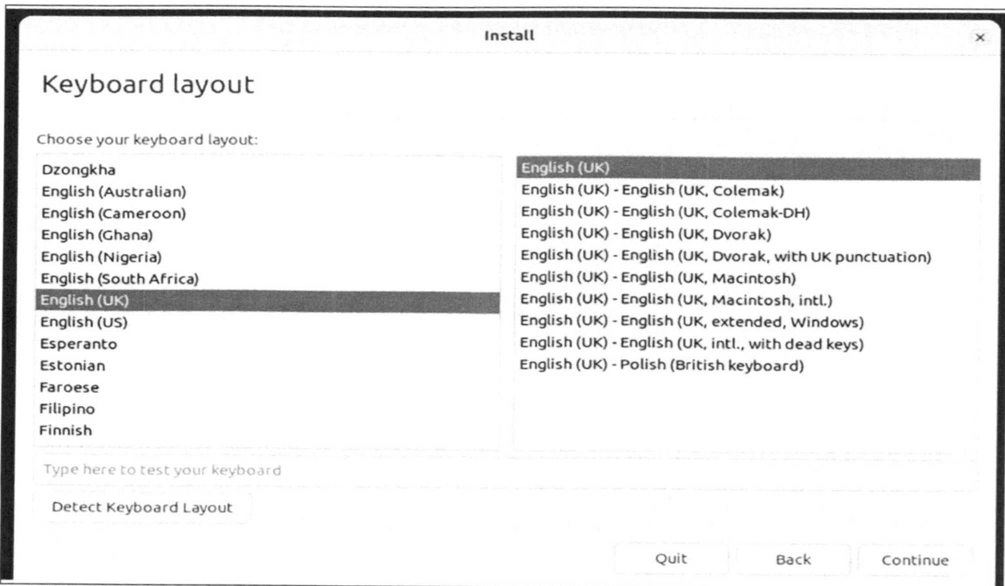

***Figure 2.24**: Selecting your language*

On the next Screen, leave it on the default (**`Normal installation`**) and click **`Continue`**.

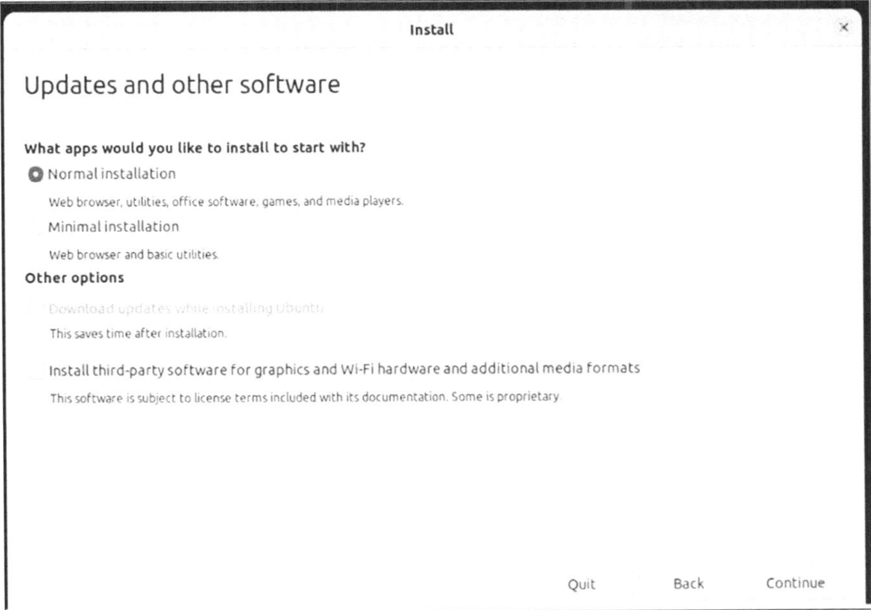

***Figure 2.25**: Ubuntu installation continues*

On the next screen, leave it on default **`Erase disk and install Ubuntu`** and click **`Install now`**.

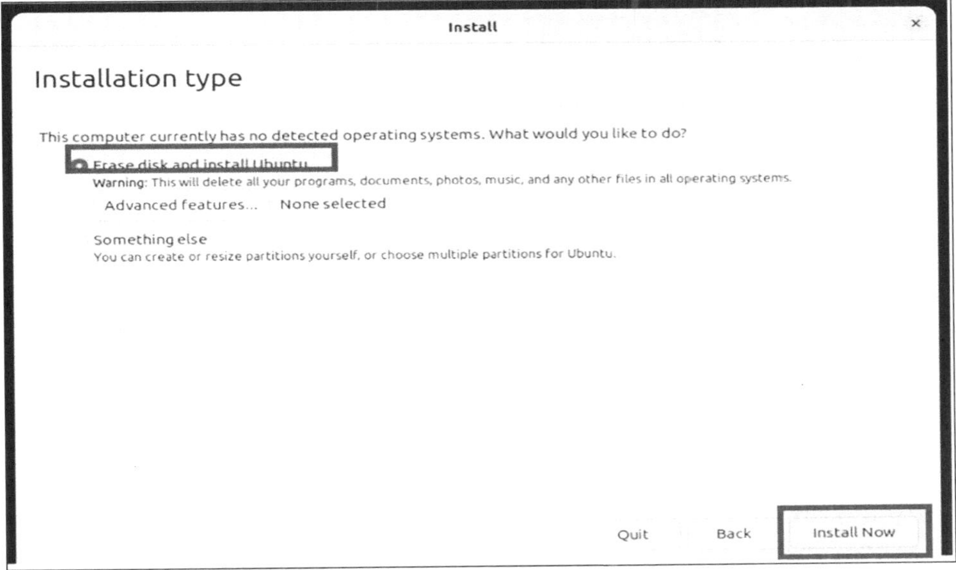

***Figure 2.26**: Choosing the Ubuntu installation type*

On the next screen as shown in *Figure 2.27*, click **Continue**.

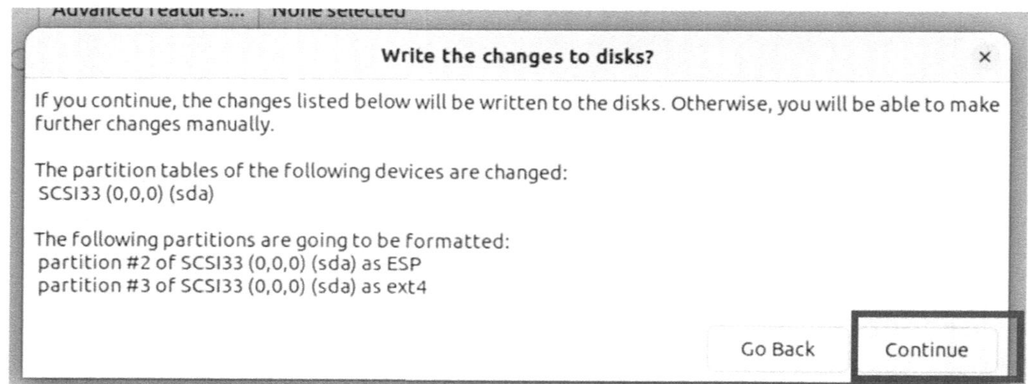

***Figure 2.27**: Ubuntu installation continues*

The next screen will automatically select your region. You can change this by hovering over the globe to your location. Click **Continue**.

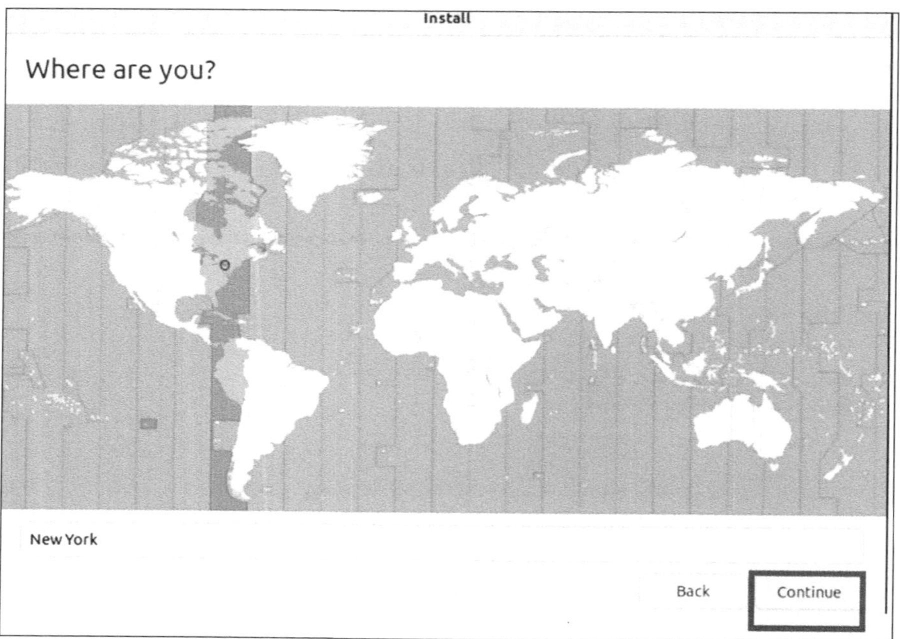

***Figure 2.28**: Choosing your region*

On the following screen, you will enter your name and your computer's name (the name you want to give to your virtual machine) and give a memorable password as it will be needed each time you want to access your virtual machine. Very important is to select **Require my password to login**. This increases the security of your virtual machine.

Figure 2.29: *Entering your login credentials*

Once you see the following screen, just be patient until the installations complete successfully:

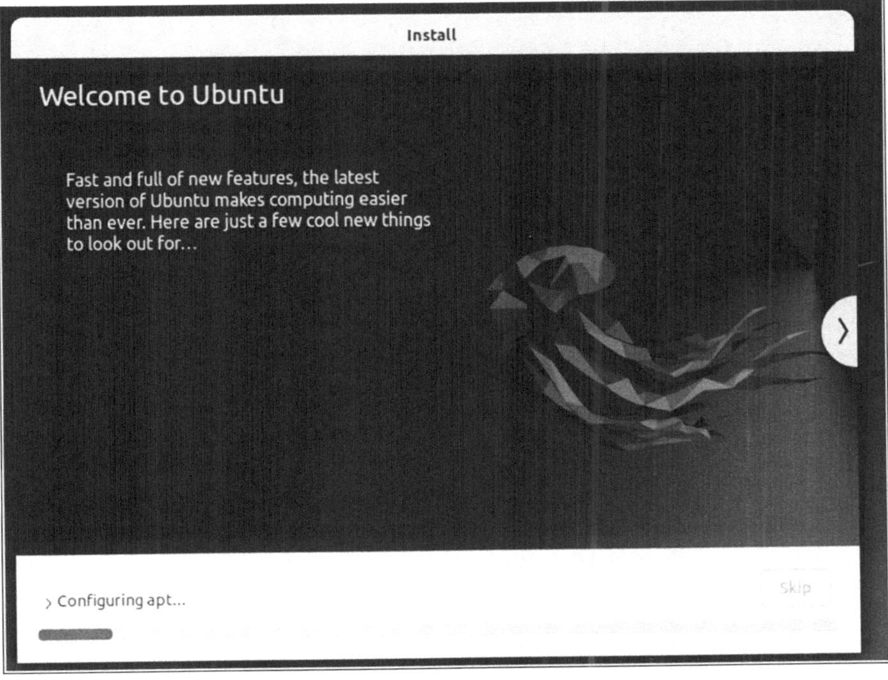

Figure 2.30: *Ubuntu is being installed*

After installation, the following screen will be displayed:

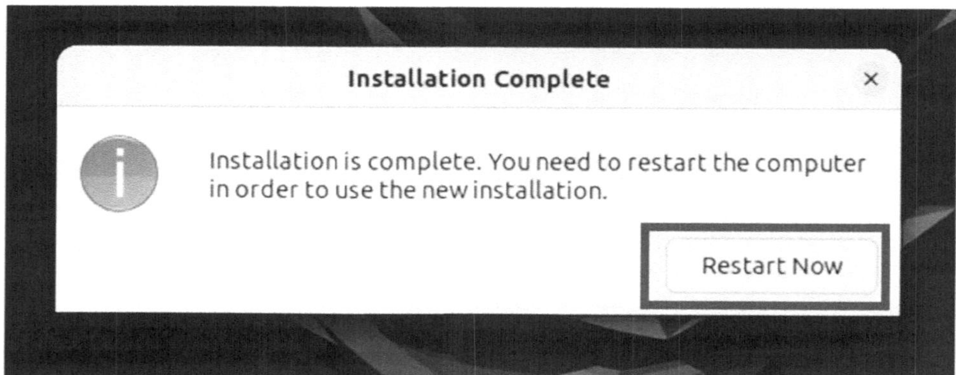

***Figure 2.31**: Ubuntu installation completed*

You will be required to press *enter* for the restart to complete.

The following screen shows that you have successfully completed the Ubuntu installation. Double-click your login name and provide the password you entered during creation to log in.

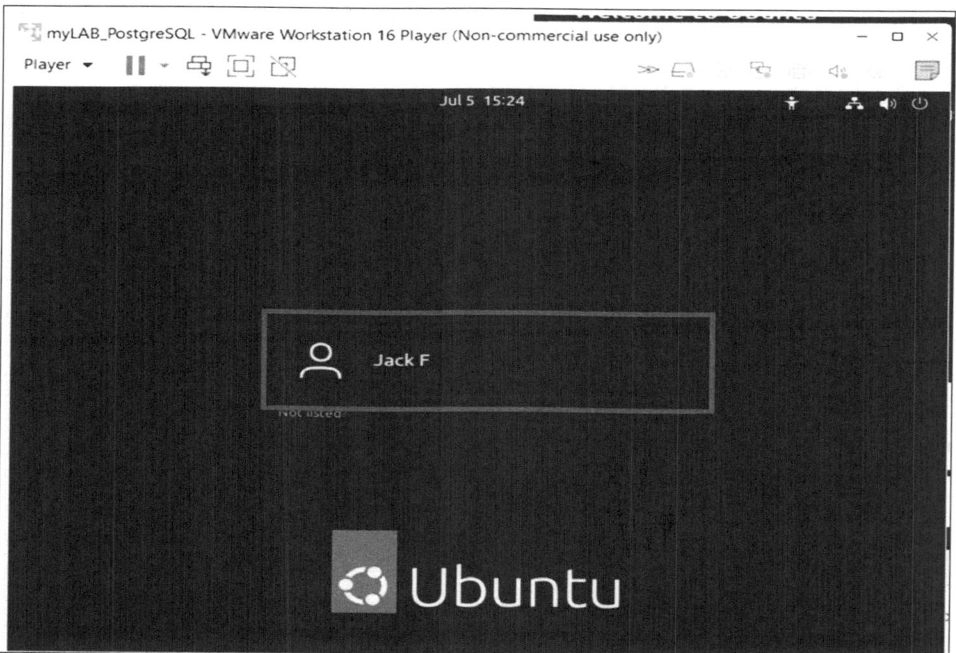

***Figure 2.32**: Ubuntu login screen*

The next few screens will just be skipped unless you know what you are doing. At this point, click **`skip`** till you get to the end.

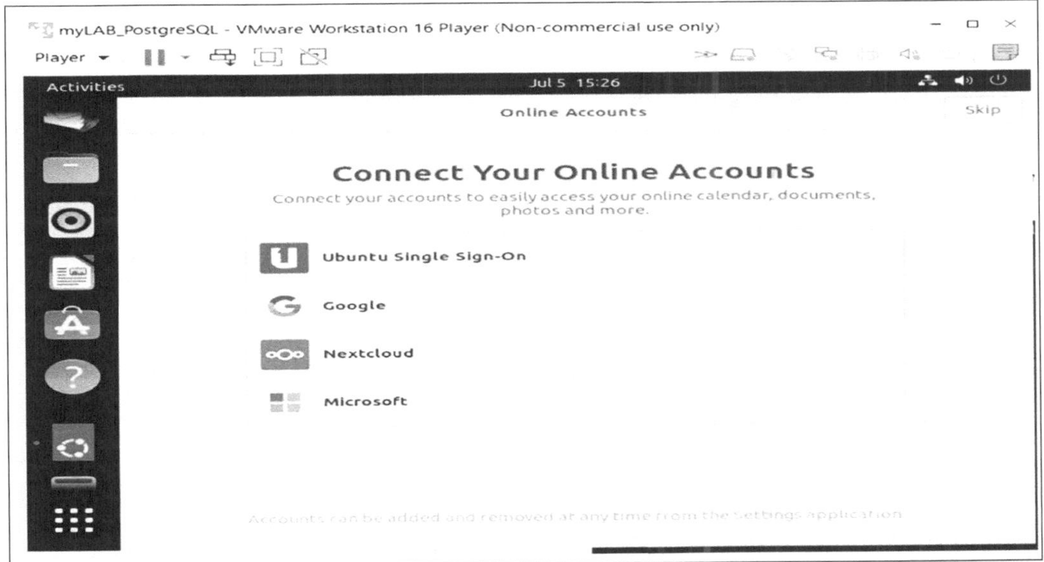

Figure 2.33: *Connecting to online accounts*

The following figure shows the home page of Ubuntu, click the highlighted dotted squares as shown:

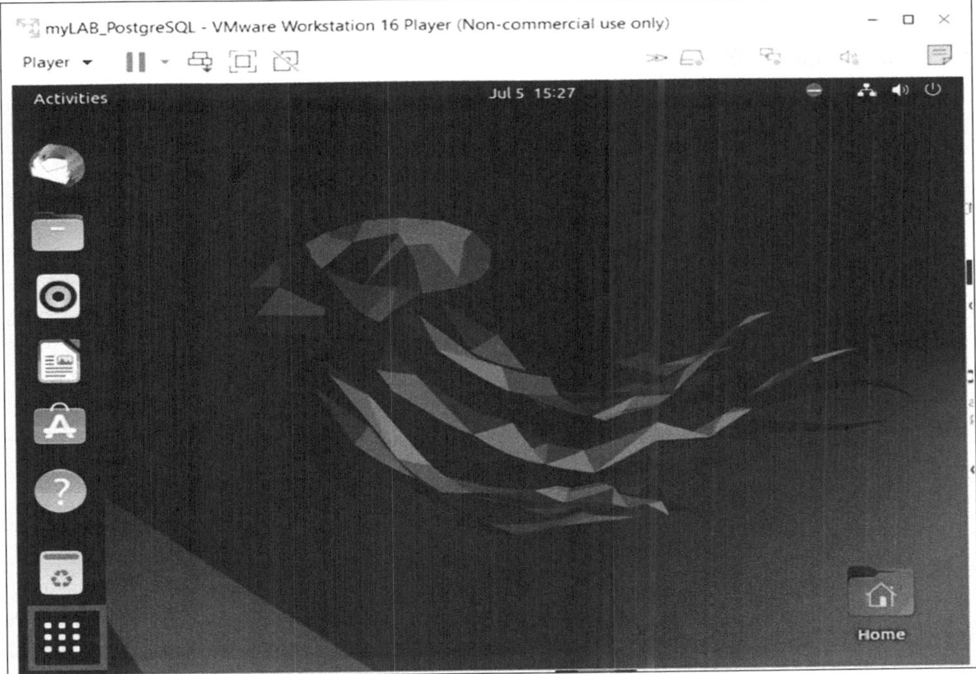

Figure 2.34: *Ubuntu home page*

Once clicked the preceding screen's square dotted shape, the following screen will open up as shown. Type the word "**`Terminal`**."

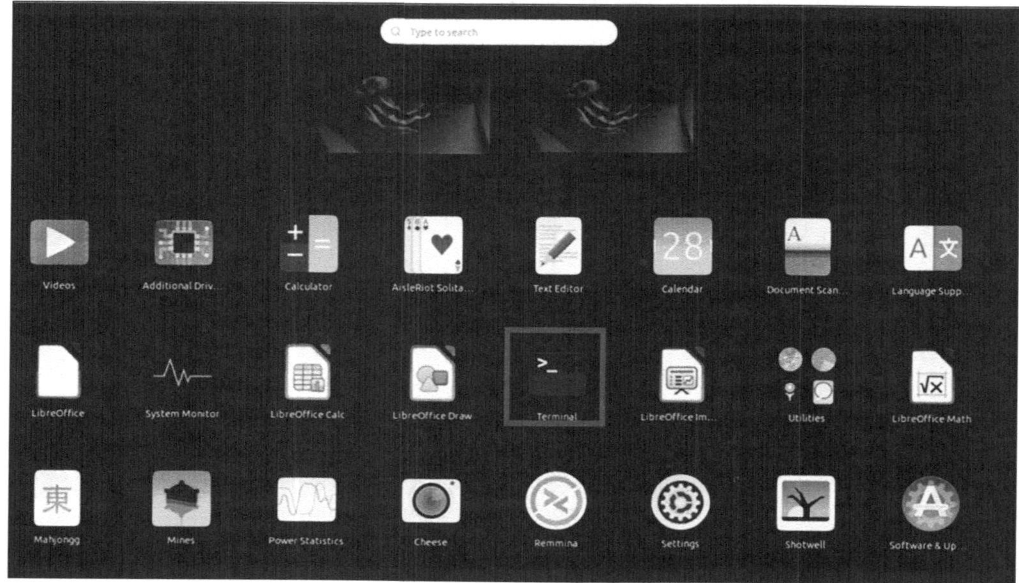

Figure 2.35: *Shows how to access a terminal*

The terminal will open as shown. It will show the username—in this case, **`'jack'`**—and the name of your virtual machine (**`jack-virtual-machine`**).

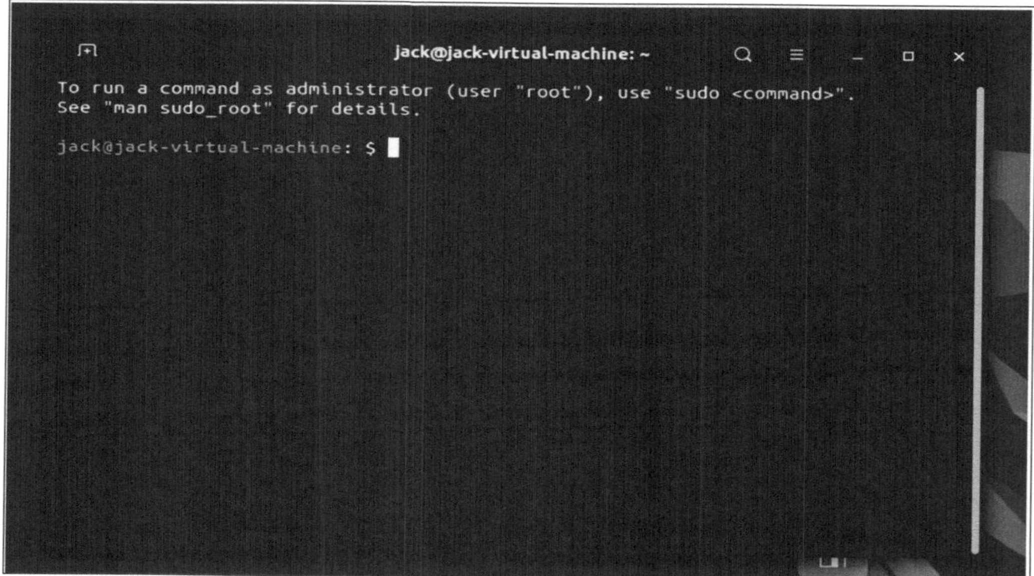

Figure 2.36: *Ubuntu terminal for commands execution*

Installing PostgreSQL

PostgreSQL, as seen in chapter 1, is an open-source relational database management system. It is also called Postgres. Postgres will be installed on Ubuntu 22.04 Linux OS.

To install Postgres, you need to open the VMware workstation by double clicking the below icon on your desktop, then click on the name of your virtual machine. In this case, it is **MyLAB_PostgreSQL**.

***Figure 2.37**: VMware workstation logo*

Choose your server as shown:

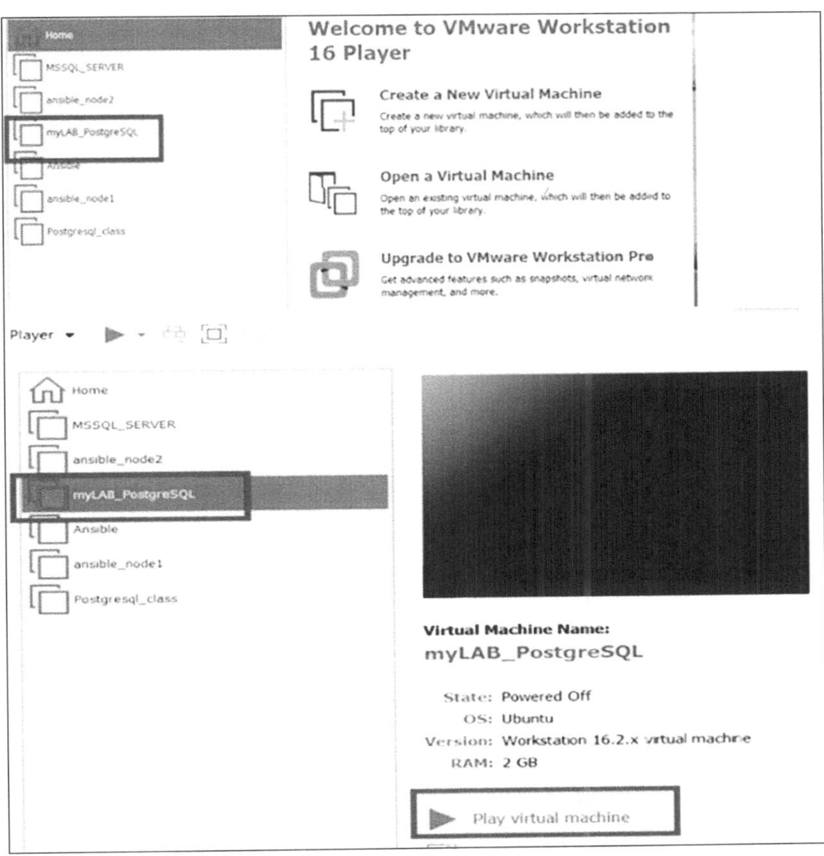

***Figure 2.38**: Powering up the Virtual machine*

Log in as mentioned earlier. Search for a terminal and you are ready to install PostgreSQL.

Perform the following steps to install PostgreSQL successfully.

1. Go to PostgreSQL official website using this link: www.postgresql.org and click **Download→**.

Figure 2.39: *PostgreSQL Download site*

On the following screen, select the operating system. In this case, it is **Linux**; then select the Linux distribution and, in this case, it is **Ubuntu**.

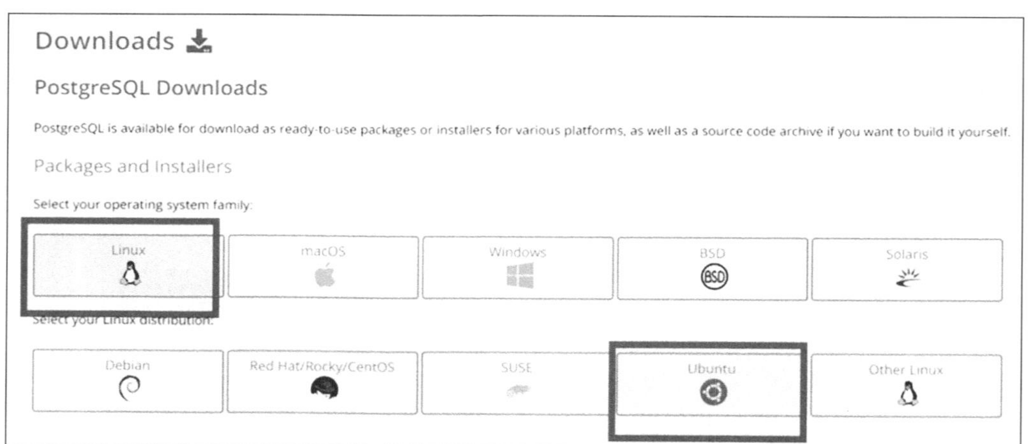

Figure 2.40: *Choosing the Operating System and distribution*

After choosing the operating system and the distribution, an installation script as shown will appear. Very important to note that to install the latest version of PostgreSQL, you can just copy and run the script all at once.

```
# Create the file repository configuration:
sudo sh -c 'echo "deb http://apt.postgresql.org/pub/repos/apt $(lsb_release -cs)-pgdg main" > /etc/apt/sources.list.d/pgdg.list'

# Import the repository signing key:
wget --quiet -O - https://www.postgresql.org/media/keys/ACCC4CF8.asc | sudo apt-key add -

# Update the package lists:
sudo apt-get update

# Install the latest version of PostgreSQL.
# If you want a specific version, use 'postgresql-12' or similar instead of 'postgresql':
sudo apt-get -y install postgresql
```

Copy Script

Figure 2.41: *PostgreSQL Installation steps*

If you want to install a version other than the latest version, the script must be run line by line. On the last line, the version to be installed must be indicated.

Copy and run the first line to create the file repository.

Note: *The red highlight shown does not indicate an error.*

```
jack@jack-virtual-machine: $ sudo sh -c "echo 'deb [signed-by=/usr/share/postgresql-common/pgdg/apt.postgresql.org.asc] https://apt.postgresql.org/pub
/repos/apt $VERSION_CODENAME-pgdg main' > /etc/apt/sources.list.d/pgdg.list"
jack@jack-virtual-machine: $
```

Figure 2.42: *PostgreSQL Installation step 1*

Copy and run the repository signing key as shown:

```
wget --quiet -O - https://www.postgresql.org/media/keys/ACCC4CF8.asc | sudo apt-key add -
```

Figure 2.43: *Import the repository signing key*

Copy and run the following for package updates.

```
sudo apt-get update
```

Figure 2.44: *To Update the package lists*

To install a specific version of PostgreSQL as shown:

```
# Install the latest version of PostgreSQL.
# If you want a specific version, use 'postgresql-12' or similar instead of 'postgresql':
```

```
sudo apt-get -y install postgresql-13
```

Figure 2.45: *Choosing PostgreSQL version*

On the last of the preceding commands, the version has been indicated as 13 (**postgresql-13**). Change the number to the required version as per your requirement.

All the preceding commands must be run inside a terminal.

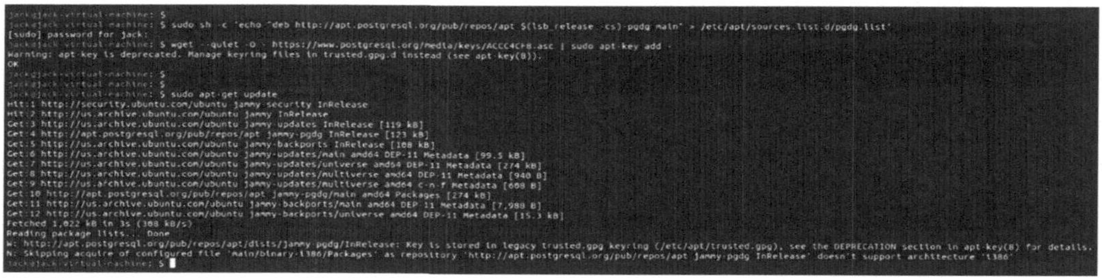

Figure 2.46: *All preceding commands shown in a terminal*

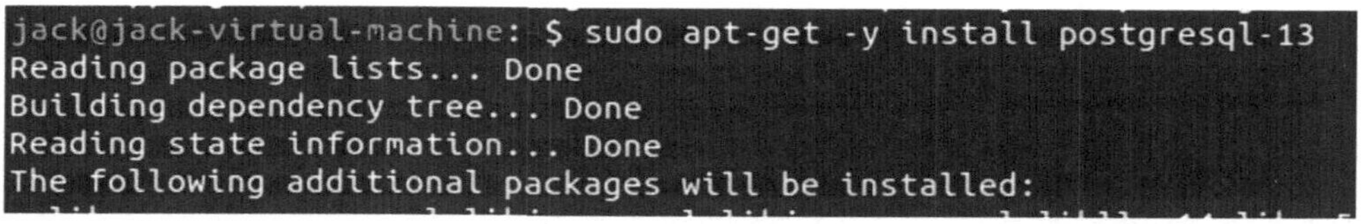

```
jack@jack-virtual-machine: $ sudo apt-get -y install postgresql-13
Reading package lists... Done
Building dependency tree... Done
Reading state information... Done
The following additional packages will be installed:
```

Figure 2.47: *Choosing the PostgreSQL version in a terminal*

The following screen shows that PostgreSQL 13 has been installed:

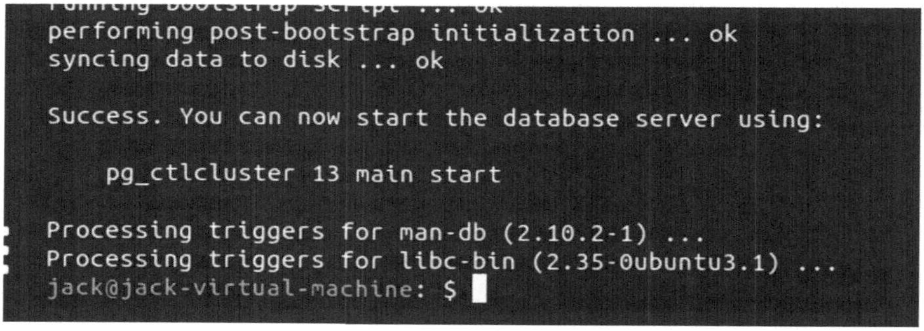

```
performing post-bootstrap initialization ... ok
syncing data to disk ... ok

Success. You can now start the database server using:

    pg_ctlcluster 13 main start

Processing triggers for man-db (2.10.2-1) ...
Processing triggers for libc-bin (2.35-0ubuntu3.1) ...
jack@jack-virtual-machine: $
```

Figure 2.48: *Screen shows PostgreSQL has been successfully installed.*

Once you see **Success**, then PostgreSQL has been successfully installed. You can now switch to user **postgres** using the command **sudo su - postgres**.

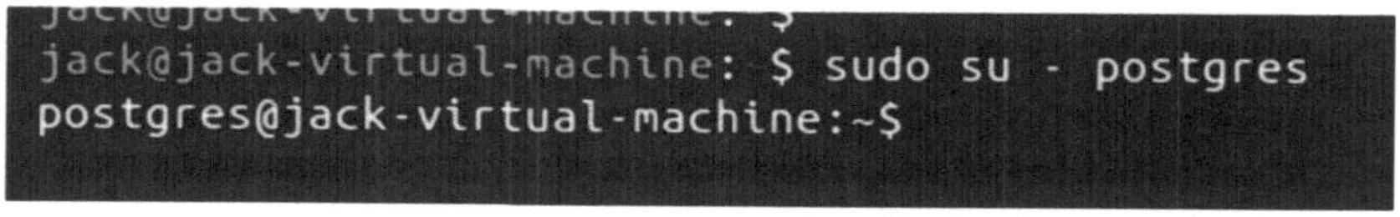

```
jack@jack-virtual-machine: $ sudo su - postgres
postgres@jack-virtual-machine:~$
```

Figure 2.49: *Switching to User postgres*

You can now login to the databases using the following command **`psql -U postgres`**.

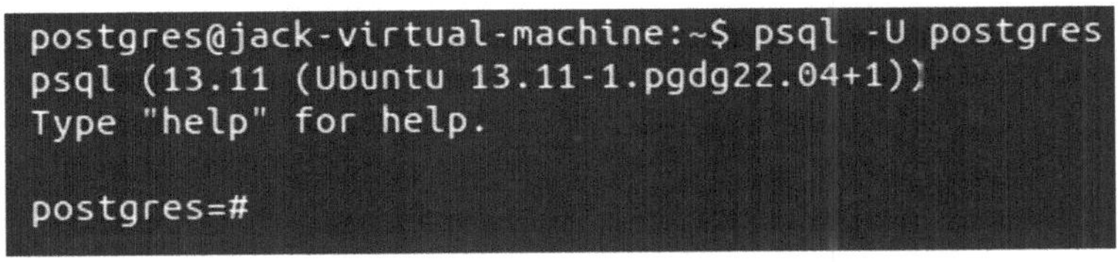

```
postgres@jack-virtual-machine:~$ psql -U postgres
psql (13.11 (Ubuntu 13.11-1.pgdg22.04+1))
Type "help" for help.

postgres=#
```

Figure 2.50: *Shows how to login to the databases as user postgres*

Conclusion

In this chapter we learned how to install VMware workstation Player, Linux Operating System (OS) Installation, PostgreSQL installation in Linux systems (Ubuntu) and finally installed PostgreSQL version 13.

In the next chapter, we will look at what SQL and its subsets are, data types and the different constraints.

CHAPTER 3
Subsets of SQL

Introduction

In this chapter, we will learn what SQL is all about and the different subsets of SQL. As this is a very important part of Postgre**SQL**, it is advisable that readers pay attention and understand all the sections of this chapter. In this chapter, we will delve into the different subsets of SQL and their respective uses.

Structure

In this chapter, we will cover the following topics:

- SQL Definition
- Data Definition Language (DDL)
- Data Manipulation Language (DML)
- Data Control Language (DCL),
- Transaction Control Language (TCL)
- Tasks

SQL Definition

SQL stands for *Structured Query Language.* It is a programming language that is used for maintaining data in relational database systems such as PostgreSQL, MySQL, MS SQL Server, and Oracle. Undisputedly, an English computer scientist employed at IBM called Edgar Frank (Ted) Codd is the father of SQL. In 1970, he published an article called **A Relational Model of Data for Large Shared Data Banks** which started the relational database era. Codd proposed two languages for the relational model: **relational calculus** and **relational algebra**. He was awarded a Turing Award in 1981, which is the highest distinction in computer science. In his article, he proposed a set of operations to extract data from tables (relations). This can be considered the first relational database queries. Following Codd's article in 1970, Raymond F. Boyce and Donald D. Chamberlain also working at IBM, further developed the SQL language.

SQL was designed for the maintenance of relational database management systems' (RDMS) data. It is used to access and manipulate data in relational databases. It helps you to create new databases, access the data stored in tables, and modify rows and columns of the tables. It helps database professionals to interact with their data easily. Different databases have specific needs and requirements. SQL has been designed with many subsets to address these database needs and requirements.

SQL has several subsets that are designed to address the requirements and needs of databases. These different subsets of SQL include **Data Definition Language (DDL)**, **Data Manipulation Language (DML), Data Control Language (DCL), Transaction Control Language (TCL), and Data Query Language (DQL)**. For learners and professionals to effectively manage and manipulate relational databases using SQL, it is very important for learners to understand the commands and uses of each of these SQL subsets. Having expert knowledge working with these subsets of SQL is paramount in the world of PostgreSQL and all other relational databases.

Data Definition Language (DDL)

Database objects such as tables, indexes, and constraints are characterized and managed using the Data Definition Language. **CREATE, DROP, ALTER,** and **TRUNCATE** keywords are DDL commands. We use the **CREATE** command to create new databases and their objects. The **DROP** command is used to remove databases and their objects. The **ALTER** keyword is used to modify existing databases and their objects. The **TRUNCATE** command is used to remove all rows from a table, effectively resetting it without deleting the table structure.

CREATE

The **CREATE** command. This section shows a syntax to create our first database, tables, and users/roles in the database.

To create a database, the command is

```
CREATE DATABASE database_name;
```

For example, to create a database called products, the command is **CREATE DATABASE products;**

All database commands must end with a semicolon (;). Let us connect back to our server and create our database.

Now let us log back into our database as user postgres.

Use the following command **psql -U postgres** as shown:

```
jack@jack-virtual-machine: $ sudo su - postgres
[sudo] password for jack:
postgres@jack-virtual-machine:~$ psql -U postgres
psql (13.11 (Ubuntu 13.11-1.pgdg22.04+1))
Type "help" for help.
```

Figure 3.1: *How to change to user postgres and log in to the database*

Type **`\l`** to see all the databases available. The default databases are **`postgres`**, **`template0`**, and **`template1`**.

```
postgres=# \l
                              List of databases
   Name    |  Owner   | Encoding |   Collate   |    Ctype    |   Access privileges
-----------+----------+----------+-------------+-------------+-----------------------
 postgres  | postgres | UTF8     | en_US.UTF-8 | en_US.UTF-8 |
 template0 | postgres | UTF8     | en_US.UTF-8 | en_US.UTF-8 | =c/postgres          +
           |          |          |             |             | postgres=CTc/postgres
 template1 | postgres | UTF8     | en_US.UTF-8 | en_US.UTF-8 | =c/postgres          +
           |          |          |             |             | postgres=CTc/postgres
(3 rows)
```

Figure 3.2: *How to change to check the list of databases*

Now let us create our first database called **`products`.** The following command will create a database called products upon execution.

```
CREATE DATABASE products;
```

You will see **`CREATE DATABASE`** response to show that the database has been created. **`Postgres=#`** shows that you are connected to postgres database.

```
postgres=# CREATE DATABASE products;
CREATE DATABASE
```

Figure 3.3: *How to create a new database*

Now let us run **`\l`** again to see that products are now shown on the list of databases.

```
postgres=# \l
                              List of databases
   Name    |  Owner   | Encoding |   Collate   |    Ctype    |   Access privileges
-----------+----------+----------+-------------+-------------+-----------------------
 postgres  | postgres | UTF8     | en_US.UTF-8 | en_US.UTF-8 |
 products  | postgres | UTF8     | en_US.UTF-8 | en_US.UTF-8 |
 template0 | postgres | UTF8     | en_US.UTF-8 | en_US.UTF-8 | =c/postgres          +
           |          |          |             |             | postgres=CTc/postgres
 template1 | postgres | UTF8     | en_US.UTF-8 | en_US.UTF-8 | =c/postgres          +
           |          |          |             |             | postgres=CTc/postgres
(4 rows)
```

Figure 3.4: *How to list all the databases*

To connect to our products database, we use the **\c** command.

```
\c products;
```

```
postgres=# \c products ;
You are now connected to database "products" as user "postgres".
products=#
```

***Figure 3.5**: How to connect to the products database*

You are now connected to the *products database* as shown:

To see the relations(tables) contain in our products database, we execute the following command: **\dt**

```
products=# \dt
Did not find any relations.
```

***Figure 3.6**: To check for tables (relations) in the database*

The result shows that there are no relations found because we have not created any tables yet.

DROP

The **DROP** command is used to remove entire database objects, such as tables, indexes, or databases. When you **DROP** a table, it deletes the table structure along with all of its data. This means the table and all related objects (such as indexes) are permanently removed from the database.

Syntax is:

```
DROP TABLE table_name;
DROP DATABASE database_name;
```

We have created the users database to be used to demonstrate the drop database command.

```
postgres=#
postgres=# CREATE DATABASE users;
CREATE DATABASE
postgres=# \l
                                    List of databases
   Name    |  Owner   | Encoding | Locale Provider |   Collate   |    Ctype    |
-----------+----------+----------+-----------------+-------------+-------------+
 dvdrental | postgres | UTF8     | libc            | en_US.UTF-8 | en_US.UTF-8 |
 postgres  | postgres | UTF8     | libc            | en_US.UTF-8 | en_US.UTF-8 |
           |          |          |                 |             |             |
           |          |          |                 |             |             |
 products  | postgres | UTF8     | libc            | en_US.UTF-8 | en_US.UTF-8 |
 template0 | postgres | UTF8     | libc            | en_US.UTF-8 | en_US.UTF-8 |
           |          |          |                 |             |             |
 template1 | postgres | UTF8     | libc            | en_US.UTF-8 | en_US.UTF-8 |
           |          |          |                 |             |             |
 users     | postgres | UTF8     | libc            | en_US.UTF-8 | en_US.UTF-8 |
(6 rows)
```

Figure 3.7: *Create database users*

Note, the database you are currently connected to cannot be dropped.

To get rid of our users database, we execute **DROP DATABASE users;** command as shown in *Figure* 3.8.

```
postgres=#
postgres=# DROP DATABASE users;
DROP DATABASE
postgres=#
postgres=# \l
                                    List of databases
   Name    |  Owner   | Encoding | Locale Provider |   Collate   |    Ctype    |
-----------+----------+----------+-----------------+-------------+-------------+
 dvdrental | postgres | UTF8     | libc            | en_US.UTF-8 | en_US.UTF-8 |
 postgres  | postgres | UTF8     | libc            | en_US.UTF-8 | en_US.UTF-8 |
           |          |          |                 |             |             |
           |          |          |                 |             |             |
 products  | postgres | UTF8     | libc            | en_US.UTF-8 | en_US.UTF-8 |
 template0 | postgres | UTF8     | libc            | en_US.UTF-8 | en_US.UTF-8 |
           |          |          |                 |             |             |
 template1 | postgres | UTF8     | libc            | en_US.UTF-8 | en_US.UTF-8 |
           |          |          |                 |             |             |
(5 rows)
```

Figure 3.8: *Drop the users database*

TRUNCATE

The **TRUNCATE** command is used to quickly remove all rows from a table, essentially resetting the table but keeping its structure intact.

When you **TRUNCATE** a table, it deletes all rows in the table, but the table structure, including indexes and constraints, remains. It is faster than a **DELETE** operation because it does not generate individual row delete statements.

Syntax is:

```
TRUNCATE TABLE table_name;
```

Key differences:

- **Scope: DROP** removes the entire table, while **TRUNCATE** only removes the data within the table but keeps the structure.
- **Speed: TRUNCATE** is faster than **DROP** because it does not involve logging individual row deletions.

For the purpose of demonstrating **TRUNCATE** and **DROP** table commands, we are going to connect to the products database and create **products_on_sales** table.

```
CREATE TABLE products_on_sales ( product_id SERIAL PRIMARY KEY,
 product_name VARCHAR(100),
category VARCHAR(50),
price NUMERIC(10, 2),
 stock_quantity INTEGER,
date_added DATE );
```

Insert the following five records into the table.

```
INSERT INTO products_on_sales (product_name, category, price,
stock_quantity, date_added)
VALUES ('Wireless Mouse', 'Electronics', 29.99, 100, '2024-01-15'),
('Running Shoes', 'Footwear', 79.99, 50, '2023-11-10'),
('Coffee Maker', 'Appliances', 49.99, 30, '2024-03-05'),
('Bluetooth Speaker', 'Electronics', 99.99, 75, '2023-12-20'),
 ('Yoga Mat', 'Fitness', 25.99, 200, '2024-02-18');
```

```
products=# CREATE TABLE products_on_sales ( product_id SERIAL PRIMARY KEY,
 product_name VARCHAR(100),
category VARCHAR(50),
price NUMERIC(10, 2),
 stock_quantity INTEGER,
date_added DATE );
CREATE TABLE
products=#
products=#  INSERT INTO products_on_sales (product_name, category, price, stock_quantity, date_added)
VALUES ('Wireless Mouse', 'Electronics', 29.99, 100, '2024-01-15'),
('Running Shoes', 'Footwear', 79.99, 50, '2023-11-10'),
('Coffee Maker', 'Appliances', 49.99, 30, '2024-03-05'),
('Bluetooth Speaker', 'Electronics', 99.99, 75, '2023-12-20'),
 ('Yoga Mat', 'Fitness', 25.99, 200, '2024-02-18');
INSERT 0 5
```

***Figure 3.9**: Create a table and insert data into it*

Query the table to confirm the records by using the

```
SELECT * FROM products_on_sales;
```

```
products=# SELECT * FROM products_on_sales;
 product_id |   product_name    |  category   | price | stock_quantity | date_added
------------+-------------------+-------------+-------+----------------+------------
          1 | Wireless Mouse    | Electronics | 29.99 |            100 | 2024-01-15
          2 | Running Shoes     | Footwear    | 79.99 |             50 | 2023-11-10
          3 | Coffee Maker      | Appliances  | 49.99 |             30 | 2024-03-05
          4 | Bluetooth Speaker | Electronics | 99.99 |             75 | 2023-12-20
          5 | Yoga Mat          | Fitness     | 25.99 |            200 | 2024-02-18
(5 rows)
```

Figure 3.10: *Query the inserted data*

The **TRUNCATE TABLE products_on_sales;** command as shown in *Figure 3.11*, delete all the data but the table structure is not deleted.

```
products=# TRUNCATE TABLE  products_on_sales;
TRUNCATE TABLE
products=#
products=# SELECT * FROM products_on_sales;
 product_id | product_name | category | price | stock_quantity | date_added
------------+--------------+----------+-------+----------------+------------
(0 rows)

products=#
products=#
products=# \dt
              List of relations
 Schema |       Name        | Type  |  Owner
--------+-------------------+-------+----------
 public | products_on_sales | table | postgres
 public | schedule          | table | postgres
(2 rows)
```

Figure 3.11: *Applying the TRUNCATE command on a table*

The **DROP TABLE products_on_sales;** command delete the whole table. The **SELECT** on the table after the **DROP** statement returns relation does not exist as shown as follows:

```
products=#
products=# DROP TABLE products_on_sales;
DROP TABLE
products=#
products=# SELECT * FROM products_on_sales;
ERROR:  relation "products_on_sales" does not exist
LINE 1: SELECT * FROM products_on_sales;
                      ^
products=# \dt
          List of relations
 Schema |   Name   | Type  |  Owner
--------+----------+-------+----------
 public | schedule | table | postgres
(1 row)
```

Figure 3.12: *Applying the DROP TABLE command on a table*

More about tables is shown as follows:

Database Tables: Tables contain columns and rows of data. Columns are referred to as fields, and rows are referred to as records. Each column in a table has a defined data type, which shows what type of data the column can hold. For example, if the column is defined with a numeric data type, it cannot take a different data type rather than the numeric data type. An example of a database table is shown:

Each row of data is usually unique, and each column usually contains one value per row.

Id integer	`First_name` `Varying character(35)`	`Last_name` `Varying character(35)`	`City` `Varying character(35)`	`State` `character(2)`
1	Deepesh	Raj	Houston	TX
2	Mariam	Chaudhry	New York	NY
3	Zainab	Kapur	New Jersey	NJ
4	Hassan	Raj	San Francisco	CA
5	Sunita	Kaur	Maryland	MD

Table 3.1: *Car Owners*

Table 3.1 contains five columns with their data types listed underneath them. We have an id column with an integer data type, a **`first_name`**, **`last_name`**, and **`city`** columns all containing varying character data types, and the last column is state with character data type. It is worth mentioning that the id column is a primary key, and it uniquely identifies each row within the table, for example, id number 5, identifies Sunita Kaur from Maryland.

It is very important to understand data types and constraints before we start creating database tables. In the next sections, learners will learn about data types and constraints.

Data Types

In PostgreSQL, there are numerous data types. In this section, we will learn the most used data types in PostgreSQL.

Numeric Data Type

This data type comprises various subtypes, each serving specific purposes:

- **`INTEGER (INT)`**: The **`INTEGER (INT)`** is the most common data type in PostgreSQL and in all relational databases. It is used to store whole numbers such as 1, 2, 50, 87, and so on. It ranges from -2,147,483,648 to 2,147,483,647. An example of data that uses the INT data type includes quantity and age.

- **SMALLINT: SMALLINT** is used to store smaller range whole numbers when less storage is required. It ranges from -32,768 to 32,767. It is suitable for storing data such as the number of pages in a book or a small set of counters.
- **BIGINT: BIGINT** is used for storing larger range whole numbers where **INTEGER** is not sufficient. It ranges from -9,223,372,036,854,775,808 to 9,223,372,036,854,775,807. It is used for data such as large counters, large ID numbers, or financial transactions.
- **DECIMAL** (or **NUMERIC**) **:** The **DECIMAL** type is used for exact numeric data with defined precision and scale. I range up to 131,072 digits before the decimal point and up to 16,383 digits after the decimal point. For instance, for storing price 9754.58, you can define the column as **NUMERIC(6,2)** where **P** (precision) is 6 and **S** (scale) is 2.
- **FLOAT4 (REAL)**: It is used for single-precision floating-point numbers where exact precision is not critical. It is suitable for scientific calculations where a rough precision is acceptable, such as sensor readings.
- **DOUBLE PRECISION**: It is used for double-precision floating-point numbers where higher precision is necessary. Appropriate for complex mathematical calculations such as financial models, where precision is key.
- **SERIAL**: The **SERIAL** data type is used for auto-incrementing integers, commonly used for unique identifiers. A typical use case for **SERIAL** is primary keys in a table, which automatically increment with each new record.

String Data Types

This data type comprises the following types of data: **CHARACTER** or **CHAR(N)**, **VARYING CHARACTER** or **VARCHAR(N)**, and **TEXT**.

- **CHARACTER or CHAR(N)**: A string of fixed length N, where N represents the fixed number of characters in the string. The maximum length for **CHAR(N)** is 10,485,760 characters. Examples of fixed-length data include state and gender. Gender can either be male (M) or female (F), thus the fixed length of N = 1.
- **VARYING CHARACTER or VARCHAR(N)**: A string of varying length with N being the maximum number of characters. The maximum length for **VARCHAR(N)** is 10,485,760 characters. Examples of data that take varying character sizes are names and emails. For instance, a first name column of VARCHAR(35) means a first name can have up to 35 characters but not more.
- **TEXT**: A data type with varying length and no maximum number of characters. The **TEXT** data type can hold up to 1 GB of text. Examples of **TEXT** data types include reviews and comments.

Time Data Type

This consists of **TIME**, **DATE**, **TIMESTAMP**, and **TIMESTAMP WITH TIME ZONE** data types.

- **TIME**: The **TIME** data type is presented in the order of **HH:MM:SS**, where **H** stands for hour, **M** stands for minutes, and **S** stands for seconds. Its use case includes opening and closing hours of a store.
- **DATE**: The **DATE** data type is in the format **YYYY-MM-DD**, where **Y** is the year in 4 digits, **M** is month, and **D** is the day. The most common use case of the **DATE** data type is the date of birth.
- **TIMESTAMP**: The **TIMESTAMP** data type is in the format **YYYY-MM-DD HH:MM:SS**. **TIMESTAMP** is typically used to store the time an order was placed or when an order was delivered.
- **TIMESTAMP WITH TIME ZONE (TIMESTAMPTZ)**: This data type includes the date and time with a time zone. It stores the timestamp in UTC and converts it to the client's local time zone upon retrieval. Its format is YYYY-MM-DD HH:MM:SS+TZ. It is widely used for logging events in systems across multiple time zones, such as user login times or transaction timestamps, ensuring the times are recorded in UTC but can be displayed in the user's local time zone.

Other common data types include **BOOLEAN** and **ENUM**.

- **BOOLEAN:** The Boolean data type is either true or false. The Boolean type is commonly used with online stores. It signifies whether there are available stocks or not.
- **ENUM:** The ENUM data type is a list of values defined by the user. It allows you to specify a fixed set of values for a column, making it easier to maintain data integrity by restricting the values that can be stored in that column.

Suppose we want to define a column for the days of the week using the **ENUM** data type.

Step 1: First, you need to create the **ENUM** type with the desired values.

```
CREATE TYPE day_of_week AS ENUM ('Sunday', 'Monday', 'Tuesday',
'Wednesday', 'Thursday', 'Friday', 'Saturday');
```

Step 2: Next, you create a table and use the **ENUM** type for a column.

```
CREATE TABLE schedule (
    id SERIAL PRIMARY KEY,
    event_name VARCHAR(100),
    event_day day_of_week
);
```

Step 3: You can now insert values into the table, using only the predefined **ENUM** values for the `event_day` column.

```
INSERT INTO schedule (event_name, event_day) VALUES ('Meeting',
'Monday');
INSERT INTO schedule (event_name, event_day) VALUES ('Workshop',
'Wednesday');
INSERT INTO schedule (event_name, event_day) VALUES ('Conference',
 'Friday');
```

Step 4: When you query the table, you will see the **ENUM** values in the **event_day** column.

```
SELECT * FROM schedule;
```

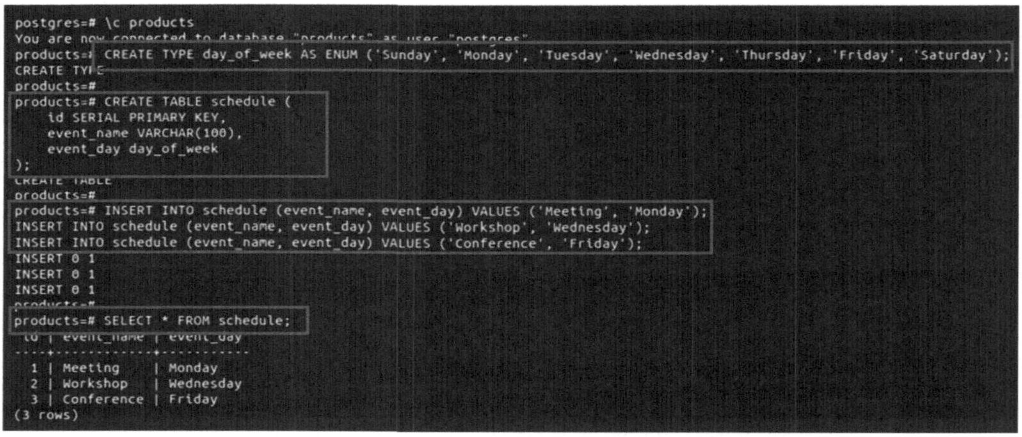

Figure 3.13: *How to apply the ENUM data type on a table*

Benefits:

- **Data Integrity: ENUM** ensures that only the specified values can be stored in the column, reducing the risk of invalid data entry.
- **Readability: ENUM** values are human-readable, making it easier to understand the data in the column.
- **Efficiency: ENUM** values are stored as integers internally, which can improve performance and storage efficiency.

Constraints

In this section, we will learn about the following constraints: **PRIMARY KEYS**, **FOREIGN KEYS**, **UNIQUE**, **NOT NULL**, and **CHECK**. Constraints are used to control the data in a database table to ensure data integrity. They define a set of valid values in the table.

- **PRIMARY KEYS**: This is a column that uniquely identifies a row or record in a table. Primary key must be unique and cannot be null (must not have missing data). There can only be one primary key per table.

Id integer	First_name Varying character(35)	Last_name Varying character(35)	City Varying character(35)	State character(2)
1	Deepesh	Raj	Houston	TX
2	Sunita	Chaudhry	New York	NY
3	Zainab	Kapur	New York	NY
4	Hassan	Raj	San Francisco	CA
5	Sunita	Kaur	Dallas	TX

Table 3.2: *Car Owners*

Referring to this car owners table previously viewed, you see that many people can have the same first name or the same last name, and many people can live in the same city and many cities can be in the same state. The only column that uniquely identifies all records without duplication is the id column. Thus, the id column is the primary key for this table as it uniquely identifies each row of data.

- **Foreign Keys:** A foreign key is used to link two tables together. A foreign key is a column where the values match the values of other tables' primary key columns. This means the values in the primary key of one table are the same as the values in the foreign key of another table. The data in the foreign key is referencing the data in the primary key column of a different table, thus forming a relationship. The table with the primary key is called the reference or parent table, while the table with the foreign key is called the child table. One table can contain many foreign keys. The referential integrity of data between the parent and child tables is maintained by foreign keys.

id Integer	make Varchar (30)	model Varchar(30)	age Integer	owner_id Integer
1	Toyota	RAV4	4	3
2	VW	Tiguan	9	5
3	Audi	A4	13	3
4	Porche	Cayenne	8	1
5	Toyota	Corolla	13	2
6	BMW	X5	16	1

Table 3.3: *Table Cars*

The **`owner_id`** column is a foreign key that references the id column in the car owner's table. To read this table and the previous car owners' table, match the owner id on the cars types table with the id on the car owners' table. From this table, it shows that the car Toyota RAV4 has an owner id of 2, which references Zainab Kapur on the car owners' table, and she is also the owner of Audi A4. Deepesh Raj is the owner of Porsche Cayenne and BMW X5. Sunita Kaur is the owner of VW Tiguan, meanwhile Sunita Chaudhry is the owner of Toyota Corolla.

- **UNIQUE** constraint ensures that a column on a table can only contain unique values. This constraint helps in restricting the entry of duplicate records in a table. This will throw an error if duplicate values are inserted into the column with a unique constraint. A very common use case is a column in a table with email addresses. No two people have the same email addresses; thus, the email address column must be applied the unique constraint. Look at the car owner table again with the email column included. The unique constraint applied to this column ensures that every entry to this column is unique.

Id integer	First_name VARCHAR(35)	Last_name VARCHAR(35)	City VARCHAR(35)	State character(2)	Emails VARCHAR(55) Unique
1	Deepesh	Raj	Houston	TX	d.raj@gmail.com
2	Sunita	Chaudhry	New York	NY	s.chaudhry@gmail.com
3	Zainab	Kapur	New York	NY	z.kapur@gmail.com
4	Hassan	Raj	San Francisco	CA	h.raj@gmail.com
5	Sunita	Kaur	Dallas	TX	s.kaur@gmail.com
6	[null]	Ramchar	[null]	CA	r.ramchar@gmail.com

Table 3.4: *Car Owners' table with Unique Constraint on email column*

- **NOT NULL**: This constraint ensures that **NULL** values are not inserted into a column. This also ensures that a column cannot be left blank. If the data is vital, such as the customer's zip code or email addresses, then you should add a **NOT NULL** constraint on the column. This constraint can be set at table creation or using the *alter* table command. Looking at owners' table, we can see that there is a sixth user entered. The **NOT NULL** constraints were applied on the state and email columns, and thus could not be left blank. Ramchar decided not to put his first name and city, but could not do the same for state and email since they had the **NOT NULL** constraints applied.
- **CHECK:** This constraint is used to check that values in a column satisfy a specific Boolean expression. For example, an age column in the car's table must have values greater than 0. This can also be used to check customers' reviews that can only accept values between 1 and 5 stars. The check constraints ensure that the data entered into the database makes sense (Data integrity).

To create a table in PostgreSQL, the following syntax is used.

```
CREATE TABLE table_name
(
column1 data_type,
column2 data_type,
column3 data_type,
column4 data_type,
 …
);
```

When using the **CREATE** command, replace **table_name** with the name of the table, **column1**, **column2**, and so on with the names of the table columns, and **data_type** with the appropriate data type as previously discussed.

To create our **car_owners**' table as earlier discussed, the command is as shown:

```
CREATE TABLE car_owner (
owner_id int primary key,
first_name varchar(35),
last_name varchar(35),
city varchar(35),
state char(2),
emails varchar(55) unique
);
```

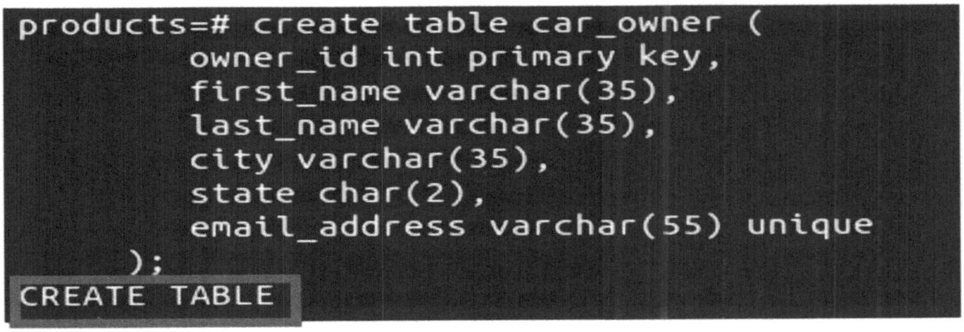

```
products=# create table car_owner (
           owner_id int primary key,
           first_name varchar(35),
           last_name varchar(35),
           city varchar(35),
           state char(2),
           email_address varchar(55) unique
       );
CREATE TABLE
```

Figure 3.14: *How to create table Car Owners inside Products database*

The **CREATE TABLE** response highlighted shows that the table has been created inside our products' database. Executing **\dt** again will now show our table. It shows that the table is in schema public, its name is **car_owner**, type is table, and the owner is postgres because we are logged in as the postgres user.

```
products=# \dt
           List of relations
 Schema |    Name    | Type  |  Owner
--------+------------+-------+----------
 public | car_owner  | table | postgres
(1 row)
```

Figure 3.15: *Using* **\dt** *command to see the Car Owner's table in Products database*

Now let us create our second table called cars using the following query:

```
CREATE TABLE cars (
car_id int primary key,
make varchar(30),
model varchar(30),
age int,
owner_id int,
);
```

Note: *The preceding query was typed on a notepad. If typing inside the database, it will turn to enter on a single line.*

```
products=# CREATE TABLE cars (
        car_id int primary key,
        make varchar(30),
        model varchar(30),
        age int,
        owner_id int
    );
CREATE TABLE
```

Figure 3.16: *Creating the cars table in products' database*

Or

```
CREATE TABLE cars (car_id int primary key, make varchar(30), model varchar(30), age int, owner_id int);
```

Figure 3.17: *Creating the cars table in products' database*

The **\dt** command will now show two databases as shown:

```
products=# \dt
          List of relations
 Schema |    Name    | Type  |  Owner
--------+------------+-------+----------
 public | car_owner  | table | postgres
 public | cars       | table | postgres
(2 rows)
```

Figure 3.18: *List of tables in Products database, showing car_owner and cars*

Data Manipulation Language (DML)

The data in the database is manipulated and modified using DML. **SELECT, UPDATE, INSERT**, and **DELETE** are the key DML commands. The **SELECT** command is used to fetch data from one, two, or more tables based on a relationship between them. New information or data is added to a table using the **INSERT** command. To modify the already present data in a table, an **UPDATE** command is used. The data in a table is removed using the **DELETE** command.

Let us now use some of the DML commands on our tables we have created. To retrieve all the data on a table, we use the following command:

```
SELECT * FROM table_name;
```

Let us perform a select statement on our two tables we created.

```
SELECT * FROM car_owner;
SELECT * FROM cars;
```

```
products=# SELECT * FROM car_owner ;
 owner_id | first_name | last_name | city | state | email_address
----------+------------+-----------+------+-------+---------------
(0 rows)
```

Figure 3.19: *Result of SELECT statement on empty car_owner table*

```
products=# SELECT * FROM cars ;
 car_id | make | model | age | owner_id
--------+------+-------+-----+----------
(0 rows)
```

Figure 3.20: *Result of SELECT statement on empty cars table*

The **SELECT** statement returned only the columns of our table because now we do not have data on our tables. We are now going to put some data into our tables using the **INSERT** keyword.

```
INSERT INTO table_name (column1, column2, column3, column4,...)
VALUES (value1, value2, value3, value4, ....);
```

OR

```
INSERT INTO table_name
VALUES (value1, value2, value3, value4, ....);
```

The following examples will be inserting data into the car_owner table using the first syntax:

```
INSERT INTO car_owner (owner_id, first_name, last_name, city,
state, email_address)
VALUES (1, 'Deepesh', 'Raj', 'Houston', 'TX', 'd.raj@gmail.com');
```

```
products=# INSERT INTO car_owner (owner_id, first_name, last_name, city, state, email_address)
VALUES (1, 'Deepesh', 'Raj', 'Houston', 'TX', 'd.raj@gmail.com');
INSERT 0 1
```

Figure 3.21: *How to insert data into car_owner table*

The "**INSERT 0 1**" shows that one piece of data has been inserted into the **car_owner** table. Note that the string data type columns data are surrounded by a single quote.

Our **SELECT** statement on **car_owner** will now show one row of data as shown in *Figure* 3.22:

```
products=# select * from car_owner ;
 owner_id | first_name | last_name |  city   | state |  email_address
----------+------------+-----------+---------+-------+-----------------
        1 | Deepesh    | Raj       | Houston | TX    | d.raj@gmail.com
(1 row)
```

Figure 3.22: *SELECT statement on car_owner's table showing the inserted data*

To insert multiple data, separate each set of data with a column (,). We are now going to insert the remaining data for the **car_owner** table at once using the following query: Type the following query very carefully before running it. If any errors, carefully check that everything is typed out correctly.

```
INSERT INTO car_owner (owner_id, first_name, last_name, city, state,
email_address)
VALUES (2, 'Sunita', 'Chaudhry', 'New York', 'NY', 's.chaudhry@gmail.
com'),
      (3, 'Zainab', 'Kapur', 'New York', 'NY', 'z.kapur@gmail.com'),
      (4, 'Hassan', 'Raj', 'San Francisco', 'CA', 'h.raj@gmail.com'),
      (5, 'Sunita', 'Kaur', 'Dallas', 'TX', 's.kaur@gmail.com'),
      (6, 'null', 'Ramchar', ' ', 'CA', 'r.ramchar@gmail.com');
```

```
products=# INSERT INTO car_owner (owner_id, first_name, last_name, city, state, email_address)
VALUES (2, 'Sunita', 'Chaudhry', 'New York', 'NY', 's.chaudhry@gmail.com'),
       (3, 'Zainab', 'Kapur', 'New York', 'NY', 'z.kapur@gmail.com'),
       (4, 'Hassan', 'Raj', 'San Francisco', 'CA', 'h.raj@gmail.com'),
       (5, 'Sunita', 'Kaur', 'Dallas', 'TX', 's.kaur@gmail.com'),
       (6, 'null', 'Ramchar', ' ', 'CA', 'r.ramchar@gmail.com');
INSERT 0 5
```

Figure 3.23: *How to insert multiple data into a table at once*

Our **SELECT *** statement will return all the six rows of data we have inserted, as shown in *Figure 3.24*:

```
products=# select * from car_owner ;
 owner_id | first_name | last_name |     city      | state |    email_address
----------+------------+-----------+---------------+-------+----------------------
        1 | Deepesh    | Raj       | Houston       | TX    | d.raj@gmail.com
        2 | Sunita     | Chaudhry  | New York      | NY    | s.chaudhry@gmail.com
        3 | Zainab     | Kapur     | New York      | NY    | z.kapur@gmail.com
        4 | Hassan     | Raj       | San Francisco | CA    | h.raj@gmail.com
        5 | Sunita     | Kaur      | Dallas        | TX    | s.kaur@gmail.com
        6 | null       | Ramchar   |               | CA    | r.ramchar@gmail.com
(6 rows)
```

Figure 3.24: *Result of SELECT statement showing multiple rows of data added*

We are now going to insert data into the cars table using the second syntax, as shown:

```
INSERT INTO cars
VALUES (1, 'Toyota', 'RAV4', 4, 3),
       (2, 'VW', 'Tiguan', 9, 5),
       (3, 'Audi', 'A4', 13, 3),
       (4, 'Porche', 'Cayenne', 8, 1),
       (5, 'Toyota', 'Corolla', 13, 2),
       (6, 'BMW', 'X5', 16, 1);
```

```
products=# INSERT INTO cars
VALUES (1, 'Toyota', 'RAV4', 4, 3),
       (2, 'VW', 'Tiguan', 9, 5),
       (3, 'Audi', 'A4', 13, 3),
       (4, 'Porche', 'Cayenne', 8, 1),
       (5, 'Toyota', 'Corolla', 13, 2),
       (6, 'BMW', 'X5', 16, 1);
INSERT 0 6
```

Figure 3.25: *Inserting data into cars' table*

Now let us see the data in the cars table using the **SELECT** statement, as shown in *Figure 3.26*:

```
products=# SELECT * FROM cars;
 car_id |  make   |  model  | age | owner_id
--------+---------+---------+-----+----------
      1 | Toyota  | RAV4    |   4 |        3
      2 | VW      | Tiguan  |   9 |        5
      3 | Audi    | A4      |  13 |        3
      4 | Porche  | Cayenne |   8 |        1
      5 | Toyota  | Corolla |  13 |        2
      6 | BMW     | X5      |  16 |        1
(6 rows)
```

Figure 3.26: *Result of SELECT statement on car's table with data*

In the next paragraph, we will see how to create a new database user. The syntax is

```
CREATE USER user_name;
```

Let us use this command to create a new user in the database called **Aisha**.

```
CREATE USER Aisha WITH PASSWORD 'password123';
```

```
postgres=# CREATE USER Aisha WITH PASSWORD 'password123';
CREATE ROLE
```

Figure 3.27: *How to create a user or role in PostgreSQL*

Now you can see that a role has been created. The concept of roles and users will be discussed in-depth in *Chapter 4, Understanding the Database*. To see the users in the database, we use **\du** command.

```
postgres=# \du
                                   List of roles
 Role name |                         Attributes                         | Member of
-----------+------------------------------------------------------------+-----------
 aisha     |                                                            | {}
 postgres  | Superuser, Create role, Create DB, Replication, Bypass RLS | {}
```

Figure 3.28: *To see all users or roles in the PostgreSQL database*

We have two users: **Aisha and postgres.** Aisha is the user we created, and postgres is a default and super user.

The other DDL commands will be discussed in subsequent chapters.

Data Control Language (DCL)

Access to the database and its objects is controlled using DCL. **GRANT** and **REVOKE** are the two DCL commands. The **GRANT** keyword is used to give access to the database and its objects to database users or roles, while the said access is removed using the **REVOKE** command.

We are now going to give the new user called Aisha some privileges. The following privileges are going to be granted to user Aisha: Connection on database products, usage on the public schema, **SELECT** permission on one specific table as well as on all tables within the schema public, grant **DELETE** and **TRUNCATE** permissions on the **car_owner** table. Finally, revoke the truncate permission from user Aisha.

1. Allow Aisha connectivity to the products database using **GRANT**.

```
GRANT CONNECT ON DATABASE products To aisha;
```

```
postgres=# GRANT CONNECT ON DATABASE products TO aisha;
GRANT
```

Figure 3.29: *How to grant access to a database to a user*

2. Allow Aisha usage on schema public.

```
GRANT USAGE ON schema public TO aisha;
```

```
postgres=# GRANT USAGE ON schema public TO aisha;
GRANT
```

Figure 3.30: *How to grant usage on a database to a user*

Now let us connect to the products database and give Aisha privileges on the tables contained there.

```
postgres=# \c products
You are now connected to database "products" as user "postgres".
```

Figure 3.31: *How to connect to a different database*

The next screen shows how to grant usage on schema public to a user.

```
products=# GRANT USAGE ON schema public TO aisha;
GRANT
```

Figure 3.32: *Grant usage on a schema to a user*

The next screen is to list all tables available in the products database.

```
products=# \dt
           List of relations
 Schema |    Name    | Type  |  Owner
--------+------------+-------+----------
 public | car_owner  | table | postgres
 public | cars       | table | postgres
(2 rows)
```

Figure 3.33: *How to see the tables in the products database*

3. To give Aisha select on the **car_owner** table, we will use the following query:

```
GRANT SELECT ON TABLE car_owner TO aisha;
```

```
products=# GRANT SELECT ON TABLE car_owner TO aisha;
GRANT
```

Figure 3.34: *How to give SELECT privilege to a user*

The next command will give Aisha select privileges on all tables in schema public.

```
GRANT SELECT ON ALL TABLES IN SCHEMA public TO aisha;
```

```
products=# GRANT SELECT ON ALL TABLES IN SCHEMA public TO aisha;
GRANT
```

Figure 3.35: *How to create a user or role in PostgreSQL*

```
GRANT DELETE, TRUNCATE ON ALL TABLES IN SCHEMA public TO aisha;
```

```
products=# GRANT  DELETE, TRUNCATE ON TABLE car_owner TO aisha;
GRANT
```

Figure 3.36: *Granting privileges to a user*

The following query is used to check individual user privileges. Change the grantee (**user_name**) and **table_name** as appropriate. Do not beat yourself up just yet if the following query is not understood. All the steps of the query will be taught in subsequent chapters.

```
SELECT privilege_type
    FROM information_schema.role_table_grants
    WHERE
        table_schema = quote_ident('public')
        AND table_name = quote_ident('car_owner')
        AND grantee = quote_ident('aisha') ;
```

```
products=# SELECT privilege_type
    FROM information_schema.role_table_grants
    WHERE
        table_schema = quote_ident('public')
        AND table_name = quote_ident('car_owner')
        AND grantee = quote_ident('aisha') ;
 privilege_type
----------------
 TRUNCATE
 DELETE
 SELECT
(3 rows)
```

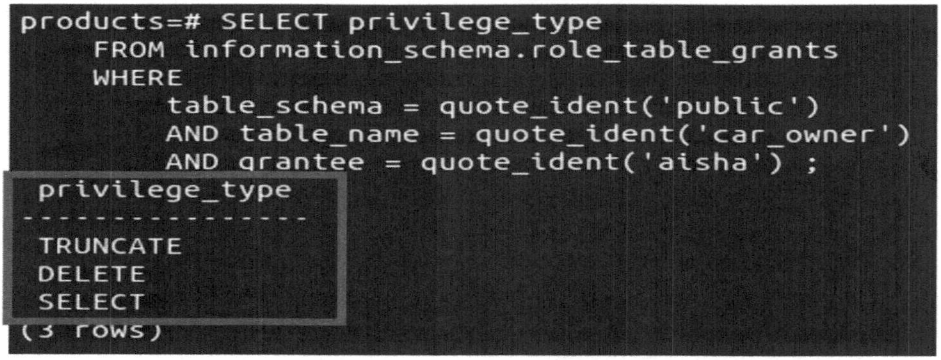

Figure 3.37: *To see all the privileges on a table*

We are now going to withdraw the **truncate** permissions from user Aisha using the **revoke** command shown:

```
REVOKE TRUNCATE ON TABLE car_owner FROM aisha;
```

```
products=# REVOKE TRUNCATE ON TABLE car_owner FROM aisha;
REVOKE
```

Figure 3.38: *How to revoke privileges from a user*

Checking the privileges for Aisha again, you can see that the truncate is now gone from the list of privileges.

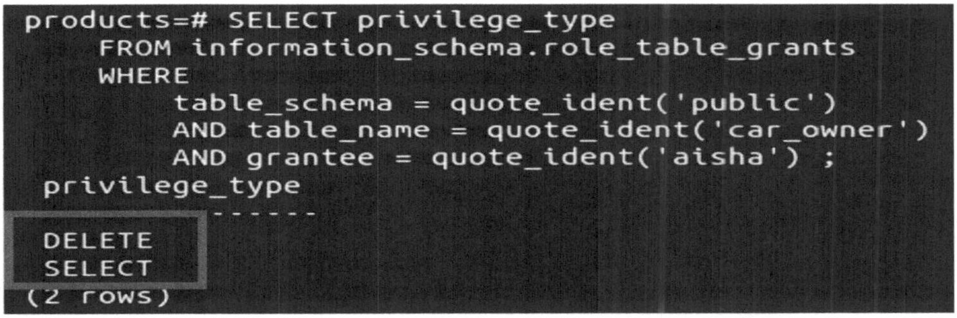

```
products=# SELECT privilege_type
    FROM information_schema.role_table_grants
    WHERE
        table_schema = quote_ident('public')
        AND table_name = quote_ident('car_owner')
        AND grantee = quote_ident('aisha') ;
 privilege_type
----------------
 DELETE
 SELECT
(2 rows)
```

Figure 3.39: *Check user's privileges*

Transaction Control Language (TCL)

All database transactions are managed by TCL. A database transaction is a single piece of work containing one or many operations. **COMMIT**, **SAVEPOINT**, and **ROLLBACK** are the three TCL commands. Changes made during a transaction are committed using the **COMMIT** command, while the **ROLLBACK** command is used to undo the changes. If there is a particular point in the transaction that you want to roll back to, then **SAVEPOINT** is the command to use.

We are going to use a bank transaction between two accounts to demonstrate **COMMIT** and **ROLLBACK**. For a transaction to be complete, it must ensure the money from account A to account B arrives and the balances in the two accounts adjusted to match the transferred amount. This means the exact amount transferred from account A is the same amount received by account B. Here we say that the transaction of PostgreSQL is atomic, consistent, isolated, and durable (ACID).

Let us create a different table called bank accounts, as shown, to practice transactions.

```
CREATE TABLE bank_accounts (
    id INT GENERATED BY DEFAULT AS IDENTITY,
    account_name VARCHAR(30) NOT NULL,
    account_balance DEC(9,2) NOT NULL,
```

```
    PRIMARY KEY(id)
);
```

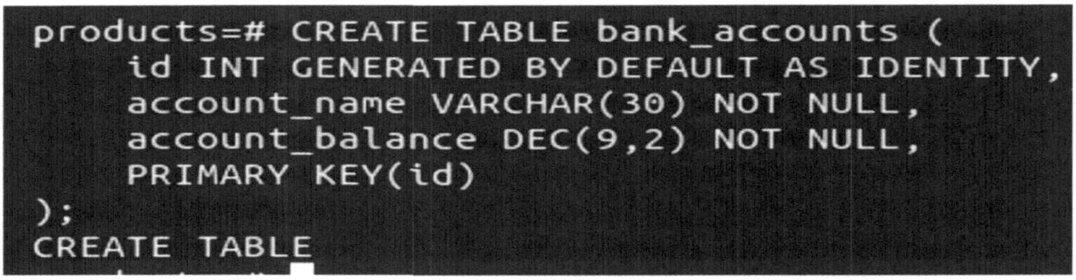

```
products=# CREATE TABLE bank_accounts (
    id INT GENERATED BY DEFAULT AS IDENTITY,
    account_name VARCHAR(30) NOT NULL,
    account_balance DEC(9,2) NOT NULL,
    PRIMARY KEY(id)
);
CREATE TABLE
```

***Figure 3.40:** Creating a bank account table for transactions*

For transactions, we always start with the **BEGIN** keyword, then followed by the transactions' queries. The first practice is to insert 10000 dollars into Laila's account through a transaction.

```
products=# BEGIN;

INSERT INTO bank_accounts(account_name,account_balance)
VALUES('Laila',10000);
BEGIN
INSERT 0 1
```

***Figure 3.41:** Inserting data into bank account table as a transaction starting with BEGIN*

Once the transaction is run, you can either commit as shown or rollback.

```
products=*# commit;
COMMIT
```

***Figure 3.42:** How to commit a transaction*

The following query will help us to see different account holders (**account_name**) and their bank balance (**account_balance**). The following **SELECT** query, is selecting specific columns (**id**, **account_name**, **account_balance**) rather than "**SELECT ***" that selects all columns.

```
SELECT
    id,
    account_name,
    account_balance
FROM
    bank_accounts;
```

```
products=# SELECT
    id,
    account_name,
    account_balance
FROM
    bank_accounts;
 id | account_name | account_balance
----+--------------+-----------------
  3 | Laila        |        10000.00
(1 row)
```

Figure 3.43: *Using SELECT statement to check content of bank account table*

The result of the above query shows that Laila is the only account holder now with an account balance of 10000.

Now, let us add **`Pramoda's`** account using the following transaction. We are going to rollback rather than commit.

```
products=# BEGIN;

INSERT INTO bank_accounts(account_name,account_balance)
VALUES('Pramoda',10000);
BEGIN
INSERT 0 1
```

Figure 3.44: *Starting another transaction with BEGIN*

The next screen shows how to rollback a transaction so that it is not saved.

```
products=*# ROLLBACK;
ROLLBACK
```

Figure 3.45: *Show how to rollback a transaction*

The following select query shows that Promoda's account was not added as we did the rollback on that transaction.

```
products=# SELECT
    id,
    account_name,
    account_balance
FROM
    bank_accounts;
 id | account_name | account_balance
----+--------------+-----------------
  3 | Laila        |        10000.00
(1 row)
```

Figure 3.46: *Using SELECT statement to see that the last transaction was rollbacked.*

Now, let us add another account details called Akram with a **COMMIT** at the end of the transaction as shown:

```
products=# BEGIN;

INSERT INTO bank_accounts(account_name,account_balance)
VALUES('Akram',10000);
BEGIN
INSERT 0 1
products=*# COMMIT;
COMMIT
```

Figure 3.47: *Adding a different account with commit*

Now we have two account holders (Laila and Akram) returned with our **SELECT** query:

```
products=# SELECT
    id,
    account_name,
    account_balance
FROM
    bank_accounts;
 id | account_name | account_balance
----+--------------+-----------------
  3 | Laila        |        10000.00
  5 | Akram        |        10000.00
(2 rows)
```

Figure 3.48: *Showing all accounts available*

The next transaction is going to deduct 1000 dollars from Laila's account to Akram's account.

```
BEGIN;
-- deduct 1000 from Laila's account
UPDATE bank_accounts
SET account_balance = account_balance - 1000
WHERE account_name = 'Akram';
-- add 1000 to Akram's account
UPDATE bank_accounts
SET account_balance = account_balance + 1000
WHERE account_name = 'Laila';
-- commit the transaction to save it
COMMIT;
```

You may not understand the preceding transaction just yet, but do not worry. The **WHERE** and **SET** will be discussed in subsequent chapters.

```
products=# BEGIN;

-- deduct 1000 from Laila's account

UPDATE bank_accounts
SET account_balance = account_balance - 1000
WHERE account_name = 'Laila';

-- add 1000 to Akram's account

UPDATE bank_accounts
SET account_balance = account_balance + 1000
WHERE account_name = 'Akram';

-- commit the transaction to save it
COMMIT;
BEGIN
UPDATE 1
UPDATE 1
COMMIT
```

Figure 3.49: *A transaction to deduct 1000 from one bank account to another*

Note that the commented lines (for example, - - - deduct 1000 from Laila's account) will not run.

The result of the following **SELECT** statement shows that 1000 has been deducted from Laila's account to Akram's account.

```
products=# SELECT
    id,
    account_name,
    account_balance
FROM
    bank_accounts;
 id | account_name | account_balance
----+--------------+-----------------
  3 | Laila        |         9000.00
  5 | Akram        |        11000.00
(2 rows)
```

Figure 3:50: *Result of the completed transaction*

Tasks

This section is to consolidate the learning from this chapter.

1. Create a Movies database

2. Connect to the movies database and create an actor table with the following columns: actor id, first name, last name, age, gender, and date of birth. The table should look like the one shown as follows:

actor_id int primary key	first_name Varchar(25)	last_name Varchar(25)	age int	gender Char(1)	date of birth (Date)

Table 3.5: *Task 1, Create a table inside the database created in step 1*

3. Products database

   Create the following tables:

   a. Customers

customer_id INT PRIMARY key	first_name Varchar(25)	last_name Varchar(25)	age int	Email_address Varchar(55)	number of complaints (Date)

Table 3.6: *Task 2*

   b. Sales with the following columns: sales id, item name, date of sales, and customer id.

Sales

sales_id SERIAL	item_name Varchar(25)	Customer_id Char(1)	date of sales (Date)

Table 3.7: *Task 3, Create sales table*

c. Create **Items** table

item_id	item_name	Unit_price	Company_id

***Table 3.8**: Task 4*

d. Create **Companies** table

Company_id	company_name

***Table 3.9**: Create company table*

Technical Interview Questions

1. What is SQL?
2. Explain the purpose of SQL in relational databases.
3. Differentiate between SQL and NoSQL.
4. What are the different sublanguages of SQL?
5. What are DDL commands in SQL?
6. Explain the purpose of the **CREATE** command.
7. How does the **ALTER** command differ from the **UPDATE** command?
8. Give an example of how to use the **DROP** command.
9. What is the **TRUNCATE** command used for, and how is it different from **DELETE**?
10. Explain Data Manipulation Language (DML).
11. What is DML in SQL, and why is it used?
12. List and explain the basic DML commands.
13. How would you insert a new record into a table?
14. Describe the differences between the **UPDATE** and **DELETE** commands.
15. What does the **SELECT** statement do?
16. What are DCL commands, and what are their main purposes?

17. Explain the **GRANT** command and its uses.
18. What is the **REVOKE** command, and how is it used?
19. What are TCL commands, and why are they important?
20. Describe the use of the **COMMIT** command.
21. What is the purpose of the **ROLLBACK** command?
22. Explain what the **SAVEPOINT** command does.
23. How does the SET **TRANSACTION** command work?
24. What are data types in SQL, and why are they important?
25. Explain the difference between **INTEGER** and **SMALLINT** data types.
26. What is the **DECIMAL** data type used for, and how is it defined?
27. Describe the use of **VARCHAR** and **CHAR** data types.
28. What are the differences between **TIMESTAMP** and **TIMESTAMPTZ**?
29. What is an **ENUM** data type, and when would you use it?

Conclusion

In this chapter, learners understood the definition and history of SQL. Learners also developed their understanding of Data Definition Language, Data Manipulation Language, Data Control Language, and Transaction Control Language.

In the next chapter, learners will learn how to create and use database schema, tables with foreign keys, and insert data into tables. Learners will also learn how to manipulate data.

Chapter 4
Understanding the Database

Introduction

In this chapter, learners will learn and become proficient in manipulating the data in the database. You will learn how to create a schema, create tables with foreign keys, insert data into a table, delete data from a table, modify tables, delete tables from a database, retrieve data from tables, and database relationships.

Structure

In this chapter, the following topics will be covered:

- Creating and Using Database Schema
- Creating Tables with Foreign Keys
- Inserting Data into Tables
- Deleting Data from Tables
- Modifying and Deleting Tables
- Retrieving Data from Tables
- Database Relationships
- Task

Creating and Using Database Schema

A Schema in PostgreSQL is the logical structure and storage of data within the database. All the tables, functions, data types, stored procedures, indexes, and object-related are held in a schema.

We define different *schemas* in a database to avoid conflicts and unnecessary interference when different people access the application.

A schema diagram illustrates the database schema of a table, its columns, data types, key constraints, and other relevant information.

A database holds one or more named *schemas*, which in turn hold tables. Schemas also hold other kinds of named objects, including data types, functions, and operators. The same object name, such as **car_owner**, can be used in different schemas without conflict.

A user connected to a database has access to objects in any of the schemas in that database if they have privileges to do so.

There are many reasons why we may want to use schemas:

- A schema allows many users to use one database without interference with each other.
- Database objects are organized into logical groups using a schema. This makes them more manageable.
- A schema is used to separate third-party applications to prevent collision with the names of other objects.

To create a schema, we use the **CREATE SCHEMA** command.

Let us create a schema called **dbschema** using the following command. Before we create a schema, let us check the available schemas using **\dn** command.

```
postgres=# \dn
  List of schemas
  Name   |  Owner
---------+----------
 public  | postgres
(1 row)
```

***Figure 4.1**: How to check available schema using \dn command*

In the next screen, we will create a new schema.

```
CREATE SCHEMA dbschema;
```

```
postgres=# CREATE SCHEMA dbschema;
CREATE SCHEMA
postgres=# \dn
   List of schemas
   Name    |  Owner
-----------+----------
 dbschema  | postgres
 public    | postgres
(2 rows)
```

***Figure 4.2:** How to create a schema*

We can also check available schemas from the **pg_catalog** table.

pg_catalog Schema

In PostgreSQL, the **pg_catalog** schema is an integral part of every database. It serves as the system catalog schema, housing essential system tables and predefined data types, functions, and operators that are built into the database system.

Components of pg_catalog:

- **System Tables:**
    - **pg_class:** This table contains metadata about tables, indexes, sequences, and other similar objects within the database.
    - **pg_attribute:** Stores detailed information about table columns, such as their names, data types, and properties.
    - **pg_proc:** Contains information about functions and stored procedures defined in the database.
    - **pg_type:** Maintains data about all data types, including built-in and user-defined types.
    - **pg_index:** Holds details about the indexes created on tables.
    - **pg_namespace:** Lists all schemas available in the database.
- **Built-in Data Types:** The **pg_catalog** schema includes definitions for all built-in data types, ensuring that they are readily available for use in database operations.
- **Functions and Operators:** This schema contains built-in functions and operators that can be utilized in SQL queries, supporting a wide range of database operations and calculations.

Significance of pg_catalog:

- **Metadata Management:** The **pg_catalog** schema is critical for managing metadata about database objects. This metadata allows the database engine to understand and manage the structure and relationships of various objects within the database.
- **Automatic Inclusion in Search Path:** The **pg_catalog** schema is automatically included in the search path, meaning that its functions, types, and operators can be accessed without explicitly mentioning the schema. This convenience simplifies the execution of SQL queries using built-in functions and operators.
- **System Integrity:** Direct modifications to the tables within the **pg_catalog** schema are not recommended, as they can impact the stability and integrity

of the entire database system. The PostgreSQL engine manages these tables, ensuring consistent and reliable database operations.

When you create a new table, the information about this table is recorded in the **pg_class** and **pg_attribute** tables within the **pg_catalog** schema.

The **pg_catalog** schema is fundamental to the internal workings of PostgreSQL, providing a robust framework for managing and accessing metadata, built-in data types, functions, and operators.

You can use the following command as well to check available schemas from **pg_catalog** table in the database.

```
SELECT *
FROM pg_catalog.pg_namespace
ORDER BY nspname;
```

```
postgres=# SELECT *
FROM pg_catalog.pg_namespace
ORDER BY nspname;
  oid  |      nspname       | nspowner |                        nspacl
-------+--------------------+----------+---------------------------------------------------------
 33140 | dbschema           |       10 |
 13199 | information_schema |       10 | {postgres=UC/postgres,=U/postgres}
    11 | pg_catalog         |       10 | {postgres=UC/postgres,=U/postgres}
    99 | pg_toast           |       10 |
  2200 | public             |       10 | {postgres=UC/postgres,=UC/postgres,aisha=U/postgres}
(5 rows)
```

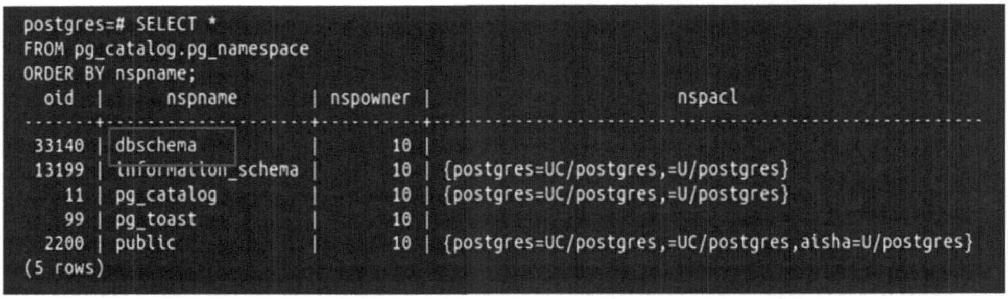

Figure 4.3: *How to check available schema using pg_catalogue*

To create or access objects in our **dbschema**, write a *schema name* with the table name separated by a dot, as shown:

```
dbschema.table_name
```

The same idea is applied to other kinds of named objects, such as procedures and functions.

The complete syntax will join the database name, schema name, and table name separated by a dot.

```
database. dbschema.table_name
```

Once you are connected to a database, using the database name is not compulsory, but if you want to write the database name, it must be the same as the database that you are connected to.

To create a table called **employees** in our new schema, we use the following syntax:

```
CREATE TABLE dbschema.employees (
   emp_id INT PRIMARY KEY,
   first_name VARCHAR(30),
```

```
last_name VARCHAR(30) NOT NULL,
department_name VARCHAR(20),
);
```

```
postgres=# CREATE TABLE dbschema.employees (
   emp_id INT PRIMARY KEY,
   first_name VARCHAR(30),
   last_name VARCHAR(30) NOT NULL,
   department_name VARCHAR(20)
   );
CREATE TABLE
```

***Figure 4.4**: How to create a table in a schema*

```
SELECT * FROM dbschema.employees;
```

```
postgres=# SELECT * FROM dbschema.employees;
 emp_id | first_name | last_name | department_name
--------+------------+-----------+-----------------
(0 rows)
```

***Figure 4.5**: Using SELECT statement to query a table inside schema*

```
INSERT INTO dbschema.employees (emp_id, first_name, last_name,
department_name)
VALUES(67, 'Aisha','Bassi','Finance'),
(106,'Fatima','Chandra', 'IT'),
(28, 'Laila', 'Kapoor', 'Sales');
```

```
postgres=# INSERT INTO dbschema.employees (emp_id, first_name, last_name, department_name)
postgres-# VALUES(67, 'Aisha','Bassi','Finance'), (106,'Fatima','Chandra', 'IT'), (28, 'Laila', 'Kapoor', 'Sales');
INSERT 0 3
```

***Figure 4.6**: How to insert data into a table inside a **dbschema** schema*

```
postgres=# SELECT * FROM dbschema.employees;
 emp_id | first_name | last_name | department_name
--------+------------+-----------+-----------------
     67 | Aisha      | Bassi     | Finance
    106 | Fatima     | Chandra   | IT
     28 | Laila      | Kapoor    | Sales
(3 rows)
```

***Figure 4.7**: Using SELECT statement to query the inserted data*

To drop a schema, it must be empty. All objects in it must be dropped first.

```
DROP SCHEMA dbschema;
```

```
postgres=# DROP SCHEMA dbschema;
ERROR:  cannot drop schema dbschema because other objects depend on it
DETAIL:  table dbschema.employees depends on schema dbschema
HINT:  Use DROP ... CASCADE to drop the dependent objects too.
```

Figure 4.8**: How to delete a schema called **dbschema

The **SCHEMA** failed to be dropped because it is not empty. If the schema is not empty, it can be dropped by adding the **CASCADE** keyword.

```
DROP SCHEMA dbschema CASCADE;
```

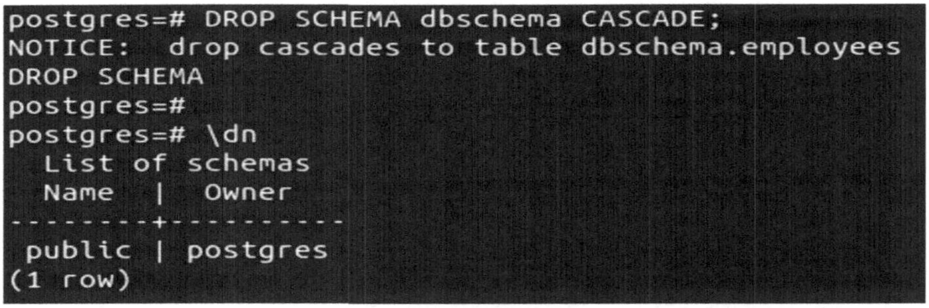

```
postgres=# DROP SCHEMA dbschema CASCADE;
NOTICE:  drop cascades to table dbschema.employees
DROP SCHEMA
postgres=#
postgres=# \dn
  List of schemas
  Name  |  Owner
--------+----------
 public | postgres
(1 row)
```

Figure 4.9: *How to delete a schema containing data*

The result of listing the schemas with the **\dn** command shows that schema *dbschema* has been deleted.

To restrict the activities of users in the database, we create a schema owned by the user.

The syntax is:

```
CREATE SCHEMA schema_name AUTHORIZATION user_name;
```

In *Chapter 3, Subsets of SQL*, we created tables **car_owners** and *cars* without specifying any schema names. By default, such tables are put into a schema named "**public**" automatically. Whenever you create a table, without specifying the schema name, it goes to schema public. All PostgreSQL databases contain a public schema by default. The following syntax will create table *cars* in the public schema:

```
CREATE TABLE cars (column1, column2,column3 …);
```

Or

```
CREATE TABLE public.cars (column1, column2,column3 …);
```

Schema Search Path

The schema search path is used by PostgreSQL to look for tables, views, and other database objects when an unqualified object name is referenced in a query. The system determines which table is referenced in the query by following a *search path*, which is a list of schemas to look in. The first matching table in the search path is taken to be the one wanted. If there is no match in the search path, an error is reported, even if matching table names exist in other schemas in the database.

In PostgreSQL, we can use the following command to see the current search path:

```
SHOW search_path;
```

```
postgres=# SHOW search_path;
   search_path
-----------------
 "$user", public
(1 row)
```

Figure 4.10: *How to check schema search path in PostgreSQL*

The result contains only the default schema public:

The first schema in the search path is the default location for creating other objects. This is why, by default, objects are created in the public schema.

For our new schema to appear in the search path, we will create the schema again since it was dropped earlier, and then use the following command:

```
CREATE SCHEMA dbschema;

SET search_path TO dbschema,public;
```

The **`SET search_path`** command sets the order in which PostgreSQL searches for database objects. The order of schemas in the search path determines the priority.

For this **`SET search_path TO dbschema,public;`** the dbschema has the highest priority.

Order of Search Path:

- **Priority:** The first schema listed (**`dbschema`**) has the highest priority.
- **Object Reference:** When you reference an object such as **`my_table`**, PostgreSQL will search for it in the order specified in the search_path.

PostgreSQL looks in **`dbschema`** for **`my_table`**. If found: It uses **`my_table`** from **`dbschema`** and stops searching further. If not found: It continues to the next schema, which is public.

The search path effectively sets a search order, and PostgreSQL uses the first matching object it finds. This mechanism allows you to control which schema's objects take precedence when multiple schemas contain objects with the same name.

```
postgres=# CREATE SCHEMA dbschema;
CREATE SCHEMA
postgres=#
postgres=# SET search_path TO dbschema,public;
SET
postgres=# SHOW search_path;
   search_path
------------------
 dbschema, public
(1 row)
```

Figure 4.11: *Recreate schema **dbschema** and set it as a default search path*

Once this is done, we can access the tables in it without schema qualification. This is because schema *dbschema* has been set as the first schema in the search path. Look at *Figure* 4.12, you will see that the table has been created and data inserted into it without using schema name as done previously.

```
postgres=# CREATE TABLE employees (
   emp_id INT PRIMARY KEY,
   first_name VARCHAR(30),
   last_name VARCHAR(30) NOT NULL,
   department_name VARCHAR(20)
   );
CREATE TABLE
postgres=# INSERT INTO employees (emp_id, first_name, last_name, department_name)
VALUES(67, 'Aisha','Bassi','Finance'), (106,'Fatima','Chandra', 'IT'), (28, 'Laila', 'Kapoor', 'Sales');
INSERT 0 3
postgres=#
```

Figure 4.12: *Using schema **dbschema** as a search path without mentioning it*

Before including the **dbschema** in the search path, table employees can only be dropped using the qualified schema syntax as shown:

```
DROP TABLE dbschema.employees;
```

Table employees can now be dropped with the following command:

```
DROP TABLE employees;
```

```
postgres=# DROP TABLE employees;
DROP TABLE
```

Figure 4.13: *Deleting a table inside dbschema without specifying the schema name*

It is worth noting that since the **dbschema** is the first element in the path, new objects would be created in it by default.

It could also have been set as shown:

```
SET search_path TO dbschema;
```

Creating Tables with Foreign Keys

In a table, a foreign key is a column that references the primary key of another table.

A child table or referencing table contains the foreign key, while a parent table is referenced by the foreign key. One table can have many foreign keys depending on the relationships it has with other tables. A foreign key is defined using the foreign key constraint. It establishes and enforces a link between data in two databases. We often call the foreign key a referencing key because it matches the primary key of another table.

The referential integrity of data between the child and parent tables is maintained by a foreign key constraint. A foreign key constraint makes sure that values in a column of the child table equal the values in a column of the parent table.

Foreign keys and primary keys are PostgreSQL constraints that enforce data validity in tables, thus preventing invalid data from entering the table.

To better understand foreign keys and primary keys, let us create the following two tables: **`clients`** and **`sales`**.

```
CREATE TABLE clients(
      customer_id int Primary key,
      first_name VARCHAR(25),
      last_name VARCHAR(25) not null,
      email VARCHAR(55) UNIQUE
);
CREATE TABLE sales (
      sales_id int Primary key,
      date_of_purchase Date,
         Email VARCHAR(55) UNIQUE,
         customer_id int,
      CONSTRAINT FK_clients_sales FOREIGN KEY(customer_id)
REFERENCES clients(customer_id)
);
```

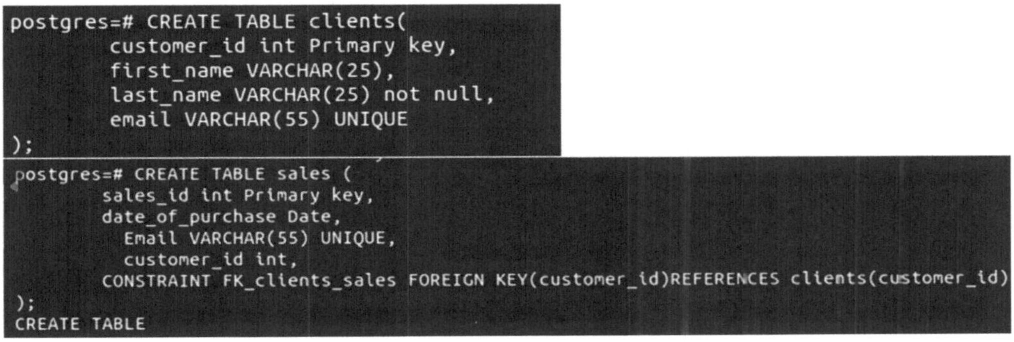

```
postgres=# CREATE TABLE clients(
       customer_id int Primary key,
       first_name VARCHAR(25),
       last_name VARCHAR(25) not null,
       email VARCHAR(55) UNIQUE
);
postgres=# CREATE TABLE sales (
       sales_id int Primary key,
       date_of_purchase Date,
         Email VARCHAR(55) UNIQUE,
         customer_id int,
       CONSTRAINT FK_clients_sales FOREIGN KEY(customer_id)REFERENCES clients(customer_id)
);
CREATE TABLE
```

Figure 4.14: *How to create a table with a foreign key*

The foreign key constraint syntax is as shown:

```
[CONSTRAINT fk_name]
FOREIGN KEY(fk_column)
REFERENCES parent_table(parent_key_columns)
[ON DELETE delete_action]
```

The first step is to specify the name for the foreign key constraint. The second step is to specify one or more foreign key columns in parentheses after the **FOREIGN KEY** keywords. The next step is to specify the parent table and parent key columns referenced by the foreign key columns in the **REFERENCES** clause.

The last step is to specify the delete action in the **ON DELETE** clause.

PostgreSQL supports the following actions for the foreign key constraints:

- **SET NULL**
- **SET DEFAULT**
- **RESTRICT**
- **NO ACTION**
- **CASCADE**

CASCADE

When a parent row is deleted or updated, all corresponding child rows are automatically deleted or updated. We are going to create the parent and child tables as shown:

```
CREATE TABLE parent (
    id SERIAL PRIMARY KEY,
    name VARCHAR(50)
);

CREATE TABLE child (
    id SERIAL PRIMARY KEY,
    parent_id INTEGER REFERENCES parent(id) ON DELETE CASCADE ON
UPDATE CASCADE,
    value VARCHAR(50)
);
```

```
products=#
products=# CREATE TABLE parent (
    id SERIAL PRIMARY KEY,
    name VARCHAR(50)
);
CREATE TABLE
products=#
products=# CREATE TABLE child (
    id SERIAL PRIMARY KEY,
    parent_id INTEGER REFERENCES parent(id) ON DELETE CASCADE ON UPDATE CASCADE,
    value VARCHAR(50)
);
CREATE TABLE
```

Figure 4.15: *How to create parent and child tables*

```
-- Insert into parent and child
INSERT INTO parent (name) VALUES ('Parent1');
INSERT INTO child (parent_id, value) VALUES (1, 'Child1');
```

```
products=#
products=#
products=# INSERT INTO parent (name) VALUES ('Parent1');
INSERT 0 1
products=#
products=# INSERT INTO child (parent_id, value) VALUES (1, 'Child1');
INSERT 0 1
products=#
```

Figure 4.16: *Inserting data into a child table*

```
-- Delete parent row
DELETE FROM parent WHERE id = 1;
```

The corresponding child row in the '**child**' table is automatically deleted when the parent row is deleted.

```
products=#
products=# DELETE FROM parent WHERE id = 1;
DELETE 1
products=#
products=# SELECT * FROM child WHERE parent_id = 1;
 id | parent_id | value
----+-----------+-------
(0 rows)
```

Figure 4.17: *Deleting data from parent table*

SET NULL

When a parent row is deleted or updated, the child rows' foreign key values are set to **NULL**.

```
CREATE TABLE parent2 (
    id SERIAL PRIMARY KEY,
    name VARCHAR(50)
);
CREATE TABLE child2 (
    id SERIAL PRIMARY KEY,
    parent_id INTEGER REFERENCES parent2(id) ON DELETE SET NULL
ON UPDATE SET NULL,
    value VARCHAR(50)
);

-- Insert into parent and child
INSERT INTO parent2 (name) VALUES ('Parent2');
INSERT INTO child2 (parent_id, value) VALUES (1, 'Child2');
```

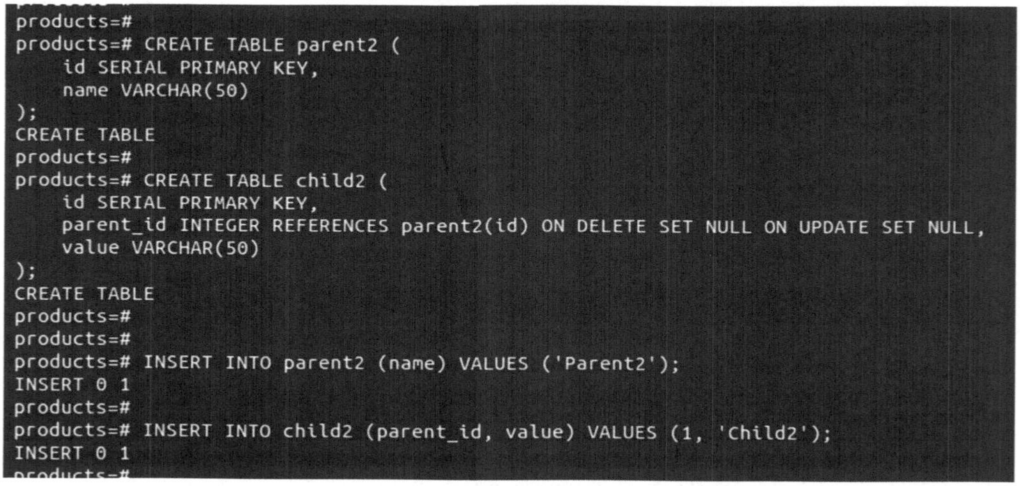

```
products=#
products=# CREATE TABLE parent2 (
    id SERIAL PRIMARY KEY,
    name VARCHAR(50)
);
CREATE TABLE
products=#
products=# CREATE TABLE child2 (
    id SERIAL PRIMARY KEY,
    parent_id INTEGER REFERENCES parent2(id) ON DELETE SET NULL ON UPDATE SET NULL,
    value VARCHAR(50)
);
CREATE TABLE
products=#
products=#
products=# INSERT INTO parent2 (name) VALUES ('Parent2');
INSERT 0 1
products=#
products=# INSERT INTO child2 (parent_id, value) VALUES (1, 'Child2');
INSERT 0 1
```

Figure 4.18: *Set the NULL on delete and update*

In the next screen, we will be checking the data in **parent2** and **child2** tables.

```
products=#
products=# SELECT * FROM parent2;
 id |  name
----+---------
  1 | Parent2
(1 row)

products=# SELECT * FROM child2;
 id | parent_id | value
----+-----------+--------
  1 |         1 | Child2
(1 row)
```

Figure 4.19: *Using SELECT statement on parent2 and child2 tables*

```
-- Delete parent row
DELETE FROM parent2 WHERE id = 1;
-- The 'parent_id' in the child row will be set to NULL
```

```
products=#
products=# DELETE FROM parent2 WHERE id = 1;
DELETE 1
products=#
products=# SELECT * FROM parent2;
 id | name
----+------
(0 rows)

products=# SELECT * FROM child2;
 id | parent_id | value
----+-----------+--------
  1 |           | Child2
(1 row)
```

Figure 4.20: *Using DELETE statement on parent2 table*

SET DEFAULT

When a parent row is deleted or updated, the child rows' foreign key values are set to their default value.

departments and **employees** tables to demonstrate the **ON DELETE SET DEFAULT** constraint:

```
CREATE TABLE departments (
    dept_id SERIAL PRIMARY KEY,
    dept_name VARCHAR(100)
);

CREATE TABLE employees (
    emp_id SERIAL PRIMARY KEY,
    emp_name VARCHAR(100),
    dept_id INTEGER DEFAULT 0 REFERENCES departments(dept_id) ON
DELETE SET DEFAULT ON UPDATE SET DEFAULT
);
-- Insert into departments
INSERT INTO departments (dept_name) VALUES ('HR');
-- Insert into employees
INSERT INTO employees (emp_name, dept_id) VALUES ('Alice', 1);
-- Delete department row
DELETE FROM departments WHERE dept_id = 1;
-- The 'dept_id' in the employees table will be set to its
default value (0)
```

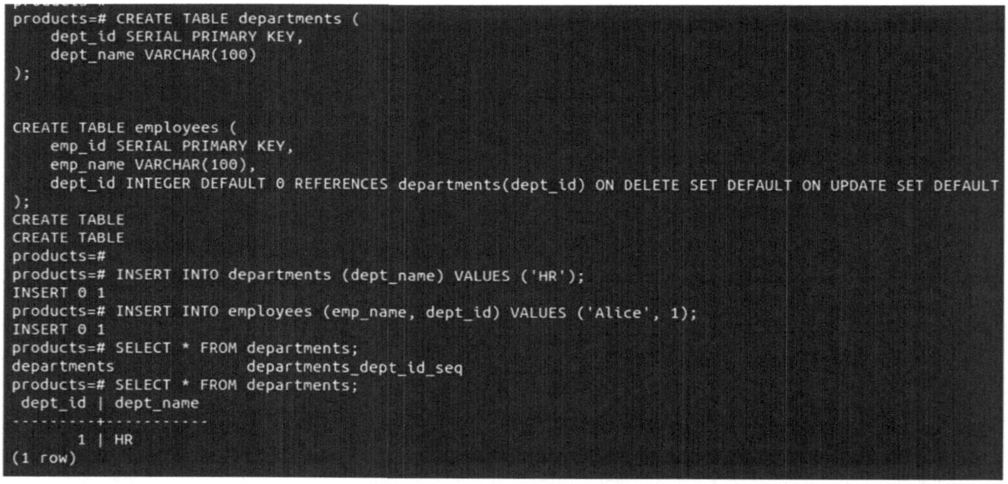

```
products=# CREATE TABLE departments (
    dept_id SERIAL PRIMARY KEY,
    dept_name VARCHAR(100)
);


CREATE TABLE employees (
    emp_id SERIAL PRIMARY KEY,
    emp_name VARCHAR(100),
    dept_id INTEGER DEFAULT 0 REFERENCES departments(dept_id) ON DELETE SET DEFAULT ON UPDATE SET DEFAULT
);
CREATE TABLE
CREATE TABLE
products=#
products=# INSERT INTO departments (dept_name) VALUES ('HR');
INSERT 0 1
products=# INSERT INTO employees (emp_name, dept_id) VALUES ('Alice', 1);
INSERT 0 1
products=# SELECT * FROM departments;
departments              departments_dept_id_seq
products=# SELECT * FROM departments;
 dept_id | dept_name
---------+-----------
       1 | HR
(1 row)
```

Figure 4.21: *On delete and update set a Default*

The error is due to a violation of the foreign key constraint in the **employees** table. Specifically, the error occurs because when you try to delete a row from the **departments** table, the foreign key constraint in the **employees** table is set to **SET DEFAULT**, and the default value (**0**) for the **dept_id** does not exist in the **departments** table.

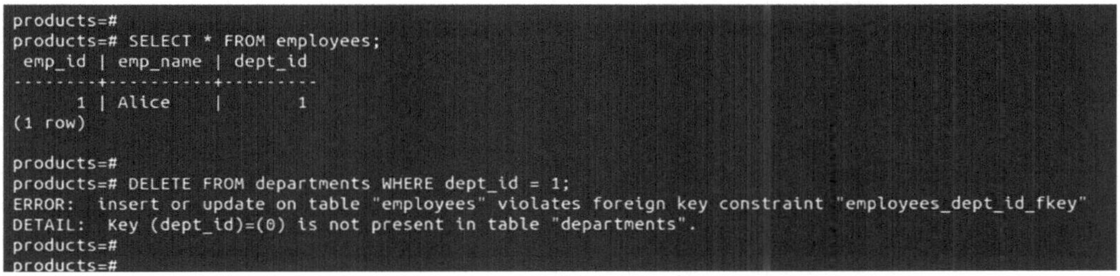

```
products=#
products=# SELECT * FROM employees;
 emp_id | emp_name | dept_id
--------+----------+---------
      1 | Alice    |       1
(1 row)

products=#
products=# DELETE FROM departments WHERE dept_id = 1;
ERROR:  insert or update on table "employees" violates foreign key constraint "employees_dept_id_fkey"
DETAIL:  Key (dept_id)=(0) is not present in table "departments".
products=#
products=#
```

Figure 4.22: *Shows an error on delete because the default value was not set*

RESTRICT

Prevents the deletion or update of an author row if there are any corresponding books rows.

```
CREATE TABLE authors (
    author_id SERIAL PRIMARY KEY,
    name VARCHAR(100)
);

CREATE TABLE books (
    book_id SERIAL PRIMARY KEY,
    title VARCHAR(100),
    author_id INTEGER REFERENCES authors(author_id) ON DELETE RE-
STRICT ON UPDATE RESTRICT
);
-- Insert into authors
INSERT INTO authors (name) VALUES ('Author1');
-- Insert into books
INSERT INTO books (title, author_id) VALUES ('Book1', 1);

-- Attempt to delete author row
DELETE FROM authors WHERE author_id = 1;
```

The delete operation failed due to the **RESTRICT** constraint.

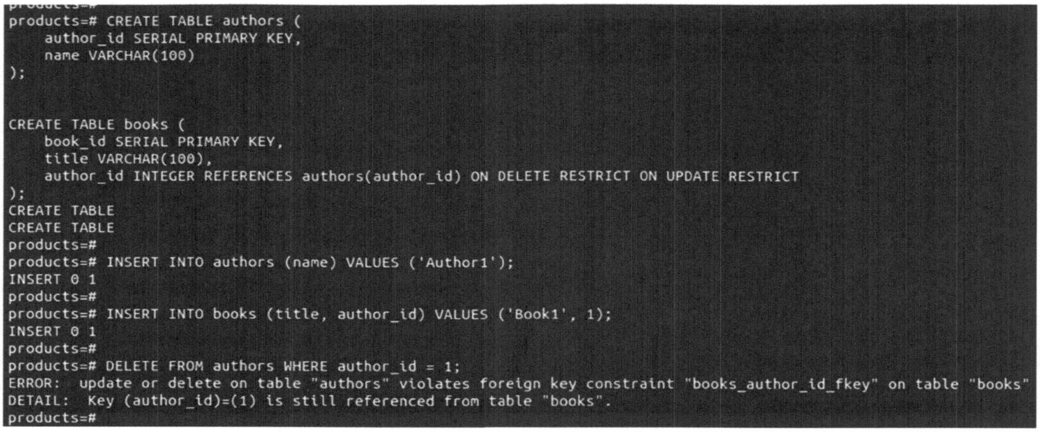

```
products=# CREATE TABLE authors (
    author_id SERIAL PRIMARY KEY,
    name VARCHAR(100)
);


CREATE TABLE books (
    book_id SERIAL PRIMARY KEY,
    title VARCHAR(100),
    author_id INTEGER REFERENCES authors(author_id) ON DELETE RESTRICT ON UPDATE RESTRICT
);
CREATE TABLE
CREATE TABLE
products=#
products=# INSERT INTO authors (name) VALUES ('Author1');
INSERT 0 1
products=#
products=# INSERT INTO books (title, author_id) VALUES ('Book1', 1);
INSERT 0 1
products=#
products=# DELETE FROM authors WHERE author_id = 1;
ERROR:  update or delete on table "authors" violates foreign key constraint "books_author_id_fkey" on table "books"
DETAIL:  Key (author_id)=(1) is still referenced from table "books".
products=#
```

Figure 4.23: *Shows RESTRICT on delete and update*

NO ACTION

Similar to **RESTRICT**, but the enforcement is deferred until the end of the transaction. If the foreign key constraint is violated by the end of the transaction, an error is raised.

```
CREATE TABLE schools (
    school_id SERIAL PRIMARY KEY,
    school_name VARCHAR(100)  );
CREATE TABLE students (
    student_id SERIAL PRIMARY KEY,
    student_name VARCHAR(100),
    school_id INTEGER REFERENCES schools(school_id) ON DELETE NO
ACTION ON UPDATE NO ACTION  );

-- Insert into schools
INSERT INTO schools (school_name) VALUES ('Buea High');
-- Insert into students
INSERT INTO students (student_name, school_id) VALUES ('Alice', 1);
-- Attempt to delete school row
 DELETE FROM schools WHERE school_id = 1;
-- The delete operation will fail due to the NO ACTION
constraint.
```

```
products=# CREATE TABLE schools (
    school_id SERIAL PRIMARY KEY,
    school_name VARCHAR(100)  );


CREATE TABLE students (
    student_id SERIAL PRIMARY KEY,
    student_name VARCHAR(100),
    school_id INTEGER REFERENCES schools(school_id) ON DELETE NO ACTION ON UPDATE NO ACTION  );
CREATE TABLE
CREATE TABLE
products=#
products=#
products=# INSERT INTO schools (school_name) VALUES ('Buea High');
INSERT 0 1
products=# INSERT INTO students (student_name, school_id) VALUES ('Alice', 1);
INSERT 0 1
products=#
products=# DELETE FROM schools WHERE school_id = 1;
ERROR:  update or delete on table "schools" violates foreign key constraint "students_school_id_fkey" on table "students"
DETAIL:  Key (school_id)=(1) is still referenced from table "students".
products=#
```

Figure 4.24: *Shows NO ACTION on delete and update*

Inserting Data into Tables

In PostgreSQL, the **INSERT** statement permits you to insert new rows into a table.

The basic syntax for the **INSERT** statement is as shown:

```
INSERT INTO table_name(column1, column2, column3 …)
VALUES (value1, value2, value3 …);
```

In this syntax, after the **INSERT INTO** is the name of the table you want to put data into, followed by a list of the column names in parentheses, separated by a comma. This is then followed by supplying a list of values in parentheses, which is also comma-separated. It is very important to make sure that columns and values are in the same order as in the column and value lists shown in the **INSERT** statement.

The **INSERT** statement also has an optional **RETURNING** clause that returns the information of the inserted row.

If you want to return the entire inserted row, you use an asterisk (*) after the **RETURNING** keyword:

```
INSERT INTO table_name(column1, column2, column3 …)
VALUES (value1, value2, column3 …)
RETURNING *;
```

If you desire to return just some information about the inserted row, then one or more columns can be specified after the **RETURNING** clause.

For example, the following statement returns the id of the inserted row:

```
INSERT INTO table_name(column1, column2, column3 …)
VALUES (value1, value2, column3 …)
RETURNING id;
```

If you desire to rename the returned value, then the AS keyword is used followed by the desired name of the output.

For example:

```
INSERT INTO table_name(column1, column2, column3 …)
VALUES (value1, value2, column3 …)
RETURNING id AS payment_id;
```

Deleting Data from Tables

In PostgreSQL, one or more rows are deleted from a table using the **DELETE** statement.

The basic syntax of a **DELETE** statement is:

```
DELETE FROM table_name
WHERE specify_condition_here;
```

After the **DELETE FROM** keywords is the table name from which the data is to be deleted. You then specify which row from the table is to be deleted using the **WHERE** clause.

It is very important to note that if the **WHERE** clause is omitted, then all rows will be deleted from the table when the **DELETE** statement is executed.

You can return the deleted row(s) to the client by using the **RETURNING** clause as follows:

```
DELETE FROM table_name
WHERE condition
RETURNING (select_list | *) ;
```

The asterisk (*) permits you to return all columns of the deleted row from the named table.

To return some specific columns, specify them after the **RETURNING** keyword as shown.

It is very important to note that the **DELETE** statement does not modify the structure of the table but removes only the data from the table. If the intention is to change the structure of the table, then the **ALTER TABLE** Statement is used.

Modifying and Deleting Tables

In PostgreSQL, the structure of a table can be modified using the **ALTER TABLE** statement.

The syntax of the **ALTER TABLE** statement is as follows:

```
ALTER TABLE table_name action;
```

There are many actions that PostgreSQL provides, some of which are adding a column, dropping a column, renaming a column, changing a column's data type, setting a column's default value, adding constraints to the table, and renaming the table.

The **ALTER TABLE ADD COLUMN** statement is used to add a new column to a table.

```
ALTER TABLE table_name
ADD COLUMN column_name datatype column_datatype;
```

The **ALTER TABLE DROP COLUMN** statement is used to drop a column from a table.

```
ALTER TABLE table_name
DROP COLUMN column_name;
```

The **ALTER TABLE RENAME COLUMN** statement is used to change the name of a column in a table, as shown:

```
ALTER TABLE table_name
RENAME COLUMN old_column_name
TO new_column_name;
```

The **ALTER TABLE ALTER COLUMN SET DEFAULT** or **DROP DEFAULT** is used to change the default value of the column, as shown:

```
ALTER TABLE table_name
ALTER COLUMN column_name
[SET DEFAULT value | DROP DEFAULT];
```

The **NOT NULL** constraint can be changed using the **ALTER TABLE ALTER COLUMN** statement, as shown:

```
ALTER TABLE table_name
ALTER COLUMN column_name
[SET NOT NULL| DROP NOT NULL];
```

A **CHECK** constraint can be added to a table using the statement **ALTER TABLE ADD CHECK**, as shown:

```
ALTER TABLE table_name
ADD CHECK expression;

ALTER TABLE table_name
ADD CONSTRAINT constraint_name constraint_definition;
```

The statement **ALTER TABLE RENAME** is used to change the name of a table, as shown.

```
ALTER TABLE table_name
RENAME TO new_table_name;
```

For practical purposes of the **ALTER TABLE** statement, we will create a database called **departments** and a table called **students**.

```
CREATE DATABASE departments;
```

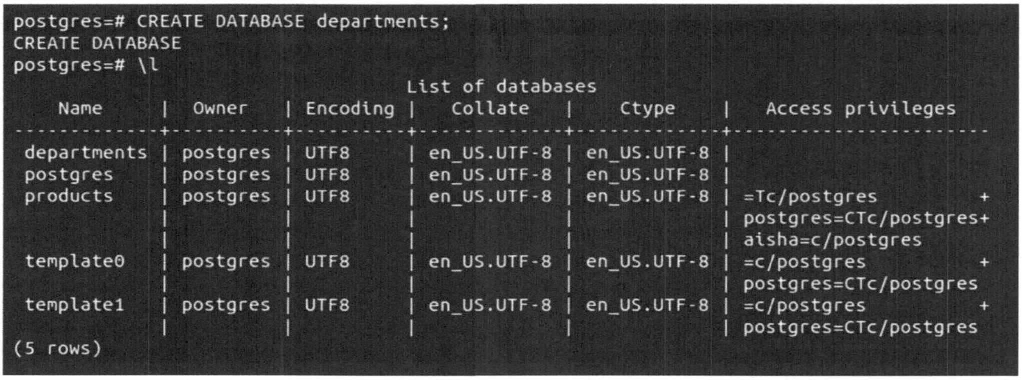

```
postgres=# CREATE DATABASE departments;
CREATE DATABASE
postgres=# \l
                                  List of databases
    Name     |  Owner   | Encoding |   Collate   |    Ctype    |   Access privileges
-------------+----------+----------+-------------+-------------+-----------------------
 departments | postgres | UTF8     | en_US.UTF-8 | en_US.UTF-8 |
 postgres    | postgres | UTF8     | en_US.UTF-8 | en_US.UTF-8 |
 products    | postgres | UTF8     | en_US.UTF-8 | en_US.UTF-8 | =Tc/postgres         +
             |          |          |             |             | postgres=CTc/postgres+
             |          |          |             |             | aisha=c/postgres
 template0   | postgres | UTF8     | en_US.UTF-8 | en_US.UTF-8 | =c/postgres          +
             |          |          |             |             | postgres=CTc/postgres
 template1   | postgres | UTF8     | en_US.UTF-8 | en_US.UTF-8 | =c/postgres          +
             |          |          |             |             | postgres=CTc/postgres
(5 rows)
```

***Figure** 4.25: Creating a new database*

```
postgres=# \c departments
You are now connected to database "departments" as user "postgres".
departments=#
```

***Figure** 4.26: Connecting to the new database*

```
DROP TABLE IF EXISTS students;
```

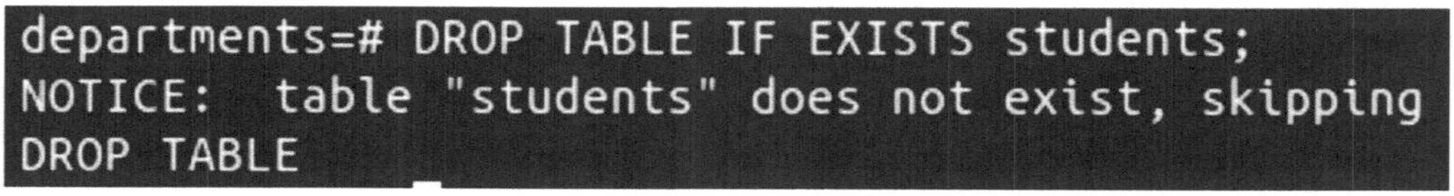

```
departments=# DROP TABLE IF EXISTS students;
NOTICE:  table "students" does not exist, skipping
DROP TABLE
```

***Figure 4.27:** Querying the existence of a table and then dropping it*

```
CREATE TABLE students (
   student_id serial PRIMARY KEY,
   sur_name VARCHAR (25) NOT NULL,
   surname VARCHAR (25) NOT NULL,
   email VARCHAR (80)
);
```

```
departments=# CREATE TABLE students (
    student_id serial PRIMARY KEY,
    sur_name VARCHAR (25) NOT NULL,
    surname VARCHAR (25) NOT NULL,
    email VARCHAR (80)
);
CREATE TABLE
departments=#
```

Figure 4.28: *Creating students' table*

To add the student's mark as a new column, we use the following statement:

```
departments=# SELECT * FROM students;
 student_id | sur_name | surname | email
------------+----------+---------+-------
(0 rows)
```

Figure 4.29: *Using SELECT statement to see columns in a table*

```
ALTER TABLE students
ADD COLUMN marks integer;
```

```
departments=# ALTER TABLE students
ADD COLUMN marks integer;
ALTER TABLE
departments=# SELECT * FROM students;
 student_id | sur_name | surname | email | marks
------------+----------+---------+-------+-------
(0 rows)
```

Figure 4.30: *Adding a new column in a table*

To remove the added marks column, use the following statement from the students' table:

```
ALTER TABLE students
DROP COLUMN marks;
```

```
departments=# ALTER TABLE students
DROP COLUMN marks;
ALTER TABLE
departments=# SELECT * FROM students;
 student_id | sur_name | surname | email
------------+----------+---------+-------
(0 rows)
```

Figure 4.31: *How to delete a column from a table*

To change the name of the surname column to **last_name**, the following statement is used:

```
ALTER TABLE students
RENAME COLUMN surname TO last_name;
```

The above query will rename surname to **last_name**. Try it out.

To add the grades as a new column to the students' table, use the following command:

```
ALTER TABLE students
ADD COLUMN grades CHAR(1);
```

```
departments=# ALTER TABLE students
ADD COLUMN grades CHAR(1);
ALTER TABLE
departments=# SELECT * FROM students;
 student_id | sur_name | surname | email | grades
------------+----------+---------+-------+--------
(0 rows)
```

Figure 4.32: *Adding grades' column to table students*

To set the default value for the grade column as **_blank** in the students' table the following statement is used:

```
ALTER TABLE students
ALTER COLUMN grades
SET DEFAULT '_blank';
```

```
departments=# ALTER TABLE students
ALTER COLUMN grades
SET DEFAULT '_blank';
ALTER TABLE
```

Figure 4.33: *Setting default values for column*

If data is inserted into the students' table without specifying a value for the grades' column, the **_blank** will be taken as the default value.

For example:

```
INSERT INTO students (student_id, first_name, last_name, email)
VALUES( 019, 'Pramoda', 'khan', 'p.khan@gmail.com');
```

```
departments=# INSERT INTO students (student_id, first_name, last_name, email, grades)
VALUES( 019, 'Pramoda', 'khan', 'p.khan@gmail.com');
ERROR:  column "first_name" of relation "students" does not exist
LINE 1: INSERT INTO students (student_id, first_name, last_name, ema...
                                          ^
```

***Figure 4.34:** Shows the first_name column does not exist*

Note: *When we created our table in Figure 4.28, we used* **sur_name** *rather than* **first_name**. *We are going to see how to correct that in the next figure (Figure 4.35).*

```
ALTER TABLE students
RENAME COLUMN sur_name TO first_name;

ALTER TABLE students
RENAME COLUMN surname TO last_name;
```

```
departments=# ALTER TABLE students
RENAME COLUMN sur_name TO first_name;
ALTER TABLE
departments=# ALTER TABLE students
RENAME COLUMN surname TO last_name;
ALTER TABLE
departments=# SELECT * FROM students;
 student_id | first_name | last_name | email | grades
------------+------------+-----------+-------+--------
(0 rows)
```

***Figure 4.35:** Renaming columns in a table*

In the next screen, we will insert data into the table.

```
departments=# INSERT INTO students (student_id, first_name, last_name, email, grades)
VALUES( 019,'Pramoda', 'khan', 'p.khan@gmail.com', 'A');
INSERT 0 1
```

***Figure 4.36:** Inserting into correct column names*

The following statement selects data from the students' table:

```
SELECT * FROM students;
```

```
departments=# SELECT * FROM students;
 student_id | first_name | last_name |       email
------------+------------+-----------+-------------------
         19 | Pramoda    | khan      | p.khan@gmail.com
(1 row)
```

Figure 4.37: *Using SELECT statement to retrieve data from a table*

The following statement adds a **CHECK** condition to the grade column so that the grade column only accepts the following values: **'A'**, **'B'**, **'C'**, **'D'**, **'E'**, and **'U'**. *Figure* 4.37 shows that the grades column no longer exists on the table. Before we run the query, we must first add the grades column again, as it was dropped previously.

```
ALTER TABLE students
ADD COLUMN grades CHAR(1);
```

```
departments=# ALTER TABLE students
ADD COLUMN grades CHAR(1);
ALTER TABLE
```

Figure 4.38: *Using alter command to add a new column in a table*

The following section will show you how to add a check constraint in a table:

```
ALTER TABLE students
ADD CHECK (grades IN ('A',  'B',  'C',  'D',  'E',  'U'));
```

```
departments=# ALTER TABLE students
ADD CHECK (grades IN ('A',  'B',  'C',  'D',  'E',  'U'));
ALTER TABLE
```

Figure 4.39: *Adding a check constraint on an existing table*

If a new row is inserted that violates the **CHECK** constraint for the grade column, PostgreSQL will throw an error message as shown in the following example:

```
INSERT INTO students (student_id, first_name, last_name, email,
grades)
VALUES (017, 'Tarjani', 'Gururani', 'n.vishnu@gmail.com', 'F');
```

```
departments=# INSERT INTO students (student_id, first_name, last_name, email, grades)
VALUES ( 017, 'Tarjani', 'Gururani', 't.gururani@gmail.com', 'F');
ERROR:  new row for relation "students" violates check constraint "students_grades_check"
DETAIL:  Failing row contains (17, Tarjani, Gururani, t.gururani@gmail.com, F).
```

Figure 4.40: *Entries violates the check constraint*

ERROR: new row for relation "**students**" violates check constraint "**students_grades_check**".

This refers to an '**F**' grade we tried to insert into the table, as it is not one of the check grades we set as constraints. Grade F is not among the grades specified in the check constraint.

The following query does not violate the check constraint; thus, there is no error. Grade **C** is among the grades specified in the check constraint.

```
INSERT INTO students (student_id, first_name, last_name, email, grades)
VALUES ( 018, 'Naveena', 'Vishnu', 'n.vishnu@gmail.com', 'C'),
                ( 012, 'Sahana', 'Arun', 's.arun@gmail.com','D');
```

```
departments=# INSERT INTO students (student_id, first_name, last_name, email, grades)
VALUES ( 018, 'Naveena', 'Vishnu', 'n.vishnu@gmail.com', 'C'),
                ( 012, 'Sahana', 'Arun', 's.arun@gmail.com','D');
INSERT 0 2
departments=# SELECT * FROM students;
 student_id | first_name | last_name |        email         | grades
------------+------------+-----------+----------------------+--------
         19 | Pramoda    | khan      | p.khan@gmail.com     |
         18 | Naveena    | Vishnu    | n.vishnu@gmail.com   | C
         12 | Sahana     | Arun      | s.arun@gmail.com     | D
(3 rows)
```

Figure 4.41: *Grade entries in accordance with the check constraint*

```
ALTER TABLE students
ADD CONSTRAINT unique_email UNIQUE (email);
```

```
departments=# ALTER TABLE students
ADD CONSTRAINT unique_email UNIQUE (email);
ALTER TABLE
```

Figure 4.42: *Adding a UNIQUE constraint on the email column*

The following statement attempts to insert the email that already exists:

```
INSERT INTO students (student_id, first_name, last_name, email, grades)
VALUES (011, 'Kaajal', 'Ramesh', 'k.ramesh@gmail.com','B');
```

```
departments=# INSERT INTO students (student_id, first_name, last_name, email, grades)
VALUES (011, 'Kallam', 'Ramesh', 'k.ramesh@gmail.com','B');
ERROR:  duplicate key value violates unique constraint "students_pkey"
DETAIL:  Key (student_id)=(11) already exists.
```

Figure 4.43: *Duplicate student_id entries violates unique constraint*

```
INSERT INTO students (student_id, first_name, last_name, email, grades)
VALUES (016, 'Kallam', 'Ramesh', 'k.ramesh@gmail.com', 'B');
```

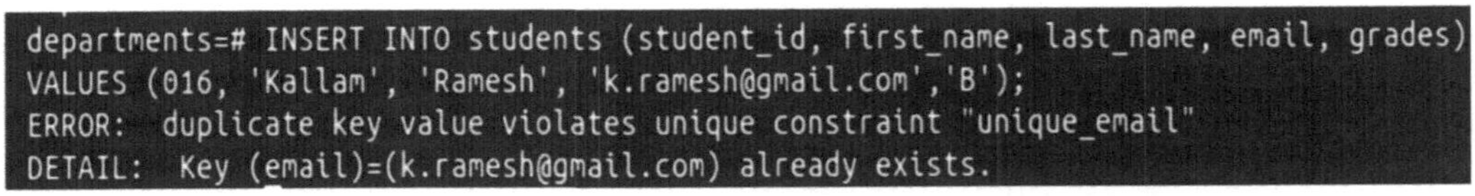

```
departments=# INSERT INTO students (student_id, first_name, last_name, email, grades)
VALUES (016, 'Kallam', 'Ramesh', 'k.ramesh@gmail.com','B');
ERROR:  duplicate key value violates unique constraint "unique_email"
DETAIL:  Key (email)=(k.ramesh@gmail.com) already exists.
```

Figure 4.44: *Duplicate email entries violates unique constraint*

It causes an error due to the **`unique_email`** constraint:

ERROR: duplicate key value violates unique constraint "**`unique_email`**"

The name of the students' table can be changed to employees using the following. Now let us change the table name from **`students`** to **`employees`**.

```
ALTER TABLE students
RENAME TO employees;
```

```
departments=# ALTER TABLE students
RENAME TO employees;
ALTER TABLE
```

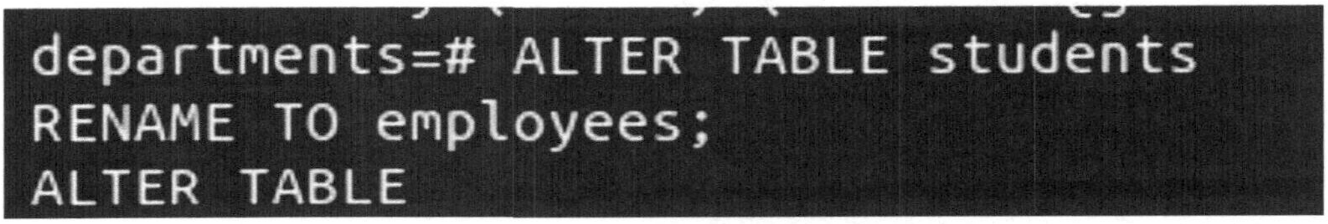

Figure 4.45: *Changing the table name from students to employees*

In the next screen, we will check that the name has actually been changed.

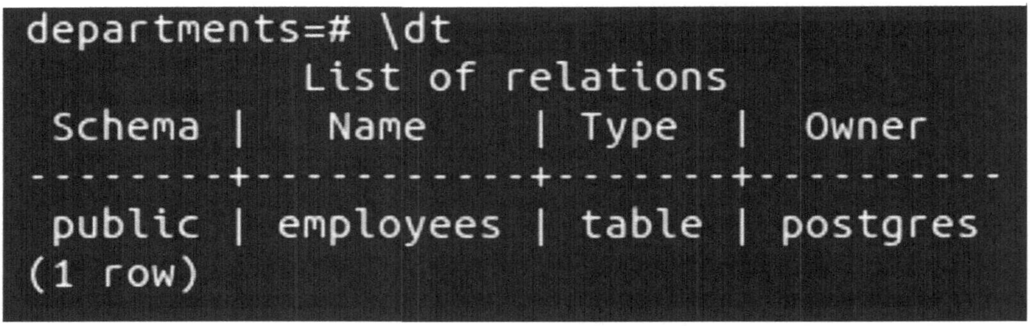

```
departments=# \dt
          List of relations
 Schema |   Name    | Type  |  Owner
--------+-----------+-------+----------
 public | employees | table | postgres
(1 row)
```

Figure 4.46: *To check that the table name has changed to employees*

With the preceding query, we saw how to change table names from **`students`** to **`employees`**. The next paragraph shows you how to change the name of a database.

The first thing is to check the number of connections.

Run the following query to check the number of connections on the database.

```
SELECT  *
FROM pg_stat_activity
WHERE datname = 'departments';
```

```
departments=# SELECT   *
FROM pg_stat_activity
WHERE datname = 'departments';
-[ RECORD 1 ]----+--------------------------------
datid            | 33119
datname          | departments
pid              | 2837
leader_pid       |
usesysid         | 10
usename          | postgres
application_name | psql
client_addr      |
client_hostname  |
client_port      | -1
backend_start    | 2023-11-22 15:22:30.602418-05
xact_start       | 2023-11-22 18:48:12.125367-05
query_start      | 2023-11-22 18:48:12.125367-05
state_change     | 2023-11-22 18:48:12.125369-05
wait_event_type  |
wait_event       |
state            | active
backend_xid      |
backend_xmin     | 748
query            | SELECT   *                          +
                 | FROM pg_stat_activity               +
                 | WHERE datname = 'departments';
backend_type     | client backend
```

Figure 4.47: *Using pg_stat_activity to check the number of connections on a table*

From the output, it is shown that there is only one connection to the database called **departments**. If the database that you want to rename has many active connections, then the users, as well as the application owners, must be informed before terminating the connections to avoid data loss.

You must terminate all the connections to the departments database by using the following statement:

```
SELECT   pg_terminate_backend (pid)
FROM      pg_stat_activity
WHERE   datname = 'departments';
```

The next step is to connect to the Postgres database as shown:

```
departments=# \c postgres
You are now connected to database "postgres" as user "postgres".
```

Figure 4.48: *Connecting to postgres database from departments*

```
postgres=# \l
                                  List of databases
    Name     |  Owner   | Encoding |   Collate   |    Ctype    |   Access privileges
-------------+----------+----------+-------------+-------------+-----------------------
 departments | postgres | UTF8     | en_US.UTF-8 | en_US.UTF-8 |
 postgres    | postgres | UTF8     | en_US.UTF-8 | en_US.UTF-8 |
 products    | postgres | UTF8     | en_US.UTF-8 | en_US.UTF-8 | =Tc/postgres         +
             |          |          |             |             | postgres=CTc/postgres+
             |          |          |             |             | aisha=c/postgres
 template0   | postgres | UTF8     | en_US.UTF-8 | en_US.UTF-8 | =c/postgres          +
             |          |          |             |             | postgres=CTc/postgres
 template1   | postgres | UTF8     | en_US.UTF-8 | en_US.UTF-8 | =c/postgres          +
             |          |          |             |             | postgres=CTc/postgres
(5 rows)
```

***Figure 4.49**: Checking the databases in postgres using \l command*

Once connected to the Postgres database, rename the **departments**' database to **sales** using the **ALTER DATABASE RENAME TO** statement as follows:

```
ALTER DATABASE departments RENAME TO sales;
```

```
postgres=# ALTER DATABASE departments RENAME TO sales;
ALTER DATABASE
```

***Figure 4.50**: How to rename a database*

Next, we will use **\l** to check that the database sales now exist rather than the database department.

```
postgres=# \l
                                  List of databases
   Name    |  Owner   | Encoding |   Collate   |    Ctype    |   Access privileges
-----------+----------+----------+-------------+-------------+-----------------------
 postgres  | postgres | UTF8     | en_US.UTF-8 | en_US.UTF-8 |
 products  | postgres | UTF8     | en_US.UTF-8 | en_US.UTF-8 | =Tc/postgres         +
           |          |          |             |             | postgres=CTc/postgres+
           |          |          |             |             | aisha=c/postgres
 sales     | postgres | UTF8     | en_US.UTF-8 | en_US.UTF-8 |
 template0 | postgres | UTF8     | en_US.UTF-8 | en_US.UTF-8 | =c/postgres          +
           |          |          |             |             | postgres=CTc/postgres
 template1 | postgres | UTF8     | en_US.UTF-8 | en_US.UTF-8 | =c/postgres          +
           |          |          |             |             | postgres=CTc/postgres
(5 rows)
```

***Figure 4.51**: Check that the database name has changed to sales*

Retrieving Data from Tables

The following skills are used to retrieve data from the database: Select, where clauses, Logical Operators, In, Not In, Like, Between, Order by, Limiting the number of results, Fetch, Distinct values, Null values, Column name alias, and Concatenation.

The first thing to learn is how to effectively use the **SELECT** statement to query data from a table.

One of the most common tasks when you work with the database is to query data from tables by using the **SELECT** statement.

One of the complex statements in PostgreSQL is the **SELECT** statement, as it has a lot of clauses that can be used to form a query.

Let us break it down into easy-to-understand short queries to enable you to learn about each clause quickly.

In the **SELECT** statement, we have the following clauses:

- Fetch distinct rows from a table using the **DISTINCT** operator.
- Sort all rows fetched using the **ORDER BY** clause.
- Filter the data being fetched using the **WHERE** clause.
- Use the **LIMIT** or **FETCH** clause to select a subset of rows from a table.
- Arrange rows into groups using the **GROUP BY** clause.
- We then filter the groups using the **HAVING** clause.
- We can retrieve data from different tables using joins such as **INNER JOIN**, **RIGHT JOIN**, **FULL JOIN**, **LEFT JOIN**, and **CROSS JOIN** clauses.
- Set operations can be performed using **UNION**, **UNION ALL**, **INTERSECT**, and **EXCEPT**.

In this section, we are going to be focusing on the **SELECT** and **FROM** clauses. We will start with the basic form of the **SELECT** statement that is used to retrieve data from a single table.

The **SELECT** statement syntax is as follows:

```
SELECT
   column_list
FROM
   table_name;
```

The **column_list** in the above query can either be a single column name or a list of column names separated by a comma (,).

Use an asterisk (*) in the place of **column_list** to select all data from all the columns of the table.

After the **FROM** keyword, we specify the table name from which we want to retrieve the data.

We are going to explore some examples of the **SELECT** statement. Let us create a customer table in the sales database with the following column names: **customer_id**, **first_name**, **last_name**, and **email**.

First, connect to the **sales** database:

```
\c sales
```

```
products=# \c sales
You are now connected to database "sales" as user "postgres".
sales=# \dt
          List of relations
 Schema |   Name    | Type  |  Owner
--------+-----------+-------+----------
 public | employees | table | postgres
(1 row)
```

Figure 4.52: *Connecting to sales database*

Now, run the following query to create the **customers**' table.

```
CREATE TABLE customers (
   Customer_id int primary key,
   first_name VARCHAR (30),
   last_name  VARCHAR (30) NOT NULL,
         store_id  CHAR (3),
         email VARCHAR (55)
);
```

```
sales=# CREATE TABLE customers (
        Customer_id int primary key,
        first_name VARCHAR (30),
        last_name  VARCHAR (30) NOT NULL,
         store_id  CHAR (3),
         email VARCHAR (55)
);
CREATE TABLE
```

Figure 4.53: *Creating customers' table inside sales database*

```
INSERT INTO customers (customer_id, first_name, last_name, store_id,
email)
VALUES (76, 'Guaresh', 'Badal', 'B09', 'g.badal@gmail.com'),
       (90, 'Aarav', 'Chadha', 'B10', 'a.chadha@gmail.com' ),
       (109, 'Pooran', 'Cheema', 'B14',  'p.cheema@gmail.com'),
    (65, 'Sanaya','Choudhry', 'B20',  's.choudhry@gmail.com),
    (23, 'Sagar', 'Ghandhi', 'B29', 's.ghandhi@gmail.com'),
    (109, 'Pooran', 'Cheema', 'B14', 'p.cheema@gmail.com'),
    (90, 'Aarav', 'Chadha', 'B10',  'a.chadha@gmail.com' ),
   (9, 'Aryan', 'Suresh', 'B14', 'a.suresh@gmail.com);
```

```
sales=# INSERT INTO customers (customer_id, first_name, last_name, store_id, email)
VALUES (76, 'Guaresh', 'Badal', 'B09', 'g.badal@gmail.com'),
       (90, 'Aarav', 'Chadha', 'B10', 'a.chadha@gmail.com' ),
       (109, 'Pooran', 'Cheema', 'B14',  'p.cheema@gmail.com'),
       (65, 'Sanaya','Choudhry', 'B20',  's.choudhry@gmail.com'),
       (23, 'Sagar', 'Ghandhi', 'B29', 's.ghandhi@gmail.com'),
       (110, 'Pooran', 'Cheema', 'B14', 'p.cheema@gmail.com'),
       (91, 'Aarav', 'Chadha', 'B10',  'a.chadha@gmail.com' ),
       (9, 'Aryan', 'Suresh','B14', 'a.suresh@gmail.com');
INSERT 0 8
```

Figure 4.54: *Inserting multiple rows of data into customers' table*

The following query will retrieve all the data from the table:

```
SELECT * FROM customers;
```

```
sales=# SELECT * FROM customers;
 customer_id | first_name | last_name | store_id |         email
-------------+------------+-----------+----------+----------------------
          76 | Guaresh    | Badal     | B09      | g.badal@gmail.com
          90 | Aarav      | Chadha    | B10      | a.chadha@gmail.com
         109 | Pooran     | Cheema    | B14      | p.cheema@gmail.com
          65 | Sanaya     | Choudhry  | B20      | s.choudhry@gmail.com
          23 | Sagar      | Ghandhi   | B29      | s.ghandhi@gmail.com
         110 | Pooran     | Cheema    | B14      | p.cheema@gmail.com
          91 | Aarav      | Chadha    | B10      | a.chadha@gmail.com
           9 | Aryan      | Suresh    | B14      | a.suresh@gmail.com
(8 rows)
```

Figure 4.55: *Retrieving all data from customers' table*

It is very important to note that **SELECT *** is widely considered a bad style in production code since adding a column to the table would change the results. Always select the required columns as shown:

```
SELECT first_name FROM customer;
```

```
sales=# SELECT first_name FROM customers;
 first_name
------------
 Guaresh
 Aarav
 Pooran
 Sanaya
 Sagar
 Pooran
 Aarav
 Aryan
(8 rows)
```

Figure 4.56: *Retrieving only first names from customers table*

Only the first names were retrieved.

The following query will retrieve the first name and last name.

```
SELECT first_name, last_name FROM customers;
```

```
sales=# SELECT first_name, last_name FROM customers;
 first_name | last_name
------------+-----------
 Guaresh    | Badal
 Aarav      | Chadha
 Pooran     | Cheema
 Sanaya     | Choudhry
 Sagar      | Ghandhi
 Pooran     | Cheema
 Aarav      | Chadha
 Aryan      | Suresh
(8 rows)
```

Figure 4.57: *Retrieving first name and last name*

```
SELECT customer_id, last_name, email FROM customers;
```

```
sales=# SELECT customer_id, last_name, email FROM customers;
 customer_id | last_name |         email
-------------+-----------+----------------------
          76 | Badal     | g.badal@gmail.com
          90 | Chadha    | a.chadha@gmail.com
         109 | Cheema    | p.cheema@gmail.com
          65 | Choudhry  | s.choudhry@gmail.com
          23 | Ghandhi   | s.ghandhi@gmail.com
         110 | Cheema    | p.cheema@gmail.com
          91 | Chadha    | a.chadha@gmail.com
           9 | Suresh    | a.suresh@gmail.com
(8 rows)
```

Figure 4.58: *Retrieving three named columns from a table*

The following query will join the first name and last name and call it a full name.

```
SELECT
      CONCAT (first_name, ' ', last_name) AS "Full_name"
FROM customers;
```

```
sales=# SELECT
        CONCAT (first_name, ' ', last_name) AS "Full_name"
FROM customers;
    Full_name
-----------------
 Guaresh Badal
 Aarav Chadha
 Pooran Cheema
 Sanaya Choudhry
 Sagar Ghandhi
 Pooran Cheema
 Aarav Chadha
 Aryan Suresh
(8 rows)
```

Figure 4.59: *Join first_name and last_name as full_name with concat command*

The AS clause is used to relabel the new output column.

A **WHERE** clause is added to a query to specify which rows are to be extracted. The **WHERE** clause normally contains an expression that is either true or false (Boolean expression). Where the **WHERE** clause is applied, the rows returned are those for which the Boolean expression is true.

The **AND**, **OR**, and **NOT** are permitted in the qualification of the Boolean operators.

For example, to retrieve the data of all customers with customer IDs greater than 80, the following query will do:

```
SELECT * FROM customer
WHERE customer_id > 80;
```

```
sales=# SELECT * FROM customers
WHERE customer_id > 80;
 customer_id | first_name | last_name | store_id |        email
-------------+------------+-----------+----------+--------------------
          90 | Aarav      | Chadha    | B10      | a.chadha@gmail.com
         109 | Pooran     | Cheema    | B14      | p.cheema@gmail.com
         110 | Pooran     | Cheema    | B14      | p.cheema@gmail.com
          91 | Aarav      | Chadha    | B10      | a.chadha@gmail.com
(4 rows)
```

Figure 4.60: *How to use the WHERE clause*

We can arrange for the results of the preceding query to be sorted in descending order based on customer ID:

```
SELECT * FROM customers
WHERE customer_id > 80
ORDER BY customer_id DESC;
```

```
sales=# SELECT * FROM customers
WHERE customer_id > 80
ORDER BY customer_id DESC;
 customer_id | first_name | last_name | store_id |        email
-------------+------------+-----------+----------+--------------------
         110 | Pooran     | Cheema    | B14      | p.cheema@gmail.com
         109 | Pooran     | Cheema    | B14      | p.cheema@gmail.com
          91 | Aarav      | Chadha    | B10      | a.chadha@gmail.com
          90 | Aarav      | Chadha    | B10      | a.chadha@gmail.com
(4 rows)
```

Figure 4.61: *How to use the WHERE clause and ORDER BY clause*

From the output, you can see that the customer IDs are arranged from biggest to smallest (descending order).

Next, let us retrieve all data with a customer ID greater than **80** and a store ID is **B14**.

```
SELECT * FROM customers
WHERE customer_id > 80
AND store_id = 'B14' ;
```

```
sales=# SELECT * FROM customers
WHERE customer_id > 80
AND store_id = 'B14' ;
 customer_id | first_name | last_name | store_id |        email
-------------+------------+-----------+----------+--------------------
         109 | Pooran     | Cheema    | B14      | p.cheema@gmail.com
         110 | Pooran     | Cheema    | B14      | p.cheema@gmail.com
(2 rows)
```

***Figure 4.62**: How to use the WHERE clause with AND operator*

Only two rows were selected that have a store ID of **B14** and a customer ID greater than **80**. These happen to be duplicate entries.

To request that duplicate rows be removed from the result of a query, execute it with a **DISTINCT** keyword as shown:

```
SELECT DISTINCT first_name, last_name, store_id, email
FROM customers
WHERE customer_id > 80
AND store_id = 'B14' ;
```

```
sales=# SELECT DISTINCT first_name, last_name, store_id, email FROM customers
WHERE customer_id > 80
AND store_id = 'B14' ;
 first_name | last_name | store_id |       email
------------+-----------+----------+--------------------
 Pooran     | Cheema    | B14      | p.cheema@gmail.com
(1 row)
```

***Figure 4.63:** How to use the WHERE clause with AND operator*

```
SELECT DISTINCT first_name, last_name, email, customer_id FROM
customers;
```

```
sales=# SELECT DISTINCT first_name, last_name, email FROM customers;
 first_name | last_name |        email
------------+-----------+----------------------
 Sanaya     | Choudhry  | s.choudhry@gmail.com
 Pooran     | Cheema    | p.cheema@gmail.com
 Sagar      | Ghandhi   | s.ghandhi@gmail.com
 Aryan      | Suresh    | a.suresh@gmail.com
 Aarav      | Chadha    | a.chadha@gmail.com
 Guaresh    | Badal     | g.badal@gmail.com
(6 rows)
```

***Figure 4.64**: How to use the DISTINCT keyword to limit the data*

With the **`DISTINCT`** keyword on the first name entries, only one Pooran and Aarav are returned.

Database Relationships

In relational databases, **database relationships** describe how data stored in different tables is related to each other. These relationships help maintain data integrity and enable efficient data retrieval. There are several types of database relationships, each serving different purposes and use cases. The primary types are:

One-to-One (1:1)

A **one-to-one** relationship occurs when a single record in one table is associated with a single record in another table. This type of relationship is used when you want to split data into different tables but maintain a direct link between them.

Example: Person and Passport

- One person has one unique passport.
- One passport is issued to one person.

Person Table:

`PersonID`	**`Name`**
1	Alice Smith
2	John Doe

Table 4.1: *Person table for one-to-one relationship*

Passport Table:

`PassportID`	**`PersonID`**	**`PassportNumber`**
P1	1	X1234567
P2	2	Y2345678

Table 4.2: *Passport table for one-to-one relationship*

The **`PersonID`** in the Person table is the primary key, while in the **`Passport`** table, it is the foreign key. In *Tables* 4.1 *and* 4.2, each **`PersonID`** in the **`Person`** table corresponds to a **`PersonID`** in the Passport table, establishing a one-to-one relationship. For example, Alice Smith with **`PersonID`** 1 is associated with **`PassportID`** P1 in the **`Passport`** table.

One-to-Many (1:M)

A **one-to-many** relationship occurs when a single record in one table is related to multiple records in another table. This is the most common type of relationship in relational databases.

Example: Author and Book

- One author can write many books.
- Each book is written by one author.

Author Table:

AuthorID	AuthorName
1	Jane Austen
2	Mark Twain

***Table 4.3:** Author's table for one-to-many relationship*

Book Table:

BookID	AuthorID	Title
1	1	Pride and Prejudice
2	1	Sense and Sensibility
3	2	The Adventures of Tom Sawyer
4	2	Adventures of Huckleberry Finn

***Table 4.4:** Book's table for one-to-many relationship*

The **`AuthorID`** in the Author table is the primary key, while in the **`Book`** table, it is the foreign key.

In *Tables 4.3 and 4.4*, each **`AuthorID`** in the Author table can have multiple corresponding **`AuthorIDs`** in the Book table, establishing a one-to-many relationship.

Many-to-One (M:1)

A **many-to-one** relationship is essentially the inverse of a one-to-many relationship. Multiple records in one table relate to a single record in another table.

Example: Employee and Department

- Many employees work in one department.
- Each department has many employees.

Employee Table:

`EmployeeID`	`Name`	`DepartmentID`
1	Alice Smith	101
2	John Doe	102
3	Mary Johnson	101
4	James Brown	103

Table 4.5: *Employee's table for many-to-one relationship*

Department Table:

`DepartmentID`	`DepartmentName`
101	Sales
102	Marketing
103	IT

Table 4.6: *Department's table for many-to-one relationship*

The **`DepartmentID`** in the Employee table is the primary key, while in the **`Department`** table, it is the foreign key. In *Tables* 4.5 *and* 4.6, each **`DepartmentID`** in the **`Department`** table corresponds to multiple **`DepartmentIDs`** in the **`Employee`** table, establishing a many-to-one relationship.

Many-to-Many (M:N)

A **many-to-many** relationship occurs when multiple records in one table are related to multiple records in another table. This type of relationship usually requires a join table (or a junction table) to manage the associations.

Example: Student and Course

- Many students can enrol in many courses.
- Each course can have many students.

Student Table:

`StudentID`	`StudentName`
1	Alice Smith
2	John Doe
3	Mary Johnson

Table 4.7: *Student's table for many-to-many relationship*

Course Table:

CourseID	CourseName
101	Mathematics
102	Literature
103	Computer Science

Table 4.8: *Course's table for many-to-many relationship*

Enrollment Table:

StudentID	CourseID
1	101
2	101
2	102
3	103
1	103

Table 4.9: *Junction table for student and course tables*

In the **Student** table, **StudentID** is the primary key, while in the **Course** table, **CourseID** is the primary key. The **Enrolment** table acts as a junction table to manage the many-to-many relationship between the **Student** and **Course** tables. Each record in the Enrolment table links a **StudentID** to a **CourseID**, allowing multiple students to enrol in multiple courses. The **StudentID** and **CourseID** are both foreign keys in the Enrolment table. Looking at the above tables, an example is a student called Alice Smith with StudentID 1 is registered to Mathematics **CourseID 101** and Computer Science **CourseID 103**. The **CourseID 101** is also registered to **StudentID** 2, which is John Doe.

Task

Create all the tables listed under database relationships, establishing the primary keys and foreign keys.

Technical Interview Questions

1. **Creating and Using Database Schema**
    a. What are the steps to create a new schema in PostgreSQL?
    b. How do you switch to a different schema in PostgreSQL?
    c. Can you explain the importance of using schemas in a database?

2. **Creating Tables with Foreign Keys**
    a. How do you create a table with a foreign key constraint in PostgreSQL?
    b. What are the cascading actions available for foreign keys, and how do they work?
3. **Inserting Data into Tables**
    a. How do you insert multiple rows into a table in a single **`INSERT`** statement?
    b. What is the purpose of the **`RETURNING`** clause in an **`INSERT`** statement?
4. **Deleting Data from Tables**
    a. How do you delete rows from a table based on a condition?
    b. What is the difference between **`TRUNCATE`** and **`DELETE`** in PostgreSQL?
5. **Modifying and Deleting Tables**
    a. How do you add a new column to an existing table in PostgreSQL?
    b. What are the steps to delete a table along with its dependent objects in PostgreSQL?
6. **Retrieving Data from Tables**
    a. How do you retrieve data from multiple tables using a **`JOIN`** operation?
    b. What is the difference between **`INNER JOIN`** and **`LEFT JOIN`**?
7. **Database Relationships**
    a. Can you explain the different types of relationships in a database?
    b. How do you define a many-to-many relationship in PostgreSQL using a join table?

Conclusion

In this chapter, we have learned how to use the PostgreSQL **`INSERT`**, **`DELETE`**, **`ADD CONSTRAINT`**, and **`ALTER TABLE`** statements to manipulate data in a table and to change the structure of an existing table. We also learned how to rename a table using **`ALTER TABLE RENAME TO`**, as well as a PostgreSQL database to a new one by using the **`ALTER DATABASE RENAME TO`** statement. We also learned the different database relationships and that these relationships are established using primary keys and foreign keys, which link the related records across different tables in the database.

In the next chapter, we will learn how to create and authenticate users to log in to the database, understand basic Linux commands, and PostgreSQL configuration.

Chapter 5
PostgreSQL Configurations and File Permissions

Introduction

In this chapter, learners will learn and understand the different configuration files of PostgreSQL. They will become proficient in using these configuration files for setting and resetting parameters. Learners will learn the basic Linux commands needed to work with PostgreSQL and its configuration.

Structure

In this chapter, the following topics will be covered:

- Understanding Linux Commands
- File Permissions and Ownership
- PostgreSQL Configuration Files
- Creating and Authenticating Users and Roles

Understanding Linux Commands

Linux is an operating system (OS) that is used to manage computer resources. It helps establish communication between software and hardware. Linux OS is very stable and secure. It is free and open source. In this section, you will learn the basic Linux commands required to work with PostgreSQL configuration files.

Let us log into our Ubuntu machine as described in *Chapter 2, Installing and Initializing PostgreSQL*. Reminder steps are as follows:

1. Double-click the vmware workstation logo

2. Double-click the virtual machine
3. Click username
4. Enter password
5. Double-click the 3 by 3 dotted square
6. Search for terminal.

Figures 5.1–5.3 is a refresher on how to log into the virtual machine on your VMWare workstation.

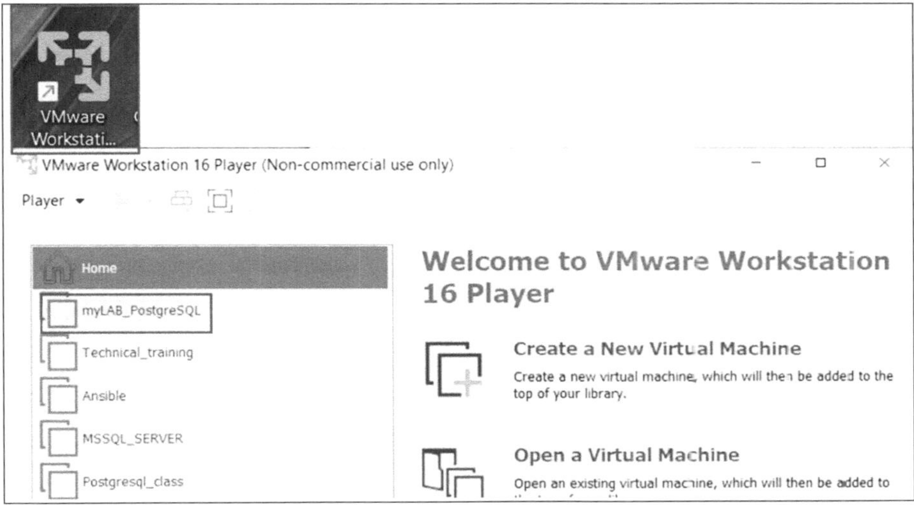

Figure 5.1: *Login step 1*

Figure 5.2: *Login step 2*

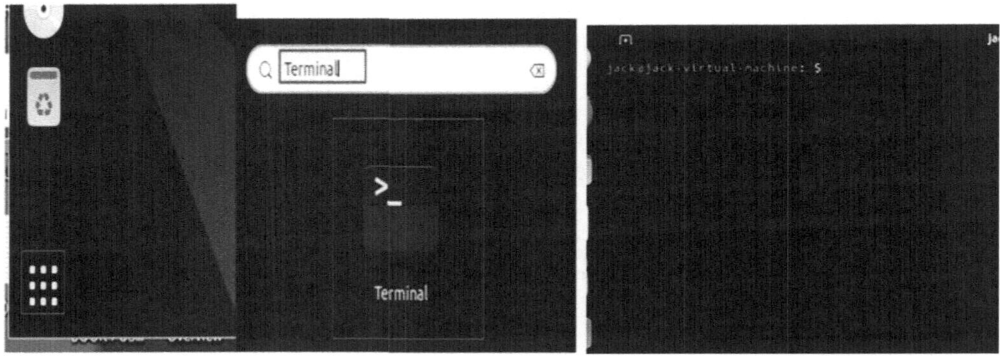

***Figure* 5.3:** *Complete the login process*

Explaining `pwd`, `ls`, and `cd` Commands

The **`pwd`** (present working directory) command returns your current or present working directory's path, like **`/home/jack`** shown as follows:

The syntax is:

```
pwd  [option]
```

There are two options supported by the **`pwd`** command. The first option **`-L`** or **`--logical`** option returns the environment variable, including symbolic links. The second option, **`-P`** or **`-physical`** returns the present directory's actual path. The **`-P`** option is often omitted and only **`pwd`** is used.

```
jack@jack-virtual-machine: $ pwd -L
/home/jack
jack@jack-virtual-machine: $ pwd -P
/home/jack
jack@jack-virtual-machine: $ pwd
/home/jack
```

***Figure* 5.4:** *Check the current working directory*

The above result shows that the present working directory is **`/home/jack.`** This means that the user is currently working inside the **`/jack`** folder, which is inside the **`/home`** directory. If other directories and files are created, they will be created inside the **`/jack`** folder (directory).

The **`ls`** command is used to list files and directories in the system. The syntax is:

```
ls [/directory/folder]
```

When **ls** is used without the path, it lists files and directories in the current environment (present working directory).

If you remove the path, the **ls** command will show the current working directory's content. You can modify the command using these options:

- **-R** will list all the files in the subdirectories.
- **-a** will show all files, including hidden files.
- **-al** will show all files in a long listing format
- **-lh** will convert all sizes to readable formats, such as **TB**, **MB**, and **GB**.

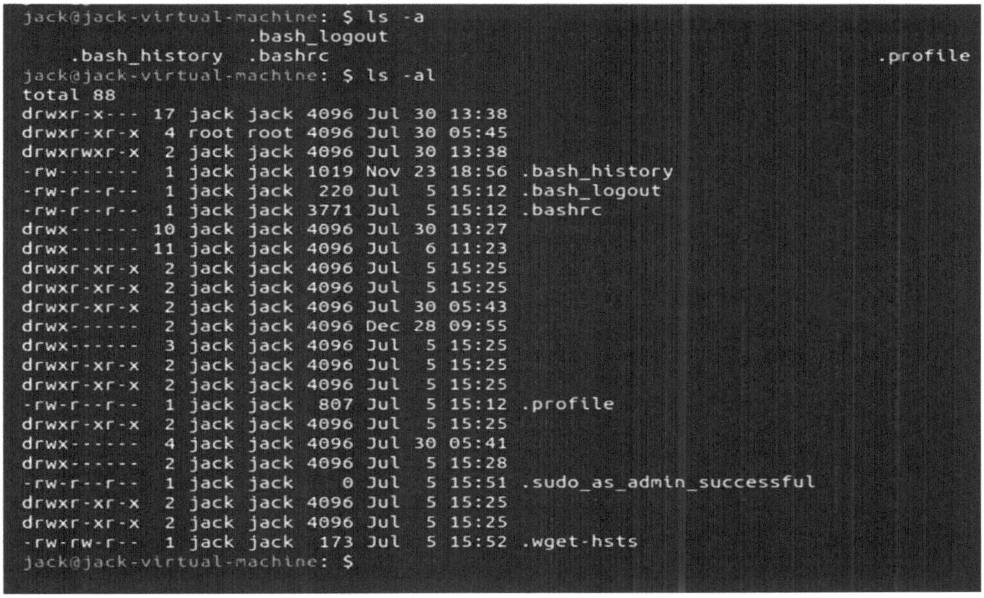

```
jack@jack-virtual-machine: $ ls -a
                    .bash_logout
    .bash_history  .bashrc                                                          .profile
jack@jack-virtual-machine: $ ls -al
total 88
drwxr-x---  17 jack jack 4096 Jul 30 13:38
drwxr-xr-x   4 root root 4096 Jul 30 05:45
drwxrwxr-x   2 jack jack 4096 Jul 30 13:38
-rw-------   1 jack jack 1019 Nov 23 18:56 .bash_history
-rw-r--r--   1 jack jack  220 Jul  5 15:12 .bash_logout
-rw-r--r--   1 jack jack 3771 Jul  5 15:12 .bashrc
drwx------  10 jack jack 4096 Jul 30 13:27
drwx------  11 jack jack 4096 Jul  6 11:23
drwxr-xr-x   2 jack jack 4096 Jul  5 15:25
drwxr-xr-x   2 jack jack 4096 Jul  5 15:25
drwxr-xr-x   2 jack jack 4096 Jul 30 05:43
drwx------   2 jack jack 4096 Dec 28 09:55
drwx------   3 jack jack 4096 Jul  5 15:25
drwxr-xr-x   2 jack jack 4096 Jul  5 15:25
drwxr-xr-x   2 jack jack 4096 Jul  5 15:25
-rw-r--r--   1 jack jack  807 Jul  5 15:12 .profile
drwxr-xr-x   2 jack jack 4096 Jul  5 15:25
drwx------   4 jack jack 4096 Jul 30 05:41
drwx------   2 jack jack 4096 Jul  5 15:28
-rw-r--r--   1 jack jack    0 Jul  5 15:51 .sudo_as_admin_successful
drwxr-xr-x   2 jack jack 4096 Jul  5 15:25
drwxr-xr-x   2 jack jack 4096 Jul  5 15:25
-rw-rw-r--   1 jack jack  173 Jul  5 15:52 .wget-hsts
jack@jack-virtual-machine: $
```

Figure 5.5: *Result of ls -l command*

The **cd** command is used to navigate from one directory to another (*Figures* 5.6–5.8). The syntax is:

```
cd /path_to_directory
```

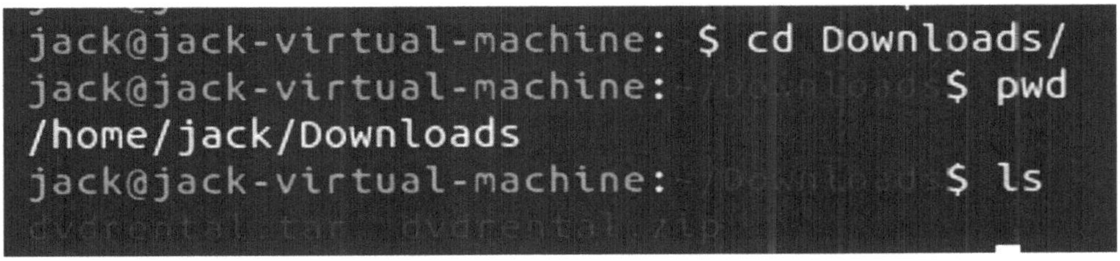

```
jack@jack-virtual-machine: $ cd Downloads/
jack@jack-virtual-machine:          $ pwd
/home/jack/Downloads
jack@jack-virtual-machine:          $ ls
```

Figure 5.6: *Change to the Downloads directory*

```
jack@jack-virtual-machine:          $ cd
jack@jack-virtual-machine: $ pwd
/home/jack
```

Figure 5.7: *Change back to the original working directory*

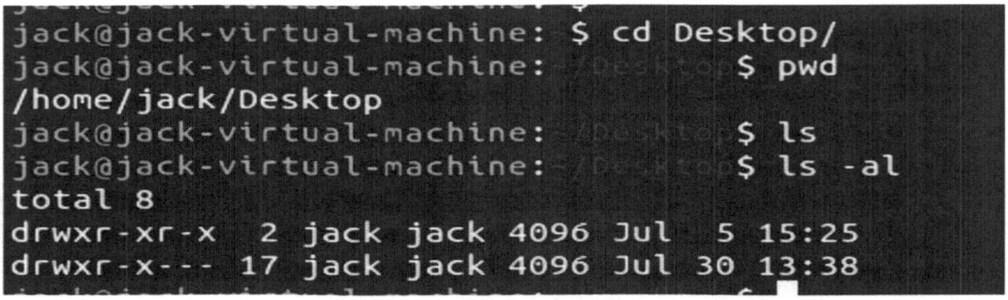

```
jack@jack-virtual-machine: $ cd Desktop/
jack@jack-virtual-machine:          $ pwd
/home/jack/Desktop
jack@jack-virtual-machine:          $ ls
jack@jack-virtual-machine:          $ ls -al
total 8
drwxr-xr-x  2 jack jack 4096 Jul  5 15:25
drwxr-x--- 17 jack jack 4096 Jul 30 13:38
```

Figure 5.8: *Change to the Desktop directory and check content*

Next, we are going to create a new directory inside the desktop directory called classwork. The syntax to create a new directory is:

```
mkdir directory_name
```

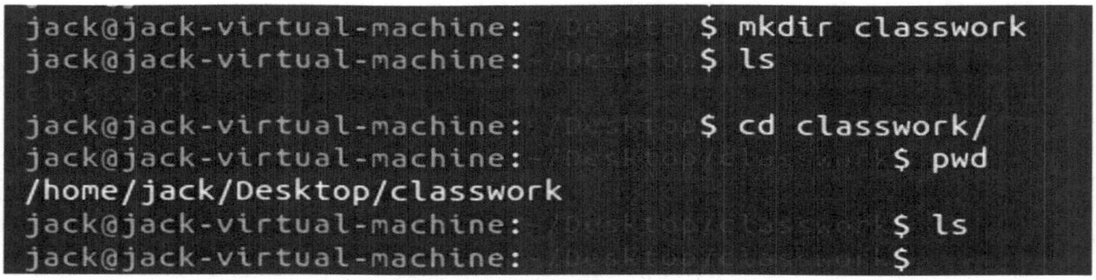

```
jack@jack-virtual-machine:          $ mkdir classwork
jack@jack-virtual-machine:          $ ls

jack@jack-virtual-machine:          $ cd classwork/
jack@jack-virtual-machine:                    $ pwd
/home/jack/Desktop/classwork
jack@jack-virtual-machine:                    $ ls
jack@jack-virtual-machine:                    $
```

Figure 5.9: *To create a new directory and cd into the new directory*

Next, we are going to create a new file called training materials inside the classwork directory using the **touch** command. The syntax to create a new file is:

```
touch file_name
touch training_materials
```

```
jack@jack-virtual-machine:                    $ touch training_materials
jack@jack-virtual-machine:                    $ ls
training_materials
jack@jack-virtual-machine:                    $ touch file1
jack@jack-virtual-machine:                    $ touch file2, file3
jack@jack-virtual-machine:                    $ ls
file1  file2,  file3  training_materials
jack@jack-virtual-machine:                    $ ls -al
total 8
drwxrwxr-x 2 jack jack 4096 Dec 28 10:55
drwxr-xr-x 3 jack jack 4096 Dec 28 10:51
-rw-rw-r-- 1 jack jack    0 Dec 28 10:55 file1
-rw-rw-r-- 1 jack jack    0 Dec 28 10:55 file2,
-rw-rw-r-- 1 jack jack    0 Dec 28 10:55 file3
-rw-rw-r-- 1 jack jack    0 Dec 28 10:54 training_materials
jack@jack-virtual-machine:                    $
```

Figure 5.10: *Creating files in Linux OS*

In the next image, we will see how to rename a file. To change the name of a file, we use the **mv** command. The syntax is:

```
mv  old_file_name   new_file_name
mv  training_materials   file4
```

The preceding command will change the file name from **training_materials** to **file4**.

```
jack@jack-virtual-machine:                    $ ls
file1  file2,  file3  training_materials
jack@jack-virtual-machine:                    $ mv training_materials file4
jack@jack-virtual-machine:                    $ ls
file1  file2,  file3  file4
jack@jack-virtual-machine:                    $ mv file1 new_file
jack@jack-virtual-machine:                    $ ls
file2,  file3  file4  new_file
jack@jack-virtual-machine:                    $
```

Figure 5.11: *To rename a file*

In the next image, we will learn how to remove files from Linux OS. We use the **rm** command to remove files. To remove multiple files, type file names with comma separated. The syntax is:

```
rm file_name
```

To remove **new_file** use the following command:

```
rm new_file
```

To remove **file2** and **file3** at the same time, use the following command:

```
rm file2, file3
```

```
jack@jack-virtual-machine:            $ ls
file1  file2,  file3  file4
jack@jack-virtual-machine:            $ mv file1 new_file
jack@jack-virtual-machine:            $ ls
file2,  file3  file4  new_file
jack@jack-virtual-machine:            $
jack@jack-virtual-machine:            $ rm new_file
jack@jack-virtual-machine:            $ ls
file2,  file3  file4
jack@jack-virtual-machine:            $ rm file2, file3
jack@jack-virtual-machine:            $ ls
file4
jack@jack-virtual-machine:            $
```

Figure 5.12: *Shows how to remove files*

There are different ways to add content to files in Linux. We can echo the content and either use a single redirection (**>**) to replace all the content of the file with the new content or use double redirection (**>>**) to add content to the existing content in the file. The **cat** command is used to see or read the content of a file. In *Figure* 5.13, the first **cat** command returns no content. After adding content with the **echo** command, the second **cat** command reviews the content of the file, as shown.

```
echo "This is the best PostgreSQL training course" > file4
cat file4
```

```
jack@jack-virtual-machine:            $ cat file4

jack@jack-virtual-machine:            $ echo "This is the best PostgreSQL training course" > file4
jack@jack-virtual-machine:            $ cat file4
"This is the best PostgreSQL training course"
jack@jack-virtual-machine:            $
```

Figure 5.13: *Adding content to file4 using echo and single redirection*

The **df** command is used to check the amount of free disk space used and available. The **-h** option presents the result in human-readable format.

```
jack@jack-virtual-machine:            $ df
Filesystem     1K-blocks     Used Available Use% Mounted on
tmpfs             190324     1788    188536   1% /run
/dev/sda3       19946096 14004784   4902772  75% /
tmpfs             951608       16    951592   1% /dev/shm
tmpfs               5120        4      5116   1% /run/lock
/dev/sda2         524252     6220    518032   2% /boot/efi
tmpfs             190320      104    190216   1% /run/user/1000
/dev/sr0         4812096  4812096         0 100% /media/jack/Ubuntu 22.04.2 LTS amd64
jack@jack-virtual-machine:            $
jack@jack-virtual-machine:            $ df -h
Filesystem      Size  Used Avail Use% Mounted on
tmpfs           186M  1.8M  185M   1% /run
/dev/sda3        20G   14G  4.7G  75% /
tmpfs           930M   16K  930M   1% /dev/shm
tmpfs           5.0M  4.0K  5.0M   1% /run/lock
/dev/sda2       512M  6.1M  506M   2% /boot/efi
tmpfs           186M  104K  186M   1% /run/user/1000
/dev/sr0        4.6G  4.6G     0 100% /media/jack/Ubuntu 22.04.2 LTS amd64
jack@jack-virtual-machine:            $
```

Figure 5.14: *To display the amount of free and used disk space*

The following steps show how to add the result of a command as content to a file:

```
jack@jack-virtual-machine:                    $ cat file4
This the best PostgreSQL training course
jack@jack-virtual-machine:                    $ df -h >> file4
jack@jack-virtual-machine:                    $
jack@jack-virtual-machine:                    $ cat file4
This the best PostgreSQL training course
Filesystem      Size  Used Avail Use% Mounted on
tmpfs           186M  1.8M  185M   1% /run
/dev/sda3        20G   14G  4.7G  75% /
tmpfs           930M   16K  930M   1% /dev/shm
tmpfs           5.0M  4.0K  5.0M   1% /run/lock
/dev/sda2       512M  6.1M  506M   2% /boot/efi
tmpfs           186M  104K  186M   1% /run/user/1000
/dev/sr0        4.6G  4.6G     0 100% /media/jack/Ubuntu 22.04.2 LTS amd64
jack@jack-virtual-machine:                    $
```

Figure 5.15: *Adding the result of a command as content to a file*

The **ps** command is used to check various processes and their ids. The following PIDs are very important in the case of terminating or killing a process.

```
ps [options]
```

The following are some options to add to **ps** command:

- **-a**: See the processes of all users rather than just the current user.
- **-u**: Gives detailed information on the individual processes.
- **-x**: Show processes that are controlled by daemons, not users.

For example, the command **ps -aux** and **ps -ef** returns a detailed list of all processes:

The **pgrep** command returns processes based on specific selection criteria.

- **-l**: Show the list of process names and PIDs.
- **-n**: To find the newest process.
- **-o**: To find the oldest process.
- **-u**: Return only processes belonging to a specific user.
- **-x**: Return processes that match exactly a given pattern.

```
pgrep -l -u root
```

The above command will return PIDs of all processes owned by the root user (*Figures 5.16 and 5.17*).

To kill a process, the syntax is:

```
Kill -9 PID
```

```
jack@jack-virtual-machine:                              $ ps
    PID TTY          TIME CMD
   3020 pts/0    00:00:00 bash
  12530 pts/0    00:00:00 ps
jack@jack-virtual-machine:                              $ ps -ef
UID          PID    PPID  C STIME TTY          TIME CMD
root           1       0  0 Dec29 ?        00:00:06 /sbin/init splash
root           2       0  0 Dec29 ?        00:00:00 [kthreadd]
root           3       2  0 Dec29 ?        00:00:00 [rcu_gp]
root           4       2  0 Dec29 ?        00:00:00 [rcu_par_gp]
root           5       2  0 Dec29 ?        00:00:00 [slub_flushwq]
root           6       2  0 Dec29 ?        00:00:00 [netns]
root           8       2  0 Dec29 ?        00:00:00 [kworker/0:0H-events_highpri]
root          10       2  0 Dec29 ?        00:00:00 [mm_percpu_wq]
root          11       2  0 Dec29 ?        00:00:00 [rcu_tasks_kthread]
root          12       2  0 Dec29 ?        00:00:00 [rcu_tasks_rude_kthread]
root          13       2  0 Dec29 ?        00:00:00 [rcu_tasks_trace_kthread]
root          14       2  0 Dec29 ?        00:00:03 [ksoftirqd/0]
root          15       2  0 Dec29 ?        00:00:14 [rcu_preempt]
root          16       2  0 Dec29 ?        00:00:00 [migration/0]
root          17       2  0 Dec29 ?        00:00:00 [idle_inject/0]
root          19       2  0 Dec29 ?        00:00:00 [cpuhp/0]
root          20       2  0 Dec29 ?        00:00:00 [cpuhp/1]
root          21       2  0 Dec29 ?        00:00:00 [idle_inject/1]
root          22       2  0 Dec29 ?        00:00:00 [migration/1]
root          23       2  0 Dec29 ?        00:00:01 [ksoftirqd/1]
root          25       2  0 Dec29 ?        00:00:00 [kworker/1:0H-events_highpri]
```

Figure 5.16: *PIDs returned from running the ps -ef command*

```
jack@jack-virtual-machine:                              $ pgrep -l -u root
1 systemd
2 kthreadd
3 rcu_gp
4 rcu_par_gp
5 slub_flushwq
6 netns
8 kworker/0:0H-events_highpri
10 mm_percpu_wq
11 rcu_tasks_kthread
12 rcu_tasks_rude_kthread
13 rcu_tasks_trace_kthread
14 ksoftirqd/0
15 rcu_preempt
16 migration/0
17 idle_inject/0
19 cpuhp/0
20 cpuhp/1
21 idle_inject/1
```

Figure 5.17: *PIDs and processes owned by the root user*

To view all currently running processes, the **top** command is used. It shows a list of all active processes and zombie processes.

```
jack@jack-virtual-machine:                    $ top

top - 10:15:18 up 12:57,  1 user,  load average: 0.37, 0.25, 0.17
Tasks: 289 total,   1 running, 288 sleeping,   0 stopped,   0 zombie
%Cpu(s):  0.3 us,  0.5 sy,  0.0 ni, 98.8 id,  0.0 wa,  0.0 hi,  0.3 si,  0.0 st
MiB Mem :   1858.6 total,     73.4 free,   1017.9 used,    767.4 buff/cache
MiB Swap:   2140.0 total,   2070.9 free,     69.1 used.    642.3 avail Mem

    PID USER      PR  NI    VIRT    RES    SHR S  %CPU  %MEM     TIME+ COMMAND
   1829 jack      20   0 4447120 218852  84724 S   2.3  11.5   3:37.27 gnome-shell
   3002 jack      20   0  559536  54176  41160 S   0.7   2.8   0:43.46 gnome-terminal-
   9583 root      20   0       0      0      0 I   0.7   0.0   0:43.66 kworker/0:2-events
    603 systemd+  20   0   14828   6784   6016 S   0.3   0.4   0:55.08 systemd-oomd
   2111 jack      20   0  144548  35412  23020 S   0.3   1.9   1:07.39 vmtoolsd
   2214 jack      20   0  226708  46212  21964 S   0.3   2.4   0:01.49 Xwayland
  12519 root      20   0       0      0      0 I   0.3   0.0   0:00.39 kworker/1:2-events
      1 root      20   0  168112  12896   7776 S   0.0   0.7   0:06.75 systemd
      2 root      20   0       0      0      0 S   0.0   0.0   0:00.11 kthreadd
      3 root       0 -20       0      0      0 I   0.0   0.0   0:00.00 rcu_gp
      4 root       0 -20       0      0      0 I   0.0   0.0   0:00.05 rcu_par_gp
      5 root       0 -20       0      0      0 I   0.0   0.0   0:00.00 slub_flushwq
      6 root       0 -20       0      0      0 I   0.0   0.0   0:00.00 netns
      8 root       0 -20       0      0      0 I   0.0   0.0   0:00.00 kworker/0:0H-events_highpri
     10 root       0 -20       0      0      0 I   0.0   0.0   0:00.00 mm_percpu_wq
     11 root      20   0       0      0      0 I   0.0   0.0   0:00.00 rcu_tasks_kthread
     12 root      20   0       0      0      0 I   0.0   0.0   0:00.00 rcu_tasks_rude_kthread
     13 root      20   0       0      0      0 I   0.0   0.0   0:00.00 rcu_tasks_trace_kthread
     14 root      20   0       0      0      0 S   0.0   0.0   0:03.73 ksoftirqd/0
```

Figure 5.18: *Overview of all running processes and their PIDs*

The next image will show how to redirect the result of a command as the content of a file. The content of **file4** will now be the result of **ps -ef** command.

```
ps -ef > file4
cat file4        #To see the content of file4 after the ps -ef
command.
```

```
jack@jack-virtual-machine:                   $ ps -ef > file4
jack@jack-virtual-machine:                   $ cat file4
UID          PID    PPID  C STIME TTY          TIME CMD
root           1       0  0 Dec29 ?        00:00:06 /sbin/init splash
root           2       0  0 Dec29 ?        00:00:00 [kthreadd]
root           3       2  0 Dec29 ?        00:00:00 [rcu_gp]
root           4       2  0 Dec29 ?        00:00:00 [rcu_par_gp]
root           5       2  0 Dec29 ?        00:00:00 [slub_flushwq]
root           6       2  0 Dec29 ?        00:00:00 [netns]
root           8       2  0 Dec29 ?        00:00:00 [kworker/0:0H-events_highpri]
root          10       2  0 Dec29 ?        00:00:00 [mm_percpu_wq]
root          11       2  0 Dec29 ?        00:00:00 [rcu_tasks_kthread]
root          12       2  0 Dec29 ?        00:00:00 [rcu_tasks_rude_kthread]
root          13       2  0 Dec29 ?        00:00:00 [rcu_tasks_trace_kthread]
root          14       2  0 Dec29 ?        00:00:03 [ksoftirqd/0]
root          15       2  0 Dec29 ?        00:00:15 [rcu_preempt]
root          16       2  0 Dec29 ?        00:00:00 [migration/0]
root          17       2  0 Dec29 ?        00:00:00 [idle_inject/0]
root          19       2  0 Dec29 ?        00:00:00 [cpuhp/0]
root          20       2  0 Dec29 ?        00:00:00 [cpuhp/1]
root          21       2  0 Dec29 ?        00:00:00 [idle_inject/1]
root          22       2  0 Dec29 ?        00:00:00 [migration/1]
root          23       2  0 Dec29 ?        00:00:01 [ksoftirqd/1]
root          25       2  0 Dec29 ?        00:00:00 [kworker/1:0H-events_highpri]
root          26       2  0 Dec29 ?        00:00:00 [kdevtmpfs]
root          27       2  0 Dec29 ?        00:00:00 [inet_frag_wq]
root          28       2  0 Dec29 ?        00:00:00 [kauditd]
root          30       2  0 Dec29 ?        00:00:00 [khungtaskd]
```

Figure 5.19: *Redirecting the result of ps -ef to a file for further commands*

Next, we will use more, head and tail to extract and view the content of a file. With the **head** and **tail** commands, we can specify how many lines we want to return.

```
jack@jack-virtual-machine:                    $ more file4
UID          PID    PPID  C STIME TTY          TIME CMD
root           1       0  0 09:50 ?        00:00:02 /sbin/init splash
root           2       0  0 09:50 ?        00:00:00 [kthreadd]
root           3       2  0 09:50 ?        00:00:00 [rcu_gp]
root           4       2  0 09:50 ?        00:00:00 [rcu_par_gp]
root           5       2  0 09:50 ?        00:00:00 [slub_flushwq]
root           6       2  0 09:50 ?        00:00:00 [netns]
root           8       2  0 09:50 ?        00:00:00 [kworker/0:0H-events_highpri]
root          10       2  0 09:50 ?        00:00:00 [mm_percpu_wq]
root          11       2  0 09:50 ?        00:00:00 [rcu_tasks_kthread]
root          12       2  0 09:50 ?        00:00:00 [rcu_tasks_rude_kthread]
root          13       2  0 09:50 ?        00:00:00 [rcu_tasks_trace_kthread]
root          14       2  0 09:50 ?        00:00:00 [ksoftirqd/0]
```

Figure 5.20: *Content of a file using the more command*

To see the top 5 lines of the file, we use **head -5 file_name**.

```
head -5 file4
```

```
jack@jack-virtual-machine:                    $ head -5 file4
UID          PID    PPID  C STIME TTY          TIME CMD
root           1       0  0 09:50 ?        00:00:02 /sbin/init splash
root           2       0  0 09:50 ?        00:00:00 [kthreadd]
root           3       2  0 09:50 ?        00:00:00 [rcu_gp]
root           4       2  0 09:50 ?        00:00:00 [rcu_par_gp]
jack@jack-virtual-machine:                    $
```

Figure 5.21: *Top 5 lines of a file using the head command*

To see the last 5 lines of the file, we use the tail **-5 file_name** command.

```
tail -5 file4
```

```
jack@jack-virtual-machine:                    $ tail -5 file4
root        3915       2  0 10:57 ?        00:00:02 [kworker/0:2-events]
root        3979       2  0 11:03 ?        00:00:00 [kworker/0:0-events]
root        4083       2  0 11:11 ?        00:00:00 [kworker/u256:1-events_power_efficient]
root        4100       2  0 11:13 ?        00:00:00 [kworker/0:1-events]
jack        4127    3020  0 11:16 pts/0    00:00:00 ps -ef
jack@jack-virtual-machine:                    $
```

Figure 5.22: *Last 5 lines of a file using the tail command*

The **grep** command is used to extract lines with specific words from a file. The syntax is:

```
grep -i "word to search" filename
grep -i daemon file4
```

```
jack@jack-virtual-machine:                    $ grep -i daemon file4
root          861       1  0 09:50 ?        00:00:00 /usr/libexec/accounts-
avahi         865       1  0 09:50 ?        00:00:00 avahi-      : running [jack-virtual-machine.loc
message+      869       1  0 09:50 ?        00:00:01 @dbus-        --system --address=systemd: --nofo
root          871       1  0 09:50 ?        00:00:02 /usr/sbin/NetworkManager --no-
root          895       1  0 09:50 ?        00:00:00 /usr/libexec/power-profiles-
avahi         941     865  0 09:50 ?        00:00:00 avahi-      : chroot helper
rtkit        1099       1  0 09:50 ?        00:00:00 /usr/libexec/rtkit-
jack         1686    1677  0 09:53 ?        00:00:02 /usr/bin/pulseaudio --       ize=no --log-target
jack         1695       1  0 09:53 ?        00:00:00 /usr/bin/gnome-keyring-       --       ize --log
jack         1704    1677  0 09:53 ?        00:00:01 /usr/bin/dbus-        --session --address=system
jack         1839    1831  0 09:53 ?        00:00:00 /usr/bin/dbus-        --config-file=/usr/share/d
unix:path=/run/user/1000/at-spi/bus
jack         1884    1677  0 09:53 ?        00:00:00 /usr/libexec/goa-
jack         2022    1677  0 09:53 ?        00:00:00 sh -c /usr/bin/ibus-        --panel disable $([
jack         2030    2022  0 09:53 ?        00:00:02 /usr/bin/ibus-        --panel disable
jack@jack-virtual-machine:                    $
```

Figure 5.23: *Search lines with the word daemon from a file*

We used the **mkdir command** to create a directory and **rmdir command** to remove a directory for the system. The syntax is:

```
mkdir directory_name
rmdir directory_name
```

Let us create and remove the same directory called **remove_me**.

```
mkdir remove_me
rmdir remove_me
```

```
jack@jack-virtual-machine:                    $ mkdir remove_me
jack@jack-virtual-machine:                    $ ls
file4
jack@jack-virtual-machine:                    $
jack@jack-virtual-machine:                    $ rmdir remove_me
jack@jack-virtual-machine:                    $
jack@jack-virtual-machine:                    $ ls
file4
jack@jack-virtual-machine:                    $
```

Figure 5.24: *To create and remove a directory*

The **man** command is used to understand all commands in Linux. The syntax is:

```
man command
Example, man df
```

```
jack@jack-virtual-machine:                              $ man df

DESCRIPTION
       This  manual page documents the GNU version of df.  df displays the am
       available on all currently mounted file systems is shown.  Disk space
       are used.

       If  an  argument  is the absolute file name of a disk device node cont
       the device node.  This version of df cannot show the space available o
       system structures.

OPTIONS
       Show information about the file system on which each FILE resides, or

       Mandatory arguments to long options are mandatory for short options to

       -a, --all
              include pseudo, duplicate, inaccessible file systems

       -B, --block-size=SIZE
              scale sizes by SIZE before printing them; e.g., '-BM' prints si

       -h, --human-readable
              print sizes in powers of 1024 (e.g., 1023M)

       -H, --si
              print sizes in powers of 1000 (e.g., 1.1G)

       -i, --inodes
              list inode information instead of block usage

       -k     like --block-size=1K

       -l, --local
              limit listing to local file systems
```

Figure 5.25: *Result of man df command*

The **history** command is used to see all the commands you have executed on the Linux OS.

```
jack@jack-virtual-machine:~/Desktop/classwork$ history
 85  touch file1
 86  touch file2, file3
 87  ls
 88  ls -al
 89  ls
 90  mv training_materials file4
 91  ls
 92  mv file1 new_file
 93  ls
 94  rm new_file
 95  ls
 96  rm file2, file3
 97  ls
 98  cat file4
 99  echo "This the best PostgreSQL training course" > file4
100  cat file4
101  df
102  df -u
103  df -du
104  df -h
105  df
106  df -h
107  ls
108  cat file4
109  df -h >> file4
110  cat file4
111  ps
112  ps -ef
```

Figure 5.26: *Result of the History command*

To check the name of the host machine, we use the **`hostname` command.** The **`who` command** shows you who is connected to the system. **`Whoami` command** also shows you who is connected.

```
jack@jack-virtual-machine:~/Desktop/classwork$ hostname
jack-virtual-machine
jack@jack-virtual-machine:~/Desktop/classwork$ who
jack     tty2         2023-12-28 09:53 (tty2)
jack@jack-virtual-machine:~/Desktop/classwork$ whoami
jack
jack@jack-virtual-machine:~/Desktop/classwork$
```

Figure 5.27: *Result of* **hostname, who,** *and* **whoami** *commands*

To count the total number of words, lines, and characters in a file, we use the **wc command** with various options, such as:

- **wc -l** display number of lines
- **wc - w** display number of words
- **wc -c** display the number of characters.

```
jack@jack-virtual-machine:                $ wc file4
  290  2468 22668 file4
jack@jack-virtual-machine:                $ wc -l file4
290 file4
jack@jack-virtual-machine:                $ wc -w file4
2468 file4
jack@jack-virtual-machine:                $ wc -c file4
22668 file4
jack@jack-virtual-machine:                $
```

Figure 5.28: *Result of* ***wc command*** *on file4*

The next image shows another way of running the **wc** command in combination with the pipe (|) symbol. The pipe symbol enables you to use the output of one command as an input of another command.

In the following image, the result of the **cat** command is accepted as the input of the **wc** command because of the pipe symbol between them.

```
cat file4 | wc
cat file4 | wc -l
```

```
jack@jack-virtual-machine:                $ cat file4 | wc
    290     2468    22668
jack@jack-virtual-machine:                $ cat file4 | wc -l
290
```

Figure 5.29: *Combining the result of two commands using the* ***|*** *symbol*

We use the **uname command** to check the name of the operating system. The **uptime command** tells you how long the system has been up and running.

```
jack@jack-virtual-machine:                $ uname
Linux
jack@jack-virtual-machine:                $ uptime
 11:54:20 up  2:03,  1 user,  load average: 0.13, 0.14, 0.16
jack@jack-virtual-machine:                $
jack@jack-virtual-machine:                $ echo $SHELL
/bin/bash
jack@jack-virtual-machine:                $
```

Figure 5.30: *Result of* ***uname****, and* ***uptime*** *command*

The **cat /etc/passwd** command is used to see the users in the Linux OS.

```
jack@jack-virtual-machine:                    $ cat /etc/passwd
root:x:0:0:root:/root:/bin/bash
daemon:x:1:1:daemon:/usr/sbin:/usr/sbin/nologin
bin:x:2:2:bin:/bin:/usr/sbin/nologin
sys:x:3:3:sys:/dev:/usr/sbin/nologin
sync:x:4:65534:sync:/bin:/bin/sync
games:x:5:60:games:/usr/games:/usr/sbin/nologin
man:x:6:12:man:/var/cache/man:/usr/sbin/nologin
lp:x:7:7:lp:/var/spool/lpd:/usr/sbin/nologin
mail:x:8:8:mail:/var/mail:/usr/sbin/nologin
news:x:9:9:news:/var/spool/news:/usr/sbin/nologin
uucp:x:10:10:uucp:/var/spool/uucp:/usr/sbin/nologin
```

Figure 5.31: *Partial result of* ***cat /etc/passwd*** *command*

To add users to the Linux OS, we need to use the **useradd command**. Let us add user Sara. We need the super user (**sudo**) privileges to add a user.

```
jack@jack-virtual-machine:                    $ useradd sara
useradd: Permission denied.
useradd: cannot lock /etc/passwd; try again later.
jack@jack-virtual-machine:                    $ sudo useradd sara
[sudo] password for jack:
jack@jack-virtual-machine:                    $ cat /etc/passwd
jack:x:1000:1000:Jack F,,,:/home/jack:/bin/bash
postgres:x:129:137:PostgreSQL administrator,,,:/var/lib/postgresql:/bin/bash
sshd:x:130:65534::/run/sshd:/usr/sbin/nologin
sara:x:1001:1001::/home/sara:/bin/sh
```

Figure 5.32: **USERADD** *command result*

The **ls** command is used to list the content of a directory, while **ls -xa*** will list all the files whose names start with **xa**. The **rm xa*** will remove all the files beginning with **xa**.

```
jack@jack-virtual-machine:                    $ ls
file4  xaa  xab  xac  xad  xae  xaf  xag  xah  xai  xaj
jack@jack-virtual-machine:                    $
jack@jack-virtual-machine:                    $ ls xa*
xaa  xab  xac  xad  xae  xaf  xag  xah  xai  xaj
jack@jack-virtual-machine:                    $ rm xa*
jack@jack-virtual-machine:                    $ ls
file4
jack@jack-virtual-machine:                    $
```

Figure 5.33: *Removing all files beginning with xa*

To create multiple versions of the same file, we use the **touch** command as shown in the following image. The **touch abc{1..9}-xyz.txt** will create 9 files, all starting with abc and ending with **-xyz.txt**. The **rm abc*** command will remove all the files beginning with abc.

```
jack@jack-virtual-machine:                    $ touch abc{1..9}-xyz.txt
jack@jack-virtual-machine:                    $ ls
abc1-xyz.txt  abc2-xyz.txt  abc3-xyz.txt  abc4-xyz.txt  abc5-xyz.txt  abc6
jack@jack-virtual-machine:                    $ ls -al
total 32
drwxrwxr-x 2 jack jack  4096 Dec 28 12:06
drwxr-xr-x 3 jack jack  4096 Dec 28 10:51
-rw-rw-r-- 1 jack jack     0 Dec 28 12:06 abc1-xyz.txt
-rw-rw-r-- 1 jack jack     0 Dec 28 12:06 abc2-xyz.txt
-rw-rw-r-- 1 jack jack     0 Dec 28 12:06 abc3-xyz.txt
-rw-rw-r-- 1 jack jack     0 Dec 28 12:06 abc4-xyz.txt
-rw-rw-r-- 1 jack jack     0 Dec 28 12:06 abc5-xyz.txt
-rw-rw-r-- 1 jack jack     0 Dec 28 12:06 abc6-xyz.txt
-rw-rw-r-- 1 jack jack     0 Dec 28 12:06 abc7-xyz.txt
-rw-rw-r-- 1 jack jack     0 Dec 28 12:06 abc8-xyz.txt
-rw-rw-r-- 1 jack jack     0 Dec 28 12:06 abc9-xyz.txt
-rw-rw-r-- 1 jack jack 22668 Dec 28 11:16 file4
jack@jack-virtual-machine:                    $ rm abc*
jack@jack-virtual-machine:                    $ ls
file4
jack@jack-virtual-machine:                    $ ls -al
total 32
drwxrwxr-x 2 jack jack  4096 Dec 28 12:07
drwxr-xr-x 3 jack jack  4096 Dec 28 10:51
-rw-rw-r-- 1 jack jack 22668 Dec 28 11:16 file4
jack@jack-virtual-machine:                    $
```

Figure 5.34: *Creating multiple versions of the same file*

The next image is another example of creating multiple files at once.

```
jack@jack-virtual-machine:                    $ touch newfile{1..5}
jack@jack-virtual-machine:                    $ ls
file4  newfile1  newfile2  newfile3  newfile4  newfile5
jack@jack-virtual-machine:                    $ ls -al
total 32
drwxrwxr-x 2 jack jack  4096 Dec 28 12:11
drwxr-xr-x 3 jack jack  4096 Dec 28 10:51
-rw-rw-r-- 1 jack jack 22668 Dec 28 11:16 file4
-rw-rw-r-- 1 jack jack     0 Dec 28 12:11 newfile1
-rw-rw-r-- 1 jack jack     0 Dec 28 12:11 newfile2
-rw-rw-r-- 1 jack jack     0 Dec 28 12:11 newfile3
-rw-rw-r-- 1 jack jack     0 Dec 28 12:11 newfile4
-rw-rw-r-- 1 jack jack     0 Dec 28 12:11 newfile5
jack@jack-virtual-machine:                    $
```

Figure 5.35: *Creating five files at the same time*

File Permissions and Ownership

Linux is a multi-user OS, so there would be many people connected and using the system at a given time. To prevent people from accessing others' confidential files and directories, it has security to prevent that called permissions and ownership. To see directories and file permissions, execute either of the following commands:

```
ls -l
ls -al
```

```
jack@jack-virtual-machine:                    $ ls -l
total 36
-rw-rw-r-- 1 jack jack 22668 Dec 28 11:16 file4
-rw-rw-r-- 1 jack jack     0 Dec 28 12:11 newfile1
-rw-rw-r-- 1 jack jack     0 Dec 28 12:11 newfile2
-rw-rw-r-- 1 jack jack     0 Dec 28 12:11 newfile3
-rw-rw-r-- 1 jack jack     0 Dec 28 12:11 newfile4
-rw-rw-r-- 1 jack jack     0 Dec 28 12:11 newfile5
drwxrwxr-x 2 jack jack  4096 Dec 29 07:57
drwxrwxr-x 2 jack jack  4096 Dec 29 07:57
drwxrwxr-x 2 jack jack  4096 Dec 29 07:57
```

***Figure* 5.36:** *Result of* **ls -l** *command*

From the result, we can deduce the following information:

- If the first character is '**-**', it means it is a file, and if it is '**d**', it means it is a directory
- The security of the system is represented by the next nine characters (**rw-rw-r--**). This will be further explained in the next paragraph.
- The next column displays the owner of the file. (Here it is `**jack**`)
- The next column displays the group to which the owner of the file belongs, which is also `**jack**`.
- The next column displays the size of the file in bytes.
- The next column displays the date and time the file was last modified.
- The last column displays the name of the file or directory. File examples are **file4**, **newfile1**, **newfile2** .... Directory examples are **subject**, **subject2**, and **subject3**.

There are three permission groups in Linux. The nine characters (**rwx rwx**) previously mentioned are divided into three sets of three. The full permission that can be granted is to read(**r**), write(**w**) and execute(**x**).

Letters	Definition
'r'	"read" the file's contents.
'w'	"write", or modify, the file's contents.
'x'	"execute" the file. This permission is given only if the file is a program.

***Figure 5.37:** Different file security permissions*

Each of the three "**rwx**" characters refers to a different operation you can perform on the file.

- **Owner or user:** The first set of three (**rwx**) are the permissions of the file owner. The permissions are exclusively applicable to the individuals who own the files or directories.
- **Groups:** The second set of three (**rwx**) permissions can be granted to a specific group of users. Only users in that group are impacted by these permissions.
- **All Users:** The last set of three permissions (**rwx**) is for the rest of the world. These permissions are universally applied to all users on the system, thus presenting the highest security risk. This set of permissions assigned to all users should be done cautiously to prevent potential security vulnerabilities.

```
---         ---            ---
rwx         rwx            rwx
user(u)     group(g)       other (o)
```

These symbols are very useful in adding or removing permissions (*Figures 5.38 and 5.39*).

Referenc e	Class	Description
u	User	Permissions apply only to the **owner** of the file or directory.
g	Group	Permissions apply only to the **group** assigned to the file or directory.
o	Others	Permissions apply to **all other users** on the system. Most critical to watch.
a	All	Applies to **all three:** owner, group, and others.

***Figure 5.38**: Classifications of different permissions*

Symbols: `+`, `-` and `=`Option in Linux File Permission	
Operators	**Definition**
`+`	Add permissions
`-`	Remove permissions
`=`	Set the permissions to the specified values

Figure 5.39: *Linux File Permissions Operators*

Looking at *Figure* 5.36, **`file4`** has the following permissions (**`rw-rw-r--`**). This means that the owner of the file has read and write permissions, the group to which the owner belongs has read and write permissions, but all other users in the system have only read permissions.

Let us consider the permissions on the **`subject`** directory (**`rwxrwxr-x`**). This shows that the owner of the directory has read(**`r`**), write(**`w`**), and execute(**`x`**) permissions. The group to which the owner belongs has read(**`r`**), write(**`w`**), and execute(**`x`**) permissions. All other users on the system have read(**`r`**) and execute(**`x`**) permissions only.

How to change the Linux file and directory Permissions

The command to change the security permissions on files and directories is '**`chmod`**'. This stands for **change mode**, as the nine security characters are collectively called the security "**`mode`**" of the file.

Let us modify the permissions on file4 and grant all users(other) the execute permissions.

```
chmod  o+x file4
```

Next, let us grant the owner(user) of file4 the execute permissions.

```
chmod u+x file4
```

```
jack@jack-virtual-machine:                    $ ls -l file4
-rw-rw-r-- 1 jack jack 22668 Dec 28 11:16 file4
jack@jack-virtual-machine:                    $ chmod o+x file4
jack@jack-virtual-machine:                    $ ls -l file4
-rw-rw-r-x 1 jack jack 22668 Dec 28 11:16 file4
jack@jack-virtual-machine:                    $ chmod u+x file4
jack@jack-virtual-machine:                    $ ls -l file4
-rwxrw-r-x 1 jack jack 22668 Dec 28 11:16 file4
jack@jack-virtual-machine:                    $
```

Figure 5.40: *Changing file permissions*

Next, let us change the permissions on the **subject3** directory by removing the execute permissions from everyone by using the following command. `chmod a-x subject3`

```
jack@jack-virtual-machine:                    $ ls -l
total 36
-rwxrw-r-x 1 jack jack 22668 Dec 28 11:16 file4
-rw-rw-r-- 1 jack jack     0 Dec 28 12:11 newfile1
-rw-rw-r-- 1 jack jack     0 Dec 28 12:11 newfile2
-rw-rw-r-- 1 jack jack     0 Dec 28 12:11 newfile3
-rw-rw-r-- 1 jack jack     0 Dec 28 12:11 newfile4
-rw-rw-r-- 1 jack jack     0 Dec 28 12:11 newfile5
drwxrwxr-x 2 jack jack  4096 Dec 29 07:57
drwxrwxr-x 2 jack jack  4096 Dec 29 07:57
drwxrwxr-x 2 jack jack  4096 Dec 29 07:57
jack@jack-virtual-machine:                    $ chmod a-x subject3
jack@jack-virtual-machine:                    $ ls -al
total 44
drwxrwxr-x 5 jack jack  4096 Dec 29 07:57
drwxr-xr-x 3 jack jack  4096 Dec 28 10:51
-rwxrw-r-x 1 jack jack 22668 Dec 28 11:16 file4
-rw-rw-r-- 1 jack jack     0 Dec 28 12:11 newfile1
-rw-rw-r-- 1 jack jack     0 Dec 28 12:11 newfile2
-rw-rw-r-- 1 jack jack     0 Dec 28 12:11 newfile3
-rw-rw-r-- 1 jack jack     0 Dec 28 12:11 newfile4
-rw-rw-r-- 1 jack jack     0 Dec 28 12:11 newfile5
drwxrwxr-x 2 jack jack  4096 Dec 29 07:57
drwxrwxr-x 2 jack jack  4096 Dec 29 07:57
drw-rw-r-- 2 jack jack  4096 Dec 29 07:57
jack@jack-virtual-machine:                    $
```

Figure 5.41: *Modifying directory permissions*

The next figure shows the numbers allocated to each permission. These are used to grant or revoke permissions to or from a file numerically.

Using chmod in absolute mode

In the absolute mode, permissions are represented in numeric form (octal system to be precise). In this system, each file permission is represented by a number.

- r (read) = 4
- w (write) = 2
- x (execute) = 1
- – (no permission) = 0

Figure 5.42: *File Permissions in numeric form*

```
rwx = 4+2+1 = 7        #All permissions (read, write, and execute)
rw- = 4+2 = 6          # Read and write permissions
r-x  =  4+1 = 5            # Read and execute
-wx =  2+1 = 3                # Write and execute
```

Let us use the absolute mode to change the permissions on **newfile2.**

```
ls -l newfile2   #This will show the permissions only for the
newfile2.
```

We are going to give the file owner (user) the read, write, and execute permissions. The group will be given only the read and execute permissions, and the others, the read and write permissions.

```
chmod 756 newfile2
```

```
jack@jack-virtual-machine: ... $ ls -l newfile2
-rw-rw-r-- 1 jack jack 0 Dec 28 12:11 newfile2
jack@jack-virtual-machine: ... $ chmod 756 newfile2
jack@jack-virtual-machine: ... $ ls -l newfile2
-rwxr-xrw- 1 jack jack 0 Dec 28 12:11 newfile2
```

Figure 5.43: *Changing file permissions using the numeric format*

Changing Linux File Ownership

The **chown** command is used to change the ownership of a Linux file. **Chown** stands for change owner.

The syntax is:

```
chown new_user_name : new_user_group file_name
```

Let us change the owner and group of **newfile3** from jack to root.

```
chown root:root newfile3
```

```
jack@jack-virtual-machine:                $ ls -l newfile3
-rw-rw-r-- 1 jack jack 0 Dec 28 12:11 newfile3
jack@jack-virtual-machine:                $ chown root:root newfile3
chown: changing ownership of 'newfile3': Operation not permitted
jack@jack-virtual-machine:                $ sudo chown root:root newfile3
[sudo] password for jack:
jack@jack-virtual-machine:                $ ls -l newfile3
-rw-rw-r-- 1 root root 0 Dec 28 12:11 newfile3
```

Figure 5.44: *Changing file ownership*

In *Figure* 5.45, we have added a new user called **kanisha** using **sudo useradd kanisha**. We then change the ownership and group from jack to **kanisha** using the command **sudo chown kanisha:kanisha newfile4**.

```
jack@jack-virtual-machine:                $
jack@jack-virtual-machine:                $ sudo useradd kanisha
jack@jack-virtual-machine:                $ ls -l newfile4
-rw-rw-r-- 1 jack jack 0 Dec 28 12:11 newfile4
jack@jack-virtual-machine:                $ sudo chown kanisha:kanisha newfile4
jack@jack-virtual-machine:                $ ls -l newfile4
-rw-rw-r-- 1 kanisha kanisha 0 Dec 28 12:11 newfile4
jack@jack-virtual-machine:                $
```

Figure 5.45: *Changing file ownership*

The **ls -l** is to do a long listing of the content of a file.

```
jack@jack-virtual-machine:                $ ls -l
total 36
-rwxrw-r-x 1 jack     jack     22668 Dec 28 11:16 file4
-rw-rw-r-- 1 jack     jack         0 Dec 28 12:11 newfile1
-rwxr-xrw- 1 kanisha  kanisha      0 Dec 28 12:11 newfile2
-rw-rw-r-- 1 root     root         0 Dec 28 12:11 newfile3
-rw-rw-r-- 1 kanisha  kanisha      0 Dec 28 12:11 newfile4
-rw-rw-r-- 1 jack     jack         0 Dec 28 12:11 newfile5
drwxrwxr-x 2 jack     jack      4096 Dec 29 07:57
drwxrwxr-x 2 jack     jack      4096 Dec 29 07:57
drw-rw-r-- 2 jack     jack      4096 Dec 29 07:57
```

Figure 5.46: *All permissions changed*

PostgreSQL Configuration Files

Understanding the different configuration files and their location is paramount when it comes to optimizing PostgreSQL performance and database administration. It helps in troubleshooting connection issues.

The smooth operation of the PostgreSQL server is determined by various configuration files. These files dictate how PostgreSQL connects with client applications and other aspects of the database. PostgreSQL stores its configuration files in the data directory, commonly referred to as the **'data cluster'**.

The specific path to this directory thus differs based on how PostgreSQL was installed. They also differ with different operating systems. Use "**SHOW data_directory;**" command from within the database to quickly find the data directory for a running PostgreSQL instance.

After you successfully log in to the Linux OS as indicated in the previous sections, switch user (**su**) to Postgres user and login to the database (**psql**). Run the SHOW data directory command as shown to see the path to the data directory.

```
sudo su - postgres
psql -U postgres
SHOW data_directory;
```

```
jack@jack-virtual-machine: $ sudo su - postgres
postgres@jack-virtual-machine:~$ psql
psql (16.1 (Ubuntu 16.1-1.pgdg22.04+1))
Type "help" for help.

postgres=# \q
postgres@jack-virtual-machine:~$ psql -U postgres
psql (16.1 (Ubuntu 16.1-1.pgdg22.04+1))
Type "help" for help.

postgres=# SHOW data_directory;
       data_directory
-----------------------------
 /var/lib/postgresql/16/main
(1 row)
```

Figure 5.47: *Finding the data directory from the database*

You can also run **psql -U postgres -c 'SHOW data_directory;'** from the terminal to check the location of the data directory.

```
postgres=# \q
postgres@jack-virtual-machine:~$ psql -U postgres -c 'SHOW data_directory;'
       data_directory
-----------------------------
 /var/lib/postgresql/16/main
(1 row)
```

Figure 5.48: *Obtaining the data directory from the terminal*

The following three important files are found inside the data directory:

- `postgresql.conf`
- `pg_hba.conf`
- `pg_ident.conf`

With Ubuntu OS, the preceding three files are not located in the data directory as shown in *Figure* 5.49.

```
postgres@jack-virtual-machine:~$ cd /var/lib/postgresql/16/main
postgres@jack-virtual-machine:~/16/main$ ls -l
total 84
drwx------ 5 postgres postgres 4096 Dec 30 17:00 base
drwx------ 2 postgres postgres 4096 Dec 30 17:04 global
drwx------ 2 postgres postgres 4096 Dec 30 17:00 pg_commit_ts
drwx------ 2 postgres postgres 4096 Dec 30 17:00 pg_dynshmem
drwx------ 4 postgres postgres 4096 Dec 30 17:05 pg_logical
drwx------ 4 postgres postgres 4096 Dec 30 17:00 pg_multixact
drwx------ 2 postgres postgres 4096 Dec 30 17:00 pg_notify
drwx------ 2 postgres postgres 4096 Dec 30 17:00 pg_replslot
drwx------ 2 postgres postgres 4096 Dec 30 17:00 pg_serial
drwx------ 2 postgres postgres 4096 Dec 30 17:00 pg_snapshots
drwx------ 2 postgres postgres 4096 Dec 30 17:00 pg_stat
drwx------ 2 postgres postgres 4096 Dec 30 17:00 pg_stat_tmp
drwx------ 2 postgres postgres 4096 Dec 30 17:00 pg_subtrans
drwx------ 2 postgres postgres 4096 Dec 30 17:00 pg_tblspc
drwx------ 2 postgres postgres 4096 Dec 30 17:00 pg_twophase
-rw------- 1 postgres postgres    3 Dec 30 17:00 PG_VERSION
drwx------ 3 postgres postgres 4096 Dec 30 17:00 pg_wal
drwx------ 2 postgres postgres 4096 Dec 30 17:00 pg_xact
-rw------- 1 postgres postgres   88 Dec 30 17:00 postgresql.auto.conf
-rw------- 1 postgres postgres  130 Dec 30 17:00 postmaster.opts
-rw------- 1 postgres postgres  109 Dec 30 17:00 postmaster.pid
```

Figure 5.49: *Files found in the data directory*

There are two commands to find these **config** files.

Use the following commands. The first one is to be run from the terminal whilst connected as a user postgres, and the second one is to be run from within the database (*Figures 5.50 and 5.51*).

```
ps -ef | grep postgres                  #Run from terminal
SHOW config_file;                       #Execute inside the database
```

```
postgres@jack-virtual-machine:~/16/main$ psql
psql (16.1 (Ubuntu 16.1-1.pgdg22.04+1))
Type "help" for help.

postgres=# SHOW config_file;
               config_file
-----------------------------------------
 /etc/postgresql/16/main/postgresql.conf
(1 row)
```

Figure 5.50: *Ubuntu config file location*

```
postgres=# \q
postgres@jack-virtual-machine:~/16/main$ cd /etc/postgresql/16/main/
postgres@jack-virtual-machine:/etc/postgresql/16/main$ ls -l
total 60
drwxr-xr-x 2 postgres postgres  4096 Dec 30 17:00 conf.d
-rw-r--r-- 1 postgres postgres   315 Dec 30 17:00 environment
-rw-r--r-- 1 postgres postgres   143 Dec 30 17:00 pg_ctl.conf
-rw-r----- 1 postgres postgres  5924 Dec 30 17:00 pg_hba.conf
-rw-r----- 1 postgres postgres  2640 Dec 30 17:00 pg_ident.conf
-rw-r--r-- 1 postgres postgres 29974 Dec 30 17:00 postgresql.conf
-rw-r--r-- 1 postgres postgres   317 Dec 30 17:00 start.conf
```

Figure 5.51: *PostgreSQL primary config files in their location*

Now that we have located the important primary files, it is worth noting that these files must always be backed up before any changes. We will be using the **`vi editor`** or any text editor of your choice to make any changes to this file. If new to text editors, please check **YouTube videos** on how to use text editors.

Now that your current location or present working directory **(`pwd`)** contains the files, we can simply type **`vi file_name`** to be able to edit it. Alternatively, if you are not in the location of the files, then the full path to the files must be used after the keyword vi.

`Postgresql.conf` File

The main configuration file for the PostgreSQL server is called **`postgresql.conf`** file. Settings related to the following are found in this file:

- Memory usage such as **`shared_buffers`**, **`maintenance_work_mem`**, **`work_mem`**, and so on.
- Write-ahead logs such as **`wal_level`**, **`wal_buffers`**, and so on.
- Authentications and connections such as **`max_connections`**, **`listen_addresses`**, and so on.

- Query tuning parameters, logging, and many other server parameters.

Always take a backup of config files before making any changes. You can use the **cp or scp commands** as shown to make another copy.

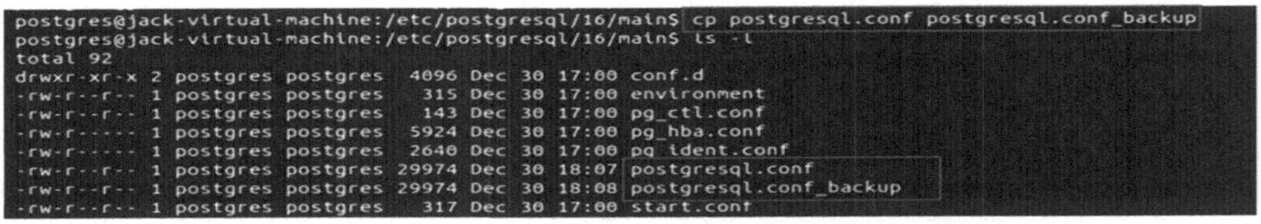

```
postgres@jack-virtual-machine:/etc/postgresql/16/main$ cp postgresql.conf postgresql.conf_backup
postgres@jack-virtual-machine:/etc/postgresql/16/main$ ls -l
total 92
drwxr-xr-x 2 postgres postgres  4096 Dec 30 17:00 conf.d
-rw-r--r-- 1 postgres postgres   315 Dec 30 17:00 environment
-rw-r--r-- 1 postgres postgres   143 Dec 30 17:00 pg_ctl.conf
-rw-r----- 1 postgres postgres  5924 Dec 30 17:00 pg_hba.conf
-rw-r----- 1 postgres postgres  2640 Dec 30 17:00 pg_ident.conf
-rw-r--r-- 1 postgres postgres 29974 Dec 30 18:07 postgresql.conf
-rw-r--r-- 1 postgres postgres 29974 Dec 30 18:08 postgresql.conf_backup
-rw-r--r-- 1 postgres postgres   317 Dec 30 17:00 start.conf
```

Figure 5.52*: Make a backup copy of postgresql.conf file*

The **postgresql.conf** file is a very big and delicate file, this is why it is always advisable to take its backup.

```
Vi postgresql.conf
Vi /etc/postgresql/16/main/postgresql.conf
```

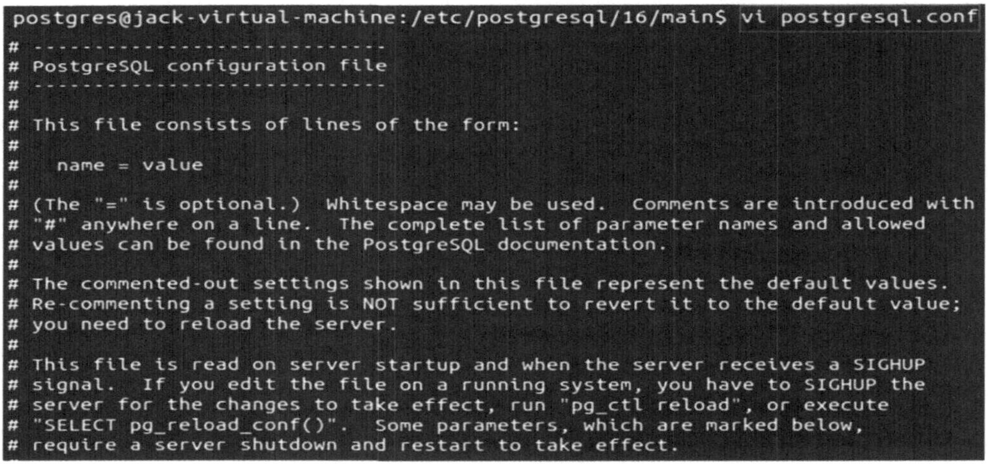

```
postgres@jack-virtual-machine:/etc/postgresql/16/main$ vi postgresql.conf
# -----------------------------
# PostgreSQL configuration file
# -----------------------------
#
# This file consists of lines of the form:
#
#   name = value
#
# (The "=" is optional.)  Whitespace may be used.  Comments are introduced with
# "#" anywhere on a line.  The complete list of parameter names and allowed
# values can be found in the PostgreSQL documentation.
#
# The commented-out settings shown in this file represent the default values.
# Re-commenting a setting is NOT sufficient to revert it to the default value;
# you need to reload the server.
#
# This file is read on server startup and when the server receives a SIGHUP
# signal.  If you edit the file on a running system, you have to SIGHUP the
# server for the changes to take effect, run "pg_ctl reload", or execute
# "SELECT pg_reload_conf()".  Some parameters, which are marked below,
# require a server shutdown and restart to take effect.
```

Figure 5.53: *Postgresql.conf file using vi editor*

To locate the section of the file you want to modify, use the **/key_word** then enter. This will take you to the region of the file you want to edit.

Let us locate the **listen_addresses**, **max_connections**, and **shared_buffers**.

Whilst inside the **postgresql.conf** file using **vi postgresql.conf**, type the following:

```
/listen_addresses
```

```
#------------------------------------------------------------------------------
# CONNECTIONS AND AUTHENTICATION
#------------------------------------------------------------------------------

# - Connection Settings -

#listen_addresses = 'localhost'         # what IP address(es) to listen on;
                                        # comma-separated list of addresses;
                                        # defaults to 'localhost'; use '*' for all
                                        # (change requires restart)
port = 5432                             # (change requires restart)
max_connections = 100                   # (change requires restart)
#reserved_connections = 0               # (change requires restart)
#superuser_reserved_connections = 3     # (change requires restart)
unix_socket_directories = '/var/run/postgresql' # comma-separated list of directories
                                        # (change requires restart)
#unix_socket_group = ''                 # (change requires restart)
#unix_socket_permissions = 0777         # begin with 0 to use octal notation
                                        # (change requires restart)
```

Figure 5.54: *Searching required sections inside postgresql.conf*

To search for shared buffers inside the **`postgresql.conf`** file, type the following:

```
/ shared_buffers
```

```
#------------------------------------------------------------------------------
# RESOURCE USAGE (except WAL)
#------------------------------------------------------------------------------

# - Memory -

shared_buffers = 128MB                  # min 128kB
                                        # (change requires restart)
#huge_pages = try                       # on, off, or try
                                        # (change requires restart)
#huge_page_size = 0                     # zero for system default
                                        # (change requires restart)
#temp_buffers = 8MB                     # min 800kB
#max_prepared_transactions = 0          # zero disables the feature
                                        # (change requires restart)
# Caution: it is not advisable to set max_prepared_transactions nonzero unless
# you actively intend to use prepared transactions.
#work_mem = 4MB                         # min 64kB
#hash_mem_multiplier = 2.0              # 1-1000.0 multiplier on hash table work_mem
#maintenance_work_mem = 64MB            # min 1MB
#autovacuum_work_mem = -1               # min 1MB, or -1 to use maintenance_work_mem
#logical_decoding_work_mem = 64MB       # min 64kB
```

Figure 5.55: *Searching shared_buffers inside postgresql.conf*

`pg_hba.conf` File

The Host-Based Authentication (**hba**) file determines which users can connect to which databases. It governs from which hosts or networks a user can connect and determines which authentication methods, such as md5, password, peer, and so on, can be used.

All connections to the PostgreSQL server are checked against the **pg_hba** file to determine if the connection should be allowed or rejected.

```
Vi pg_hba.conf
Vi /etc/postgresql/16/main/ pg_hba.conf
```

```
postgres@jack-virtual-machine:/etc/postgresql/16/main$ vi pg_hba.conf
# Database administrative login by Unix domain socket
local   all             postgres                                peer

# TYPE  DATABASE        USER            ADDRESS                 METHOD

# "local" is for Unix domain socket connections only
local   all             all                                     peer
# IPv4 local connections:
host    all             all             127.0.0.1/32            scram-sha-256
# IPv6 local connections:
host    all             all             ::1/128                 scram-sha-256
# Allow replication connections from localhost, by a user with the
# replication privilege.
local   replication     all                                     peer
host    replication     all             127.0.0.1/32            scram-sha-256
host    replication     all             ::1/128                 scram-sha-256
```

Figure 5.56: *Editable section of pg_hba.conf file*

pg_ident.conf File

This file is essential for mapping OS-level usernames to PostgreSQL usernames. This is particularly useful when using peer or **ident** authentication methods.

Creating and Authenticating Users and Roles

In PostgreSQL, we manage database access by granting permissions to *roles*. A role is either a database user or a group of database users. Database objects such as tables and functions can be owned by roles. The role owning these objects can assign privileges on these objects to other roles, thus controlling who has access to which objects.

It is a very common practice in PostgreSQL to grant *membership* in a role to another role or user, thus allowing the member role to use privileges assigned to another role.

A database role does have several attributes defining its privileges, and thus, this helps the role to interact with the client authentication system.

login privilege: For database connections, only roles that have the **LOGIN** attribute are permitted. A database user is considered the same as a role with the **LOGIN** attribute.

A role with login privilege can be created using either:

```
CREATE ROLE role_name LOGIN;
CREATE USER user_name;
```

CREATE USER and **CREATE ROLE** are the same, except that the **CREATE USER** includes **LOGIN** by default, while **CREATE ROLE** does not.

In PostgreSQL, user accounts are represented by roles. Roles can log in equally as users do and are called login roles. Group roles are roles that contain other roles.

Once a role is created, it is valid in the cluster (all the databases in that server).

Now, let us use the following statement to create a role called **dbadmin**.

```
CREATE ROLE dbadmin;
```

```
postgres=# CREATE ROLE dbadmin;
CREATE ROLE
```

Figure 5.57: *Create a database role*

To see all the roles in the current PostgreSQL database server, use the **pg_roles** system catalog table:

```
SELECT rolname FROM pg_roles;
```

```
postgres=# SELECT rolname FROM pg_roles;
          rolname
---------------------------
 postgres
 pg_monitor
 pg_read_all_settings
 pg_read_all_stats
 pg_stat_scan_tables
 pg_read_server_files
 pg_write_server_files
 pg_execute_server_program
 pg_signal_backend
 aisha
 dbadmin
(11 rows)
```

Figure 5.58: *To see all roles in postgresql database*

All the roles starting with **pg_** are system roles.

The **\du** command will give a list of all existing roles in the current PostgreSQL database server.

```
\du
```

```
postgres=# CREATE USER aisha WITH ENCRYPTED PASSWORD 'aisha@123';
CREATE ROLE
postgres=# \du
                              List of roles
 Role name |                         Attributes
-----------+------------------------------------------------------------
 aisha     |
 dbadmin   | Cannot login
 postgres  | Superuser, Create role, Create DB, Replication, Bypass RLS
```

***Figure 5.59**: The \du result*

Looking at the list of attributes from the output, our newly created **dbadmin** do not have login privileges as we did not include the keyword **LOGIN** when the role dbadmin was created.

Now, let us create a different role called **administrator** with login privileges.

```
CREATE ROLE administrator LOGIN;
```

```
postgres=# CREATE ROLE administrator LOGIN;
CREATE ROLE
postgres=# \du
                                   List of roles
   Role name   |                         Attributes                         | Member of
---------------+------------------------------------------------------------+-----------
 administrator |                                                            | {}
 aisha         |                                                            | {}
 dbadmin       | Cannot login                                               | {}
 postgres      | Superuser, Create role, Create DB, Replication, Bypass RLS | {}
```

***Figure 5.60:** Roles with login privileges*

The following query statement will create a role called **Narmada** that has login privileges and a **password**:

```
CREATE ROLE Narmada WITH LOGIN PASSWORD 'Welcome@2023';
```

```
postgres=# CREATE ROLE Narmada LOGIN PASSWORD 'Welcome@2023';
CREATE ROLE
postgres=# \du
                                   List of roles
   Role name   |                         Attributes                         | Member of
---------------+------------------------------------------------------------+-----------
 administrator |                                                            | {}
 aisha         |                                                            | {}
 dbadmin       | Cannot login                                               | {}
 narmada       |                                                            | {}
 postgres      | Superuser, Create role, Create DB, Replication, Bypass RLS | {}
```

***Figure 5.61**: New roles added*

Very important to note is that the password is in single quotes ('). Now, role **Narmada** can be used to log in to the PostgreSQL database server as follows:

```
psql -U Narmada -W
```

```
postgres@jack-virtual-machine:~$ psql -U Narmada -W
Password:
```

***Figure 5.62**: New user login with password*

The preceding command is run on the Linux OS, not inside the database. This is to gain access to the database. When entering the password, it is invisible (take note of the fact that you do not see the password when it is being entered). The **-W** prompts for a password to be entered.

```
postgres@jack-virtual-machine:/etc/postgresql/16/main$ psql -U Narmada -W
Password:
psql: error: connection to server on socket "/var/run/postgresql/.s.PGSQL.5432" failed: FATAL:  Peer authentication failed for user "Narmada"
```

***Figure 5.63**: User login failed*

After entering the password, the following error message pops up.

Error: *Connection to server on socket* "**/var/run/postgresql/.s.PGSQL.5432**" *failed:* FATAL: *Peer authentication failed for user "Narmada"*

Take note of the above error as we are going to fix that next. This error means that user "Narmada" has not been authenticated in an authentication file called **pg_hba.conf** (**hba** = host-based authentication). On the Linux command prompt, type the following command:

```
Ps -ef | grep postgres
```

```
jack@jack-virtual-machine: $ sudo su - postgres
[sudo] password for jack:
postgres@jack-virtual-machine:~$ ps -ef | grep postgres
postgres    1095       1  0 08:18 ?        00:00:00 /usr/lib/postgresql/16/bin/postgres -D /var/lib/postgresql/16/main -c config_file=/etc/postgresql/16/main/postgresql.conf
postgres    1114    1095  0 08:19 ?        00:00:00 postgres: 16/main: checkpointer
postgres    1115    1095  0 08:19 ?        00:00:00 postgres: 16/main: background writer
postgres    1124    1095  0 08:19 ?        00:00:00 postgres: 16/main: walwriter
postgres    1125    1095  0 08:19 ?        00:00:00 postgres: 16/main: autovacuum launcher
postgres    1126    1095  0 08:19 ?        00:00:00 postgres: 16/main: archiver
postgres    1127    1095  0 08:19 ?        00:00:00 postgres: 16/main: logical replication launcher
root        3098    3049  0 08:32 pts/0    00:00:00 sudo su - postgres
root        3099    3098  0 08:32 pts/1    00:00:00 sudo su - postgres
root        3100    3099  0 08:32 pts/1    00:00:00 su - postgres
postgres    3101    3100  0 08:32 pts/1    00:00:00 -bash
postgres    3183    3101  0 08:32 pts/1    00:00:00 ps -ef
postgres    3184    3101  0 08:32 pts/1    00:00:00 grep postgres
postgres@jack-virtual-machine:~$
```

***Figure 5.64**: Accessing config files*

We are going to cd to that location to see all the configuration files.

```
postgres@jack-virtual-machine:~$ cd /etc/postgresql/16/main/
postgres@jack-virtual-machine:/etc/postgresql/16/main$ ls -l
total 92
drwxr-xr-x 2 postgres postgres  4096 Dec 30 17:00 conf.d
-rw-r--r-- 1 postgres postgres   315 Dec 30 17:00 environment
-rw-r--r-- 1 postgres postgres   143 Dec 30 17:00 pg_ctl.conf
-rw-r----- 1 postgres postgres  5933 Dec 30 20:28 pg_hba.conf
-rw-r----- 1 postgres postgres  2640 Dec 30 17:00 pg_ident.conf
-rw-r--r-- 1 postgres postgres 29965 Dec 30 20:05 postgresql.conf
-rw-r--r-- 1 postgres postgres 29974 Dec 30 18:08 postgresql.conf_backup
-rw-r--r-- 1 postgres postgres   317 Dec 30 17:00 start.conf
```

***Figure 5.65**: Locating pg_hba.conf file*

We need to edit the file using a text editor such as **vi**, **vim nano**, and so on.

```
vi pg_hba.conf
```

We are going to edit **pg_hba.conf** file. We will be replacing peer with **scram-sha-256** since user "Narmada" was granted the password. Also, make sure Narmada is spelled correctly as shown with the **\du command** in the database.

Before changes.

```
# Database administrative login by Unix domain socket
local   all             postgres                                peer

# TYPE  DATABASE        USER            ADDRESS                 METHOD

# "local" is for Unix domain socket connections only
local   all             all                                     peer
# IPv4 local connections:
host    all             all             127.0.0.1/32            scram-sha-256
```

After the changes.

```
# Database administrative login by Unix domain socket
local   all             postgres                                peer

# TYPE  DATABASE        USER            ADDRESS                 METHOD

# "local" is for Unix domain socket connections only
local   all             all                                     scram-sha-256
# IPv4 local connections:
host    all             all             127.0.0.1/32            scram-sha-256
```

Figure 5.66: *Modifying pg_hba.conf file*

Whenever changes are made in the **pg_hba.conf**, the database must be reloaded for it to take effect.

Log in to the database and reload it using the following query:

```
SELECT pg_reload_conf();
```

```
postgres@jack-virtual-machine:/etc/postgresql/16/main$ psql
psql (16.1 (Ubuntu 16.1-1.pgdg22.04+1))
Type "help" for help.

postgres=# SELECT pg_reload_conf();
 pg_reload_conf
----------------
 t
(1 row)

postgres=#
```

Figure 5.67: *Reload database following changes in pg_hba.conf file*

Figure 5.68 shows how to log in to the database as a different user from postgres.

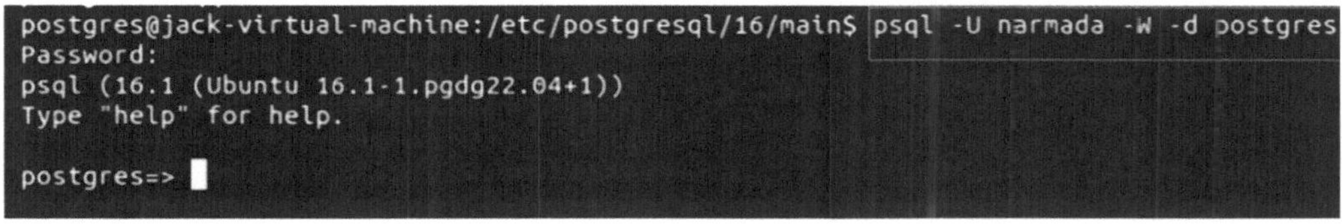

```
postgres@jack-virtual-machine:/etc/postgresql/16/main$ psql -U narmada -W -d postgres
Password:
psql (16.1 (Ubuntu 16.1-1.pgdg22.04+1))
Type "help" for help.

postgres=>
```

Figure 5.68: *User login to database with password*

- **superuser status:** In PostgreSQL, except for the right to log in, a database **superuser** bypasses all permission checks. It is best practice to work as a non-superuser because of the danger of its privileges.

  The query to create a new database superuser is shown:

  `CREATE ROLE role_name SUPERUSER.`

  Only a role that is already a superuser can run the above query.

- **Database creation:** To be able to create a database, a role must be granted the create database permission. To create a role with such privileges, use the following query.

  `CREATE ROLE role_name CREATEDB.`

- **Role creation:** A role must be granted permission to be able to create more roles. Use the following query to create such a role.

  `CREATE ROLE role_name CREATEROLE.`

  The role with **CREATEROLE** privileges can alter and drop roles that have been created by them. The **ALTER ROLE** command can be used to change the role password.

  CREATEROLE cannot grant or revoke the **BYPASSRLS** privilege.

- **password:** It is worth noting that a password is only of importance if the client authentication method requires the user to supply a password when connecting to the database.

  A password must be specified upon role creation as shown:

  `CREATE ROLE role_name PASSWORD 'enterpassword_here'.`

- **Privileges inheritance:** A role inherits the privileges of roles it is a member of, by default. However, to create a role that does not inherit privileges by default, use the

  `CREATE ROLE role_name NOINHERIT.`

  Alternatively, inheritance can be overridden for individual grants by using **WITH INHERIT TRUE** or **WITH INHERIT FALSE**.

- **Connection limit:** The number of concurrent connections a role can make can be specified using the connection limit. The value of **-1** (the default) means no limit. Specify the connection limit when creating the role.

  CREATE ROLE **role_name CONNECTION LIMIT 'integer'**.

  Once a role is created, its attributes can be modified with an **ALTER ROLE** statement.

  ALTER ROLE **role_name** [WITH] **option;**

```
postgres=# \du
                                   List of roles
   Role name    |                         Attributes
----------------+------------------------------------------------------------
 administrator |
 aisha         |
 dbadmin       | Cannot login
 narmada       |
 postgres      | Superuser, Create role, Create DB, Replication, Bypass RLS
```

Figure 5.69: *User dbadmin with no login privileges*

We used the following alter commands to grant **dbadmin** user login and create database privileges.

```
ALTER ROLE dbadmin WITH LOGIN;
ALTER ROLE dbadmin WITH CREATEDB;
```

```
postgres=# ALTER ROLE dbadmin WITH LOGIN;
ALTER ROLE
postgres=# ALTER ROLE dbadmin WITH CREATEDB;
ALTER ROLE
postgres=# \du
                                   List of roles
   Role name    |                         Attributes
----------------+------------------------------------------------------------
 administrator |
 aisha         |
 dbadmin       | Create DB
 narmada       |
 postgres      | Superuser, Create role, Create DB, Replication, Bypass RLS
```

Figure 5.70: *Showing user dbadmin privileges*

Conclusion

In this chapter, we learned the basic Linux commands necessary to be a successful database administrator and or database engineer. We learned how to manipulate file permissions and file ownership. We also learned how to edit PostgreSQL configuration files and use these files to authenticate users and roles.

In the next chapter, we are going to learn one of the most needed and essential skills of a PostgreSQL DBA or engineer, which is backup and restore. Every good PostgreSQL environment must have a good backup and recovery plan. This is how we recover from incidents affecting data such as server class or accidental deletion of data.

CHAPTER 6
Backup and Restore

Introduction

PostgreSQL databases usually house valuable data. An essential aspect of managing this data is taking regular backups and having a recovery plan. This is very important for the data integrity and availability. Backups are crucial for accidental/incidental loss of data in the database. This chapter teaches learners how to effectively backup and restore a PostgreSQL database and its objects. It is very important for learners to understand the underlying methods for backing up and restoring PostgreSQL database objects. Learners will learn the logical and physical backup methods. It is crucial to understand these backup types before choosing which of them to use.

Structure

In this chapter, we will discuss the following topics:

- Downloading and Installing a Sample Database
- Logical Backups (`Pg_dump` and `pg_dumpall`)
- Physical Backups (`pg_basebackups`)

Downloading and Installing a Sample Database

The sample database we will be using is called DVD rental. The DVD rental database is a true representation of a DVD rental business store. There are 15 tables inside the database.

There are 15 tables in our sample database which are listed below:

1. **actor** – stores actors' data including first name and last name.
2. **film** – stores film data such as title, release year, length, rating, etc
3. **film_actor** – stores the relationships between films and actors.
4. **category** – stores film's categories data.
5. **film_category**– stores the relationships between films and categories.
6. **store** – contains the store data including manager staff and address.
7. **inventory** – stores inventory data.
8. **rental** – stores rental data.
9. **payment** – stores customer's payments.
10. **staff** – stores staff data.
11. **customer** – stores customers' data.
12. **address** – stores address data for staff and customers
13. **city** – stores the city names.
14. **country** – stores the country names.
15. **language** – stores the different languages.

Figure 6.1: *Description of tables inside the dvdrental sample database*

To successfully download the sample database, we first need to log in to the Linux Operating system as indicated in *Figures* 2.32–2.36 in *Chapter* 2, *Installing and Initializing PostgreSQL*.

Once logged in, double-click to open Firefox and search for the PostgreSQL sample database as indicated in *Figure* 6.2.

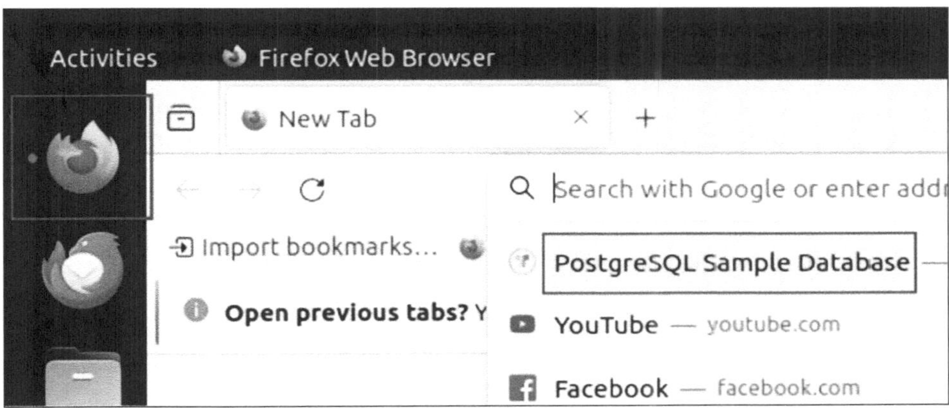

Figure 6.2: *Using Firefox to search for Postgresql Sample database*

The link to the sample database should be as follows:

https://www.postgresqltutorial.com/postgresql-getting-started/postgresql-sample-database/

Scroll down to the **Download the PostgreSQL sample database** section as shown in *Figure 6.3*.

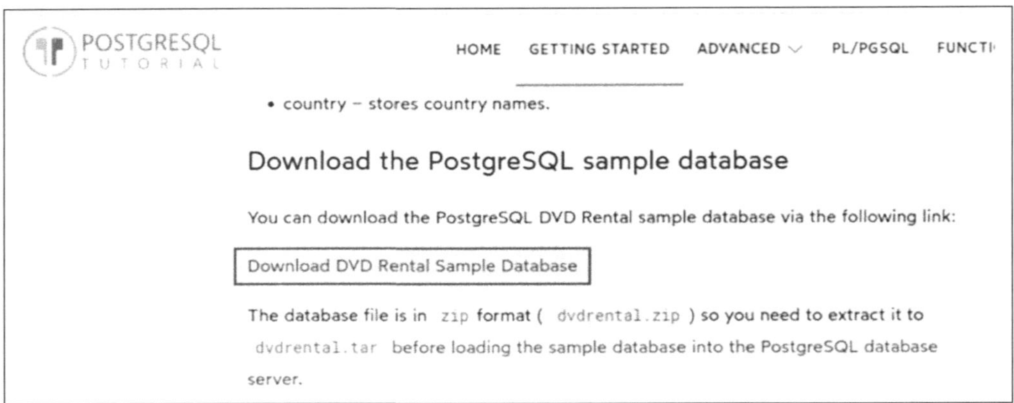

Figure 6.3: *Downloading DVD rental sample database*

Figure 6.4 shows that DVD rental sample download is complete.

Figure 6.4: *DVD rental sample download completed*

Once the download is complete, head back to your Ubuntu operating system, and **ls** to see the **Downloads** directory. **Cd** into the Downloads directory and **ls** to see the sample **dvdrental** database as a zip file (**dvdrental.zip**).

```
jack@jack-virtual-machine: $ ls
jack@jack-virtual-machine: $ cd Downloads/
jack@jack-virtual-machine:              $ ls
jack@jack-virtual-machine:              $
```

Figure 6.5: *DVD rental inside the Downloads directory*

Switch the user to Postgres and run the **pwd** command to see the path where the sample database will be copied to.

```
sudo su - postgres           #To switch to user postgres
pwd                          # Get the path where the database
                             will be copied to
exit                         # to switch back to OS user.
sudo cp /home/jack/Downloads/dvdrental.zip  /var/lib/postgresql
                             #cp = copy command
#/home/jack/Downloads/dvdrental.zip  = current file location
                              #/var/lib/postgresql = destination
                              of the file
```

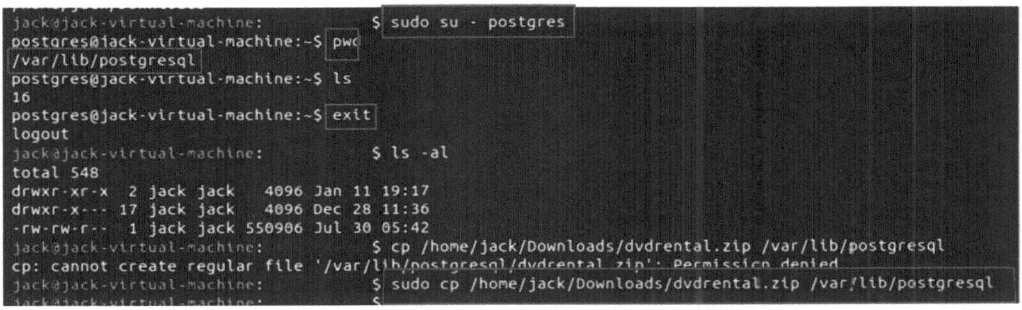

```
jack@jack-virtual-machine:          $ sudo su - postgres
postgres@jack-virtual-machine:~$ pwd
/var/lib/postgresql
postgres@jack-virtual-machine:~$ ls
16
postgres@jack-virtual-machine:~$ exit
logout
jack@jack-virtual-machine:          $ ls -al
total 548
drwxr-xr-x  2 jack jack   4096 Jan 11 19:17
drwxr-x--- 17 jack jack   4096 Dec 28 11:36
-rw-rw-r--  1 jack jack 550906 Jul 30 05:42
jack@jack-virtual-machine:          $ cp /home/jack/Downloads/dvdrental.zip /var/lib/postgresql
cp: cannot create regular file '/var/lib/postgresql/dvdrental.zip': Permission denied
jack@jack-virtual-machine:          $ sudo cp /home/jack/Downloads/dvdrental.zip /var/lib/postgresql
```

Figure 6.6: *How to copy from one directory to another*

The next screen is to list using the **ls** command to see the **dvdrental.zip** file.

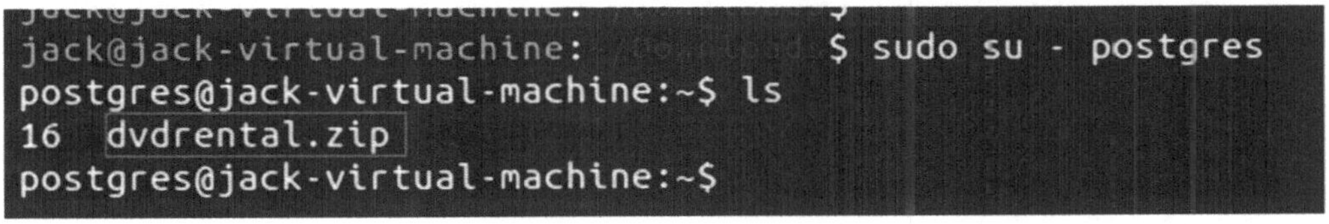

```
jack@jack-virtual-machine:          $ sudo su - postgres
postgres@jack-virtual-machine:~$ ls
16  dvdrental.zip
postgres@jack-virtual-machine:~$
```

Figure 6.7: *Shows the dvdrental zip file inside Postgresql directory*

To unzip the DVD rental file, we will use the following command:

```
unzip dvdrental.zip
```

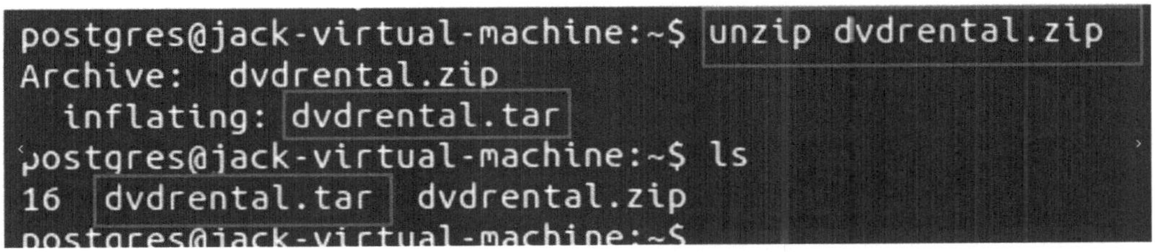

```
postgres@jack-virtual-machine:~$ unzip dvdrental.zip
Archive:  dvdrental.zip
  inflating: dvdrental.tar
postgres@jack-virtual-machine:~$ ls
16  dvdrental.tar  dvdrental.zip
postgres@jack-virtual-machine:~$
```

Figure 6.8: *How to unzip the dvd rental file to .tar file*

We are now going to log into the database and check all the databases available using the following commands:

```
psql -U postgres
\l
```

```
postgres@jack-virtual-machine:~$ psql -U postgres
psql (16.1 (Ubuntu 16.1-1.pgdg22.04+1))
Type "help" for help.

postgres=# \l
                                          List of databases
   Name    |  Owner   | Encoding | Locale Provider |   Collate   |    Ctype    |
-----------+----------+----------+-----------------+-------------+-------------+
 postgres  | postgres | UTF8     | libc            | en_US.UTF-8 | en_US.UTF-8 |
           |          |          |                 |             |             |
           |          |          |                 |             |             |
 template0 | postgres | UTF8     | libc            | en_US.UTF-8 | en_US.UTF-8 |
           |          |          |                 |             |             |
 template1 | postgres | UTF8     | libc            | en_US.UTF-8 | en_US.UTF-8 |
           |          |          |                 |             |             |
(3 rows)
```

Figure 6.9: *Shows there is no dvdrental database*

Next, we will create the dvdrental database using the following command:

```
CREATE DATABASE dvdrental;
```

```
postgres=# CREATE DATABASE dvdrental;
CREATE DATABASE
postgres=# \l
                                          List of databases
   Name    |  Owner   | Encoding | Locale Provider |   Collate   |    Ctype
-----------+----------+----------+-----------------+-------------+-------------
 dvdrental | postgres | UTF8     | libc            | en_US.UTF-8 | en_US.UTF-8
 postgres  | postgres | UTF8     | libc            | en_US.UTF-8 | en_US.UTF-8
           |          |          |                 |             |
           |          |          |                 |             |
 template0 | postgres | UTF8     | libc            | en_US.UTF-8 | en_US.UTF-8
           |          |          |                 |             |
 template1 | postgres | UTF8     | libc            | en_US.UTF-8 | en_US.UTF-8
           |          |          |                 |             |
(4 rows)

postgres=#
```

Figure 6.10: *How to create the dvdrental database*

The next step is to connect to the newly created dvdrental database and check for the availability of tables (relations).

```
postgres=# \c dvdrental
You are now connected to database "dvdrental" as user "postgres".
dvdrental=# \dt
Did not find any relations.
dvdrental=#
```

Figure 6.11: *Shows there are no tables in the newly created dvdrental database*

The next step is to import data from our downloaded sample DVD rental database (**`dvdrental.tar`**) into our newly created dvdrental database using the **`pg_restore`** utility as follows.

Before executing the **`pg_restore`** command, we must first exit the database to the OS.

```
\q
pg_restore -U postgres -d dvdrental < /var/lib/postgresql/dvdrental.
tar
```

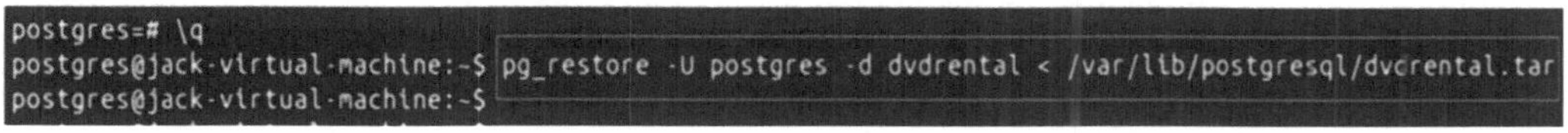

```
postgres=# \q
postgres@jack-virtual-machine:~$ pg_restore -U postgres -d dvdrental < /var/lib/postgresql/dvdrental.tar
postgres@jack-virtual-machine:~$
```

Figure 6.12: *Inserting data into dvdrental database using the pg_restore utility*

Next, we will connect to postgres and then to dvdrental database to check that there are now tables and also data inside those tables using the following commands.

```
psql -U postgres
\c dvdrental
\dt
```

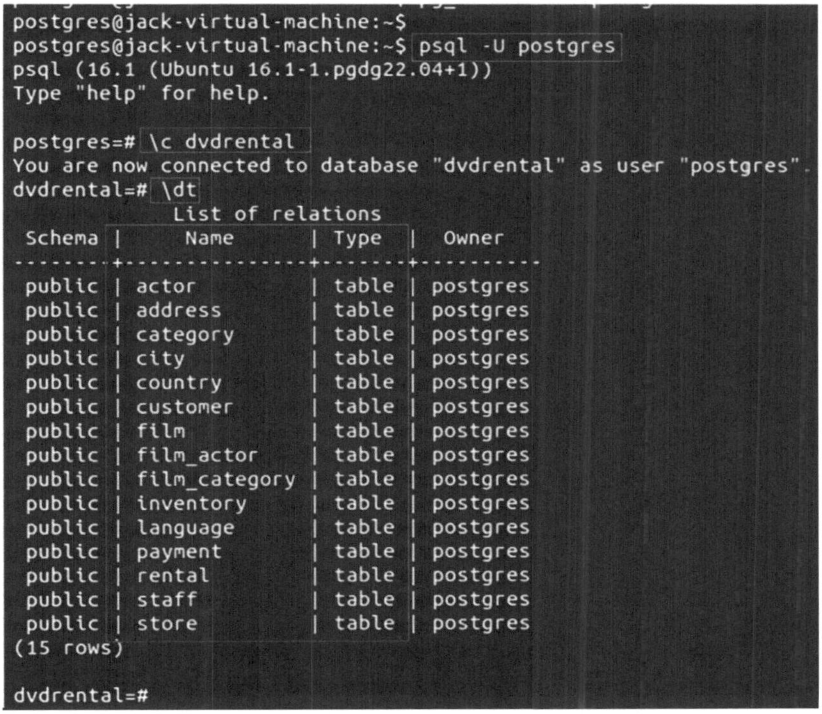

```
postgres@jack-virtual-machine:~$
postgres@jack-virtual-machine:~$ psql -U postgres
psql (16.1 (Ubuntu 16.1-1.pgdg22.04+1))
Type "help" for help.

postgres=# \c dvdrental
You are now connected to database "dvdrental" as user "postgres".
dvdrental=# \dt
             List of relations
 Schema |     Name      | Type  |  Owner
--------+---------------+-------+----------
 public | actor         | table | postgres
 public | address       | table | postgres
 public | category      | table | postgres
 public | city          | table | postgres
 public | country       | table | postgres
 public | customer      | table | postgres
 public | film          | table | postgres
 public | film_actor    | table | postgres
 public | film_category | table | postgres
 public | inventory     | table | postgres
 public | language      | table | postgres
 public | payment       | table | postgres
 public | rental        | table | postgres
 public | staff         | table | postgres
 public | store         | table | postgres
(15 rows)

dvdrental=#
```

Figure 6.13: *Shows tables available inside dvdrental database*

We are going to use a **`SELECT`** statement to see the data inside the actor's table. The data has been restored using the **`pg_restore`** command shown in *Figure 6.12*.

```
dvdrental=# SELECT * FROM actor;
 actor_id | first_name  |  last_name   |      last_update
----------+-------------+--------------+------------------------
        1 | Penelope    | Guiness      | 2013-05-26 14:47:57.62
        2 | Nick        | Wahlberg     | 2013-05-26 14:47:57.62
        3 | Ed          | Chase        | 2013-05-26 14:47:57.62
        4 | Jennifer    | Davis        | 2013-05-26 14:47:57.62
        5 | Johnny      | Lollobrigida | 2013-05-26 14:47:57.62
        6 | Bette       | Nicholson    | 2013-05-26 14:47:57.62
        7 | Grace       | Mostel       | 2013-05-26 14:47:57.62
        8 | Matthew     | Johansson    | 2013-05-26 14:47:57.62
        9 | Joe         | Swank        | 2013-05-26 14:47:57.62
       10 | Christian   | Gable        | 2013-05-26 14:47:57.62
       11 | Zero        | Cage         | 2013-05-26 14:47:57.62
       12 | Karl        | Berry        | 2013-05-26 14:47:57.62
       13 | Uma         | Wood         | 2013-05-26 14:47:57.62
       14 | Vivien      | Bergen       | 2013-05-26 14:47:57.62
       15 | Cuba        | Olivier      | 2013-05-26 14:47:57.62
       16 | Fred        | Costner      | 2013-05-26 14:47:57.62
       17 | Helen       | Voight       | 2013-05-26 14:47:57.62
```

Figure 6.14: *Using the SELECT statement to view data in the actor's table*

Note: *There are 200 rows of data inside the actor's table.*

Logical Backup

Logical backups represent database data in a text-based format, making it easily readable by humans. This approach is beneficial as it allows for the restoration of individual rows of data if needed. Logical backups can be used to restore specific rows, entire tables, schemas, or even the full database. However, using this method to restore an entire database can be time-consuming.

In PostgreSQL, the tool used for creating logical backups is called **pg_dump**. This utility can back up individual rows, tables, or an entire database. To back up the entire cluster, the **pg_dumpall** utility is used. Typically, **pg_dump** generates consistent backups.

This backup method supports various output formats. By default, the **pg_dump** utility generates backups in plain text format, which can be restored using the **psql** command. Other supported formats include tar, custom binary, and compressed formats, which can be restored using the **pg_restore** command.

It is important to note that any changes made during the backup process are not captured, meaning the backup represents the data as it existed at the time the **pg_dump** command was run.

In this section, we will back up and restore the whole database, a single table, data only, and an entire cluster.

Backing Up and Restoring a Whole Database

We have previously demonstrated the process of downloading and restoring the DVD rental sample database. The restoration was carried out on our PostgreSQL server using the **pg_restore** utility. The choice between using **psql** or **pg_restore** for restoration depends on the format in which the database backup is stored.

In this instance, we will create a backup of the dvdrental database as a tar file utilizing the **pg_dump** utility.

For Practical purposes, we will use the DVD rental sample database (**dvdrental**) installed earlier.

Let us create a backup directory to host the backups using the **mkdir** command.

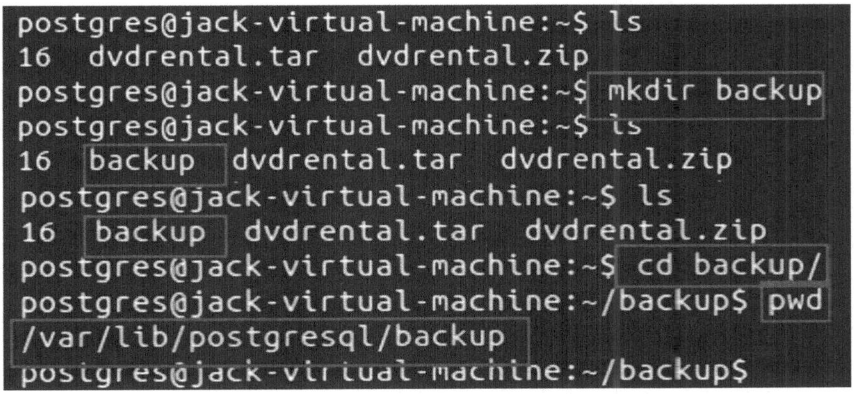

```
postgres@jack-virtual-machine:~$ ls
16  dvdrental.tar  dvdrental.zip
postgres@jack-virtual-machine:~$ mkdir backup
postgres@jack-virtual-machine:~$ ls
16  backup  dvdrental.tar  dvdrental.zip
postgres@jack-virtual-machine:~$ ls
16  backup  dvdrental.tar  dvdrental.zip
postgres@jack-virtual-machine:~$ cd backup/
postgres@jack-virtual-machine:~/backup$ pwd
/var/lib/postgresql/backup
postgres@jack-virtual-machine:~/backup$
```

Figure 6.15: *Location of the backups*

We are now going to take a backup of the whole database using the **pg_dump** utility.

The syntax for **pg_dump** is as follows:

```
pg_dump -d dbname -U user_name -Ft> /path_to_backup/dumpfile_name

pg_dump dbname | gzip > /path_to_backup_location/filename.gz
                                                #For large databases
```

Where:

```
dbname = database name
user_name = name of postgres user
Ft = format is .tar
/path_to_backup = absolute path to the backup directory

dumpfile_name = Name of the backup file
```

```
pg_dump -Ft  -d dvdrental -U postgres > /var/lib/postgresql/
backup/ dvdrental.tar
```

```
postgres@jack-virtual-machine:~/backup$ pg_dump -d dvdrental -U postgres -Ft > /var/lib/postgresql/backup/dvdrental.tar
postgres@jack-virtual-machine:~/backup$
postgres@jack-virtual-machine:~/backup$ ls
dvdrental.tar
postgres@jack-virtual-machine:~/backup$
```

***Figure 6.16**: A backup of the dvdrental database using pg_dump utility*

Incident 1: We are now going to simulate a situation at work where someone accidentally drops the whole database. This means that the whole database and its data will be lost.

Log in to the database as a Postgres user, and drop the dvdrental database.

```
psql -U postgres
DROP DATABASE dvdrental;
```

```
postgres@jack-virtual-machine:~/backup$ psql -U postgres
psql (16.1 (Ubuntu 16.1-1.pgdg22.04+1))
Type "help" for help.

postgres=# \l
                                        List of databases
   Name    |  Owner   | Encoding | Locale Provider |   Collate   |    Ctype    |
-----------+----------+----------+-----------------+-------------+-------------+
 dvdrental | postgres | UTF8     | libc            | en_US.UTF-8 | en_US.UTF-8 |
 postgres  | postgres | UTF8     | libc            | en_US.UTF-8 | en_US.UTF-8 |
           |          |          |                 |             |             |
           |          |          |                 |             |             |
 template0 | postgres | UTF8     | libc            | en_US.UTF-8 | en_US.UTF-8 |
           |          |          |                 |             |             |
 template1 | postgres | UTF8     | libc            | en_US.UTF-8 | en_US.UTF-8 |
           |          |          |                 |             |             |
(4 rows)

postgres=# DROP DATABASE dvdrental;
DROP DATABASE
postgres=#
```

***Figure 6.17**: The dvdrental database accidentally dropped*

The next screenshot shows that the dvdrental database does not exist anymore because it has been dropped.

```
postgres=#
postgres=# \l
                                                List of databases
   Name    |  Owner   | Encoding | Locale Provider |   Collate   |    Ctype    |
-----------+----------+----------+-----------------+-------------+-------------+
 postgres  | postgres | UTF8     | libc            | en_US.UTF-8 | en_US.UTF-8 |
           |          |          |                 |             |             |
           |          |          |                 |             |             |
 template0 | postgres | UTF8     | libc            | en_US.UTF-8 | en_US.UTF-8 |
           |          |          |                 |             |             |
 template1 | postgres | UTF8     | libc            | en_US.UTF-8 | en_US.UTF-8 |
           |          |          |                 |             |             |
(3 rows)
```

Figure 6.18: *Shows the dvdrental database no longer exists after the dropped incident above*

Restoring the `dvdrental` Database Using the Backup File

A crucial task for a PostgreSQL DBA is to restore the database from the available backups. For this restoration process, we will utilize the **`pg_restore`** utility.

The process is:

1. Create a new dvdrental database
2. Run the **`pg_restore`** command provided above to restore the backup database file to the actual database.

```
postgres=# CREATE DATABASE dvdrental;
CREATE DATABASE
postgres=# \l
                                                List of databases
   Name    |  Owner   | Encoding | Locale Provider |   Collate   |    Ctype    |
-----------+----------+----------+-----------------+-------------+-------------+
 dvdrental | postgres | UTF8     | libc            | en_US.UTF-8 | en_US.UTF-8 |
 postgres  | postgres | UTF8     | libc            | en_US.UTF-8 | en_US.UTF-8 |
           |          |          |                 |             |             |
           |          |          |                 |             |             |
 template0 | postgres | UTF8     | libc            | en_US.UTF-8 | en_US.UTF-8 |
           |          |          |                 |             |             |
 template1 | postgres | UTF8     | libc            | en_US.UTF-8 | en_US.UTF-8 |
           |          |          |                 |             |             |
(4 rows)
```

Figure 6.19: *A new dvdrental database has been created*

We will use the **`pg_restore`** utility to restore all the data to the dvdrental database as it was before the incident.

Syntax:

```
pg_restore -U userName -d databaseName  <  /path_to_backup/
fileName.tar
pg_restore -U postgres -d dvdrental  <  /var/lib/postgresql/
backup/dvdrental.tar
```

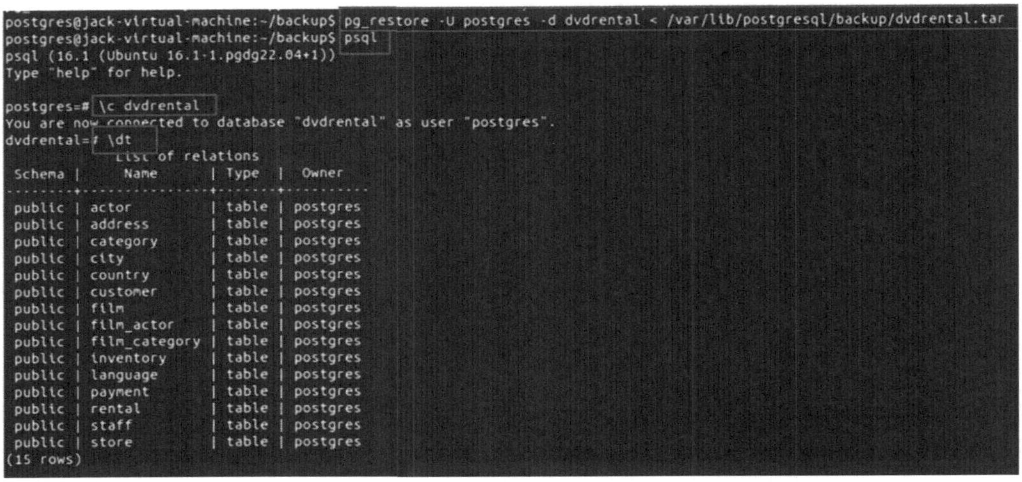

```
postgres@jack-virtual-machine:~/backup$ pg_restore -U postgres -d dvdrental < /var/lib/postgresql/backup/dvdrental.tar
postgres@jack-virtual-machine:~/backup$ psql
psql (16.1 (Ubuntu 16.1-1.pgdg22.04+1))
Type "help" for help.

postgres=# \c dvdrental
You are now connected to database "dvdrental" as user "postgres".
dvdrental=# \dt
             List of relations
 Schema |      Name      | Type  |  Owner
--------+----------------+-------+----------
 public | actor          | table | postgres
 public | address        | table | postgres
 public | category       | table | postgres
 public | city           | table | postgres
 public | country        | table | postgres
 public | customer       | table | postgres
 public | film           | table | postgres
 public | film_actor     | table | postgres
 public | film_category  | table | postgres
 public | inventory      | table | postgres
 public | language       | table | postgres
 public | payment        | table | postgres
 public | rental         | table | postgres
 public | staff          | table | postgres
 public | store          | table | postgres
(15 rows)
```

***Figure 6.20**: Shows dvdrental restored with all its 15 tables available*

Another method of backup and recovery would be to save the **dumpfile** as **.dump** or **.sql** and restore it using the **psql** tool. Let us see an example:

```
pg_dump -Fc  -d dvdrental -U postgres > /var/lib/postgresql/
backup/dvdrental.dump

pg_dump -d dvdrental -U postgres > /var/lib/postgresql/backup/
dvdrental.sql
```

```
postgres@jack-virtual-machine:~/backup$ ls
dvdrental.tar
postgres@jack-virtual-machine:~/backup$ cd
postgres@jack-virtual-machine:~$
postgres@jack-virtual-machine:~$ pwd
/var/lib/postgresql
postgres@jack-virtual-machine:~$ cd backup/
postgres@jack-virtual-machine:~/backup$ pwd
/var/lib/postgresql/backup
postgres@jack-virtual-machine:~/backup$ ls
dvdrental.tar
dvdrental=# \q
postgres@jack-virtual-machine:~/backup$ pg_dump -Fc  -d dvdrental -U postgres > /var/lib/postgresql/backup/dvdrental.dump
postgres@jack-virtual-machine:~/backup$
postgres@jack-virtual-machine:~/backup$ ls
dvdrental.dump  dvdrental.tar
postgres@jack-virtual-machine:~/backup$ pg_dump -d dvdrental -U postgres > /var/lib/postgresql/backup/dvdrental.sql
postgres@jack-virtual-machine:~/backup$ ls
dvdrental.dump  dvdrental.sql  dvdrental.tar
postgres@jack-virtual-machine:~/backup$
```

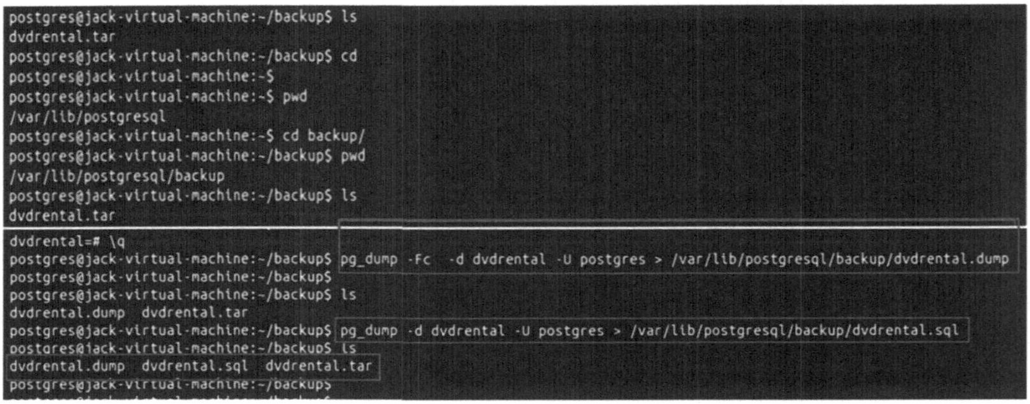

***Figure 6.21**: Shows all the three backups taken*

We are going to repeat incident 1, and drop the **dvdrental** database again, and then restore it with either the **.dump** or the **.sql** backup file using psql tool for **.sql** file.

```
postgres@jack-virtual-machine:~/backup$ psql -U postgres
psql (16.1 (Ubuntu 16.1-1.pgdg22.04+1))
Type "help" for help.

postgres=# DROP DATABASE dvdrental;
DROP DATABASE
postgres=# \l
                                          List of databases
   Name    |  Owner   | Encoding | Locale Provider |   Collate   |    Ctype    |
-----------+----------+----------+-----------------+-------------+-------------+
 postgres  | postgres | UTF8     | libc            | en_US.UTF-8 | en_US.UTF-8 |
           |          |          |                 |             |             |
           |          |          |                 |             |             |
 template0 | postgres | UTF8     | libc            | en_US.UTF-8 | en_US.UTF-8 |
           |          |          |                 |             |             |
 template1 | postgres | UTF8     | libc            | en_US.UTF-8 | en_US.UTF-8 |
           |          |          |                 |             |             |
(3 rows)

postgres=# CREATE DATABASE dvdrental;
CREATE DATABASE
postgres=# \l
                                          List of databases
   Name    |  Owner   | Encoding | Locale Provider |   Collate   |    Ctype    |
-----------+----------+----------+-----------------+-------------+-------------+
 dvdrental | postgres | UTF8     | libc            | en_US.UTF-8 | en_US.UTF-8 |
 postgres  | postgres | UTF8     | libc            | en_US.UTF-8 | en_US.UTF-8 |
           |          |          |                 |             |             |
           |          |          |                 |             |             |
 template0 | postgres | UTF8     | libc            | en_US.UTF-8 | en_US.UTF-8 |
           |          |          |                 |             |             |
 template1 | postgres | UTF8     | libc            | en_US.UTF-8 | en_US.UTF-8 |
           |          |          |                 |             |             |
(4 rows)
```

***Figure 6.22**: dvdrental database dropped and recreated*

Restoring a Whole Database

Restoring a whole database can be done using the following commands:

```
Pg_restore -U postgres -d dvdrental  <  /var/lib/postgresql/back-
up/dvdrental.dump
psql -U postgres -d dvdrental  <  /var/lib/postgresql/backup/dvd-
rental.sql
```

```
postgres@jack-virtual-machine:~/backup$ psql -U postgres -d dvdrental < /var/lib/postgresql/dvdrental.tar
SET
SET
SET
SET
SET
 set_config
------------
```

```
postgres@jack-virtual-machine:~/backup$ psql -U postgres
psql (16.1 (Ubuntu 16.1-1.pgdg22.04+1))
Type "help" for help.

postgres=# \c dvdrental
You are now connected to database "dvdrental" as user "postgres".
dvdrental=# \dt
             List of relations
 Schema |     Name      | Type  |  Owner
--------+---------------+-------+----------
 public | actor         | table | postgres
 public | address       | table | postgres
 public | category      | table | postgres
 public | city          | table | postgres
 public | country       | table | postgres
 public | customer      | table | postgres
 public | film          | table | postgres
 public | film_actor    | table | postgres
 public | film_category | table | postgres
 public | inventory     | table | postgres
 public | language      | table | postgres
 public | payment       | table | postgres
 public | rental        | table | postgres
 public | staff         | table | postgres
 public | store         | table | postgres
(15 rows)
```

***Figure 6.23**: dvdrental database restored using **psql** utility*

Backing Up and Restoring a Whole Table

In this section, we are going to:

- backup a whole table
- delete the table from the database and
- then restore it

Before we drop the actor table, let us take its backup.

```
pg_dump -d dvdrental -t actor > /var/lib/postgresql/backup/actor.sql
```

where:

`-d` option specifies the database

`-t` option specifies the table to be backed up

```
postgres@jack-virtual-machine:~/backup$ ls
dvdrental.dump  dvdrental.sql  dvdrental.tar
postgres@jack-virtual-machine:~/backup$ pg_dump -d dvdrental -t actor >  /var/lib/postgresql/backup/actor.sql
postgres@jack-virtual-machine:~/backup$ ls
actor.sql  dvdrental.dump  dvdrental.sql  dvdrental.tar
postgres@jack-virtual-machine:~/backup$
```

***Figure 6.24:** A single table (actor) backed up*

Incident 2: A colleague accidentally deleted the actor table with all its data from the dvdrental database.

```
dvdrental=#
dvdrental=# DROP TABLE actor;
ERROR:  cannot drop table actor because other objects depend on it
DETAIL:  view actor_info depends on table actor
view film_list depends on table actor
view nicer_but_slower_film_list depends on table actor
constraint film_actor_actor_id_fkey on table film_actor depends on table actor
HINT:  Use DROP ... CASCADE to drop the dependent objects too.
dvdrental=#
```

Figure 6.25: *ERROR message and HINT while dropping table actor*

The next screenshot shows how we used the **CASCADE** as suggested in the **HINT** to then drop the table.

```
dvdrental=#
dvdrental=# DROP TABLE actor CASCADE;
NOTICE:  drop cascades to 4 other objects
DETAIL:  drop cascades to view actor_info
drop cascades to view film_list
drop cascades to view nicer_but_slower_film_list
drop cascades to constraint film_actor_actor_id_fkey on table film_actor
DROP TABLE
dvdrental=#
```

Figure 6.26: *Table actor dropped with CASCADE*

We are going to use the **\dt** command to see that the actor table has been deleted and no longer exists.

```
dvdrental=# \dt
             List of relations
 Schema |      Name       | Type  |  Owner
--------+-----------------+-------+----------
 public | address         | table | postgres
 public | category        | table | postgres
 public | city            | table | postgres
 public | country         | table | postgres
 public | customer        | table | postgres
 public | film            | table | postgres
 public | film_actor      | table | postgres
 public | film_category   | table | postgres
 public | inventory       | table | postgres
 public | language        | table | postgres
 public | payment         | table | postgres
 public | rental          | table | postgres
 public | staff           | table | postgres
 public | store           | table | postgres
(14 rows)
```

Figure 6.27: *List of tables without table actor*

Restoring a Table

As a PostgreSQL DBA or Engineer, you must restore the table and its data to their original state. This is done using the following query:

```
psql -U postgres -d dvdrental -t actor  <  /var/lib/postgresql/
backup/actor.sql
```

```
dvdrental=# \q
postgres@jack-virtual-machine:~/backup$ ls
actor.sql  data_only.sql  dvdrental.dump  dvdrental.sql  dvdrental.tar  julia.sql
postgres@jack-virtual-machine:~/backup$
postgres@jack-virtual-machine:~/backup$ psql -U postgres -d dvdrental -t actor  <  /var/lib/postgresql/backup/actor.sql
psql: warning: extra command-line argument "actor" ignored
SET
SET
SET
SET
SET


SET
SET
SET
SET
SET
SET
CREATE TABLE
ALTER TABLE
COPY 200
ALTER TABLE
CREATE INDEX
CREATE TRIGGER
postgres@jack-virtual-machine:~/backup$
```

Figure 6.28: *Table actor being restored using the psql tool*

From the result shown in *Figure* 6.28, table actor has been restored using the backup file called **`actor.sql`**. At the bottom right of the screenshot, it shows that the table has been created and 200 rows of data restored to it.

```
postgres@jack-virtual-machine:~/backup$ psql
psql (16.1 (Ubuntu 16.1-1.pgdg22.04+1))
Type "help" for help.

postgres=# \c dvdrental
You are now connected to database "dvdrental" as user "postgres".
dvdrental=# \dt
             List of relations
 Schema |     Name      | Type  |  Owner
--------+---------------+-------+----------
 public | actor         | table | postgres
 public | address       | table | postgres
 public | category      | table | postgres
 public | city          | table | postgres
 public | country       | table | postgres
 public | customer      | table | postgres
 public | film          | table | postgres
 public | film_actor    | table | postgres
 public | film_category | table | postgres
 public | inventory     | table | postgres
 public | language      | table | postgres
 public | payment       | table | postgres
 public | rental        | table | postgres
 public | staff         | table | postgres
 public | store         | table | postgres
(15 rows)

dvdrental=#
```

Figure 6.29: *Shows that the table actor has been restored*

The **SELECT** statement shown with **LIMIT 5** will retrieve the first 5 rows of records.

```
dvdrental=# SELECT * FROM actor LIMIT 5;
 actor_id | first_name |  last_name   |      last_update
----------+------------+--------------+------------------------
        1 | Penelope   | Guiness      | 2013-05-26 14:47:57.62
        2 | Nick       | Wahlberg     | 2013-05-26 14:47:57.62
        3 | Ed         | Chase        | 2013-05-26 14:47:57.62
        4 | Jennifer   | Davis        | 2013-05-26 14:47:57.62
        5 | Johnny     | Lollobrigida | 2013-05-26 14:47:57.62
(5 rows)
```

Figure 6.30: *SELECT statement on the restored table actor*

The **SELECT** statement shows that the data has been restored.

Backing Up and Restoring Data Only

In this section, we are going to see how to back up data only from a table before an incident happens.

To backup data only, the command is as follows.

```
pg_dump -d dvdrental -t actor -a  >  /var/lib/postgresql/backup/
data_only.sql
pg_dump -d dvdrental -t actor --data  >  /var/lib/postgresql/backup/
data_only.sql
```

where **-a** or **--data** option is for data only.

```
postgres@jack-virtual-machine:~/backup$ pg_dump -d dvdrental -t actor -a  >  /var/lib/postgresql/backup/data_only.sql
postgres@jack-virtual-machine:~/backup$ ls
actor.sql  data_only.sql  dvdrental.dump  dvdrental.sql  dvdrental.tar  julia.sql
postgres@jack-virtual-machine:~/backup$ cat julia
cat: julia: No such file or directory
postgres@jack-virtual-machine:~/backup$ tail -5 data_only.sql

--
-- PostgreSQL database dump complete
--

postgres@jack-virtual-machine:~/backup$ tail -10 data_only.sql
198     Mary    Keitel  2013-05-26 14:47:57.62
199     Julia   Fawcett 2013-05-26 14:47:57.62
200     Thora   Temple  2013-05-26 14:47:57.62
\.


--
-- PostgreSQL database dump complete
--
```

Figure 6.31: *Backing up data only from the table actor to the backup directory*

Incident 3: A colleague accidentally deleted all rows of data from the actor table using the **TRUNCATE** command, as shown in *Figure* 6.33. As a PostgreSQL DBA or Engineer, you must restore the data to the table.

```
dvdrental=# select * from actor;
 actor_id | first_name  |  last_name   |      last_update
----------+-------------+--------------+------------------------
        1 | Penelope    | Guiness      | 2013-05-26 14:47:57.62
        2 | Nick        | Wahlberg     | 2013-05-26 14:47:57.62
        3 | Ed          | Chase        | 2013-05-26 14:47:57.62
        4 | Jennifer    | Davis        | 2013-05-26 14:47:57.62
        5 | Johnny      | Lollobrigida | 2013-05-26 14:47:57.62
        6 | Bette       | Nicholson    | 2013-05-26 14:47:57.62
        7 | Grace       | Mostel       | 2013-05-26 14:47:57.62
        8 | Matthew     | Johansson    | 2013-05-26 14:47:57.62
        9 | Joe         | Swank        | 2013-05-26 14:47:57.62
```

Figure 6.32: *Data in the table actor before an incident shown below*

The **TRUNCATE** command will delete all the rows of data from the table.

```
dvdrental=# TRUNCATE TABLE actor;
TRUNCATE TABLE
dvdrental=#
dvdrental=# SELECT * FROM actor;
 actor_id | first_name | last_name | last_update
----------+------------+-----------+-------------
(0 rows)

dvdrental=#
```

Figure 6.33: *All data deleted using a truncate command*

Restoring Data only to a Table

We are going to restore the deleted/truncated data using the following query:

```
psql -U postgres -d dvdrental -t actor -a <  /var/lib/postgresql/
backup/data_only.sql
```

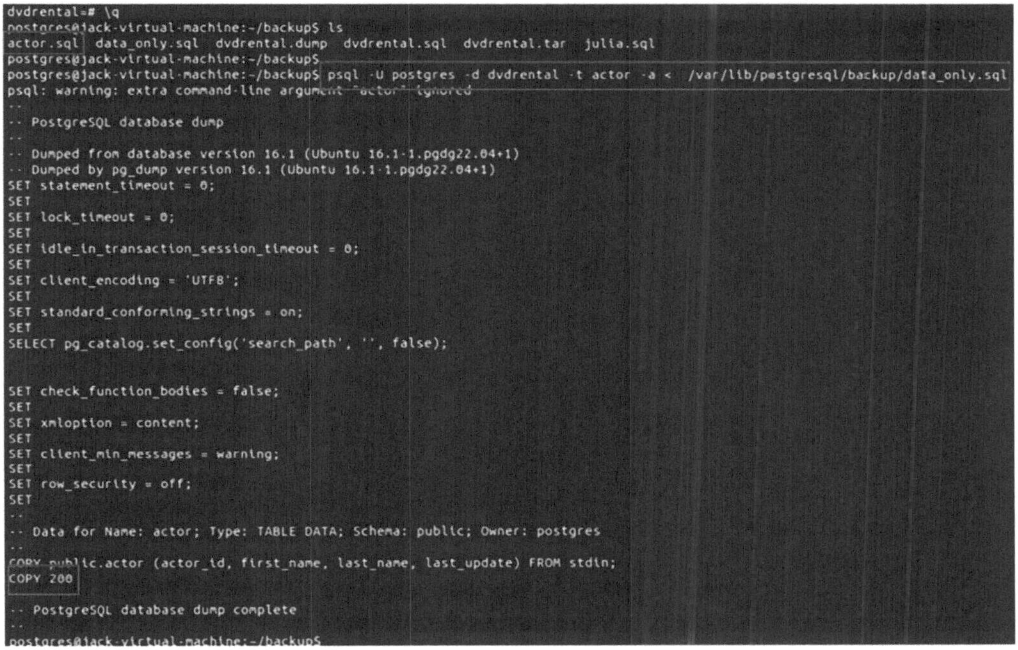

```
dvdrental=# \q
postgres@jack-virtual-machine:~/backup$ ls
actor.sql  data_only.sql  dvdrental.dump  dvdrental.sql  dvdrental.tar  julia.sql
postgres@jack-virtual-machine:~/backup$
postgres@jack-virtual-machine:~/backup$ psql -U postgres -d dvdrental -t actor -a <  /var/lib/postgresql/backup/data_only.sql
psql: warning: extra command-line argument "actor" ignored
--
-- PostgreSQL database dump
--
-- Dumped from database version 16.1 (Ubuntu 16.1-1.pgdg22.04+1)
-- Dumped by pg_dump version 16.1 (Ubuntu 16.1-1.pgdg22.04+1)
SET statement_timeout = 0;
SET
SET lock_timeout = 0;
SET
SET idle_in_transaction_session_timeout = 0;
SET
SET client_encoding = 'UTF8';
SET
SET standard_conforming_strings = on;
SET
SELECT pg_catalog.set_config('search_path', '', false);


SET check_function_bodies = false;
SET
SET xmloption = content;
SET
SET client_min_messages = warning;
SET
SET row_security = off;
SET
--
-- Data for Name: actor; Type: TABLE DATA; Schema: public; Owner: postgres
--
COPY public.actor (actor_id, first_name, last_name, last_update) FROM stdin;
COPY 200
-- PostgreSQL database dump complete
--
postgres@jack-virtual-machine:~/backup$
```

Figure 6.34: *All data restored to table actor*

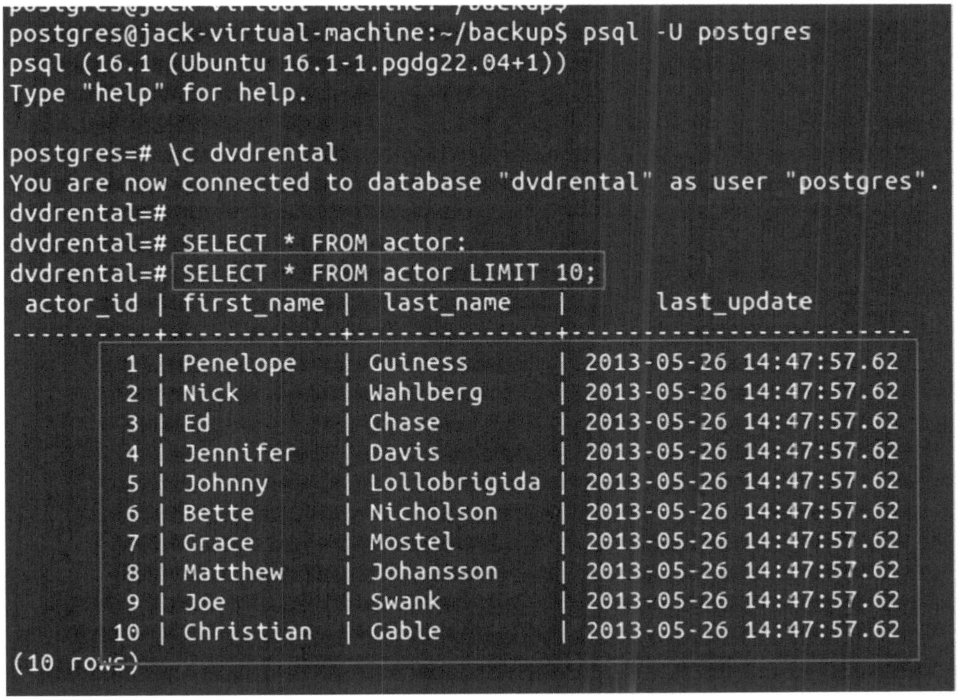

```
postgres@jack-virtual-machine:~/backup$ psql -U postgres
psql (16.1 (Ubuntu 16.1-1.pgdg22.04+1))
Type "help" for help.

postgres=# \c dvdrental
You are now connected to database "dvdrental" as user "postgres".
dvdrental=#
dvdrental=# SELECT * FROM actor;
dvdrental=# SELECT * FROM actor LIMIT 10;
 actor_id | first_name |  last_name   |      last_update
----------+------------+--------------+------------------------
        1 | Penelope   | Guiness      | 2013-05-26 14:47:57.62
        2 | Nick       | Wahlberg     | 2013-05-26 14:47:57.62
        3 | Ed         | Chase        | 2013-05-26 14:47:57.62
        4 | Jennifer   | Davis        | 2013-05-26 14:47:57.62
        5 | Johnny     | Lollobrigida | 2013-05-26 14:47:57.62
        6 | Bette      | Nicholson    | 2013-05-26 14:47:57.62
        7 | Grace      | Mostel       | 2013-05-26 14:47:57.62
        8 | Matthew    | Johansson    | 2013-05-26 14:47:57.62
        9 | Joe        | Swank        | 2013-05-26 14:47:57.62
       10 | Christian  | Gable        | 2013-05-26 14:47:57.62
(10 rows)
```

Figure 6.35: *SELECT statement to show the restored data*

pg_dumpall

The **pg_dump** utility is designed to back up a single database at a time, but it does not include roles and tablespaces in its backup. This is because roles and tablespaces span across multiple databases rather than being confined to one.

To properly back up the entire cluster, we use **pg_dumpall**. This tool creates a backup for each database within the cluster and also includes data such as cluster-wide roles and tablespace definitions. The syntax for using **pg_dumpall** is:

```
pg_dumpall > /path_to_backup_location/dumpfile
```

```
pg_dumpall > /var/lib/postgresql/backup/cluster.sql
```

```
postgres@jack-virtual-machine:~/backup$
postgres@jack-virtual-machine:~/backup$ pg_dumpall > /var/lib/postgresql/backup/cluster.sql
postgres@jack-virtual-machine:~/backup$
postgres@jack-virtual-machine:~/backup$ ls
actor.sql  cluster.sql  data_only.sql  dvdrental.dump  dvdrental.sql  dvdrental.tar  julia.sql
postgres@jack-virtual-machine:~/backup$
```

Figure 6.36: *Backup of the entire cluster using pg_dumpall*

As it is very unlikely to drop the entire cluster, we are not going to drop it either. Talking of backing the entire cluster, the next section will show learners another way to do that.

Physical Backups

This type of backup captures a snapshot of the database at a specific point in time. When performing backups, it is crucial to consider that database changes might occur concurrently. For the backup to be reliable, it must be consistent. Physical backups are particularly effective for restoring entire datasets. The speed of these backups depends on hardware capabilities, including disk performance and network speed, if data is transferred over a network. There are two main types of physical backups: **pg_basebackup** and **pg_backrest**. It is essential to evaluate the Recovery Point Objective (RPO) and Recovery Time Objective (RTO) before deciding which backup method to use.

- **Recovery Point Objective (RPO):** RPO defines the maximum acceptable amount of data loss in case of a disaster. It indicates how much data an organization is willing to lose, measured from the last backup to the point of failure. This metric helps DBAs determine the appropriate frequency for data backups.
- **Recovery Time Objective (RTO):** RTO specifies the target time within which the system must be restored after a failure. For example, if the RTO is set at 1.5 hours, the expectation is that normal business operations will resume within this period following a disaster. If this timeframe is exceeded during

an actual disaster, the RTO should be reassessed, and disaster recovery plans should be updated accordingly.

When calculating RTO, consider factors such as:

- The cost of downtime per hour.
- The priority and importance of individual systems.
- The steps involved in disaster recovery.
- Available resources and budget.

pg_basebackup

This tool is a powerful and widely utilized backup utility in PostgreSQL. It is designed to perform consistent online backups of the file system. These consistent backups are essential for setting up a secondary/slave server during streaming replication and for point-in-time recovery. While **pg_dump** is used to back up a single database, **pg_basebackup** is employed to back up the entire cluster. It does not support backing up individual databases or database objects. The backups created by **pg_basebackup** can be in tar or plain text formats.

Using tar and compressed formats is beneficial for minimizing the disk space required to back up and store the data directory, tablespaces, and WAL segments.

It is crucial to understand the client's requirements before initiating the backup procedure. Different organizations may have varying backup and recovery objectives based on their operational needs.

We will follow these steps to perform a successful base backup:

1. Create a backup directory and an archive location.
2. Set up continuous WAL archiving.
3. Execute the base backup.

Creating a Backup Directory and Archive Location

Log into the OS, and switch to root user to create a backup directory under the root directory (/).

1. Create a backup location called **pgbackup** under the root directory (/).
2. Change the ownership of the **pgbackup** to user postgres.

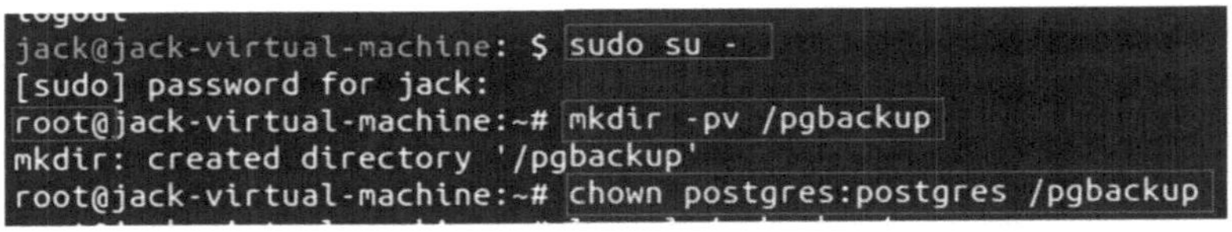

```
jack@jack-virtual-machine: $ sudo su -
[sudo] password for jack:
root@jack-virtual-machine:~# mkdir -pv /pgbackup
mkdir: created directory '/pgbackup'
root@jack-virtual-machine:~# chown postgres:postgres /pgbackup
```

Figure 6.37: *Creating the pgbackup directory under the root directory (/)*

Next, we must exit from the root user and go back to user **postgres** to create our archive location under the **pgbackup** location created above.

```
exit
sudo su - postgres
mkdir -pv /pgbackup/archivedir
cd /pgbackup/
ls
```

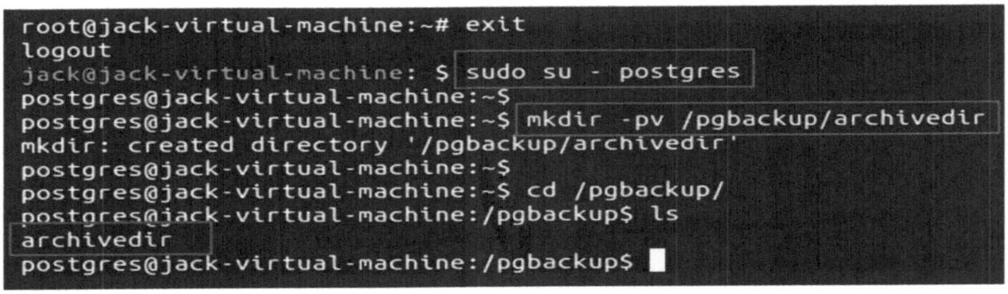

```
root@jack-virtual-machine:~# exit
logout
jack@jack-virtual-machine: $ sudo su - postgres
postgres@jack-virtual-machine:~$
postgres@jack-virtual-machine:~$ mkdir -pv /pgbackup/archivedir
mkdir: created directory '/pgbackup/archivedir'
postgres@jack-virtual-machine:~$
postgres@jack-virtual-machine:~$ cd /pgbackup/
postgres@jack-virtual-machine:/pgbackup$ ls
archivedir
postgres@jack-virtual-machine:/pgbackup$
```

Figure 6.38: *Archive directory created under the pgbackup directory*

Setting up Continuous WAL Archiving

Next, we will modify the **postgresql.conf** file to enable continuous WAL archiving. The following adjustments need to be made:

1. Enable archive mode for the database.
2. Set the WAL level to replica.
3. Configure **wal_keep_segments** and **wal_sender**.
4. Specify the archive command.
5. Trigger a log switch to verify the archive command.
6. Turn on archive mode.

To locate the archive mode setting within the **postgresql.conf** file, use the search term **/archive_mode**.

For us to make changes in the **postgresql.conf** file, we must first find its location.

From the Linux OS, run the following command.

```
ps -ef | grep postgres
```

From within the database, run the following command.

```
show config_file;
```

The results of which are shown in the following image:

```
jack@jack-virtual-machine: $ sudo su - postgres
[sudo] password for jack:
postgres@jack-virtual-machine:~$ ps -ef | grep postgres
postgres    1095       1  0 08:18 ?        00:00:00 /usr/lib/postgresql/16/bin/postgres -D /var/lib/postgresql/16/main -c config_f
ile=/etc/postgresql/16/main/postgresql.conf
postgres    1114    1095  0 08:19 ?        00:00:00 postgres: 16/main: checkpointer
postgres    1115    1095  0 08:19 ?        00:00:00 postgres: 16/main: background writer
postgres    1124    1095  0 08:19 ?        00:00:00 postgres: 16/main: walwriter
postgres    1125    1095  0 08:19 ?        00:00:00 postgres: 16/main: autovacuum launcher
postgres    1126    1095  0 08:19 ?        00:00:00 postgres: 16/main: archiver
postgres    1127    1095  0 08:19 ?        00:00:00 postgres: 16/main: logical replication launcher
root        3098    3049  0 08:32 pts/0    00:00:00 sudo su - postgres
root        3099    3098  0 08:32 pts/1    00:00:00 sudo su - postgres
root        3100    3099  0 08:32 pts/1    00:00:00 su - postgres
postgres    3101    3100  0 08:32 pts/1    00:00:00 -bash
postgres    3183    3101  0 08:32 pts/1    00:00:00 ps -ef
postgres    3184    3101  0 08:32 pts/1    00:00:00 grep postgres
postgres@jack-virtual-machine:~$
```

Figure 6.39: *Using ps -ef | grep postgres to find the location of postgresql.conf.*

The location is `/etc/postgresql/16/main/postgresql.conf`

```
postgres@jack-virtual-machine:/pgbackup$ psql
psql (16.1 (Ubuntu 16.1-1.pgdg22.04+1))
Type "help" for help.

postgres=# show config_file;
              config_file
-----------------------------------------
 /etc/postgresql/16/main/postgresql.conf
(1 row)
```

Figure 6.40: *Finding the location of postgresql.conf from within the database*

To make the necessary changes, we must edit the **postgresql.conf** file using a text editor. The command is:

Either

```
Vi /etc/postgresql/16/main/postgresql.conf
```

Or

```
cd  /etc/postgresql/16/main/
vi  postgresql.conf
```

We must always make a copy of the **postgresql.conf** file before editing it.

```
postgres=# \q
postgres@jack-virtual-machine:/pgbackup$ cd
postgres@jack-virtual-machine:~$
postgres@jack-virtual-machine:~$ cd  /etc/postgresql/16/main/
postgres@jack-virtual-machine:/etc/postgresql/16/main$ ls
conf.d  environment  pg_ctl.conf  pg_hba.conf  pg_ident.conf  postgresql.conf  postgresql.conf_backup  start.conf
postgres@jack-virtual-machine:/etc/postgresql/16/main$ cp postgresql.conf postgresql.conf_bkup
postgres@jack-virtual-machine:/etc/postgresql/16/main$ ls
conf.d  environment  pg_ctl.conf  pg_hba.conf  pg_ident.conf  postgresql.conf  postgresql.conf_backup  postgresql.conf_bkup
postgres@jack-virtual-machine:/etc/postgresql/16/main$
```

Figure 6.41: *Take a backup of postgresql.conf before editing*

As our current location is **/etc/postgresql/16/main**, which contains all the data files including **postgresql.conf** file, we can just vi into it without specifying the absolute path. The following are two commands, with and without the absolute path to the postgresql.conf file.

```
Vi /etc/postgresql/16/main/postgresql.conf
```

Or

```
Vi postgresql.conf          #Only if you are in that location as
                            shown below.
```

Postgresql.conf is a very big and delicate file. To search for something, click the forward slash (/) , and then type the keyword you want to search.

For instance, for us to look for archive mode, we type **/archive_mode** as shown in the following image:

```
postgres@jack-virtual-machine:/etc/postgresql/16/main$ vi postgresql.conf

# The default values of these variables are driven from the -D command-line
# option or PGDATA environment variable, represented here as ConfigDir.

data_directory = '/var/lib/postgresql/16/main'          # use data in another directory
                                        # (change requires restart)
hba_file = '/etc/postgresql/16/main/pg_hba.conf'        # host-based authentication file
                                        # (change requires restart)
ident_file = '/etc/postgresql/16/main/pg_ident.conf'    # ident configuration file
/archive_mode
```

Figure 6.42: *Searching for items to modify inside postgresql.conf file*

By default, the archive mode is off, as shown. We must change it to on and then uncomment by removing the hash sign at the start.

```
# - Archiving -

#archive_mode = off             # enables archiving; off, on, or always
                                # (change requires restart)
#archive_library = ''           # library to use to archive a WAL file
                                # (empty string indicates archive_command should
                                # be used)
#archive_command = ''           # command to use to archive a WAL file
                                # placeholders: %p = path of file to archive
                                #               %f = file name only
                                # e.g. 'test ! -f /mnt/server/archivedir/%f && cp %p /mnt/server/archivedir/%
#archive_timeout = 0            # force a WAL file switch after this
                                # number of seconds; 0 disables
```

Figure 6.43: *Default archive mode inside postgresql.conf file*

To save changes made to **postgresql.conf** file, the process is as follows:

1. Press the escape (**Esc**) key
2. Click the column (**:**) key
3. Press the **w** and **q** keys

4. Finally press **!** key
5. Then *Enter*.

To quit the **vi editor** without saving the changes, follow the same process omitting the **w** key.

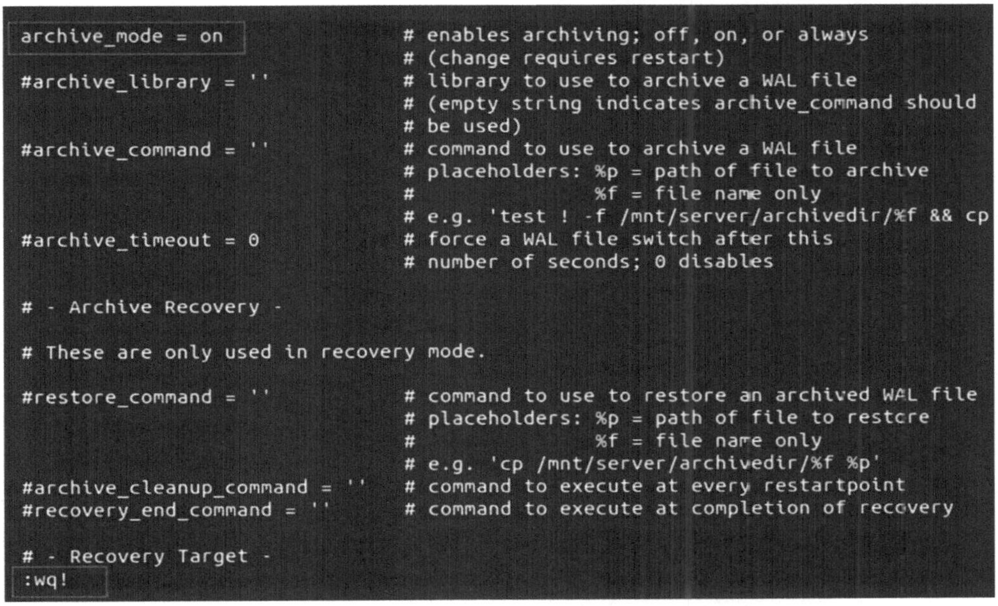

```
archive_mode = on               # enables archiving; off, on, or always
                                # (change requires restart)
#archive_library = ''           # library to use to archive a WAL file
                                # (empty string indicates archive_command should
                                # be used)
#archive_command = ''           # command to use to archive a WAL file
                                # placeholders: %p = path of file to archive
                                #               %f = file name only
                                # e.g. 'test ! -f /mnt/server/archivedir/%f && cp
#archive_timeout = 0            # force a WAL file switch after this
                                # number of seconds; 0 disables

# - Archive Recovery -

# These are only used in recovery mode.

#restore_command = ''           # command to use to restore an archived WAL file
                                # placeholders: %p = path of file to restore
                                #               %f = file name only
                                # e.g. 'cp /mnt/server/archivedir/%f %p'
#archive_cleanup_command = ''   # command to execute at every restartpoint
#recovery_end_command = ''      # command to execute at completion of recovery

# - Recovery Target -
:wq!
```

Figure 6.44: *How to save changes made to postgresql.conf file via the vi-text editor*

Next, we are going to set the **archive** command, making sure that it is pointing to the archive directory we created.

To obtain the path to the archive directory, follow the steps:

1. **cd   #will take you all the way back**
2. **cd /pgbackup   #This will take you to the pbbackup directory**
3. **cd archivedir   #This will then take you to the archive directory**
4. **pwd   # This will show you your present working directory or current directory**

Or

```
cd          #will take you all the way back
```

The following command will take you to the archive directory which is inside the **pgbackup** directory. The **pwd** is to display your current working directory.

```
cd /pgbackup/archivedir   #
pwd
```

```
postgres@jack-virtual-machine:/etc/postgresql/16/main$ cd
postgres@jack-virtual-machine:~$ cd /pgbackup/
postgres@jack-virtual-machine:/pgbackup$ ls
archivedir
postgres@jack-virtual-machine:/pgbackup$ cd archivedir/
postgres@jack-virtual-machine:/pgbackup/archivedir$ pwd
/pgbackup/archivedir
```

Figure 6.45: *How to find the location (path) to the archive directory*

We are going to configure the archive command to stream wals to the archive directory. The syntax is as shown, surrounded by single quotes.

```
cp %p /pgbackup/archivedir/%f
```

Make sure it is uncommended by removing the hash (#) at the start of it.

```
# - Archiving -

archive_mode = on               # enables archiving; off, on, or always
                                # (change requires restart)
#archive_library = ''           # library to use to archive a WAL file
                                # (empty string indicates archive_comma
                                # be used)
archive_command = 'cp %p /pgbackup/archivedir/%f'          # comma
                                # placeholders: %p = path of file to ar
                                #               %f = file name only
                                # e.g. 'test ! -f /mnt/server/archivedi
#archive_timeout = 0            # force a WAL file switch after this
                                # number of seconds; 0 disables
```

Figure 6.46: *How to set up the archive command in Postgresql.conf file*

The **`wal_level`** must be set to replica.

```
#------------------------------------------------------------------------------
# WRITE-AHEAD LOG
#------------------------------------------------------------------------------

# - Settings -

wal_level = replica                 # minimal, replica, or logical
                                    # (change requires restart)
#fsync = on                         # flush data to disk for crash safety
                                    # (turning this off can cause
                                    # unrecoverable data corruption)
```

Figure 6.47: *Wal level is set to replica*

We will be setting the Wal senders to 3 as we do not have a big CPU. This value will be different depending on the size of the database.

The **max_wal_senders** parameter defines the maximum number of concurrent connections that a streaming base backup client can have. It limits the number of WAL sender processes that can run simultaneously.

```
#------------------------------------------------------------------------------
# REPLICATION
#------------------------------------------------------------------------------

# - Sending Servers -

# Set these on the primary and on any standby that will send replication data.

max_wal_senders = 3             # max number of walsender processes
                                # (change requires restart)
#max_replication_slots = 10     # max number of replication slots
                                # (change requires restart)
#wal_keep_size = 0              # in megabytes; 0 disables
```

Figure 6.48: *How to configure max wal senders*

We have made all the changes in the **postgresql.conf** file needed for the **pg_basebackup** to run successfully. Now, **SAVE** or write (**:wq!**) the changes and restart the cluster for the changes to take effect.

The command to restart the cluster is as shown. The same command can be used to check status, stop, and even reload the database by just changing the last parameter.

```
/usr/lib/postgresql/16/bin/pg_ctl -D /etc/postgresql/16/main/ restart
/usr/lib/postgresql/16/bin/pg_ctl -D /etc/postgresql/16/main/ status
/usr/lib/postgresql/16/bin/pg_ctl -D /etc/postgresql/16/main/ stop
/usr/lib/postgresql/16/bin/pg_ctl -D /etc/postgresql/16/main/ reload
```

```
postgres@jack-virtual-machine:/var/log/postgresql$ /usr/lib/postgresql/16/bin/pg_ctl -D /etc/postgresql/16/main/ restart
2024-01-23 14:55:32.668 EST [18968] LOG:  received fast shutdown request
waiting for server to shut down....2024-01-23 14:55:32.669 EST [18968] LOG:  aborting any active transactions
2024-01-23 14:55:32.676 EST [18968] LOG:  background worker "logical replication launcher" (PID 18975) exited with exit c
2024-01-23 14:55:32.676 EST [18969] LOG:  shutting down
2024-01-23 14:55:32.875 EST [18969] LOG:  checkpoint starting: shutdown immediate
.2024-01-23 14:55:33.780 EST [18969] LOG:  checkpoint complete: wrote 3 buffers (0.0%); 0 WAL file(s) added, 0 removed, 2
01 s, average=0.001 s; distance=16384 kB, estimate=16384 kB; lsn=0/8000028, redo lsn=0/8000028
2024-01-23 14:55:33.805 EST [18968] LOG:  database system is shut down
 done
server stopped
waiting for server to start....2024-01-23 14:55:33.936 EST [18995] LOG:  starting PostgreSQL 16.1 (Ubuntu 16.1-1.pgdg22.0
.0, 64-bit
2024-01-23 14:55:33.939 EST [18995] LOG:  listening on IPv4 address "0.0.0.0", port 5432
2024-01-23 14:55:33.939 EST [18995] LOG:  listening on IPv6 address "::", port 5432
2024-01-23 14:55:33.940 EST [18995] LOG:  listening on Unix socket "/var/run/postgresql/.s.PGSQL.5432"
2024-01-23 14:55:33.945 EST [18998] LOG:  database system was shut down at 2024-01-23 14:55:33 EST
2024-01-23 14:55:33.960 EST [18995] LOG:  database system is ready to accept connections
 done
server started
postgres@jack-virtual-machine:/var/log/postgresql$
```

Figure 6.49: *How to restart the server*

It is worth noting here that the **-D** is the path to the cluster directory, which is the postgres data directory.

For us to verify that our archive command was set correctly and that it is sending wals to the archive location, we will force a log switch using the **pg_switch_wal()** function. **SELECT pg_switch_wal** will generate wal files and send them to the archive location. We have force-switched the logs three times, as shown.

The query is;

```
SELECT pg_switch_wal();
```

```
postgres@jack-virtual-machine:/var/log/postgresql$ psql
psql (16.1 (Ubuntu 16.1-1.pgdg22.04+1))
Type "help" for help.

postgres=# SELECT pg_switch_wal();
 pg_switch_wal
---------------
 0/9000078
(1 row)

postgres=# SELECT pg_switch_wal();
 pg_switch_wal
---------------
 0/A000078
(1 row)

postgres=# SELECT pg_switch_wal();
 pg_switch_wal
---------------
 0/B000078
(1 row)

postgres=#
```

Figure 6.50: *How to force switch wals by calling the pg_switch_wal function*

The next screen shows the available WAL files in our archive directory.

```
postgres@jack-virtual-machine:~$
postgres@jack-virtual-machine:~$ cd /pgbackup/
postgres@jack-virtual-machine:/pgbackup$ ls
archivedir  basebackup
postgres@jack-virtual-machine:/pgbackup$ cd archivedir/
postgres@jack-virtual-machine:/pgbackup/archivedir$ ls -al
total 131084
drwxrwxr-x 2 postgres postgres     4096 Jan 23 15:40 .
drwxr-xr-x 4 postgres postgres     4096 Jan 23 15:40 ..
-rw------- 1 postgres postgres 16777216 Jan 23 14:54 000000010000000000000006
-rw------- 1 postgres postgres 16777216 Jan 23 14:55 000000010000000000000007
-rw------- 1 postgres postgres 16777216 Jan 23 15:00 000000010000000000000008
-rw------- 1 postgres postgres 16777216 Jan 23 15:01 000000010000000000000009
-rw------- 1 postgres postgres 16777216 Jan 23 15:01 00000001000000000000000A
-rw------- 1 postgres postgres 16777216 Jan 23 15:02 00000001000000000000000B
-rw------- 1 postgres postgres 16777216 Jan 23 15:40 00000001000000000000000C
-rw------- 1 postgres postgres 16777216 Jan 23 15:40 00000001000000000000000D
-rw------- 1 postgres postgres      338 Jan 23 15:40 00000001000000000000000D.00000028.backup
postgres@jack-virtual-machine:/pgbackup/archivedir$
```

Figure 6.51: *WAL files available in the archive directory*

From the *Figure* 6.51, we can see the WAL files available in the archive location.

Very important to note is that if the WALs are not present in the said location, check the logs.

How to check Postgresql Logs

To access the postgresql logs, cd to **/var/log/postgresql/** and cat the most recent log file.

```
cd /var/log/postgresql/
```

```
postgres@jack-virtual-machine:~$
postgres@jack-virtual-machine:~$ cd /var/log/postgresql/
postgres@jack-virtual-machine:/var/log/postgresql$ ls -l
total 8292
-rw-r----- 1 postgres adm    3110 Jan 23 14:46 postgresql-16-main.log
-rw-r----- 1 postgres adm 8476993 Jan 21 15:26 postgresql-16-main.log.1
-rw-r----- 1 postgres adm     798 Jan 14 06:46 postgresql-16-main.log.2.gz
-rw-r----- 1 postgres adm    1868 Jan  8 10:10 postgresql-16-main.log.3.gz
postgres@jack-virtual-machine:/var/log/postgresql$
```

Figure 6.52: *How to locate Postgresql logs for troubleshooting*

If there are no WAL files in the archive directory, the most recent log file will indicate the issue. During troubleshooting, consider the following steps:

- Ensure that the archivedir is owned by the **postgres** user.
- Verify that the archive command is correctly pointing to the intended backup directory.
- Confirm that the server was restarted after modifying the **postgresql.conf** file.

Research more on continuous archiving and point-in-time recovery with Postgresql.

Now that our WAL files have been archived correctly and are available in the archive directory, we are good to run a base backup. But before we take the backup, let us create two tablespaces to help us access the successful nature of the backup.

```
postgres@jack-virtual-machine:~$ exit
logout
jack@jack-virtual-machine: $ mkdir -pv /pgdata
mkdir: cannot create directory '/pgdata': Permission denied
jack@jack-virtual-machine: $ sudo mkdir -pv /pgdata
[sudo] password for jack:
mkdir: created directory '/pgdata'
jack@jack-virtual-machine: $ sudo chown postgres:postgres /pgdata
jack@jack-virtual-machine: $ sudo su - postgres
postgres@jack-virtual-machine:~$ mkdir /pg
pgbackup/ pgdata/
postgres@jack-virtual-machine:~$ mkdir /pg
pgbackup/ pgdata/
postgres@jack-virtual-machine:~$ mkdir /pgdata/tablespaces
postgres@jack-virtual-machine:~$ cd /pgdata/tablespaces
postgres@jack-virtual-machine:/pgdata/tablespaces$ pwd
/pgdata/tablespaces
postgres@jack-virtual-machine:/pgdata/tablespaces$
```

Figure 6.53: *Creating a directory for tablespaces*

The **CREATE TABLESPACE** command, as shown in *Figure* 6.54, is used to create table spaces if the public table space is not to be used.

```
postgres=#
postgres=# CREATE TABLESPACE tb_dvd
OWNER postgres
LOCATION '/pgdata/tablespaces';
CREATE TABLESPACE
postgres=#
postgres=# \db
          List of tablespaces
    Name    |  Owner   |      Location
------------+----------+---------------------
 pg_default | postgres |
 pg_global  | postgres |
 tb_dvd     | postgres | /pgdata/tablespaces
(3 rows)

postgres=# \q
postgres@jack-virtual-machine:/pgdata/tablespaces$ mkdir /pgdata/sporttb
postgres@jack-virtual-machine:/pgdata/tablespaces$ psql
psql (16.1 (Ubuntu 16.1-1.pgdg22.04+1))
Type "help" for help.

postgres=# CREATE TABLESPACE tbs_sportdb
OWNER postgres
LOCATION '/pgdata/sporttb';
CREATE TABLESPACE
postgres=#
postgres=# \db
          List of tablespaces
    Name     |  Owner   |      Location
-------------+----------+---------------------
 pg_default  | postgres |
 pg_global   | postgres |
 tb_dvd      | postgres | /pgdata/tablespaces
 tbs_sportdb | postgres | /pgdata/sporttb
(4 rows)
```

Figure 6.54: *Creating a directory for tablespaces*

Taking the base backup

Taking the base backup requires the **-D** option again, but this time points to the path to the location where the backup is to be stored. The command to take the base backup is as shown:

pg_basebackup -Ft -p5432 -v -D /pgbackup/basebackup

a. **Pg_basebackup**

b. **-Ft** the format is tar

c. **-p5432** we are backing up the default cluster

d. **-v** verbose (will initiate a detailed report)

e. **-D /pgbackup/basebackup** directory is the location where the backup will be stored. This basebackup directory can be any other name of choice.

```
postgres@jack-virtual-machine:/pgdata/tablespaces$ cd
postgres@jack-virtual-machine:~$ cd /pgbackup/
postgres@jack-virtual-machine:/pgbackup$ ls
archivedir
postgres@jack-virtual-machine:/pgbackup$ pg_basebackup -Ft -p5432 -v -D /pgbackup/basebackup
pg_basebackup: initiating base backup, waiting for checkpoint to complete
pg_basebackup: checkpoint completed
pg_basebackup: write-ahead log start point: 0/D000028 on timeline 1
pg_basebackup: starting background WAL receiver
pg_basebackup: created temporary replication slot "pg_basebackup_19333"
pg_basebackup: write-ahead log end point: 0/D000100
pg_basebackup: waiting for background process to finish streaming ...
pg_basebackup: syncing data to disk ...
pg_basebackup: renaming backup_manifest.tmp to backup_manifest
pg_basebackup: base backup completed
postgres@jack-virtual-machine:/pgbackup$
postgres@jack-virtual-machine:/pgbackup$ ls
archivedir  basebackup
postgres@jack-virtual-machine:/pgbackup$
```

Figure 6.55: *Taking the basebackup using the pg_basebackup command*

The following screen shows the files that have been backed up:

```
postgres@jack-virtual-machine:/pgbackup/basebackup$ ls -l
total 54312
-rw------- 1 postgres postgres     1536 Jan 23 15:40 19540.tar
-rw------- 1 postgres postgres     1536 Jan 23 15:40 19541.tar
-rw------- 1 postgres postgres   192036 Jan 23 15:40 backup_manifest
-rw------- 1 postgres postgres 38630912 Jan 23 15:40 base.tar
-rw------- 1 postgres postgres 16778752 Jan 23 15:40 pg_wal.tar
postgres@jack-virtual-machine:/pgbackup/basebackup$
```

Figure 6.56: *Showing the content of the recently taken backup*

Figure 6.56 shows four tar files created by the **base backup** command. The top two files correspond to the two tablespaces. The **base.tar** file contains the backup of the **pg_default** tablespace along with all cluster directories and files. The **pg_wal.tar** file includes the backup of all the WAL files required for this backup set to remain consistent. These four tar files together form a complete and consistent backup. This set of files can be used to restore the backup to a different location and initiate a new cluster, as it contains all the essential files.

Restoring the Base Backup to Create a New Cluster

Let us create a location for our new cluster under **pgdata** and call it **restorebasebkp**. This is where we will be restoring our **base backup** set to start a new cluster.

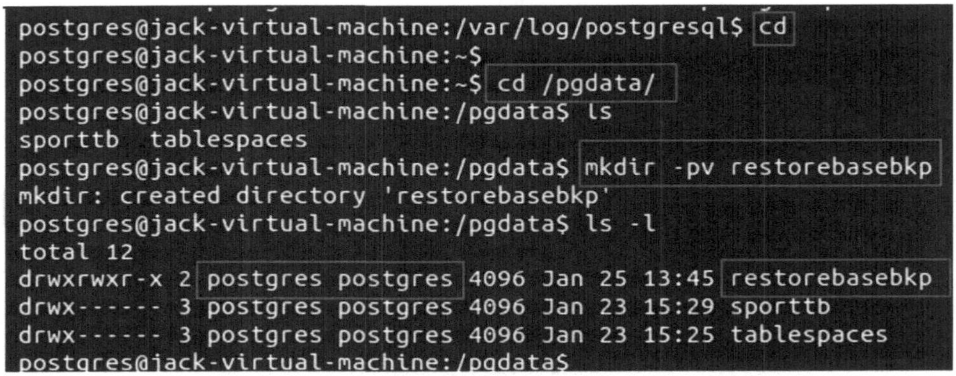

```
postgres@jack-virtual-machine:/var/log/postgresql$ cd
postgres@jack-virtual-machine:~$
postgres@jack-virtual-machine:~$ cd /pgdata/
postgres@jack-virtual-machine:/pgdata$ ls
sporttb  tablespaces
postgres@jack-virtual-machine:/pgdata$ mkdir -pv restorebasebkp
mkdir: created directory 'restorebasebkp'
postgres@jack-virtual-machine:/pgdata$ ls -l
total 12
drwxrwxr-x 2 postgres postgres 4096 Jan 25 13:45 restorebasebkp
drwx------ 3 postgres postgres 4096 Jan 23 15:29 sporttb
drwx------ 3 postgres postgres 4096 Jan 23 15:25 tablespaces
postgres@jack-virtual-machine:/pgdata$
```

Figure 6.57: *Creating the restorebasebkp directory for the restoration*

With the restoration directory now set up, we will proceed to restore the tar files into it individually. Before executing the commands, ensure that the restoration directory is set as your current directory. This means you should navigate to that location using the **cd** command.

```
cd
cd /pgdata/restorebasebkp
```

The command to restore the files is:

```
tar xvf  /pgbackup/basebackup/base.tar
tar xvf  /pgbackup/basebackup/pg_wal.tar
tar xvf  /pgbackup/basebackup/19540.tar
tar xvf  /pgbackup/basebackup/19541.tar
```

```
postgres@jack-virtual-machine:/pgdata/restorebasebkp$ cd
postgres@jack-virtual-machine:~$
postgres@jack-virtual-machine:~$ cd /pgdata/restorebasebkp/
postgres@jack-virtual-machine:/pgdata/restorebasebkp$ tar xvf  /pgbackup/basebackup/base.tar
base/1/2704
base/1/3601_vm
base/1/3502
base/1/13424
base/1/2690
base/1/3764_fsm
base/1/2679
base/1/3609
base/1/13426_vm
pg_xact/
pg_xact/0000
pg_replslot/
pg_logical/
pg_logical/mappings/
pg_logical/replorigin_checkpoint
pg_logical/snapshots/
pg_wal/
./pg_wal/archive_status/
global/pg_control
postgres@jack-virtual-machine:/pgdata/restorebasebkp$
```

Figure 6.58: *base.tar restored successfully*

The following **ls -l** command will display all the files that have been extracted:

```
postgres@jack-virtual-machine:/pgdata/restorebasebkp$
postgres@jack-virtual-machine:/pgdata/restorebasebkp$ ls -l
total 84
-rw------- 1 postgres postgres  225 Jan 23 15:40 backup_label
drwx------ 6 postgres postgres 4096 Jan 23 15:40 base
drwx------ 2 postgres postgres 4096 Jan 25 13:56 global
drwx------ 2 postgres postgres 4096 Dec 30 17:00 pg_commit_ts
drwx------ 2 postgres postgres 4096 Dec 30 17:00 pg_dynshmem
drwx------ 4 postgres postgres 4096 Jan 23 15:40 pg_logical
drwx------ 4 postgres postgres 4096 Dec 30 17:00 pg_multixact
drwx------ 2 postgres postgres 4096 Dec 30 17:00 pg_notify
drwx------ 2 postgres postgres 4096 Jan 23 15:40 pg_replslot
drwx------ 2 postgres postgres 4096 Dec 30 17:00 pg_serial
drwx------ 2 postgres postgres 4096 Dec 30 17:00 pg_snapshots
drwx------ 2 postgres postgres 4096 Jan 23 14:55 pg_stat
drwx------ 2 postgres postgres 4096 Dec 30 17:00 pg_stat_tmp
drwx------ 2 postgres postgres 4096 Dec 30 17:00 pg_subtrans
drwx------ 2 postgres postgres 4096 Jan 23 15:29 pg_tblspc
drwx------ 2 postgres postgres 4096 Dec 30 17:00 pg_twophase
-rw------- 1 postgres postgres    3 Dec 30 17:00 PG_VERSION
drwx------ 3 postgres postgres 4096 Jan 25 13:56 pg_wal
drwx------ 2 postgres postgres 4096 Dec 30 17:00 pg_xact
-rw------- 1 postgres postgres   88 Dec 30 17:00 postgresql.auto.conf
-rw------- 1 postgres postgres   48 Jan 23 15:40 tablespace_map
postgres@jack-virtual-machine:/pgdata/restorebasebkp$
```

Figure 6.59: *Files restored from base.tar*

The following commands will extract the tablespaces and the **pw_wal** files:

```
postgres@jack-virtual-machine:/pgdata/restorebasebkp$
postgres@jack-virtual-machine:/pgdata/restorebasebkp$ tar xvf  /pgbackup/basebackup/19540.tar
PG_16_202307071/
postgres@jack-virtual-machine:/pgdata/restorebasebkp$ tar xvf  /pgbackup/basebackup/19541.tar
PG_16_202307071/
postgres@jack-virtual-machine:/pgdata/restorebasebkp$ tar xvf  /pgbackup/basebackup/pg_wal.tar
00000001000000000000000D
postgres@jack-virtual-machine:/pgdata/restorebasebkp$
```

Figure 6.60: *All Files restored*

With the physical backup of the entire database cluster and files restored using the tar xvf command, we can now proceed to do point-in-time recovery.

We are going to stop the database cluster using the following command:

```
sudo systemctl status postgresql@16-main
sudo systemctl stop postgresql@16-main
```

```
postgres@jack-virtual-machine:/pgdata/restorebasebkp$ exit
logout
jack@jack-virtual-machine: $ sudo systemctl status postgresql@16-main
● postgresql@16-main.service - PostgreSQL Cluster 16-main
     Loaded: loaded (/lib/systemd/system/postgresql@.service; enabled-runti
     Active: active (running) since Sat 2024-02-03 19:20:52 EST; 11min ago
    Process: 968 ExecStart=/usr/bin/pg_ctlcluster --skip-systemctl-redirect
   Main PID: 1033 (postgres)
      Tasks: 7 (limit: 2139)
     Memory: 31.0M
        CPU: 732ms
     CGroup: /system.slice/system-postgresql.slice/postgresql@16-main.servic
             └─1033 /usr/lib/postgresql/16/bin/postgres -D /var/lib/postgres
```

Figure 6.61: *How to check the status of Postgresql cluster*

The next screenshot shows how to stop **postgresql** and also check its status.

```
jack@jack-virtual-machine: $ sudo systemctl stop postgresql@16-main
jack@jack-virtual-machine: $ sudo systemctl status postgresql@16-main
○ postgresql@16-main.service - PostgreSQL Cluster 16-main
     Loaded: loaded (/lib/systemd/system/postgresql@.service; enabled-ru
     Active: inactive (dead) since Sat 2024-02-03 19:32:28 EST; 3s ago
    Process: 968 ExecStart=/usr/bin/pg_ctlcluster --skip-systemctl-redir
    Process: 2694 ExecStop=/usr/bin/pg_ctlcluster --skip-systemctl-redir
   Main PID: 1033 (code=exited, status=0/SUCCESS)
        CPU: 878ms
```

Figure 6.62: *How to check the stop Postgresql cluster*

We will now move the **pg_wal** directory to a different location, as it might contain unarchived WAL files that are important for recovery. The mv command will move the **pg_wal** from **/var/lib/postgresql/16/main** to **~/** location.

Note that you can use any location.

```
sudo mv /var/lib/postgresql/16/main/pg_wal ~/
```

```
jack@jack-virtual-machine: $ sudo mv /var/lib/postgresql/16/main/pg_wal ~/
jack@jack-virtual-machine: $
jack@jack-virtual-machine: $ ls ~/
```

Figure 6.63: *Moving the pg_wal to a different location*

Now that the database has been stopped, we will remove all the files in PostgreSQL's data directory.

Incident 4: Imagine that this was at work, and a colleague accidentally removed or deleted the postgresql main directory.

```
sudo rm -rf /var/lib/postgresql/16/main
```

```
jack@jack-virtual-machine: $
jack@jack-virtual-machine: $ sudo rm -rf /var/lib/postgresql/16/main
```

Figure 6.64: *The main directory has been removed accidentally*

We will recreate the **main** directory and assign ownership to Postgres using the following commands:

Note that this newly created main directory is empty.

```
sudo mkdir /var/lib/postgresql/16/main
sudo chown postgres:postgres /var/lib/postgresql/16/main
```

```
jack@jack-virtual-machine: $
jack@jack-virtual-machine: $ sudo mkdir /var/lib/postgresql/16/main
jack@jack-virtual-machine: $
jack@jack-virtual-machine: $ sudo chown postgres:postgres /var/lib/postgresql/13/main
chown: cannot access '/var/lib/postgresql/13/main': No such file or directory
jack@jack-virtual-machine: $ sudo chown postgres:postgres /var/lib/postgresql/16/main
jack@jack-virtual-machine: $
```

Figure 6.65: *Creating an empty main directory and assigning ownership to postgres*

```
jack@jack-virtual-machine: $
jack@jack-virtual-machine: $ sudo su - postgres
postgres@jack-virtual-machine:~$ cd /var/lib/postgresql/16/main/
postgres@jack-virtual-machine:~/16/main$ ls
postgres@jack-virtual-machine:~/16/main$
```

Figure 6.66: *Shows that the main directory is empty*

We will now copy all the files from the physical restore location (**/pgdata/restorebasebkp**) to the new empty main directory using the **cp** command:

```
jack@jack-virtual-machine: $ sudo su - postgres
[sudo] password for jack:
postgres@jack-virtual-machine:~$ cp -a /pgdata/restorebasebkp/. /var/lib/postgresql/13/main/
cp: cannot create directory '/var/lib/postgresql/13/main/': No such file or directory
postgres@jack-virtual-machine:~$ cp -a /pgdata/restorebasebkp/. /var/lib/postgresql/16/main/
postgres@jack-virtual-machine:~$
postgres@jack-virtual-machine:~$ ls -al /var/lib/postgresql/16/main/
total 16480
drwxrwxr-x 20 postgres postgres     4096 Jan 25 14:13 .
drwxr-xr-x  3 postgres postgres     4096 Feb  3 19:43 ..
-rw-------  1 postgres postgres 16777216 Jan 23 15:40 00000001000000000000000D
-rw-------  1 postgres postgres      225 Jan 23 15:40 backup_label
drwx------  6 postgres postgres     4096 Jan 23 15:40 base
drwx------  2 postgres postgres     4096 Jan 25 13:56 global
drwx------  2 postgres postgres     4096 Jan 23 15:29 PG_16_202307071
drwx------  2 postgres postgres     4096 Dec 30 17:00 pg_commit_ts
drwx------  2 postgres postgres     4096 Dec 30 17:00 pg_dynshmem
drwx------  4 postgres postgres     4096 Jan 23 15:40 pg_logical
drwx------  4 postgres postgres     4096 Dec 30 17:00 pg_multixact
drwx------  2 postgres postgres     4096 Dec 30 17:00 pg_notify
drwx------  2 postgres postgres     4096 Jan 23 15:40 pg_replslot
drwx------  2 postgres postgres     4096 Dec 30 17:00 pg_serial
drwx------  2 postgres postgres     4096 Dec 30 17:00 pg_snapshots
drwx------  2 postgres postgres     4096 Jan 23 14:55 pg_stat
drwx------  2 postgres postgres     4096 Dec 30 17:00 pg_stat_tmp
drwx------  2 postgres postgres     4096 Dec 30 17:00 pg_subtrans
drwx------  2 postgres postgres     4096 Jan 23 15:29 pg_tblspc
drwx------  2 postgres postgres     4096 Dec 30 17:00 pg_twophase
-rw-------  1 postgres postgres        3 Dec 30 17:00 PG_VERSION
drwx------  3 postgres postgres     4096 Jan 25 13:56 pg_wal
drwx------  2 postgres postgres     4096 Dec 30 17:00 pg_xact
-rw-------  1 postgres postgres       88 Dec 30 17:00 postgresql.auto.conf
-rw-------  1 postgres postgres       48 Jan 23 15:40 tablespace_map
postgres@jack-virtual-machine:~$
```

Figure 6.67: *Copying backup files to the main directory*

The outdated WAL files in the **pg_wal** directory will be substituted with the WAL files that were relocated before the incident. This is necessary because some files may not have been archived prior to the server shutdown.

We will first remove the **pg_wal** file from the **/var/lib/postgresql/16/main** directory and then copy the previously relocated **pg_wal** back as follows:

```
postgres@jack-virtual-machine:~$ rm -rf /var/lib/postgresql/16/main/pg_wal
postgres@jack-virtual-machine:~$
```

```
postgres@jack-virtual-machine:~$ exit
logout
jack@jack-virtual-machine: $ sudo cp -a ~/pg_wal /var/lib/postgresql/16/main/pg_wal
[sudo] password for jack:
```

Figure 6.68: *Remove pg_wal and replace it with the previously relocated one*

Now that the data directory has been properly restored, we need to configure the recovery settings to ensure that all archived WAL files are accurately restored. The recovery settings can be found in the **postgresql.conf** file, located in the **/etc/**

postgresql/16/main/ directory. We will use the vi text editor to make the necessary changes.

Inside the file, locate the **restore_command** and remove the **#** symbol at the beginning of the line to uncomment it, similar to how we uncommented the **archive_command.** The **archive_command** directs the files to the archive location, while the **restore_ command** will retrieve the files back.

```
postgres@jack-virtual-machine:~$
postgres@jack-virtual-machine:~$ vi /etc/postgresql/16/main/postgresql.conf

restore_command = 'cp /pgbackup/archivedir/%f %p'          # command to use to restore
                                # placeholders: %p = path of file to restore
                                #               %f = file name only
                                # e.g. 'cp /mnt/server/archivedir/%f %p'
#archive_cleanup_command = ''   # command to execute at every restartpoint
#recovery_end_command = ''      # command to execute at completion of recovery

# - Recovery Target -

# Set these only when performing a targeted recovery.

recovery_target = 'immediate'           # 'immediate' to end recovery as soon as a
                                # consistent state is reached
                                # (change requires restart)
#recovery_target_name = ''      # the named restore point to which recovery will proceed
                                # (change requires restart)
#recovery_target_time = ''      # the time stamp up to which recovery will proceed
                                # (change requires restart)
#recovery_target_xid = ''       # the transaction ID up to which recovery will proceed
                                # (change requires restart)
#recovery_target_lsn = ''       # the WAL LSN up to which recovery will proceed
                                # (change requires restart)
#recovery_target_inclusive = on # Specifies whether to stop:
                                # just after the specified recovery target (on)
                                # just before the recovery target (off)
                                # (change requires restart)
```

Figure 6.69: *How to set the restore command with path to archive directory*

Point In Time Recovery (PITR)

Point-in-Time Recovery (PITR) allows the replaying of WAL files to a specific point in time, which is crucial for disaster recovery. For PITR to function correctly, two key elements are required: a complete base backup and WAL files stored in the archive directory. **pg_basebackup** is typically used for this purpose.

The options **recovery_target** and **restore_command** are essential for configuring PITR. The **restore_command** specifies the location from which WAL files should be copied or replayed. Both **restore_command** and **archive_command** can use the same placeholders.

The **recovery_target_time** parameter instructs the server on when to stop the recovery or replay process, halting as soon as the specified timestamp is reached. The

recovery_target_inclusive parameter controls whether to stop replaying WALs just before or after the **recovery_target_time** is reached.

```
restore_command = 'cp /pgbackup/archivedir/%f %p'
recovery_target= 'immediate'
recovery_target_inclusive = on
```

We will specify the recovery target as **'immediate'**. The recovery target defines the point at which the database cluster will attempt to recover before exiting recovery mode. This target is typically a timestamp but can also be a transaction ID or log sequence number.

It is important to note that the recovery target should be set to a point after the physical backup was taken. PostgreSQL must be instructed to start in recovery mode before initiating this process, which is done by creating an empty file named **recovery.signal** in the data directory of the cluster.

Use the **touch** command to create the file:

```
[sudo] password for jack:
jack@jack-virtual-machine: $ sudo touch /var/lib/postgresql/16/main/recovery.signal
jack@jack-virtual-machine: $
```

Figure 6.70: *Creating a recovery.signal file*

We are going to restart the database cluster by running:

```
sudo systemctl start postgresql@16-main
```

```
jack@jack-virtual-machine: $ sudo systemctl start postgresql@16-main
Job for postgresql@16-main.service failed because the service did not take the steps required by its unit configuration.
See "systemctl status postgresql@16-main.service" and "journalctl -xeu postgresql@16-main.service" for details.
jack@jack-virtual-machine: $ cd /var/lib/postgresql/
jack@jack-virtual-machine:                   $ ls
```

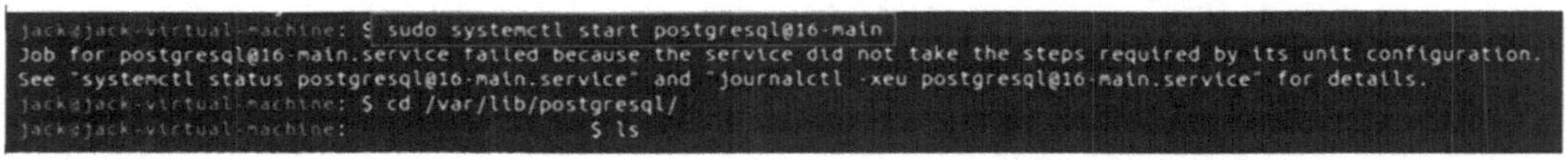

Figure 6.71: *Postgresql failed to start*

Next, we are going to troubleshoot to find out why postgresql did not start. We will change directory (**cd**) to where Postgresql logs are located. Postgresql logs are found in **/var/log/postgresql**. The log file is called **postgresql-16-main.log.** To see the content of this file, we use the cat command as shown:

```
cd  /var/log/postgresql
cat postgresql-16-main.log
```

At the bottom of the log file, it is shown that it failed due to wrong permissions. It even gives suggestions on what the permissions on the file should be.

Figure 6.72: Troubleshooting using Postgresql logs.

Next, we will grant the right permissions on the main directory using the **chmod** command, as earlier discussed.

```
sudo chmod 0700 /var/lib/postgresql/16/main
```

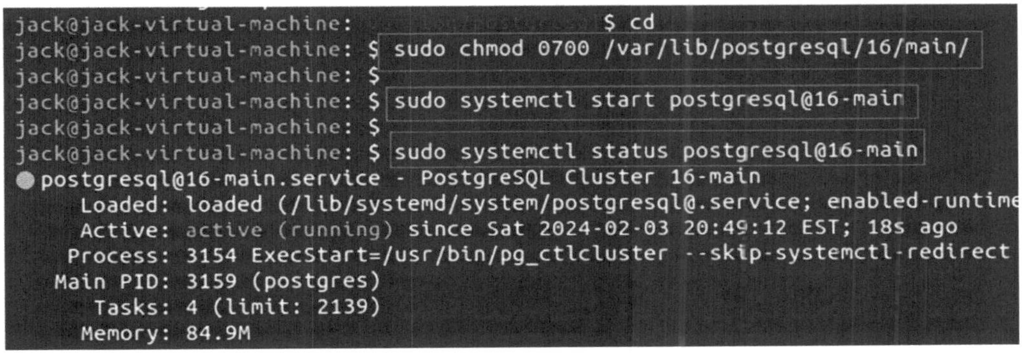

Figure 6.73: Granting the right permissions on the file and starting Postgresql

Conclusion

In this chapter, we explored various backup and restoration techniques in PostgreSQL. We also highlighted the significance of Point-in-Time Recovery (PITR) for data recovery following a server crash or failure, which might cause an invalid or corrupted data directory and result in downtime.

One of the critical responsibilities of a PostgreSQL DBA is maintaining up-to-date full backups along with WAL files. This practice ensures a smooth restoration process. The duration of the recovery process depends on the recency of the full backup and the number of WAL files generated since then.

In the next chapter, we will be looking at Replication and High Availability (HA).

Chapter 7
Replication and High Availability

Introduction

In this chapter, learners will learn and understand the different ways to copy and make available the database data in more than one location. Different replication methods will be discussed and demonstrated in this chapter. Take note that replication is one type of High Availability (HA).

They will become proficient in setting up the different replication methods and making organization data highly available in different locations for security, reliability, and availability.

Replication is the process of copying data from one server to one or more servers.

The **primary or master** server is where the data is being copied from, and the **replica or standby** is where the data is being copied to.

Write and read operations are carried out in the primary server, but the replicas accept only read transactions. The primary server and the replica servers are collectively called a database cluster.

Structure

In this chapter, the following topics will be covered:

- Streaming Replication (also called Physical Replication)
- Logical Replication
- Failover (Promoting a replica to a master using `pg_promote`)

Streaming Replication

In streaming replication, the primary server streams Write Ahead Logs (WAL) files to the connected replicas as the WALs are generated. This process does not wait for the WAL file to be filled.

Write Ahead Log

A WAL is a way of writing both actions and changes to a transaction log before any updates or changes to the database. This method is used to ensure data integrity and reliability in the database. The WALs are very useful in recovering the database and its data to a particular point in time (point-in-time recovery) in the case of a database crash or failure resulting in data loss. As there are changes to the database, the WAL files keep building up. The purpose of this is to be able to recreate a database from scratch following a crash by replaying the built-up WAL files. The directory where these WAL logs are stored is called **pg_wal**.

Once streaming replication is configured, the WAL files are streamed and replicated in real time from the primary server to the replica(s).

Replication Parameters

The following are the parameters that need to be configured in **postgresql.conf** file for the replication to work effectively:

- **Wal_level:** The quantity of information that can be written the WAL file is determined by the **wal_level** parameter. To enable replication, this parameter must either be set to **logical** or **replica**. The default is a replica.
- **Max_wal_senders:** This parameter shows the maximum number of active WAL sender processes. This value is set depending on the number of replicas to be configured in the replication.
- **Wal_kep_segments:** The **pg_wal** directory maintains the minimum number of WAL segments needed to support the replication. Here we specify the minimum number of WAL segments.
- **Hot_standby:** This parameter must be switched to on for the replica to accept read-only queries.
- **Max_standby_streaming_delay:** This must be set to a value appropriate for the replication. This specifies the maximum delay in applying the WAL files on the replica to prevent it from lagging too far behind the master server.
- **Synchronous_standby_names:** Here we set the names of the replicas that we want to use for synchronous replication. These replicas will confirm that they have received the WAL data before the transaction is committed in the primary.

- **`Listen_addresses`:** Here we list the IP addresses that PostgreSQL will receive incoming connections. Make sure the primary and replica IP addresses are configured if they are not the default local hosts (**`localhost`**). If this is set to * means all IP addresses are allowed.
- **`Port`:** This ensures that the port PostgreSQL listens on for TCP connections is open and accessible between the replicas and the primary servers.
- **`Archive_mode`:** This is to be switched on to enable WAL archiving for the replication. WAL files will be sent to the designated archive location once this is enabled.
- archive_mode = on
- **`Archive_command`:** This is configured to point to the path of the location where the WAL files will be archived.

  ```
  archive_command = 'cp %p /path/to/archive_location/%f'
  ```

For practical purposes, we are going to install two different virtual machines, **primary** and **replica.** We are then going to install PostgreSQL version 15 on them. It is worth noting that the primary and replica must contain the same version of PostgreSQL.

Refer to *Chapter 2, Installing and Initializing PostgreSQL* on how to create virtual machines and install PostgreSQL.

Two virtual machines, namely a **replica** and **primary server** have been created on the VMware workstation, as shown in *Figure 7.1*.

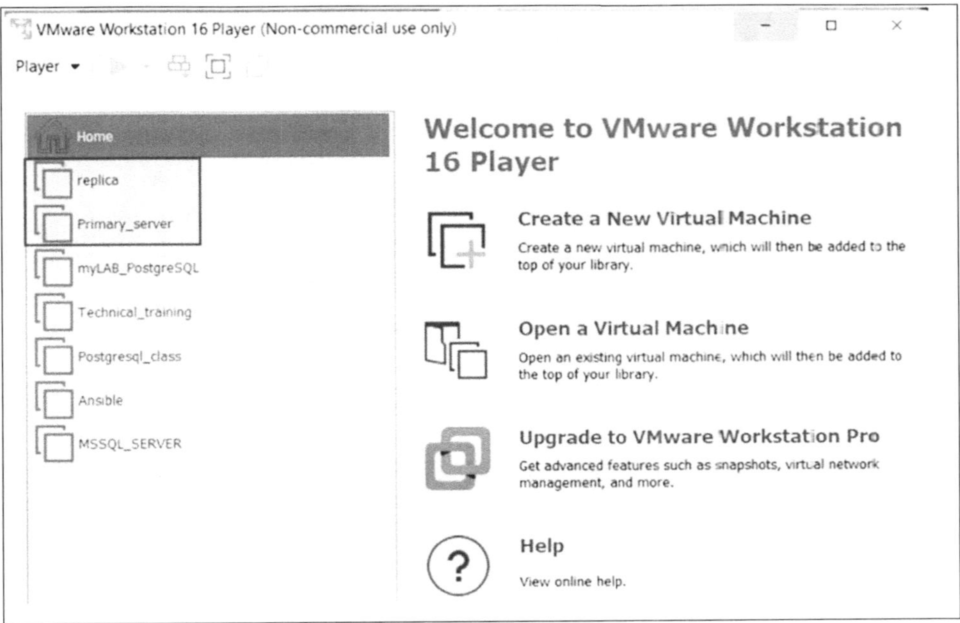

Figure 7.1: *Primary and replica virtual machines*

The next step is to log in to the primary and replica machines and install PostgreSQL version 15. Refer to *Chapter 2* on how to install PostgreSQL.

```
jack@jack-primary-server: $ sudo su - postgres
postgres@jack-primary-server:~$ psql -U postgres
psql (15.6 (Ubuntu 15.6-1.pgdg22.04+1))
Type "help" for help.

postgres=# show data_directory
postgres-# ;
       data_directory
-----------------------------
 /var/lib/postgresql/15/main
(1 row)
```

Figure 7.2: *Primary server with PostgreSQL v15 installed*

The next image shows the replica server with PostgreSQL version 15 installed.

```
jack@jack-replica-server: $ sudo su - postgres
postgres@jack-replica-server:~$ psql -U postgres
psql (15.6 (Ubuntu 15.6-1.pgdg22.04+1))
Type "help" for help.

postgres=# show config_file;
              config_file
-----------------------------------------
 /etc/postgresql/15/main/postgresql.conf
(1 row)
```

Figure 7.3: *Replica server with PostgreSQL v15 installed*

Next, we are going to use the **mkdir** command to create the archive directory where the WAL files will be sent.

```
jack@jack-primary-server: $ sudo mkdir /var/lib/archivedir
[sudo] password for jack:
jack@jack-primary-server: $ sudo chown postgres:postgres /var/lib/archivedir
jack@jack-primary-server: $
```

Figure 7.4: *WAL archive directory created in the primary server*

To find the IP address of your server, use the following commands. If the first one does not return the IP address, then the second command will install the appropriate tools. Then, re-run the first command to obtain the IP address.

This process should be repeated on the replica as well to obtain its IP address.

```
ifconfig -a
sudo apt install net-tools
ifconfig -a
```

```
jack@jack-primary-server: $ ifconfig -a
Command 'ifconfig' not found, but can be installed with:
sudo apt install net-tools
jack@jack-primary-server: $ sudo apt install net-tools
[sudo] password for jack:
Reading package lists... Done
Building dependency tree... Done
Reading state information... Done
The following NEW packages will be installed:
  net-tools
0 upgraded, 1 newly installed, 0 to remove and 382 not upgraded.
Need to get 204 kB of archives.
After this operation, 819 kB of additional disk space will be used.
Get:1 http://ca.archive.ubuntu.com/ubuntu jammy/main amd64 net-tools amd64 1.60
Fetched 204 kB in 0s (496 kB/s)
Selecting previously unselected package net-tools.
(Reading database ... 201092 files and directories currently installed.)
Preparing to unpack .../net-tools_1.60+git20181103.0eebece-1ubuntu5_amd64.deb
Unpacking net-tools (1.60+git20181103.0eebece-1ubuntu5) ...
Setting up net-tools (1.60+git20181103.0eebece-1ubuntu5) ...
Processing triggers for man-db (2.10.2-1) ...
jack@jack-primary-server: $ ifconfig -a
ens33: flags=4163<UP,BROADCAST,RUNNING,MULTICAST>  mtu 1500
        inet 192.168.222.166  netmask 255.255.255.0  broadcast 192.168.222.255
        inet6 fe80::47fc:3b5b:81b7:2251  prefixlen 64  scopeid 0x20<link>
        ether 00:0c:29:fa:0e:40  txqueuelen 1000  (Ethernet)
        RX packets 535619  bytes 806584910 (806.5 MB)
```

Figure 7.5: *How to obtain an IP address*

Primary server IP = **192.168.222.166**

Replica server IP = **192.168.222.167**

Setting Up the Primary Server for Streaming Replication

Step 1: Configure the Primary Database to Accept Connections

In this first step, we will configure the primary database listen addresses in **postgresql.conf** file to allow your replica database(s) to connect. By default, the listening address for PostgreSQL connections is **localhost (127.0.0.1)**.

```
Vi /etc/postgresql/15/main/postgresql.conf
```

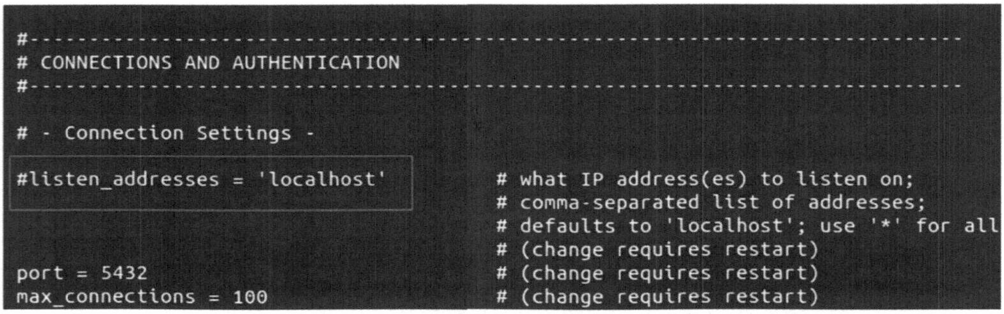

```
#------------------------------------------------------------------------------
# CONNECTIONS AND AUTHENTICATION
#------------------------------------------------------------------------------

# - Connection Settings -

#listen_addresses = 'localhost'         # what IP address(es) to listen on;
                                        # comma-separated list of addresses;
                                        # defaults to 'localhost'; use '*' for all
                                        # (change requires restart)
port = 5432                             # (change requires restart)
max_connections = 100                   # (change requires restart)
```

Figure 7.6: *Listen address before configuration*

To change this, you will first edit the **listen_addresses** configuration parameter on the primary database.

a. Change the listen address to primary **ip** OR change to '*****'. Execute the **ifconfig -a** to get the primary IP address. For the sake of practice, we will change to *****, which will allow all connections.

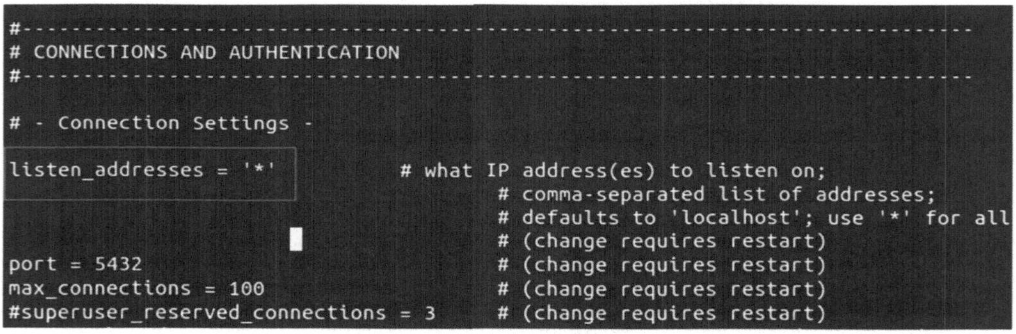

```
#------------------------------------------------------------------------------
# CONNECTIONS AND AUTHENTICATION
#------------------------------------------------------------------------------

# - Connection Settings -

listen_addresses = '*'          # what IP address(es) to listen on;
                                        # comma-separated list of addresses;
                                        # defaults to 'localhost'; use '*' for all
                                        # (change requires restart)
port = 5432                             # (change requires restart)
max_connections = 100                   # (change requires restart)
#superuser_reserved_connections = 3     # (change requires restart)
```

Figure 7.7: *Listen address configured to receive all connections*

Make the following changes in the **postgresql.conf** file of the primary server:

```
listen_addresses = '*'
archive_mode = on
max_wal_senders = 3
max_wal_size = 1GB
wal_level = replica
hot_standby = on
archive_command = 'cp %p /var/lib/archivedir/%f'
```

Next is to create a replicator user with replication privileges as shown.

```
CREATE ROLE replicator_user WITH REPLICATION PASSWORD
'repuser123' LOGIN;
# USE ANY REPLICATOR NAME AND PASSWORD OF YOUR CHOICE
```

```
postgres@jack-primary-server:~$ psql
psql (15.6 (Ubuntu 15.6-1.pgdg22.04+1))
Type "help" for help.

postgres=# CREATE ROLE replicator_user WITH REPLICATION PASSWORD 'repuser123' LOGIN;
CREATE ROLE
postgres=#
postgres=#
```

Figure 7.8: *Creating a replicator user*

Step 2: Next step is to edit the `pg_hba.conf` file to authenticate the replicator user as shown.

```
vi /etc/postgresql/15/main/pg_hba.conf
#host    replication     replicator_name    your-replica-IP/32
scram-sha-256
host    replication     replicator_user    192.168.222.167/32
scram-sha-256
```

```
# DO NOT DISABLE!
# If you change this first entry you will need to make sure that the
# database superuser can access the database using some other method.
# Noninteractive access to all databases is required during automatic
# maintenance (custom daily cronjobs, replication, and similar tasks).
#
# Database administrative login by Unix domain socket
local   all             postgres                                peer

# TYPE  DATABASE        USER            ADDRESS                 METHOD

# "local" is for Unix domain socket connections only
local   all             all                                     peer
# IPv4 local connections:
host    all             all             127.0.0.1/32            scram-sha-256
# IPv6 local connections:
host    all             all             ::1/128                 scram-sha-256
# Allow replication connections from localhost, by a user with the
# replication privilege.
local   replication     all                                     peer
host    replication     all             127.0.0.1/32            scram-sha-256
host    replication     all             ::1/128                 scram-sha-256
host    replication     replicator_user    192.168.222.167/32   scram-sha-256
```

Figure 7.9: *Replicator user configured in pg_hba.conf*

Step 3: The next step is to restart the primary server for the changes above to take effect. Restart using the following command:

```
sudo systemctl restart postgresql@15-main
```

```
postgres@jack-primary-server:~$ exit
logout
jack@jack-primary-server: $ sudo systemctl restart postgresql@15-main
jack@jack-primary-server: $
jack@jack-primary-server: $ sudo systemctl status postgresql@15-main
● postgresql@15-main.service - PostgreSQL Cluster 15-main
     Loaded: loaded (/lib/systemd/system/postgresql@.service; enabled-ru
     Active: active (running) since Sun 2024-02-11 14:49:20 EST; 15s ago
    Process: 8812 ExecStart=/usr/bin/pg_ctlcluster --skip-systemctl-redi
   Main PID: 8818 (postgres)
      Tasks: 7 (limit: 2142)
     Memory: 20.8M
        CPU: 375ms
     CGroup: /system.slice/system-postgresql.slice/postgresql@15-main.se
             └─8818 /usr/lib/postgresql/15/bin/postgres -D /var/lib/post
```

Figure 7.10: *How to restart the server and check the status*

Setting Up the Replica Server for Streaming Replication

Backup the data directory of the replica. You can use the show **data_directory** from within the database to locate the data directory and **cp** command to make a copy.

```
SHOW data_directory;
/var/lib/postgresql/15/main
```

```
jack@jack-replica-server: $ sudo su - postgres
postgres@jack-replica-server:~$ psql
psql (15.6 (Ubuntu 15.6-1.pgdg22.04+1))
Type "help" for help.

postgres=# show data_directory;
       data_directory
-----------------------------
 /var/lib/postgresql/15/main
(1 row)

postgres=# \q
postgres@jack-replica-server:~$ cp -r /var/lib/postgresql/15/main /var/lib/postgresql/15/main_old
postgres@jack-replica-server:~$
```

Figure 7.11: *Taking a backup of the data directory*

We will be deleting the content of the data directory, as the data directory of the primary will replicate to the replica. Use one of the following commands to completely remove the data directory's content.

```
sudo -u postgres rm -r /var/lib/postgresql/15/main/*
```

Or

```
rm -r /var/lib/postgresql/15/main/*
```

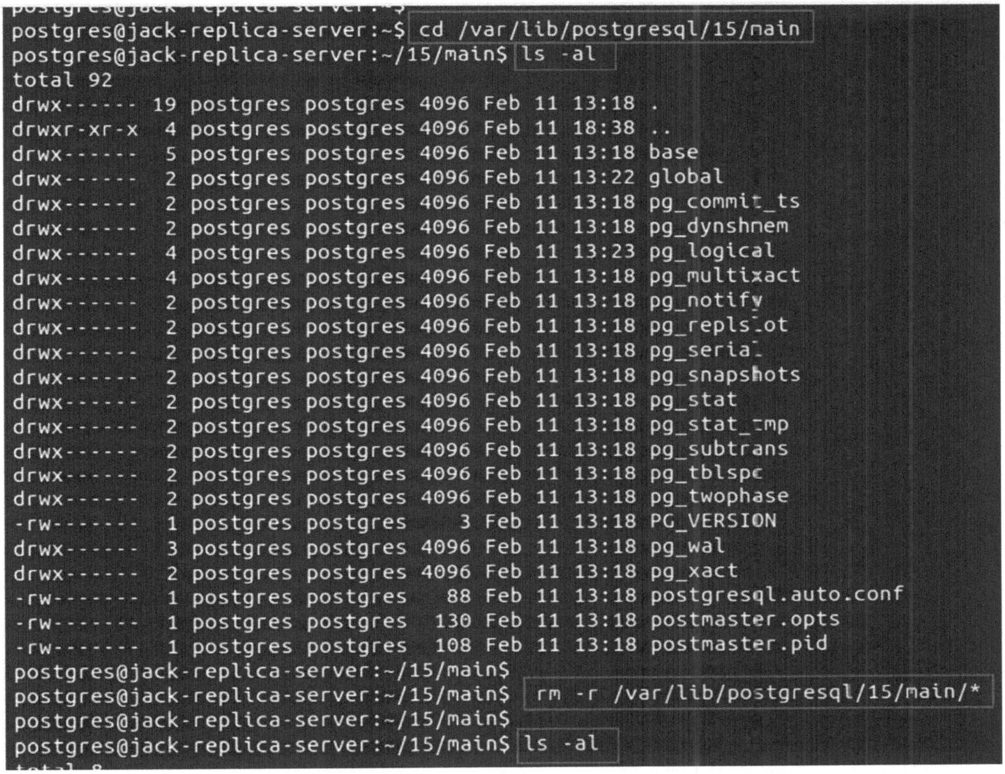

```
postgres@jack-replica-server:~$ cd /var/lib/postgresql/15/main
postgres@jack-replica-server:~/15/main$ ls -al
total 92
drwx------ 19 postgres postgres 4096 Feb 11 13:18 .
drwxr-xr-x  4 postgres postgres 4096 Feb 11 18:38 ..
drwx------  5 postgres postgres 4096 Feb 11 13:18 base
drwx------  2 postgres postgres 4096 Feb 11 13:22 global
drwx------  2 postgres postgres 4096 Feb 11 13:18 pg_commit_ts
drwx------  2 postgres postgres 4096 Feb 11 13:18 pg_dynshmem
drwx------  4 postgres postgres 4096 Feb 11 13:23 pg_logical
drwx------  4 postgres postgres 4096 Feb 11 13:18 pg_multixact
drwx------  2 postgres postgres 4096 Feb 11 13:18 pg_notify
drwx------  2 postgres postgres 4096 Feb 11 13:18 pg_replslot
drwx------  2 postgres postgres 4096 Feb 11 13:18 pg_serial
drwx------  2 postgres postgres 4096 Feb 11 13:18 pg_snapshots
drwx------  2 postgres postgres 4096 Feb 11 13:18 pg_stat
drwx------  2 postgres postgres 4096 Feb 11 13:18 pg_stat_tmp
drwx------  2 postgres postgres 4096 Feb 11 13:18 pg_subtrans
drwx------  2 postgres postgres 4096 Feb 11 13:18 pg_tblspc
drwx------  2 postgres postgres 4096 Feb 11 13:18 pg_twophase
-rw-------  1 postgres postgres    3 Feb 11 13:18 PG_VERSION
drwx------  3 postgres postgres 4096 Feb 11 13:18 pg_wal
drwx------  2 postgres postgres 4096 Feb 11 13:18 pg_xact
-rw-------  1 postgres postgres   88 Feb 11 13:18 postgresql.auto.conf
-rw-------  1 postgres postgres  130 Feb 11 13:18 postmaster.opts
-rw-------  1 postgres postgres  108 Feb 11 13:18 postmaster.pid
postgres@jack-replica-server:~/15/main$
postgres@jack-replica-server:~/15/main$ rm -r /var/lib/postgresql/15/main/*
postgres@jack-replica-server:~/15/main$
postgres@jack-replica-server:~/15/main$ ls -al
```

***Figure 7.12**: Deleting all the data from the data directory*

We are now going to take a base backup of the primary server from the replica using the replicator user we created in the primary server. The syntax is:

```
# pg_basebackup -h primary-ip-addr -p 5432 -U replicator_user -D
/var/lib/postgresql/15/main/ -Fp -Xs -R
```

- **-h** or **--host**: host specify the IP address of the primary server.
- **-D**: This option specifies the directory where the base backup will be stored.
- **-U**: This option specifies the user (replicator user) to connect to the PostgreSQL server.
- **-Fp**: indicates that it is a plain format (This is the default format).
- **-Xs**: WAL files are streaming while the **pg_basebackup** is in progress
- **-R**: This option to write a **recovery.conf** file is useful for setting up a standby server in the target directory.

```
pg_basebackup -h 192.168.222.166 -p 5432 -U replicator_user -D /
var/lib/postgresql/15/main/ -Fp -      Xs -R
```

```
postgres@jack-replica-server:~/15/main$ cd
postgres@jack-replica-server:~$ pg_basebackup -h 192.168.222.166 -p 5432 -U replicator_user -D /var/lib/postgresql/15/main/ -Fp -Xs -R
Password:
postgres@jack-replica-server:~$
postgres@jack-replica-server:~$
```

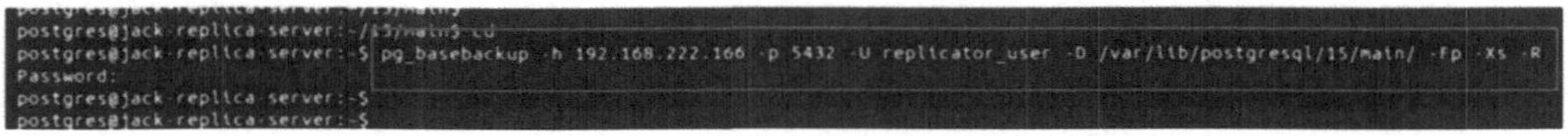

Figure 7.13: *Taking a base backup of primary from the replica to start the replication*

The next screen lists all the files from the **pg_basebackup**.

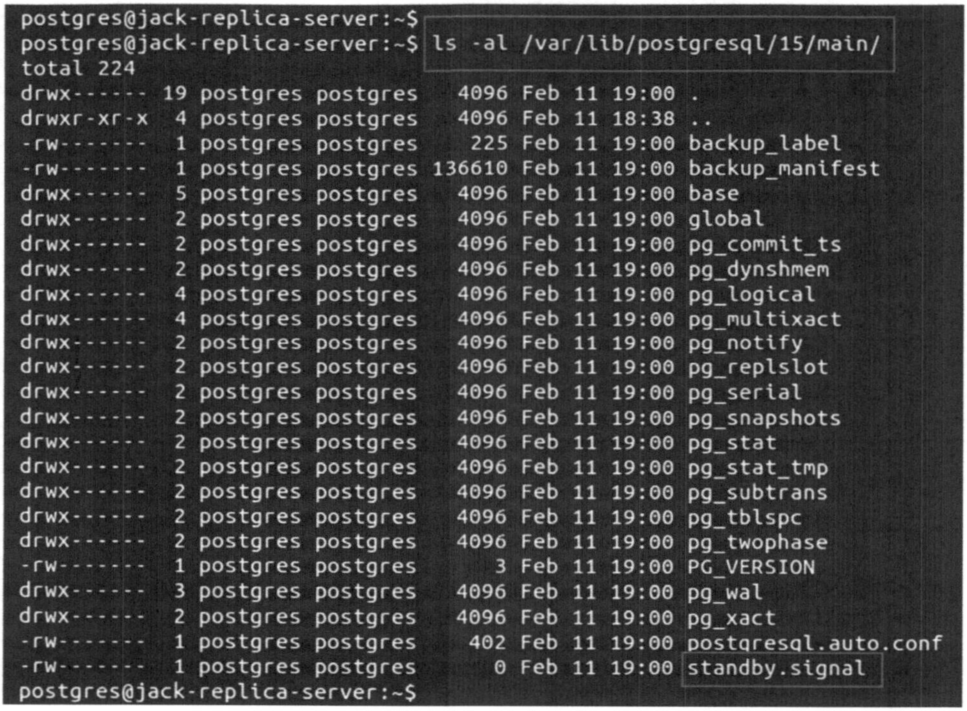

```
postgres@jack-replica-server:~$
postgres@jack-replica-server:~$ ls -al /var/lib/postgresql/15/main/
total 224
drwx------ 19 postgres postgres   4096 Feb 11 19:00 .
drwxr-xr-x  4 postgres postgres   4096 Feb 11 18:38 ..
-rw-------  1 postgres postgres    225 Feb 11 19:00 backup_label
-rw-------  1 postgres postgres 136610 Feb 11 19:00 backup_manifest
drwx------  5 postgres postgres   4096 Feb 11 19:00 base
drwx------  2 postgres postgres   4096 Feb 11 19:00 global
drwx------  2 postgres postgres   4096 Feb 11 19:00 pg_commit_ts
drwx------  2 postgres postgres   4096 Feb 11 19:00 pg_dynshmem
drwx------  4 postgres postgres   4096 Feb 11 19:00 pg_logical
drwx------  4 postgres postgres   4096 Feb 11 19:00 pg_multixact
drwx------  2 postgres postgres   4096 Feb 11 19:00 pg_notify
drwx------  2 postgres postgres   4096 Feb 11 19:00 pg_replslot
drwx------  2 postgres postgres   4096 Feb 11 19:00 pg_serial
drwx------  2 postgres postgres   4096 Feb 11 19:00 pg_snapshots
drwx------  2 postgres postgres   4096 Feb 11 19:00 pg_stat
drwx------  2 postgres postgres   4096 Feb 11 19:00 pg_stat_tmp
drwx------  2 postgres postgres   4096 Feb 11 19:00 pg_subtrans
drwx------  2 postgres postgres   4096 Feb 11 19:00 pg_tblspc
drwx------  2 postgres postgres   4096 Feb 11 19:00 pg_twophase
-rw-------  1 postgres postgres      3 Feb 11 19:00 PG_VERSION
drwx------  3 postgres postgres   4096 Feb 11 19:00 pg_wal
drwx------  2 postgres postgres   4096 Feb 11 19:00 pg_xact
-rw-------  1 postgres postgres    402 Feb 11 19:00 postgresql.auto.conf
-rw-------  1 postgres postgres      0 Feb 11 19:00 standby.signal
postgres@jack-replica-server:~$
```

Figure 7.14: *Shows the data of the primary server in the replica data directory*

The **standby.signal** file shows that the server will start up as a hot standby.

The next step is to restart the replica server using the following command:

```
sudo systemctl restart postgresql@15-main
```

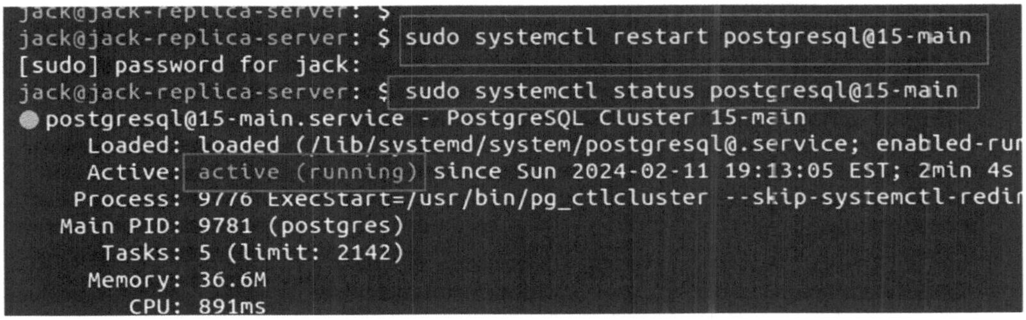

```
jack@jack-replica-server: $
jack@jack-replica-server: $ sudo systemctl restart postgresql@15-main
[sudo] password for jack:
jack@jack-replica-server: $ sudo systemctl status postgresql@15-main
● postgresql@15-main.service - PostgreSQL Cluster 15-main
     Loaded: loaded (/lib/systemd/system/postgresql@.service; enabled-ru
     Active: active (running) since Sun 2024-02-11 19:13:05 EST; 2min 4s
    Process: 9776 ExecStart=/usr/bin/pg_ctlcluster --skip-systemctl-redi
   Main PID: 9781 (postgres)
      Tasks: 5 (limit: 2142)
     Memory: 36.6M
        CPU: 891ms
```

***Figure 7.15**: Restarting the replica server and checking the status*

Now that the replica has been restarted, it shows that it is replicating.

The next step is to confirm that there is replication between the primary server and the replica. Connect to the primary database and run the following command:

```
sudo -u postgres
psql
SELECT client_addr, state FROM pg_stat_replication;
```

```
jack@jack-primary-server: $
jack@jack-primary-server: $ sudo su - postgres
[sudo] password for jack:
postgres@jack-primary-server:~$ psql -U postgres
psql (15.6 (Ubuntu 15.6-1.pgdg22.04+1))
Type "help" for help.

postgres=# SELECT client_addr, state FROM pg_stat_replication;
   client_addr   |   state
-----------------+-----------
 192.168.222.167 | streaming
(1 row)
```

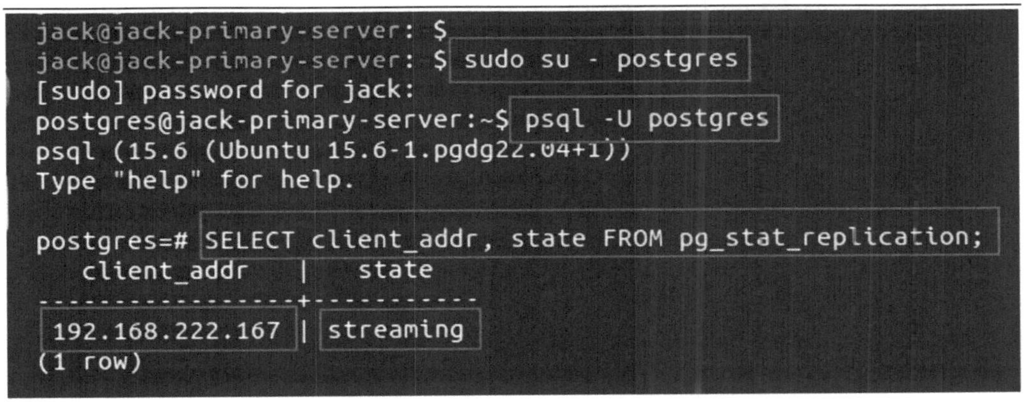

***Figure 7.16**: Checking the replication status on the primary server*

From the replica, run the following command to check the replication status:

```
SELECT * FROM pg_stat_wal_receiver;
```

```
jack@jack-replica-server: $ sudo su - postgres
postgres@jack-replica-server:~$
postgres@jack-replica-server:~$ psql -U postgres
psql (15.6 (Ubuntu 15.6-1.pgdg22.04+1))
Type "help" for help.

postgres=# SELECT * FROM pg_stat_wal_receiver;
postgres=# \x
Expanded display is on.
postgres=# SELECT * FROM pg_stat_wal_receiver;
-[ RECORD 1 ]---------+-----------------------------------------
---------------------------------------------------------------
pid                   | 9785
status                | streaming
receive_start_lsn     | 0/3000000
receive_start_tli     | 1
written_lsn           | 0/3000148
flushed_lsn           | 0/3000148
received_tli          | 1
last_msg_send_time    | 2024-02-11 19:28:35.004671-05
last_msg_receipt_time | 2024-02-11 19:28:35.014431-05
latest_end_lsn        | 0/3000148
latest_end_time       | 2024-02-11 19:13:03.393892-05
slot_name             |
sender_host           | 192.168.222.166
sender_port           | 5432
conninfo              | user=replicator_user password=**
```

Figure 7.17: *Checking the replication status on the replica server*

The status on both the primary server and the replica server shows that they are streaming, which means that the primary is replicating onto the replica. Any data inserted into the primary or any changes will automatically appear in the replica. This is one aspect of high availability.

```
select pg_is_in_recovery();
```

We will create a database called department in the primary server. We will then confirm that it has also been created in the replica server.

```
jack@jack-primary-server: $ sudo su - postgres
[sudo] password for jack:
postgres@jack-primary-server:~$ psql -U postgres
psql (15.6 (Ubuntu 15.6-1.pgdg22.04+1))
Type "help" for help.

postgres=# SELECT client_addr, state FROM pg_stat_replication;
   client_addr   |   state
-----------------+-----------
 192.168.222.167 | streaming
(1 row)

postgres=# CREATE DATABASE department;
CREATE DATABASE
postgres=# \l
                                                         List of databases
    Name     |  Owner   | Encoding |   Collate    |    Ctype     | ICU
-------------+----------+----------+--------------+--------------+----
 department  | postgres | UTF8     | en_CA.UTF-8  | en_CA.UTF-8  |
 postgres    | postgres | UTF8     | en_CA.UTF-8  | en_CA.UTF-8  |
 template0   | postgres | UTF8     | en_CA.UTF-8  | en_CA.UTF-8  |
             |          |          |              |              |
 template1   | postgres | UTF8     | en_CA.UTF-8  | en_CA.UTF-8  |
             |          |          |              |              |
(4 rows)
```

Figure 7.18: *Creating database department in the primary server*

The department database created in the primary server is also available in the replica, as shown in *Figure 7.19*.

```
jack@jack-replica-server: $
jack@jack-replica-server: $ sudo su - postgres
[sudo] password for jack:
postgres@jack-replica-server:~$ psql -U postgres
psql (15.6 (Ubuntu 15.6-1.pgdg22.04+1))
Type "help" for help.

postgres=# \l
                                                List
    Name     |  Owner   | Encoding |   Collate    |    Ct
-------------+----------+----------+--------------+------
 department  | postgres | UTF8     | en_CA.UTF-8  | en_CA
 postgres    | postgres | UTF8     | en_CA.UTF-8  | en_CA
 template0   | postgres | UTF8     | en_CA.UTF-8  | en_CA
             |          |          |              |
 template1   | postgres | UTF8     | en_CA.UTF-8  | en_CA
```

Figure 7.19: *Database department has been replicated in the replica server*

There are two modes of streaming reapplication, namely, asynchronous and synchronous modes. Asynchronous replication is copying data from primary to replica after the data is written to the primary storage, while in asynchronous replication, data is being written to the primary and replica at the same time.

Logical Replication

In PostgreSQL, logical replication is a feature that is used to copy data synchronously from one database to another. This process is necessary for load balancing and high availability. It is also used when you want to replicate data in real-time without server downtime.

With logical replication, we need to create a publication on the source database. A subscription is created on the destination database. It is based on the **publisher and subscriber model.** On the publisher, which is the logical-master server, we define the data to be replicated, while on the subscriber, which is the logical-replica server, the replicated data are applied. Here, the subscriber connects to the publisher and applies the replicated data.

With logical replication, we can replicate specific columns, rows, and tables based on our set replication rules. This is very useful in situations where we need to replicate only a subset of our data.

Pre-requisite: Create two virtual machines, namely, **logical master and logical replica**. Install PostgreSQL version 14 on these machines. Obtain their IP addresses.

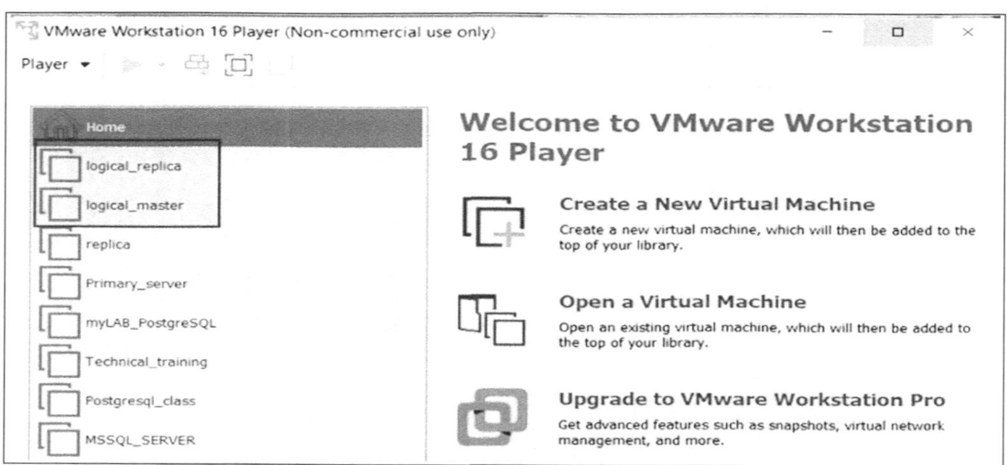

Figure 7.20: *Two virtual machines created for logical replication*

To obtain the IP addresses, run the following commands on the Ubuntu Linux OS. When run the first time, you will be prompted to install the net-tools with the second command shown as follows. This must be done on both the master and replica servers.

1. **`ifconfig -a`**
2. **`sudo apt install net-tools`**
3. **`ifconfig -a`**
4. **`master IP = 192.168.222.195 (master IP address)`**
5. **`replica IP = 192.168.222.196 (replica IP address)`**

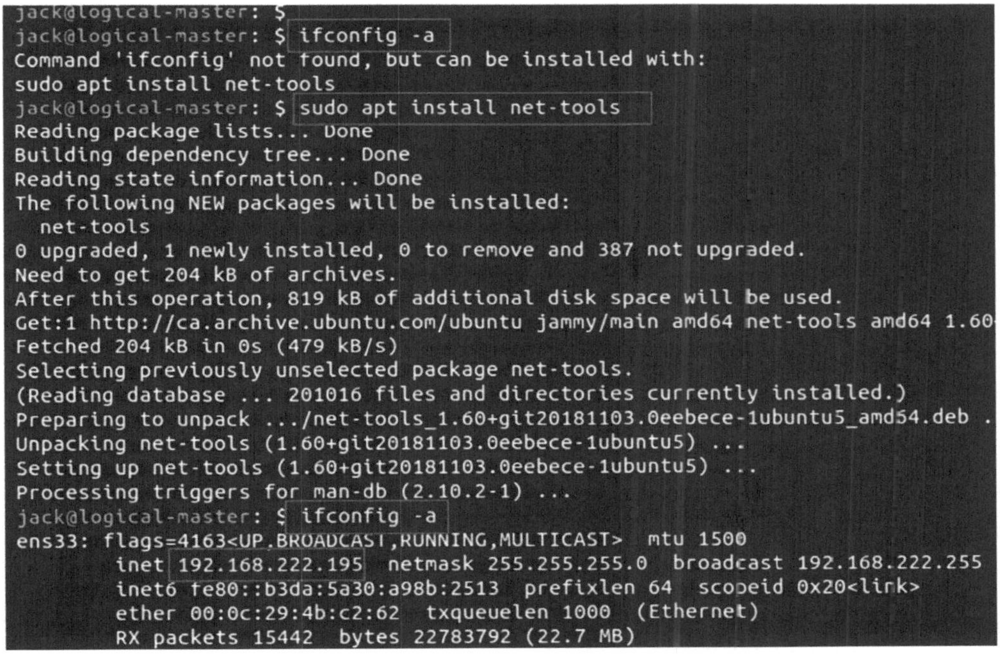

```
jack@logical-master: $
jack@logical-master: $ ifconfig -a
Command 'ifconfig' not found, but can be installed with:
sudo apt install net-tools
jack@logical-master: $ sudo apt install net-tools
Reading package lists... Done
Building dependency tree... Done
Reading state information... Done
The following NEW packages will be installed:
  net-tools
0 upgraded, 1 newly installed, 0 to remove and 387 not upgraded.
Need to get 204 kB of archives.
After this operation, 819 kB of additional disk space will be used.
Get:1 http://ca.archive.ubuntu.com/ubuntu jammy/main amd64 net-tools amd64 1.60
Fetched 204 kB in 0s (479 kB/s)
Selecting previously unselected package net-tools.
(Reading database ... 201016 files and directories currently installed.)
Preparing to unpack .../net-tools_1.60+git20181103.0eebece-1ubuntu5_amd64.deb .
Unpacking net-tools (1.60+git20181103.0eebece-1ubuntu5) ...
Setting up net-tools (1.60+git20181103.0eebece-1ubuntu5) ...
Processing triggers for man-db (2.10.2-1) ...
jack@logical-master: $ ifconfig -a
ens33: flags=4163<UP,BROADCAST,RUNNING,MULTICAST>  mtu 1500
        inet 192.168.222.195  netmask 255.255.255.0  broadcast 192.168.222.255
        inet6 fe80::b3da:5a30:a98b:2513  prefixlen 64  scopeid 0x20<link>
        ether 00:0c:29:4b:c2:62  txqueuelen 1000  (Ethernet)
        RX packets 15442  bytes 22783792 (22.7 MB)
```

Figure 7.21: *How to obtain the IP address of a machine*

Logical-master Configuration

Step 1: Configure **`Postgreql.conf`** file and **`pg_hba.conf`** file for Logical Replication.

On logical master, using a vi editor, open **`/etc/postgresql/14/main/postgresql.conf`**, which is the main PostgreSQL server configuration file as shown below:

```
sudo vi /etc/postgresql/14/main/postgresql.conf
```

Or

```
vi /etc/postgresql/14/main/postgresql.conf
```

Find the following line:

```
#listen_addresses = 'localhost'         # what IP address(es) to listen on;
```

Uncomment the line by removing the #. To enable connections on the private network, add your **logical_master_ip_address** after uncommenting.

```
listen_addresses = 'localhost, 192.168.222.195'
```

Adding to the loopback interface, makes the **logical-master** virtual machine listen for incoming connections on the private network.

Next, uncomment the following line, and change it to logical:

```
#wal_level = replica                    # minimal, replica, or logical
```

wal_level = logical

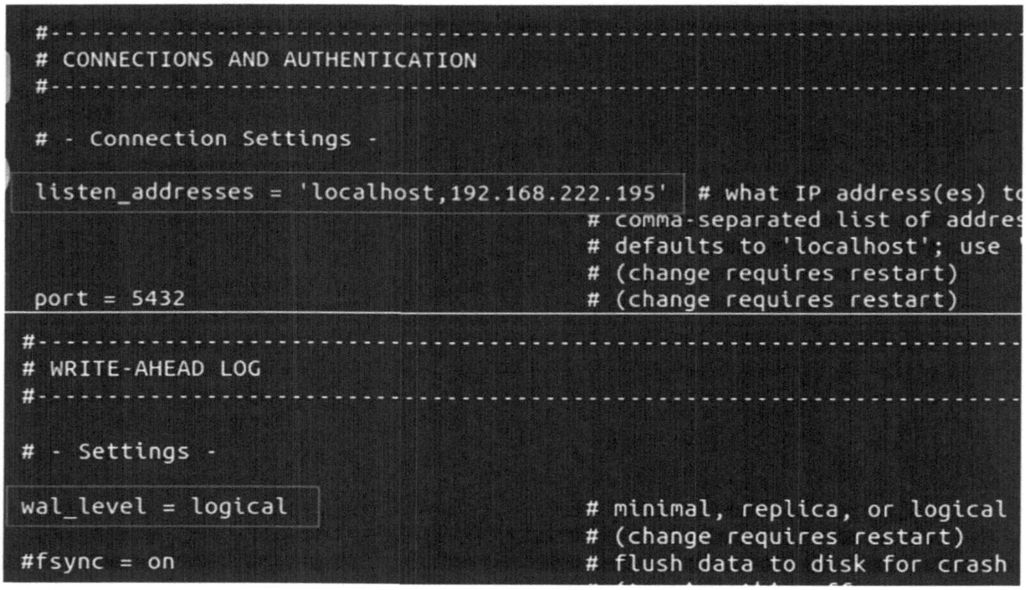

```
#------------------------------------------------------------------------------
# CONNECTIONS AND AUTHENTICATION
#------------------------------------------------------------------------------

# - Connection Settings -

listen_addresses = 'localhost,192.168.222.195'  # what IP address(es) t
                                        # comma-separated list of addre
                                        # defaults to 'localhost'; use
                                        # (change requires restart)
port = 5432                             # (change requires restart)
#------------------------------------------------------------------------------
# WRITE-AHEAD LOG
#------------------------------------------------------------------------------

# - Settings -

wal_level = logical                     # minimal, replica, or logical
                                        # (change requires restart)
#fsync = on                             # flush data to disk for crash
```

***Figure 7.22: logical-master** postgresql.conf configuration for logical replication*

pg_hba.conf File

Next, we will edit **/etc/postgresql/14/main/pg_hba.conf**. This file controls the hosts, and authenticates access to databases:

```
sudo vi /etc/postgresql/14/main/pg_hba.conf
```

Or

```
vi /etc/postgresql/14/main/pg_hba.conf
```

Copy and paste the following line with the replica ip address:

```
#host         all            all           logical_replica_ip_address/32
scram-sha-256

host          all            all           192.168.222.196/32
scram-sha-256
```

This line will enable incoming connections from the logical replica server.

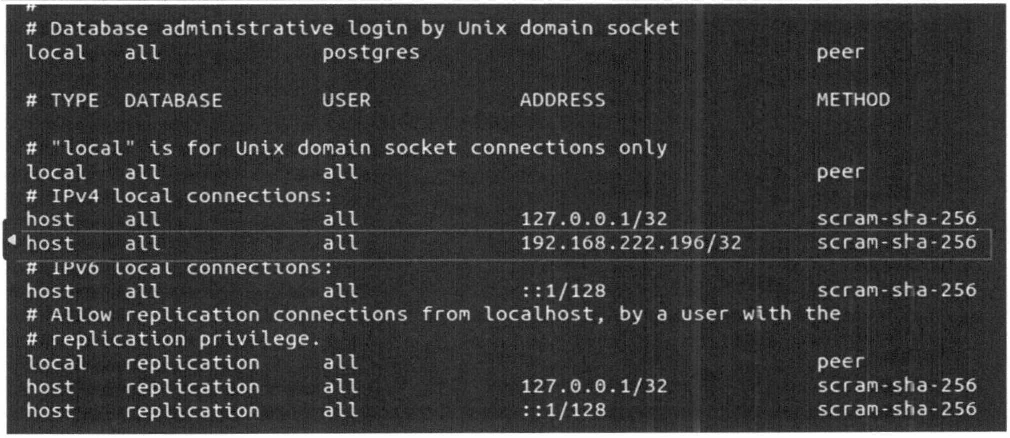

```
# Database administrative login by Unix domain socket
local   all             postgres                                peer

# TYPE  DATABASE        USER            ADDRESS                 METHOD

# "local" is for Unix domain socket connections only
local   all             all                                     peer
# IPv4 local connections:
host    all             all             127.0.0.1/32            scram-sha-256
host    all             all             192.168.222.196/32      scram-sha-256
# IPv6 local connections:
host    all             all             ::1/128                 scram-sha-256
# Allow replication connections from localhost, by a user with the
# replication privilege.
local   replication     all                                     peer
host    replication     all             127.0.0.1/32            scram-sha-256
host    replication     all             ::1/128                 scram-sha-256
```

Figure 7.23: Authenticating replica *IP address in master's pg_hba.conf file*

Finally, restart the PostgreSQL services on the logical-master server for the changes to take effect.

```
sudo systemctl restart postgresql
```

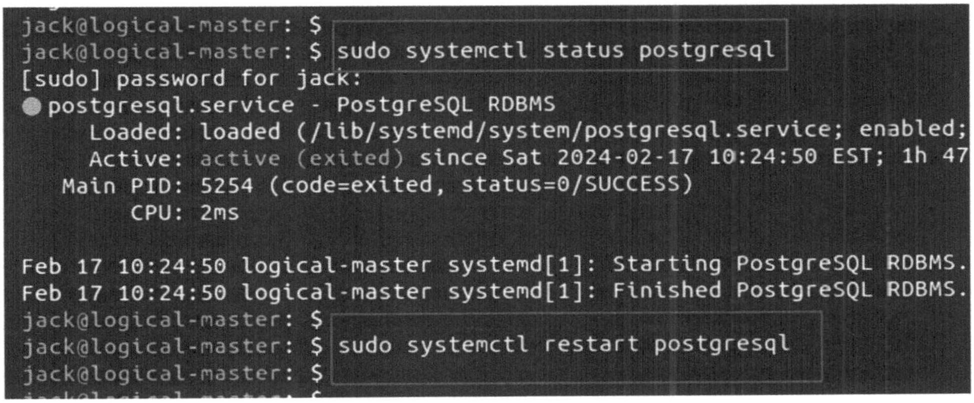

```
jack@logical-master: $
jack@logical-master: $ sudo systemctl status postgresql
[sudo] password for jack:
● postgresql.service - PostgreSQL RDBMS
     Loaded: loaded (/lib/systemd/system/postgresql.service; enabled;
     Active: active (exited) since Sat 2024-02-17 10:24:50 EST; 1h 47
   Main PID: 5254 (code=exited, status=0/SUCCESS)
        CPU: 2ms

Feb 17 10:24:50 logical-master systemd[1]: Starting PostgreSQL RDBMS.
Feb 17 10:24:50 logical-master systemd[1]: Finished PostgreSQL RDBMS.
jack@logical-master: $
jack@logical-master: $ sudo systemctl restart postgresql
jack@logical-master: $
```

Figure 7.24: *Restarting PostgreSQL server after changes to configuration files*

We are next going to set the firewall rules to allow traffic from logical-replica to port **5432** on logical-master server:

sudo ufw allow from **logical_replica _ip_address** to any port **5432**

```
sudo ufw enable

sudo ufw allow from 192.168.222.196 to any port 5432

sudo ufw status
```

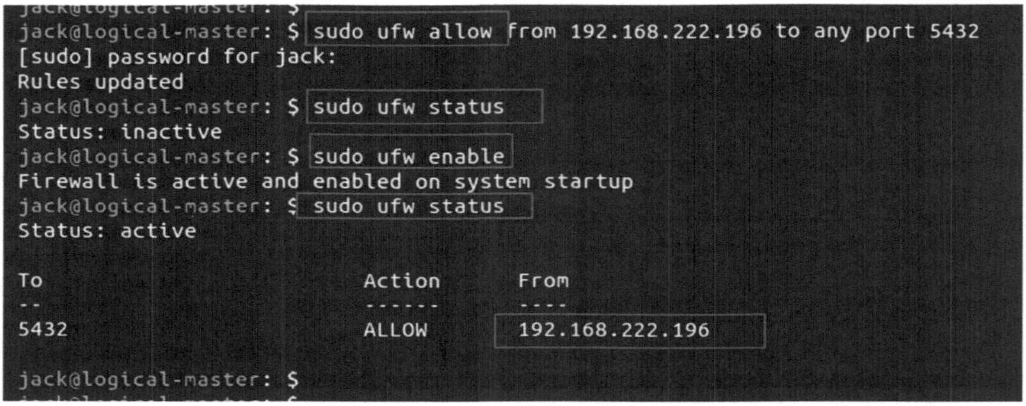

```
jack@logical-master: $ sudo ufw allow from 192.168.222.196 to any port 5432
[sudo] password for jack:
Rules updated
jack@logical-master: $ sudo ufw status
Status: inactive
jack@logical-master: $ sudo ufw enable
Firewall is active and enabled on system startup
jack@logical-master: $ sudo ufw status
Status: active

To                         Action      From
--                         ------      ----
5432                       ALLOW       192.168.222.196

jack@logical-master: $
```

***Figure** 7.25: Enable and allow firewall in master server*

Step 2: Set up a User, Database, and Table for replication

On both master and replica: First, open the psql prompt as the postgres user. On both the logical-master and logical-replica, create products database with the sales table as shown. This will be used to test that our replication is well set up and master server is replicating to the replica server.

```
CREATE DATABASE products;

\c products;

CREATE TABLE sales
(
    id SERIAL,
    product_name varchar(50),
    sales_price DECIMAL,
    CONSTRAINT sales_pkey PRIMARY KEY (id)
);
```

```
jack@logical-master: $ sudo su - postgres
[sudo] password for jack:
postgres@logical-master:~$
postgres@logical-master:~$ psql -U postgres
psql (14.11 (Ubuntu 14.11-1.pgdg22.04+1))
Type "help" for help.

postgres=# CREATE DATABASE products;
CREATE DATABASE
postgres=#
postgres=# \c products
You are now connected to database "products" as user "postgres".
products=#
products=# CREATE TABLE sales
(
    id SERIAL,
    product_name varchar(50),
    sales_price DECIMAL,
    CONSTRAINT sales_pkey PRIMARY KEY (id)
);
CREATE TABLE
products=#
```

Figure 7.26: *Creating the products database and table sales*

On logical-master: We are going to create a new user role with the **REPLICATION** option and a login password.

It is very important to assign the **REPLICATION** attribute to any role used for replication. We will call our user Aryan. You can give your user a different name and different password.

```
CREATE ROLE aryan WITH REPLICATION LOGIN PASSWORD 'aryan123£';

GRANT ALL PRIVILEGES ON DATABASE products TO aryan;

GRANT ALL PRIVILEGES ON ALL TABLES IN SCHEMA public TO aryan;
```

```
products=#
products=# CREATE ROLE aryan WITH REPLICATION LOGIN PASSWORD 'aryan123£';
CREATE ROLE
products=#
products=# GRANT ALL PRIVILEGES ON DATABASE products TO aryan;
GRANT
products=#
products=# GRANT ALL PRIVILEGES ON ALL TABLES IN SCHEMA public TO aryan;
GRANT
products=#
products=#
```

Figure 7.27: *Creating a replication user with password and privileges*

Step 3: Set up a Publication on logical master server

On logical-master server, create a publication called **`my_publication`**:

```
\c products;

CREATE PUBLICATION sales_publication;

ALTER PUBLICATION sales_publication ADD TABLE sales;
```

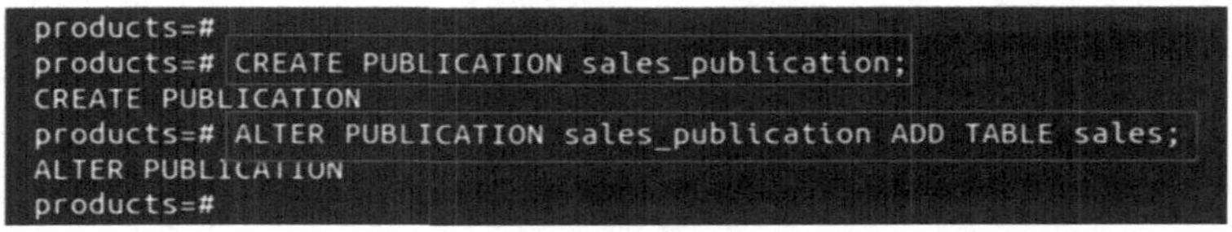

```
products=#
products=# CREATE PUBLICATION sales_publication;
CREATE PUBLICATION
products=# ALTER PUBLICATION sales_publication ADD TABLE sales;
ALTER PUBLICATION
products=#
```

Figure 7.28: *Creating a publication on master server*

Step 4: Creating a Subscription on a replica server

On **`logical_replica`**

```
\c products

CREATE SUBSCRIPTION sales_subscription CONNECTION
'host=192.168.222.195 port=5432 password= aryan123£ user=aryan
dbname=products' PUBLICATION sales_publication;
```

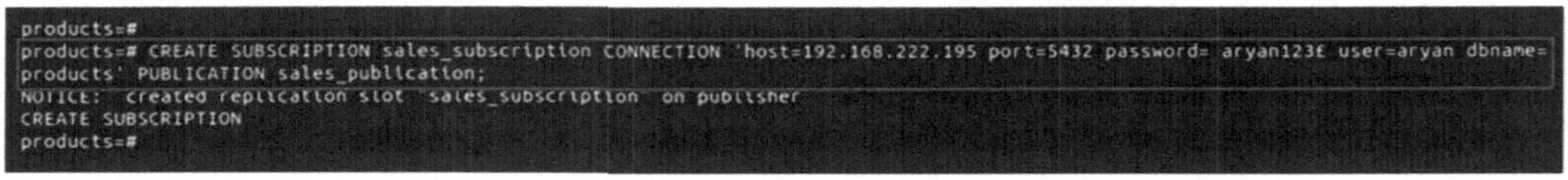

```
products=#
products=# CREATE SUBSCRIPTION sales_subscription CONNECTION 'host=192.168.222.195 port=5432 password= aryan123£ user=aryan dbname=
products' PUBLICATION sales_publication;
NOTICE:  created replication slot "sales_subscription" on publisher
CREATE SUBSCRIPTION
products=#
```

Figure 7.29: *Creating a subscription on replica server*

Note: *The IP in the subscription is logical-master server.*

Step 5: Testing the replication between master and replica.

On logical-master server, insert the following data into the sales table:

```
\c products

INSERT INTO sales (product_name, sales_price)
VALUES ('Spoons', 3.60), ('Tea Mug', 7.80),
('plate', 3.90);
```

```
products=#
products=# INSERT INTO sales (product_name, sales_price)
VALUES ('Spoons', 3.60), ('Tea Mug', 7.80),
('plate', 3.90);
INSERT 0 3
products=#
```

Figure 7.30: *Inserting data into the sales table in the master server*

On logical-replica, run the following query to fetch all the entries on this table:

```
\c products

SELECT * FROM sales;
```

```
postgres@logical-replica:~$
postgres@logical-replica:~$ psql -U postgres
psql (14.11 (Ubuntu 14.11-1.pgdg22.04+1))
Type "help" for help.

postgres=# \c products;
You are now connected to database "products" as user "postgres".
products=# SELECT * FROM sales;
 id | product_name | sales_price
----+--------------+-------------
  4 | Spoons       |        3.60
  5 | Tea Mug      |        7.80
  6 | plate        |        3.90
(3 rows)

products=#
```

Figure 7.31: *Retrieving replicated data from the sales table in the replica server*

The preceding output from a **SELECT** statement on the sales table shows that there is replication between the master and replica.

Failover (Promoting a Replica to a New Master Using pg_promote)

Failover is when an organization's system can continue to function in the case of a failure of the primary server. When the primary server fails, the replica is promoted to assume the primary server's function. The process of failing over can either be automated using tools such as Enterprise Failover Manager (EFM) or by running a function called **pg_promote**. The replica server should begin the failover procedure when the primary server fails.

If the replica server fails, then no failover is needed. If the replica server can be restarted, then the recovery process will be restarted immediately. If the replica server cannot be restarted, then a new replica instance will be created.

When the replica server has been promoted to become the new primary after the failure of the primary server, we must have a mechanism that will inform the failed primary server that they are no longer a primary server when it restarts. This process is called STONITH, which means to shoot the failed primary server in the head. This is very important to avoid a situation where both servers think they are both masters, which can eventually lead to data loss.

Most often, failover systems use the primary, replica, and a witness server, which act as a heartbeat mechanism to continually monitor the connectivity between the primary and replica servers. The setup complexity of this mechanism is not for chapter.

We are going to simulate a situation where the primary server fails, and then we promote the replica server. To achieve this, we are going to stop the primary server.

Before stopping the primary server, we will use the following command to check if the replica is replicating and also to check that it is a read-only mode.

```
SELECT pg_is_in_recovery();

CREATE DATABASE  dvdrental;
```

```
postgres=# SELECT pg_is_in_recovery();
 pg_is_in_recovery
-------------------
 t
(1 row)

postgres=# CREATE DATABASE dvdrental;
ERROR:  cannot execute CREATE DATABASE in a read-only transaction
postgres=#
postgres=#
```

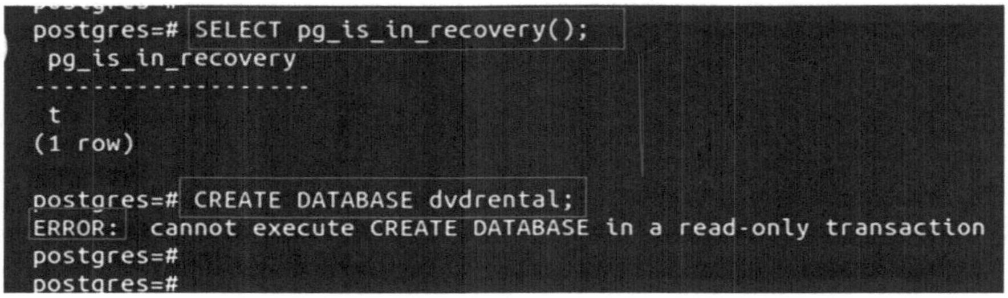

Figure 7.32: *This is a replica server*

We are unable to create a database as this is a replica server. The error above says that the server is in a read-only transaction, we can execute the **CREATE COMMAND**.

We will stop the primary server to simulate a crash that causes the server to be off. We will run this command on the primary server to stop it.

```
sudo systemctl stop postgresql@15-main
```

```
jack@jack-primary-server: $ sudo systemctl stop postgresql@15-main
[sudo] password for jack:
jack@jack-primary-server: $ sudo systemctl status postgresql@15-main
○ postgresql@15-main.service - PostgreSQL Cluster 15-main
     Loaded: loaded (/lib/systemd/system/postgresql@.service; enabled
     Active: inactive (dead) since Fri 2024-03-08 13:32:44 EST; 11s a
    Process: 105170 ExecStart=/usr/bin/pg_ctlcluster --skip-systemctl
    Process: 105778 ExecStop=/usr/bin/pg_ctlcluster --skip-systemctl-
   Main PID: 105176 (code=exited, status=0/SUCCESS)
        CPU: 3.282s
```

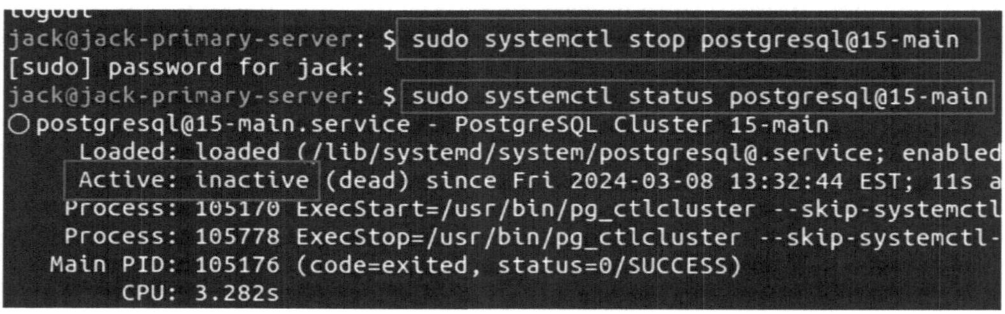

Figure 7.33: *Stopping the primary server*

Once the primary server is down from the above stoppage, we are ready to promote our replica to become the new master/primary server using the following command:

```
SELECT pg_promote();
```

Once, run, the replica will stop from being in the recovery mode, thus showing a status of false (f) from running the following command:

```
SELECT pg_is_in_recovery();
```

```
postgres=#
postgres=# SELECT pg_promote();
 pg_promote
------------
 t
(1 row)

postgres=# SELECT pg_is_in_recovery();
 pg_is_in_recovery
-------------------
 f
(1 row)

postgres=#
```

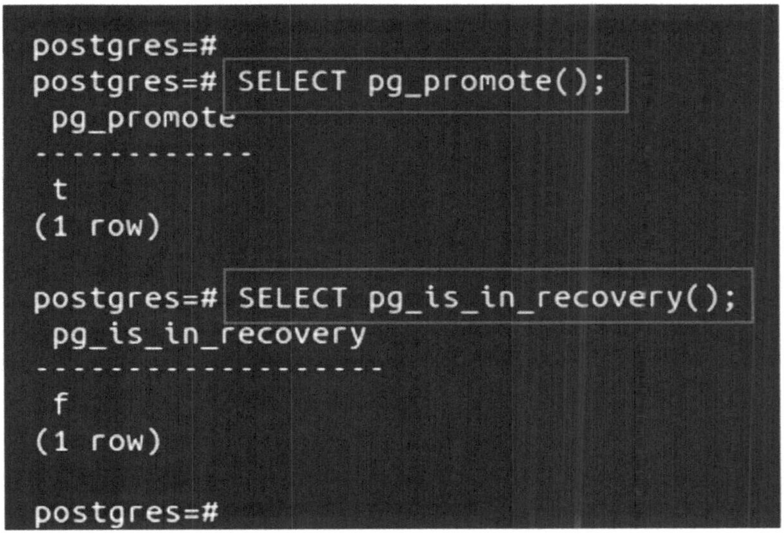

Figure 7.34: *Executing pg_promote on replica server*

The **CREATE** command can now be executed in the replica server as it is now the new primary server and accepting write statements.

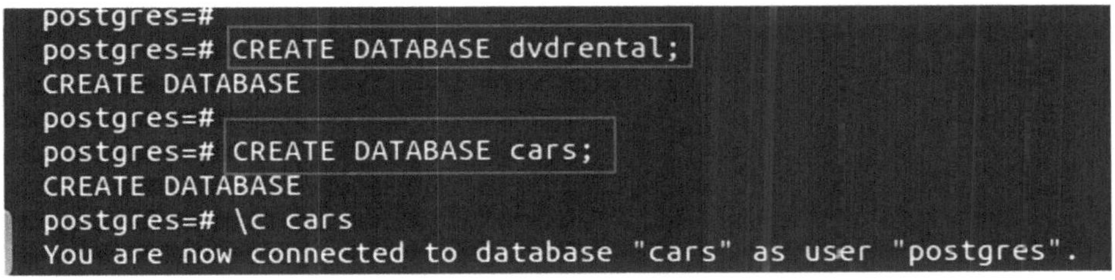

```
postgres=#
postgres=# CREATE DATABASE dvdrental;
CREATE DATABASE
postgres=#
postgres=# CREATE DATABASE cars;
CREATE DATABASE
postgres=# \c cars
You are now connected to database "cars" as user "postgres".
```

Figure 7.35: *Creating a database in the new primary server*

Let us now create a table and insert data into it after connecting to database cars.

```
CREATE TABLE car_make (
 car_id int,
 car_make varchar(30),
 colour  varchar(15)
);

INSERT INTO car_make VALUES
(65,'TOYOTA', 'BLACK'),
 (75,'VW', 'WHITE'),
(75,'BMW', 'RED');
```

```
cars=# CREATE TABLE car_make (
 car_id int,
 car_make varchar(30),
 colour  varchar(15)
);
CREATE TABLE
cars=# INSERT INTO car_make VALUES
(65,'TOYOTA', 'BLACK'),
 (75,'VW', 'WHITE'),
(75,'BMW', 'RED');
INSERT 0 3
cars=#
```

Figure 7.36: *More write operations in the new primary server*

The new primary server (which was formally a replica) is now receiving write operations such as **CREATE** and **INSERT** statements.

Conclusion

In this chapter, we learned set up streaming replication (also called Physical replication), Logical replication, and Failover by promoting a replica to a new primary using **pg_promote**.

In the next chapter, we will look at upgrading our PostgreSQL instance from one major version to another.

CHAPTER 8
Upgrade

Introduction

In this chapter, learners will learn to upgrade the PostgreSQL database from one version to another. This could be either a minor version or a major version and usually involves updating the database to a newer version. Depending on the version you want to upgrade from and to, the process thus differs in complexity.

Learners will become proficient in evaluating their employers' environment and recommending and implementing the necessary upgrades.

Structure

In this chapter, the following topics will be covered:

- Reasons Why Organizations Upgrade Their Databases
- Types of Upgrades

Reasons Why Organizations Upgrade Their Databases

Older versions and or outdated versions of PostgreSQL can potentially expose the organization to different inefficiencies and risks. Due to this and other reasons shown as follows, organizations regularly upgrade their PostgreSQL versions.

- **Security Enhancements:** With each new release of PostgreSQL, there are security fixes, including vulnerability patches. Databases become more unprotected or vulnerable when outdated.

  When it comes to data security, there are compliance regulations such as those of General Data Protection Regulation (GDPR) that organizations must abide by.

- **Performance Improvements:** Database performance is usually improved with the newer versions of PostgreSQL. This includes better query performance due to optimizations that improve the query execution times and indexing efficiency.

In high-load environments, performance is improved due to enhanced concurrency with newer versions.

Newer versions usually come with advanced techniques for indexing, which includes B-tree indexing improvements, thus significantly optimizing query performance.

- **New Features:** There are usually new features such as new data types, SQL, and functions, with major PostgreSQL releases. With these, developers can write better queries.

  Large datasets are handled more efficiently with improved partitioning and sharding capabilities.

- **Bug Fixes and stability:** Organizations benefit from a stable and more reliable system as they upgrade to newer versions. Bugs identified in older versions are usually fixed in newer versions.

  After a certain point in the life of PostgreSQL, they no longer receive updates, which include bug fixes and security patches, thus exposing the system to vulnerabilities and making it unsupported.

- **Scalability:** Newer versions usually have improved high-availability features and improved replication, thus improved scalability and failover strategies.

- **Compatibility with modern tools:** To ensure that their systems are compatible with modern tools, organizations regularly upgrade their PostgreSQL databases to be up to date with new technologies.

  PostgreSQL versions that are supported are strongly recommended by cloud providers for managed database services.

- **Community and Vendor Support:** Newer versions of PostgreSQL are supported by an active community. This includes better tutorials, documentation, and more experts to troubleshoot and fix issues.

  Many of the third-party vendors like to support newer PostgreSQL versions, thus organizations upgrade to receive their full support.

- **Regulatory compliance:** Data protection laws require software to be kept up to date so that personal data is secure, thus, organizations are mandated to upgrade PostgreSQL to newer versions for compliance.

- **Future-Proofing:** When PostgreSQL is upgraded regularly, it ensures the alignment of the infrastructure with the latest developments. This enhances future upgrades.

For organizations, upgrading PostgreSQL is not just a technical necessity but a critical decision that affects security, performance, compliance, and overall operational efficiency.

Types of Upgrades

It is very important to understand the PostgreSQL versioning system before performing an upgrade.

The first two numbers indicate the major version, and the last number(s) indicates the minor version. For instance, 15.6 means the major version is 15 and the minor version is 6.

A major version upgrade involves many steps as compared to a minor version upgrade.

There are two different types of upgrades, namely, minor upgrade and major upgrade.

Minor Upgrade

A minor upgrade involves updating the PostgreSQL version from one release to a new version within the same major version. Upgrading PostgreSQL from 14.3 to 14.7, for instance, is a minor upgrade and requires very minimal downtime. This involves updating within the same major release and usually focuses on security patches and bug fixes.

Minor PostgreSQL releases are always compatible with other releases within the same major version and do not change the internal storage format. 13.6 is compatible with 13.1 and 13.16.

We recommend that users always run the current minor release associated with their major version.

Releases

Version	Current minor	Supported	First Release	Final Release
16	16.4	Yes	September 14, 2023	November 9, 2028
15	15.8	Yes	October 13, 2022	November 11, 2027
14	14.13	Yes	September 30, 2021	November 12, 2026
13	13.16	Yes	September 24, 2020	November 13, 2025
12	12.20	Yes	October 3, 2019	November 14, 2024

Figure 8.1: *Available PostgreSQL major and current minor versions*

Before performing an upgrade, we first need to take a backup of the database. Refer to *Chapter 6, Backup and Restore*, on how to perform Postgresql backup.

```
postgres@jfozao-virtual-machine:~$
postgres@jfozao-virtual-machine:~$ mkdir upgrade_backup
postgres@jfozao-virtual-machine:~$ cd upgrade_backup/
postgres@jfozao-virtual-machine:~/upgrade_backup$ pwd
/var/lib/postgresql/upgrade_backup
postgres@jfozao-virtual-machine:~/upgrade_backup$ pg_dump -U postgres -d dvdrental -F c -f dvdrental_backup.dump
postgres@jfozao-virtual-machine:~/upgrade_backup$
postgres@jfozao-virtual-machine:~/upgrade_backup$ ls -al
total 656
drwxrwxr-x 2 postgres postgres   4096 Aug 31 07:05 .
drwxr-xr-x 7 postgres postgres   4096 Aug 31 07:04 ..
-rw-rw-r-- 1 postgres postgres 661411 Aug 31 07:05 dvdrental_backup.dump
postgres@jfozao-virtual-machine:~/upgrade_backup$
```

Figure 8.2: *Taking the Postgresql database backup*

The next step is to log into the database as a postgres user and check the current Postgresql version.

```
postgres@jfozao-virtual-machine:~/upgrade_backup$
postgres@jfozao-virtual-machine:~/upgrade_backup$ psql -U postgres
psql (13.11 (Ubuntu 13.11-1.pgdg22.04+1))
Type "help" for help.

postgres=# SELECT version();
                                                    version
---------------------------------------------------------------------------
 PostgreSQL 13.11 (Ubuntu 13.11-1.pgdg22.04+1) on x86_64-pc-linux-gnu, comp
(1 row)

postgres=#
postgres=#
```

Figure 8.3: *How to check the Postgresql version*

Next, we will be upgrading from Postgresql 13.11 to 13.16 which is the latest release for Postgresql 13. Before we proceed with the minor version upgrade, we are going to stop the Postgresql service after checking the status.

```
postgres=#
postgres=# \q
postgres@j....-virtual-machine:~/upgrade_backup$ sudo systemctl status postgresql
[sudo] password for postgres:
sudo: a password is required
postgres@jfozao-virtual-machine:~/upgrade_backup$ exit
logout
jfozao@jfozao-virtual-machine: $ sudo systemctl status postgresql
[sudo] password for jfozao:
● postgresql.service - PostgreSQL RDBMS
     Loaded: loaded (/lib/systemd/system/postgresql.service; enabled; vendor preset: enabled)
     Active: active (exited) since Sat 2024-08-31 06:48:28 EDT; 38min ago
   Main PID: 1234 (code=exited, status=0/SUCCESS)
        CPU: 2ms

Aug 31 06:48:28 jfozao-virtual-machine systemd[1]: Starting PostgreSQL RDBMS...
Aug 31 06:48:28 jfozao-virtual-machine systemd[1]: Finished PostgreSQL RDBMS.
jfozao@jfozao-virtual-machine: $
jfozao@jfozao-virtual-machine: $ sudo systemctl stop postgresql
jfozao@jfozao-virtual-machine: $
jfozao@jfozao-virtual-machine: $ sudo systemctl status postgresql
○ postgresql.service - PostgreSQL RDBMS
     Loaded: loaded (/lib/systemd/system/postgresql.service; enabled; vendor preset: enabled)
     Active: inactive (dead) since Sat 2024-08-31 07:27:01 EDT; 9s ago
   Main PID: 1234 (code=exited, status=0/SUCCESS)
        CPU: 2ms

Aug 31 06:48:28 jfozao-virtual-machine systemd[1]: Starting PostgreSQL RDBMS...
Aug 31 06:48:28 jfozao-virtual-machine systemd[1]: Finished PostgreSQL RDBMS.
Aug 31 07:27:01 jfozao-virtual-machine systemd[1]: postgresql.service: Deactivated successfully
Aug 31 07:27:01 jfozao-virtual-machine systemd[1]: Stopped PostgreSQL RDBMS.
jfozao@jfozao-virtual-machine: $
```

Figure 8.4: *How to check status and then stop Postgresql service*

Update PostgreSQL Packages

The exact steps to update the PostgreSQL packages depend on your operating system. The operating system needs to be updated before an upgrade is performed.

```
jfozao@jfozao-virtual-machine: $
jfozao@jfozao-virtual-machine: $
jfozao@jfozao-virtual-machine: $ sudo apt-get update
Get:1 http://security.ubuntu.com/ubuntu jammy-security InRelease [129 kB]
Hit:2 http://ca.archive.ubuntu.com/ubuntu jammy InRelease
Get:3 http://apt.postgresql.org/pub/repos/apt jammy-pgdg InRelease [129 kB]
Get:4 http://ca.archive.ubuntu.com/ubuntu jammy-updates InRelease [128 kB]
Get:5 http://security.ubuntu.com/ubuntu jammy-security/main amd64 Packages [1,769 kB]
Get:6 http://ca.archive.ubuntu.com/ubuntu jammy-backports InRelease [127 kB]
Get:7 http://security.ubuntu.com/ubuntu jammy-security/main i386 Packages [531 kB]
Get:8 http://security.ubuntu.com/ubuntu jammy-security/main Translation-en [290 kB]
Get:9 http://security.ubuntu.com/ubuntu jammy-security/main amd64 DEP-11 Metadata [43.2 kB]
Get:10 http://security.ubuntu.com/ubuntu jammy-security/main DEP-11 48x48 Icons [16.9 kB]
Get:11 http://security.ubuntu.com/ubuntu jammy-security/main DEP-11 64x64 Icons [26.5 kB]
Get:12 http://security.ubuntu.com/ubuntu jammy-security/main amd64 c-n-f Metadata [13.2 kB]
Get:13 http://security.ubuntu.com/ubuntu jammy-security/restricted amd64 Packages [2,308 kB
```

Figure 8.5: *Updating the operating system*

The following step is the first step to perform the minor upgrade.

```
jfozao@jfozao-virtual-machine: $
jfozao@jfozao-virtual-machine: $ sudo apt-get install postgresql-13
Reading package lists... Done
Building dependency tree... Done
Reading state information... Done
The following package was automatically installed and is no longer required:
  libllvm14
Use 'sudo apt autoremove' to remove it.
The following packages will be upgraded:
  postgresql-13
1 upgraded, 0 newly installed, 0 to remove and 377 not upgraded.
Need to get 15.8 MB of archives.
After this operation, 209 kB of additional disk space will be used.
Get:1 http://apt.postgresql.org/pub/repos/apt jammy-pgdg/main amd64 postgresql-13
Fetched 15.8 MB in 2s (6,892 kB/s)
Preconfiguring packages ...
(Reading database ... 201113 files and directories currently installed.)
Preparing to unpack .../postgresql-13_13.16-1.pgdg22.04+1_amd64.deb ...
Unpacking postgresql-13 (13.16-1.pgdg22.04+1) over (13.11-1.pgdg22.04+1) ...
Setting up postgresql-13 (13.16-1.pgdg22.04+1) ...
Processing triggers for postgresql-common (250.pgdg22.04+1) ...
Building PostgreSQL dictionaries from installed myspell/hunspell packages...
  en_ca
  en_us
Removing obsolete dictionary files:
```

Figure 8.6: *Performing the version 13 minor upgrade*

Once the packages have been updated, restart the PostgreSQL service and check the status to be sure it is active:

```
jfozao@jfozao-virtual-machine: $
jfozao@jfozao-virtual-machine: $ sudo systemctl start postgresql
jfozao@jfozao-virtual-machine: $
jfozao@jfozao-virtual-machine: $ sudo systemctl status postgresql
● postgresql.service - PostgreSQL RDBMS
     Loaded: loaded (/lib/systemd/system/postgresql.service; enabled; vend
     Active: active (exited) since Sat 2024-08-31 07:36:37 EDT; 5min ago
    Process: 6220 ExecStart=/bin/true (code=exited, status=0/SUCCESS)
   Main PID: 6220 (code=exited, status=0/SUCCESS)
        CPU: 2ms

Aug 31 07:36:37 jfozao-virtual-machine systemd[1]: Starting PostgreSQL RDB
Aug 31 07:36:37 jfozao-virtual-machine systemd[1]: Finished PostgreSQL RDB
jfozao@jfozao-virtual-machine: $
```

Figure 8.7: *Starting the PostgreSQL services after the minor upgrade*

After the service has restarted, log in to the database and verify that the upgrade was successful by checking the PostgreSQL version again:

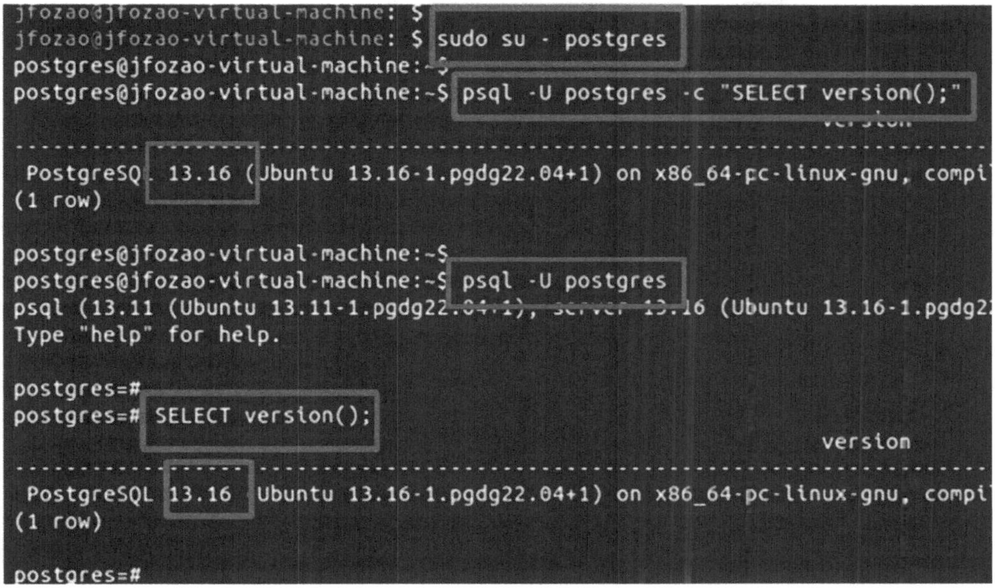

Figure 8.8: Checking the PostgreSQL version post the minor upgrade

Next, we will perform the Post upgrade task to update the statistics. This is achieved by running the **ANALYZE** command in the database. This thus improves performance as the statistics are up to date for query planner.

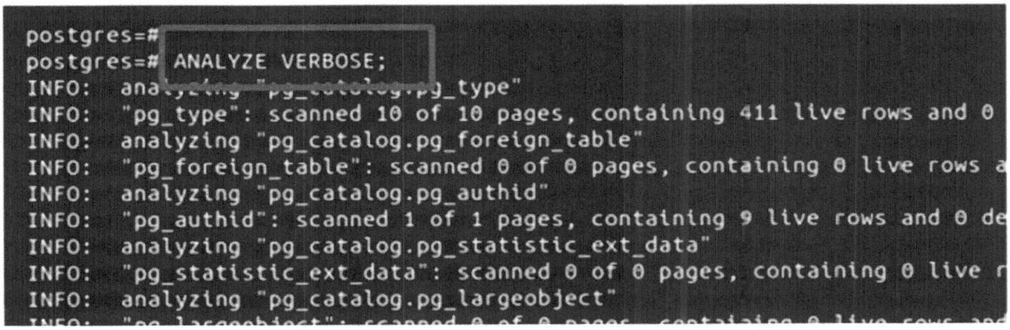

Figure 8.9: Updating statistics post the minor upgrade

Major Upgrade

This involves upgrading PostgreSQL from one major release to another, for example, from 13.x to 16.x. There are significant changes to the database when doing a major upgrade, thus it requires careful planning as it is more complex than a minor upgrade.

For major upgrade projects, we shall be using the **pg_upgrade** utility. This is to ease the upgrade of a major version of PostgreSQL database cluster to another. It facilitates data migration and configuration from older version (current version) to the newer version.

For the **pg_upgrade** utility to work effectively in upgrading the major version to another, the following needs to be done.

Prepare the environment by stopping the old server (current version server), installing the new server, and making sure that the old cluster is compatible with the new one.

pg_upgrade utility copies the old cluster's data to the new cluster's data directory. It also copies the old bin directory to the new cluster. Once the checks and upgrade are completed successfully, we will need to analyze and optimize the new cluster by running the **vacuumdb** command provided at the end of the upgrade process. We also need to perform necessary testing of the application to ensure everything with the new version works as expected.

The **pg_upgrade** utility is fast and very beneficial for large databases as it avoids the need for a dump and restore of data. It requires minimal downtime for the major upgrade. It minimizes the database performance and availability impact.

The first task here is to take ***the backup of the databases*** as we previously did in *Figure* 8.2 before the minor version upgrade.

After taking the backup, log in to the database as user Postgres and check the current version.

```
jfozao@jfozao-virtual-machine: $
jfozao@jfozao-virtual-machine: $ sudo su - postgres
postgres@jfozao-virtual-machine:~$
postgres@jfozao-virtual-machine:~$ psql -U postgres
psql (13.11 (Ubuntu 13.11-1.pgdg22.04+1), server 13.16 (Ub
Type "help" for help.

postgres=# SELECT version();

--------------------------------------------------------------
 PostgreSQL 13.16 (Ubuntu 13.16-1.pgdg22.04+1) on x86_64-p
(1 row)
```

Figure 8.10: *Checking the PostgreSQL version before starting the upgrade*

Our task here is to upgrade the current version of 13.x as shown above to version 16.x, which is the latest version of PostgreSQL as of writing this chapter.

If you are inside the database, quit and then exit from the postgres user. Update the operating system and then install the PostgreSQL 16 server.

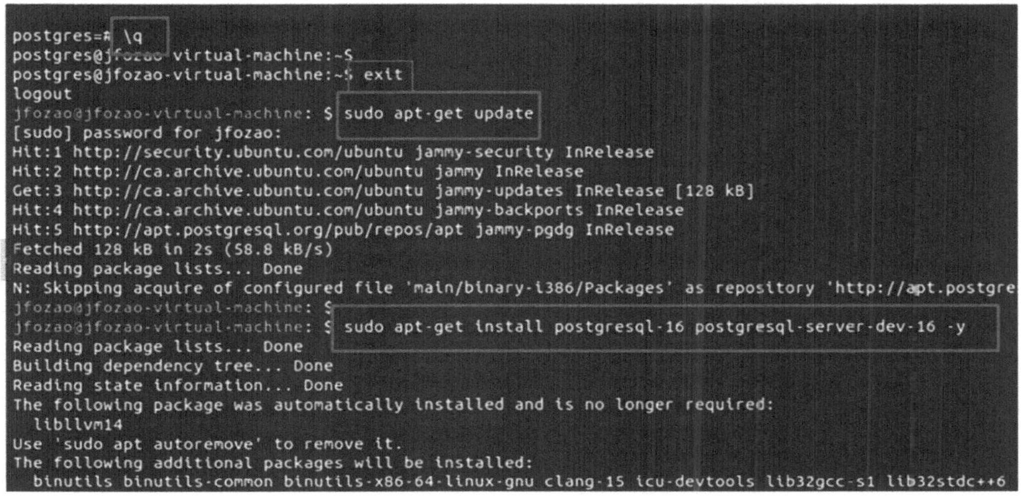

```
postgres=# \q
postgres@jfozao-virtual-machine:~$
postgres@jfozao-virtual-machine:~$ exit
logout
jfozao@jfozao-virtual-machine: $ sudo apt-get update
[sudo] password for jfozao:
Hit:1 http://security.ubuntu.com/ubuntu jammy-security InRelease
Hit:2 http://ca.archive.ubuntu.com/ubuntu jammy InRelease
Get:3 http://ca.archive.ubuntu.com/ubuntu jammy-updates InRelease [128 kB]
Hit:4 http://ca.archive.ubuntu.com/ubuntu jammy-backports InRelease
Hit:5 http://apt.postgresql.org/pub/repos/apt jammy-pgdg InRelease
Fetched 128 kB in 2s (58.8 kB/s)
Reading package lists... Done
N: Skipping acquire of configured file 'main/binary-i386/Packages' as repository 'http://apt.postgre
jfozao@jfozao-virtual-machine: $
jfozao@jfozao-virtual-machine: $ sudo apt-get install postgresql-16 postgresql-server-dev-16 -y
Reading package lists... Done
Building dependency tree... Done
Reading state information... Done
The following package was automatically installed and is no longer required:
  libllvm14
Use 'sudo apt autoremove' to remove it.
The following additional packages will be installed:
  binutils binutils-common binutils-x86-64-linux-gnu clang-15 icu-devtools lib32gcc-s1 lib32stdc++6
```

Figure 8.11: *Installing the new PostgreSQL server*

The end of the installation file is shown in *Figure* 8.12. As it is a huge file, only the end of the file is shown.

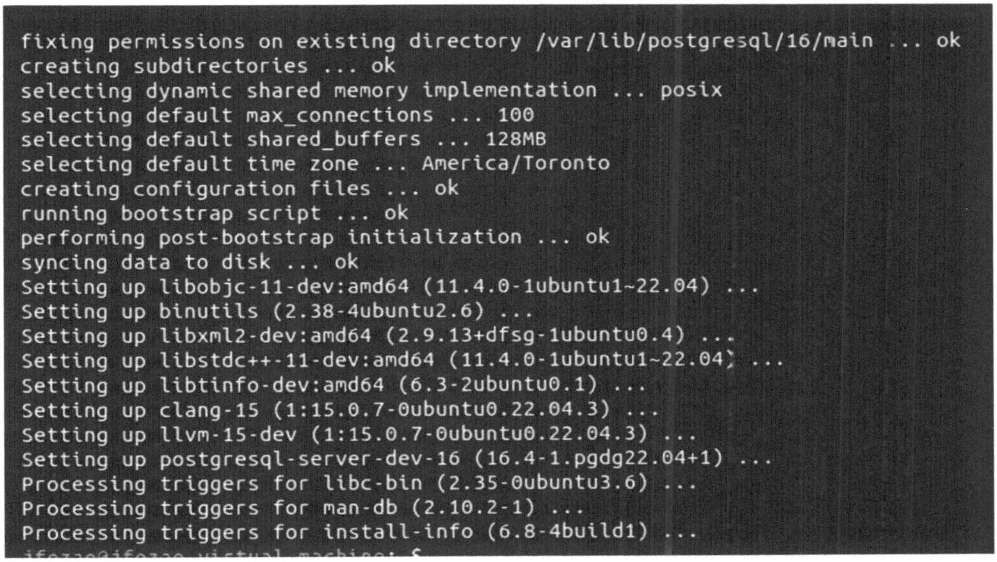

```
fixing permissions on existing directory /var/lib/postgresql/16/main ... ok
creating subdirectories ... ok
selecting dynamic shared memory implementation ... posix
selecting default max_connections ... 100
selecting default shared_buffers ... 128MB
selecting default time zone ... America/Toronto
creating configuration files ... ok
running bootstrap script ... ok
performing post-bootstrap initialization ... ok
syncing data to disk ... ok
Setting up libobjc-11-dev:amd64 (11.4.0-1ubuntu1~22.04) ...
Setting up binutils (2.38-4ubuntu2.6) ...
Setting up libxml2-dev:amd64 (2.9.13+dfsg-1ubuntu0.4) ...
Setting up libstdc++-11-dev:amd64 (11.4.0-1ubuntu1~22.04) ...
Setting up libtinfo-dev:amd64 (6.3-2ubuntu0.1) ...
Setting up clang-15 (1:15.0.7-0ubuntu0.22.04.3) ...
Setting up llvm-15-dev (1:15.0.7-0ubuntu0.22.04.3) ...
Setting up postgresql-server-dev-16 (16.4-1.pgdg22.04+1) ...
Processing triggers for libc-bin (2.35-0ubuntu3.6) ...
Processing triggers for man-db (2.10.2-1) ...
Processing triggers for install-info (6.8-4build1) ...
```

Figure 8.12: *The completion of the new PostgreSQL server installation*

Now that the server installation is completed, our next task is to stop the currently running PostgreSQL service.

To stop the PostgreSQL service, the first command is to check the status of the PostgreSQL service, the next command is to stop the service, and the final command is to check the status again that it is inactive.

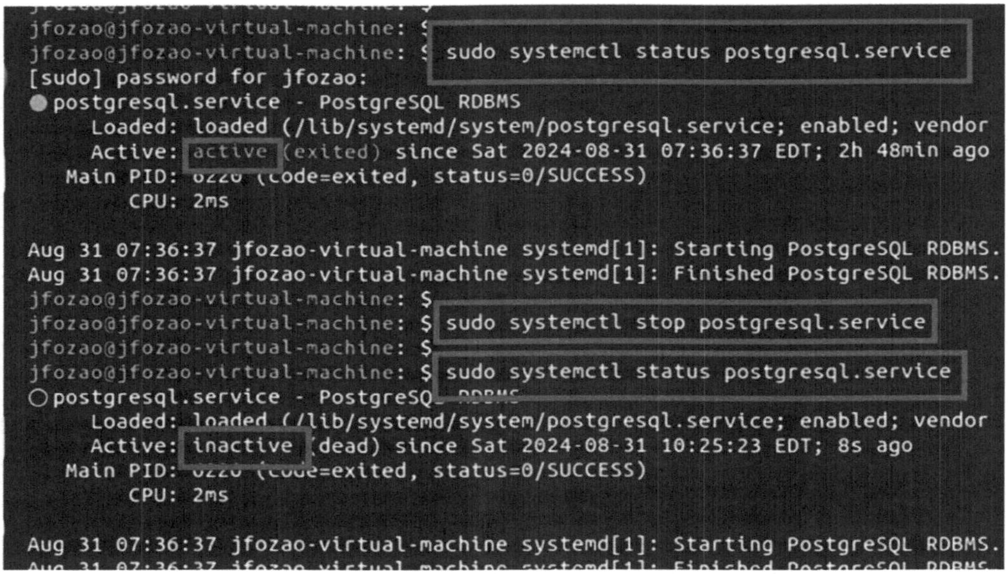

Figure 8.13: *How to stop the PostgreSQL service*

Our next task is to check the current version (in this case 13) and the version we will be upgrading to (in this case 16) for compatibility. The two servers ***must pass*** the compatibility check for the upgrade to be successful.

Login as a postgres user and run the compatibility checks. The **`--check`** or **`-c`** at the end of the command block shows that this command only check for compatibility. The command is provided in the code script attached.

Use the following syntax and change the versions as appropriate:

```
  /usr/lib/postgresql/16/bin/pg_upgrade \
--old-datadir=/var/lib/postgresql/13/main \
--new-datadir=/var/lib/postgresql/16/main \
--old-bindir=/usr/lib/postgresql/13/bin \
--new-bindir=/usr/lib/postgresql/16/bin \
--old-options '-c config_file=/etc/postgresql/13/main/
  postgresql.conf' \
--new-options '-c config_file=/etc/postgresql/16/main/
  postgresql.conf' \
--check
```

```
jfozao@jfozao-virtual-machine: $
jfozao@jfozao-virtual-machine: $ sudo su - postgres
postgres@jfozao-virtual-machine:~$
postgres@jfozao-virtual-machine:~$
postgres@jfozao-virtual-machine:~$ /usr/lib/postgresql/16/bin/pg_upgrade \
  --old-datadir=/var/lib/postgresql/13/main \
  --new-datadir=/var/lib/postgresql/16/main \
  --old-bindir=/usr/lib/postgresql/13/bin \
  --new-bindir=/usr/lib/postgresql/16/bin \
  --old-options '-c config_file=/etc/postgresql/13/main/postgresql.conf' \
  --new-options '-c config_file=/etc/postgresql/16/main/postgresql.conf' \
  --check
```

Figure 8.14: *Performing the compatibility checks before the upgrade*

The different parts of the command are explained as follows. Change the versions accordingly as per your environment requirements.

- **`/usr/lib/postgresql/16/bin/pg_upgrade`**: This is the path to the **`pg_upgrade`** executable from PostgreSQL 16. The tool is responsible for performing the upgrade from one major version of PostgreSQL to another.
- **`--old-datadir=/var/lib/postgresql/13/main`**: Shows the data directory of the old PostgreSQL version (13). This directory contains the database files for PostgreSQL 13.
- **`--new-datadir=/var/lib/postgresql/16/main`**: Shows the data directory where the new PostgreSQL version (16) will store its database files.
- **`--old-bindir=/usr/lib/postgresql/13/bin`**: Shows the binary directory of the old PostgreSQL version (13). This is where the executables (such as **`postgres`**, **`pg_ctl`**, and so on) for PostgreSQL 13 are located.
- **`--new-bindir=/usr/lib/postgresql/16/bin`**: Specifies the binary directory of the new PostgreSQL version (16). This directory contains the executables for PostgreSQL 16.
- **`--old-options '-c config_file=/etc/postgresql/13/main/postgresql.conf'`**: This shows the configuration file to be used by the old server during the upgrade process.
- **`--new-options '-c config_file=/etc/postgresql/16/main/postgresql.conf'`**: This shows the configuration file for the new server during the upgrade process.
- **`--check`**: This flag tells **`pg_upgrade`** to perform a "**`dry run`**" of the upgrade. It checks for potential issues that might prevent the upgrade from being successful but does not actually perform the upgrade. This is an important step to identify any incompatibilities or problems before proceeding with the actual upgrade.

The result of the compatibility checks is shown as follows. All checks must be marked as "**OK**" for the upgrade to be successful. This is also indicated at the end as "**Clusters are compatible**".

```
Performing Consistency Checks
-----------------------------
Checking cluster versions                                   ok
Checking database user is the install user                  ok
Checking database connection settings                       ok
Checking for prepared transactions                          ok
Checking for system-defined composite types in user tables  ok
Checking for reg* data types in user tables                 ok
Checking for contrib/isn with bigint-passing mismatch       ok
Checking for incompatible "aclitem" data type in user tables ok
Checking for user-defined encoding conversions              ok
Checking for user-defined postfix operators                 ok
Checking for incompatible polymorphic functions             ok
Checking for presence of required libraries                 ok
Checking database user is the install user                  ok
Checking for prepared transactions                          ok
Checking for new cluster tablespace directories             ok

*Clusters are compatible*
postgres@jfozao-virtual-machine:~$
```

Figure 8.15: *Compatibility checks result*

To perform the upgrade from versions 13 to 16, we will take off the **--check** from the command block and then execute it to upgrade our Postgresql database.

```
postgres@jfozao-virtual-machine:~$

postgres@jfozao-virtual-machine:~$
postgres@jfozao-virtual-machine:~$ /usr/lib/postgresql/16/bin/pg_upgrade \
  --old-datadir=/var/lib/postgresql/13/main \
  --new-datadir=/var/lib/postgresql/16/main \
  --old-bindir=/usr/lib/postgresql/13/bin \
  --new-bindir=/usr/lib/postgresql/16/bin \
  --old-options '-c config_file=/etc/postgresql/13/main/postgresql.conf' \
  --new-options '-c config_file=/etc/postgresql/16/main/postgresql.conf' \
```

Figure 8.16: *Performing the major upgrade from version 13 to 16*

The result of the upgrade is shown in two parts. The query first performs the consistency checks as shown in *Figure* 8.17. Look and understand the sentence at the end of the checks. The sentence means that if any of the checks do not return an ok status, then you must **reinstall** the new server.

```
Performing Consistency Checks
-----------------------------
Checking cluster versions                                     ok
Checking database user is the install user                    ok
Checking database connection settings                         ok
Checking for prepared transactions                            ok
Checking for system-defined composite types in user tables    ok
Checking for reg* data types in user tables                   ok
Checking for contrib/isn with bigint-passing mismatch         ok
Checking for incompatible "aclitem" data type in user tables  ok
Checking for user-defined encoding conversions                ok
Checking for user-defined postfix operators                   ok
Checking for incompatible polymorphic functions               ok
Creating dump of global objects                               ok
Creating dump of database schemas
                                                              ok
Checking for presence of required libraries                   ok
Checking database user is the install user                    ok
Checking for prepared transactions                            ok
Checking for new cluster tablespace directories               ok

If pg_upgrade fails after this point, you must re-initdb the
new cluster before continuing.
```

Figure 8.17: *First part of the upgrade result*

The upgrade is then performed by the query as shown in *Figure* 8.18. All items upgraded are marked with a status of "**`ok`**", and at the end, it is also shown the successful statement of "**`upgrade complete`**". You are also prompted to optimize the statistics by running the **`vacuumdb`** command provided at the end.

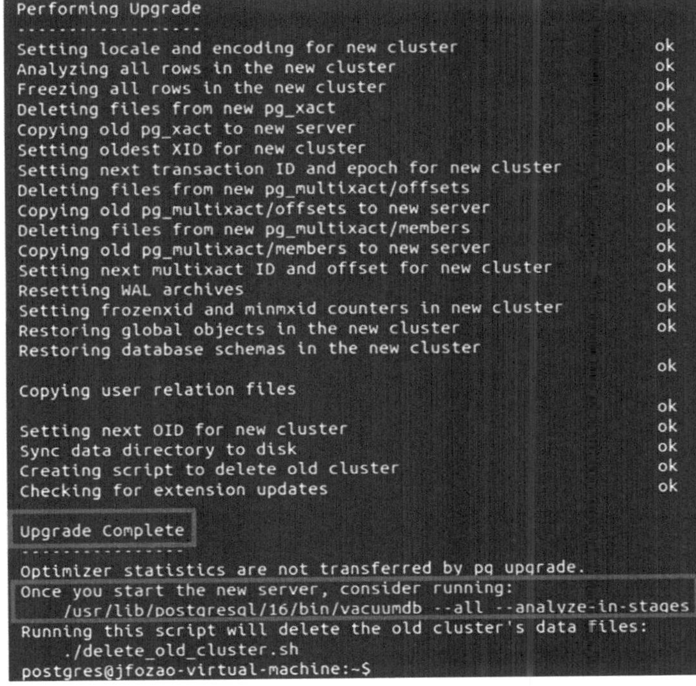

```
Performing Upgrade
------------------
Setting locale and encoding for new cluster                   ok
Analyzing all rows in the new cluster                         ok
Freezing all rows in the new cluster                          ok
Deleting files from new pg_xact                               ok
Copying old pg_xact to new server                             ok
Setting oldest XID for new cluster                            ok
Setting next transaction ID and epoch for new cluster         ok
Deleting files from new pg_multixact/offsets                  ok
Copying old pg_multixact/offsets to new server                ok
Deleting files from new pg_multixact/members                  ok
Copying old pg_multixact/members to new server                ok
Setting next multixact ID and offset for new cluster          ok
Resetting WAL archives                                        ok
Setting frozenxid and minmxid counters in new cluster         ok
Restoring global objects in the new cluster                   ok
Restoring database schemas in the new cluster
                                                              ok
Copying user relation files
                                                              ok
Setting next OID for new cluster                              ok
Sync data directory to disk                                   ok
Creating script to delete old cluster                         ok
Checking for extension updates                                ok

Upgrade Complete
----------------
Optimizer statistics are not transferred by pg_upgrade.
Once you start the new server, consider running:
    /usr/lib/postgresql/16/bin/vacuumdb --all --analyze-in-stages
Running this script will delete the old cluster's data files:
    ./delete_old_cluster.sh
postgres@jfozao-virtual-machine:~$
```

Figure 8.18: *Last part of the upgrade result*

The -j Option

The **-j** (short for "**jobs**") option in **pg_upgrade** is a feature that enables **parallel processing** during the upgrade. This means that multiple database objects (such as tables and indexes) can be processed simultaneously, instead of one at a time.

Parallel processing uses multiple threads to carry out the upgrade tasks in parallel, which can drastically reduce the overall time needed, particularly for databases with a large number of objects or high data volume.

When you run **pg_upgrade** with the **-j** option, you specify the number of threads that should be used.

The syntax is as shown as follows:

```
pg_upgrade -b /path/to/old/bin -B /path/to/new/bin -d /path/to/
old/data -D /path/to/new/data -j 4
```

The **-j 4** option tells **pg_upgrade** to use **4 threads** to process the upgrade tasks in parallel.

Each thread will independently work on upgrading different database objects. This is especially helpful for systems with **multiple CPU cores**, as it takes advantage of the hardware's ability to process multiple tasks concurrently. We did not use the **-j** option as our database was not big.

The --link Option

When you use the **--link** option during **pg_upgrade**, it creates **hard links** for the database files instead of duplicating (copying) them. A hard link essentially creates another directory entry that points to the same underlying data on disk. This means:

1. There is no physical duplication of data, saving disk space.
2. The process is faster because files aren't physically copied.

The syntax is as shown as follows:

```
pg_upgrade -b /path/to/old/bin -B /path/to/new/bin -d /path/to/old/data
-D /path/to/new/data -j 4 --link
```

```
  /usr/lib/postgresql/16/bin/pg_upgrade \
--old-datadir=/var/lib/postgresql/13/main \
--new-datadir=/var/lib/postgresql/16/main \
--old-bindir=/usr/lib/postgresql/13/bin \
--new-bindir=/usr/lib/postgresql/16/bin \
--old-options '-c config_file=/etc/postgresql/13/main/
  postgresql.conf' \
```

```
--new-options '-c config_file=/etc/postgresql/16/main/
  postgresql.conf' \
-j 4 --link
```

If you use the **--link** option, once the upgrade completes, any changes to the files in the new database cluster will also affect the old cluster (since both use the same hard-linked files). This means you cannot safely roll back to the old cluster unless you back it up beforehand. Always backup your database before using **--link**, even though it saves disk space. If anything goes wrong during or after the upgrade, you'll have a recovery path. Like with any upgrade process, test the **--link** upgrade in a staging environment before applying it to production.

Using **--link** is highly efficient but best suited for cases where you are confident in the upgrade process and the new cluster's stability. If you are unsure or need to preserve the old cluster independently, it may be better to avoid **--link** and rely on the traditional copy-based approach.

We are going to switch back to the regular user and then swap the ports for the old and new PostgreSQL versions. This is to make sure that our upgraded version (new version) is running on port 5432.

```
postgres@jfozao-virtual-machine:~$
postgres@jfozao-virtual-machine:~$ exit
logout
jfozao@jfozao-virtual-machine: $
jfozao@jfozao-virtual-machine: $ sudo vi /etc/postgresql/16/main/postgresql.conf
[sudo] password for jfozao:
jfozao@jfozao-virtual-machine: $
```

Figure 8.19: *Accessing the Postgresql.conf file via vi text editor*

Once inside the text editor, scroll down to find the port number or search using the forward slash "/".

```
#------------------------------------------------------------------------------
# CONNECTIONS AND AUTHENTICATION
#------------------------------------------------------------------------------

# - Connection Settings -

#listen_addresses = 'localhost'         # what IP address(es) to listen on;
                                        # comma-separated list of addresses;
                                        # defaults to 'localhost'; use '*' fo
                                        # (change requires restart)
port = 5433                             # (change requires restart)
max_connections = 100                   # (change requires restart)
#reserved_connections = 0               # (change requires restart)
#superuser_reserved_connections = 3     # (change requires restart)
```

Figure 8.20: *Locating port 5433 to be changed*

Change "**port = 5433**" to "**port = 5432**". This is very important, and remember to save before quitting using: **wq!** as shown in *Figure 8.21*.

```
# CONNECTIONS AND AUTHENTICATION
#------------------------------------------------------------------------------

# - Connection Settings -

#listen_addresses = 'localhost'         # what IP address(es) to list
                                        # comma-separated list of add
                                        # defaults to 'localhost'; us
                                        # (change requires restart)
port = 5432                             # (change requires restart)
max_connections = 100                   # (change requires restart)
#reserved_connections = 0               # (change requires restart)
#superuser_reserved_connections = 3     # (change requires restart)
unix_socket_directories = '/var/run/postgresql' # comma-separated lis
                                        # (change requires restart)
#unix_socket_group = ''                 # (change requires restart)
#unix_socket_permissions = 0777         # begin with 0 to use octal n
                                        # (change requires restart)
#bonjour = off                          # advertise server via Bonjou
                                        # (change requires restart)
#bonjour_name = ''                      # defaults to the computer na
                                        # (change requires restart)

# - TCP settings -
# see "man tcp" for details
:wq!
```

Figure 8.21: *Port number of new server changed to 5432*

Repeat the same process to change the port for the old version (Postgresql 13) from 5432 to 5433. Save and quit using wq!

```
jfozao@jfozao-virtual-machine: $ sudo vi /etc/postgresql/13/main/postgresql.conf
jfozao@jfozao-virtual-machine: $

#------------------------------------------------------------------------------
# CONNECTIONS AND AUTHENTICATION
#------------------------------------------------------------------------------

# - Connection Settings -

#listen_addresses = 'localhost'         # what IP address(es) to listen on;
                                        # comma-separated list of addresses;
                                        # defaults to 'localhost'; use '*' for al
                                        # (change requires restart)
port = 5433                     # (change requires restart)
max_connections = 100                   # (change requires restart)
#superuser_reserved_connections = 3     # (change requires restart)
:wq!
```

Figure 8.22: *Port number of old server changed to 5433*

Start the PostgreSQL service, and check the status to make sure it is active.

```
jfozao@jfozao-virtual-machine: $
jfozao@jfozao-virtual-machine: $ sudo systemctl start postgresql.service
jfozao@jfozao-virtual-machine: $
jfozao@jfozao-virtual-machine: $ sudo systemctl status postgresql.service
● postgresql.service - PostgreSQL RDBMS
     Loaded: loaded (/lib/systemd/system/postgresql.service; enabled; vendor
     Active: active (exited) since Sat 2024-08-31 11:01:02 EDT; 3min 41s ago
    Process: 12042 ExecStart=/bin/true (code=exited, status=0/SUCCESS)
   Main PID: 12042 (code=exited, status=0/SUCCESS)
        CPU: 4ms

Aug 31 11:01:02 jfozao-virtual-machine systemd[1]: Starting PostgreSQL RDBMS
```

Figure 8.23: *Starting the PostgreSQL services on the new server*

Next is to switch to Postgres user and log in to the database again and check that the version has changed to 16.x.

```
jfozao@jfozao-virtual-machine: $
jfozao@jfozao-virtual-machine: $ sudo su - postgres
postgres@jfozao-virtual-machine:~$
postgres@jfozao-virtual-machine:~$ psql -U postgres
psql (16.4 (Ubuntu 16.4-1.pgdg22.04+1))
Type "help" for help.

postgres=#
postgres=# SELECT version();

--------------------------------------------------------
 PostgreSQL 16.4 (Ubuntu 16.4-1.pgdg22.04+1) on x86_64-
(1 row)

postgres=#
```

Figure 8.24: *Confirming the major version upgrade*

We have successfully upgraded from PostgreSQL version 13 to version 16.

Run the recommended **`vacuumdb`** command that was provided at the end of the upgrade results on *Figure* 8.17 to gather statistics. This is a very important task to be done after the upgrade.

```
postgres=#
postgres=# \q
postgres@jfozao-virtual-machine:~$
postgres@jfozao-virtual-machine:~$ /usr/lib/postgresql/16/bin/vacuumdb --all --analyze-in-stages
vacuumdb: processing database "dvdrental": Generating minimal optimizer statistics (1 target)
vacuumdb: processing database "postgres": Generating minimal optimizer statistics (1 target)
vacuumdb: processing database "template1": Generating minimal optimizer statistics (1 target)
vacuumdb: processing database "dvdrental": Generating medium optimizer statistics (10 targets)
vacuumdb: processing database "postgres": Generating medium optimizer statistics (10 targets)
vacuumdb: processing database "template1": Generating medium optimizer statistics (10 targets)
vacuumdb: processing database "dvdrental": Generating default (full) optimizer statistics
vacuumdb: processing database "postgres": Generating default (full) optimizer statistics
vacuumdb: processing database "template1": Generating default (full) optimizer statistics
postgres@jfozao-virtual-machine:~$
```

Figure 8.25: *Updating PostgreSQL statistics after the major upgrade*

The next step is to check all the old PostgreSQL packages that are still installed using the command shown in *Figure* 8.26.

```
postgres@jfozao-virtual-machine:~$
postgres@jfozao-virtual-machine:~$ apt list --installed | grep postgresql

WARNING: apt does not have a stable CLI interface. Use with caution in scripts.

postgresql-13/jammy-pgdg,now 13.16-1.pgdg22.04+1 amd64 [installed]
postgresql-16/jammy-pgdg,now 16.4-1.pgdg22.04+1 amd64 [installed]
postgresql-client-13/now 13.11-1.pgdg22.04+1 amd64 [installed,upgradable to: 13.16-1.pgdg22.04+1]
postgresql-client-16/jammy-pgdg,now 16.4-1.pgdg22.04+1 amd64 [installed,automatic]
postgresql-client-common/jammy-pgdg,now 262.pgdg22.04+1 all [installed,automatic]
postgresql-common/jammy-pgdg,now 262.pgdg22.04+1 all [installed,automatic]
postgresql-server-dev-16/jammy-pgdg,now 16.4-1.pgdg22.04+1 amd64 [installed]
postgres@jfozao-virtual-machine:~$
```

Figure 8.26: *How to check for the old packages still installed*

Remove the old PostgreSQL packages (from the listing above). Exit from Postgres user and execute as a super user either with **`sudo`** or as **`root`**.

The next steps are to be executed after all the tests have been performed to check that the new version is working effectively as intended. The amount of time for testing is environment dependent. Before deleting the old version, this must be discussed and agreed upon with the DBA team, and the new server testing time must have elapsed as per the environment.

```
postgres@jfozao-virtual-machine:~$ exit
logout
jfozao@jfozao-virtual-machine: $ sudo apt-get remove postgresql-13 postgresql-server-dev-11
[sudo] password for jfozao:
Reading package lists... Done
Building dependency tree... Done
Reading state information... Done
Package 'postgresql-server-dev-11' is not installed, so not removed
The following package was automatically installed and is no longer required:
  libllvm14
Use 'sudo apt autoremove' to remove it.
The following packages will be REMOVED:
  postgresql-13
0 upgraded, 0 newly installed, 1 to remove and 370 not upgraded.
After this operation, 50.0 MB disk space will be freed.
Do you want to continue? [Y/n] y
(Reading database ... 213930 files and directories currently installed.)
Removing postgresql-13 (13.16-1.pgdg22.04+1) ...
Processing triggers for postgresql-common (262.pgdg22.04+1) ...
Building PostgreSQL dictionaries from installed myspell/hunspell packages...
  en_ca
  en_us
Removing obsolete dictionary files:
jfozao@jfozao-virtual-machine: $
```

Figure 8.27: *Removing the old Postgresql 13 packages still installed*

Next, remove the old configuration, and then switch to user Postgres and delete the old cluster using the last command "**./delete_old_cluster.sh**" provided at the end of the upgraded result.

```
jfozao@jfozao-virtual-machine: $
jfozao@jfozao-virtual-machine: $ sudo rm -rf /etc/postgresql/13/
jfozao@jfozao-virtual-machine: $
jfozao@jfozao-virtual-machine: $ sudo su - postgres
postgres@jfozao-virtual-machine:~$
postgres@jfozao-virtual-machine:~$ ./delete_old_cluster.sh
postgres@jfozao-virtual-machine:~$
```

Figure 8.28: *Deleting the old Postgresql cluster*

Conclusion

In this chapter, we learned how to upgrade from one minor version to another within the same major version. We also learned how to upgrade from one major version to another using the **pg_upgrade** utility. We also learned that **pg_dump** and **pg_dumpall** utilities are not the right tools for upgrading large databases but **pg_upgrade** utility.

In the next chapter, we will look at PostgreSQL database performance tuning.

CHAPTER 9
Performance Tuning

Introduction

In this chapter, learners will understand the importance of performance tuning and how to adjust various parameters to optimize their PostgreSQL database for speed and responsiveness. Performance tuning is the process of optimizing the PostgreSQL database to enhance its responsiveness, speed, and efficiency by modifying its configuration to achieve better performance.

Learners will learn how to identify and optimize slow or long-running queries. They will also read and comprehend the "`EXPLAIN`" and "`EXPLAIN ANALYZE`" plans, which are known as PostgreSQL query execution plans.

For a PostgreSQL system to run efficiently, it needs to be properly tuned for performance. This requires a thorough understanding of how the PostgreSQL database operates, as well as knowledge of the different configuration parameters, their functions, and the appropriate values needed for optimal configuration.

As a database administrator or application developer, ensuring the database performs well is paramount. A poorly performing database often leads to slow query execution times, long-running queries, and a poor user experience.

While PostgreSQL is a powerful database, like any other database, it can become slow if not properly tuned and maintained. This can lead to frustration for end users who expect quick data retrieval. Performance tuning involves adjusting the database's configuration settings to enhance its performance. This requires a comprehensive understanding of the database's internal workings, the function of each configuration parameter, and the values necessary for optimal performance.

Structure

In this chapter, the following topics will be covered: All these topics are geared toward addressing the most common database issues, such as slow and long-running queries.

- Common Issues Leading to Performance Tuning
- Factors Affecting PostgreSQL Performance
- Vacuum and Analyze

- Execution Plans
- Index Tuning
- Memory Tuning
- Storage Tuning
- Operating System Tuning
- Network Tuning
- Application Tuning
- Configuration Tuning
- Reporting and Logging
- Table Bloat (Dead Tuples)
- Transaction ID (XID) Wraparound
- Visibility Map (VM)

Common Issues Leading to Performance Tuning

Several common issues can necessitate performance tuning in a PostgreSQL database:

- **Slow Queries:** Often the primary performance issue, slow queries can stem from various factors such as suboptimal database design, inefficient queries, or hardware limitations.
- **Deadlocks:** These occur when two or more transactions are mutually waiting for each other to complete, causing the database to become unresponsive and potentially needing a restart.
- **High CPU usage**: PostgreSQL uses the CPU to execute queries and process data. If the database is running complex queries or processing large amounts of data, this can lead to high CPU usage and performance problems.
- **High memory usage**: PostgreSQL uses memory to cache data and indexes. If the database has too much data or indexes in memory, this can lead to high memory usage and performance problems.

Factors Affecting PostgreSQL Performance

Different factors affect PostgreSQL performance. Some of these are database configuration, hardware configuration, database workload, query optimization, maintenance tasks, and database design.

Database Configuration

The PostgreSQL configuration file called **postgresql.conf** contains a variety of settings that can impact the performance of the database server. It is necessary to know what the impact of each setting is and to understand how to configure them appropriately in order that resources will be used effectively by the server. It is important to regularly monitor performance and make changes to the configuration settings to achieve optimal performance over time.

The following are the most important performance-tuning settings that can be tweaked in **postgresql.conf** file:

- **shared_buffers:** This is the amount of memory used by the server for caching.
- **max_connections:** This is used by the server to control the maximum number of concurrent connections.
- **effective_cache_size:** This is used by the query planner to estimate the amount of data that is held in the operating system cache.
- **autovacuum_vacuum_scale_factor:** This is used to control the behavior of the autovacuum daemon.
- **max_wal_size:** This is used to control the maximum size of write-ahead logs.
- **checkpoint_timeout:** This is used to control checkpoint frequency.
- **work_mem:** This is the amount of memory used for sorting and hashing operations by the query planner.
- **maintenance_work_mem:** This is the amount of memory used for maintenance operations such as **REINDEX**, **VACUUM**, and **ANALYZE**.
- **random_page_cost:** This is the cost of reading a random page from disk.
- **wal_buffers:** This is the amount of memory used for the write-ahead logs.

Hardware Configuration

Network, central processing unit (CPU), memory, and disk space are some of the hardware configurations that can affect PostgreSQL performance.

- **Network:** Remote connections to the database server can be affected by the network bandwidth and latency. This can hamper the PostgreSQL performance of remote client's connection to the database. A low-latency and fast network can improve remote clients' performance.
- **CPU:** The number of CPU cores is very important for database performance. The ability of the database server to process queries faster is highly dependent on the number of CPU cores, thus, the more the cores, the faster queries are processed and the better the database performance. This is because

PostgreSQL uses the CPU to execute queries and process data. If it is a complex query that the database is running or it is processing a large amount of data, this can lead to high CPU usage and thus performance issues.

- **Memory:** Cached data in memory is accessed faster than data from disk. The more available memory, the more frequently accessed data is cached, thus improving query performance and other operations that rely on the amount of cached data in memory. Increasing the memory can increase the amount of data being cached, thus allowing the database to access the data faster. As PostgreSQL caches data and indexes in memory, this can lead to high memory usage and performance issues if the database has too much data or indexes in the memory.
- **Concurrent Connections:** The performance of the database is affected depending on the number of concurrent connections the hardware can handle. If the hardware is underutilized because it is not operating at its full capacity with the number of connections, it can hinder its performance.

Database Workload

The performance of the PostgreSQL database can be impacted significantly by its workload. The more the number of concurrent transactions and the larger the amount of data, the poorer the performance will be. Contention of resources such as shared buffers and locks is usually caused by a high number of concurrent connections. This can slow down performance. Furthermore, a huge amount of data can make query retrieval of records very difficult, thus slowing down the system.

- **Application workload:** Application design is very important for database performance. A lot of pressure can be exerted on the database if there is a high number of simple queries executed regularly, thus slowing down performance.
- **Connection pooling:** This is a technique that is used by organizations to manage a database with a large number of connections. With this, multiple applications can share a connection pool, thus minimizing the creation and closing of connections overhead. However, this can lead to the contention of resources such as CPU and memory, if the number of connections becomes too high. **`PgBouncer`** and **`pgpool`** are tools that can be used to manage connection pooling and workload and improve the performance of the database.

Database Design

The design of a database is crucial for performance as it significantly impacts query speed. One fundamental aspect of database design is the use of indexes. Properly created indexes allow for quick data location and retrieval from a table. Without an index, the entire table must be scanned to find the required data, which can be very slow for large tables. Proper use of indexes in PostgreSQL can greatly improve query performance.

Query performance can also be enhanced by using the correct data types for each column. For example, a column that stores whole numbers should use the integer data type rather than a floating-point data type, as this requires less space and thus improves performance. Similarly, a date column will perform better and faster with the date data type rather than the text data type.

Database normalization is another important factor for performance, as it improves data integrity and eliminates redundancy. A normalized database is well-organized, which enhances query performance and data retrieval.

Designing databases effectively is a critical step in optimizing performance. In PostgreSQL, partitioning data into multiple logically separate tables instead of relying on a single large table can result in immediate and significant query performance improvements.

Another technique that can enhance PostgreSQL performance is creating partial indexes. This strategy depends on the specific use case and the queries frequently executed. The goal is to filter as much data as possible, reducing the amount of data to be processed. Therefore, creating indexes on columns commonly used as filters in frequent queries is advisable.

However, it is essential to use indexes judiciously. Overusing indexes can have adverse effects, as creating and maintaining indexes is a resource-intensive task. Having too many indexes can degrade the overall performance of the database.

Query Optimization

This is the process of improving the performance of SQL queries by identifying and addressing bottlenecks in the query execution process. In PostgreSQL, the query planner and optimizer perform the query optimization. The query planner and optimizer analyze a query and generate a query execution plan that is designed to choose the plan with a low cost of execution.

Query optimization is a very important aspect of PostgreSQL performance improvement. This improves query planning and data retrieval efficiency from hardware. To optimize a query or tune it effectively, we first need to understand how PostgreSQL executes a query on the data.

The **`EXPLAIN`** and **`EXPLAIN ANALYZE`** commands enable us to optimize query performance.

EXPLAIN: The **`EXPLAIN`** statement in PostgreSQL is a great tool that is used to analyze how the database is going to execute a query. When a query is run with an **`EXPLAIN`**,

PostgreSQL gives information relating to the query execution plan. This includes things such as:

- **Node Types**: Things such as joins, scans, and sorts are some of the different stages in the execution.
- **Cost Estimates**: To gauge the efficiency of the query, the execution plan includes an estimated startup and total costs for each operation
- **Row Estimates**: The number of rows processed at each step is estimated here.

`EXPLAIN` can be used with different options:

- **`EXPLAIN ANALYZE`**: This will execute the query and return the actual execution details. It will provide real execution time, and the actual rows are returned.

  The **`EXPLAIN`** statement will help to identify bottlenecks and to understand how indexes and joins will be utilized. We can then optimize the query by making necessary changes regarding indexes and joins.

  The **`EXPLAIN`** and or **`EXPLAIN ANALYZE`** when executed, gives a clear idea as to how the query is going to be run by PostgreSQL, and whether it is going to use an index scan or a table scan. This helps us to make the necessary changes (optimize). We either alter the query for better performance or update the stats by running analyze again. They enable us to see the execution plan that the query planner and optimizer have generated for a query. Thus, we can identify the different parts of the query taking the most time and resources.

- **`ANALYZE:`** The **`ANALYZE`** command in PostgreSQL is crucial for updating statistics about data distribution in the database. These statistics are essential for the query planner to formulate the most efficient execution plans for queries. When a query is executed, PostgreSQL transforms it and sends it to the planner, which uses these statistics to develop a cost-effective execution plan. Keeping statistics up to date is vital for producing accurate and efficient execution plans. Without regular updates, the statistics become stale, leading to suboptimal plans and poor performance.

  The statistics gathered by the **`ANALYZE`** command cover various aspects such as the distribution of data within a table, the table's size, the occurrence of null values, and the frequency of distinct values. This information enables the query planner to estimate the number of rows a query will return and the selectivity of each condition in the query.

Understanding data distribution allows the query planner to determine the most efficient ways to access the data. For example, if a table has many distinct values in a column, the planner may decide to use an index for more efficient data retrieval. Conversely, if a column has many null values, a sequential scan might be chosen for better performance.

The statistics obtained through the **ANALYZE** command are particularly valuable in environments where data distribution changes over time, such as frequently updated tables or those with high data turnover. These statistics allow the query planner to adjust the query execution plan accordingly, reflecting the current data distribution.

Here's how to use the **ANALYZE** command in PostgreSQL:

```
ANALYZE VERBOSE [tablename];
```

The **VERBOSE** option provides additional details about the statistics gathered by the **ANALYZE** command.

INDEXES for Performance: In PostgreSQL, indexes are used to facilitate data retrieval. Data is retrieved more quickly when indexes are created. Thus, indexing in PostgreSQL is a process used to improve query speed and performance. They provide a faster lookup mechanism, thus enabling PostgreSQL not to scan the entire table for data retrieval, which can be very slow for tables with large datasets. It is a pointer to data in a table.

Indexes are very useful in speeding up searches, joins, and filtering queries.

There are several indexes in PostgreSQL, namely, **B-tree**, **GIN** (Generalized Inverted Index), **GiST** (Generalized Search Tree), **BRIN** (Block Range Indexes) and **Hash** indexes.

The default index structure in PostgreSQL is B-tree, which is the most used index. It is good for equality (=), range queries (<, <=, >, >=), and ordering (**ORDER BY**). Complex data types such as arrays, JSONB, and full-text search are indexed using GIN indexes. GiST indexes are effective for complex data types such as geometrical data, network addresses, and full-text search. The BRIN indexes are mostly used where data is stored in blocks, especially large and naturally ordered data. Hash indexes are used with queries with equality comparisons (=). They are less commonly used than the B-tree as they do not support range queries.

In PostgreSQL, indexes are usually created on columns that are used in the **WHERE** clause, **JOINS**, **ORDER BY**, and **GROUP BY** operations.

When an index is created on a column in a table, a sorted data structure that allows the database engine to quickly locate specific values is also created by PostgreSQL.

For instance, on a table without indexes, PostgreSQL will scan every row in that table to find rows that meet the query condition. This is called a sequential scan. With an index created on a specified column in a table, relevant rows are directly located, thus improving query performance. This is called an index scan.

We use the **CREATE INDEX** statement to create an index in PostgreSQL:

```
CREATE INDEX index_name ON table_name (column_name);
```

Indexing for Performance improvement

Indexes are great in improving the performance of read queries, but they can slow down the write operations, such as inserts, updates, and deletes because PostgreSQL must update the index as well as the table. Too many indexes thus increase the overhead for maintenance

To view all the indexes on a table, use the following query:

```
SELECT  indexname, indexdef
FROM  pg_indexes
WHERE  tablename = 'your_table_name';
```

When an index is no longer in use, it can be removed using the following query:

```
DROP INDEX index_name;
```

As indexing is very important in PostgreSQL for improving query performance, it is worth noting that creating indexes on a table must be a balancing act due to the impacts they can have on the write operations.

REINDEXING in PostgreSQL

The **REINDEX** command in PostgreSQL is a valuable tool that allows you to rebuild one or more indexes in a database. When executed, it replaces the existing version of the index with a new one, enhancing database performance by improving the efficiency of queries that utilize the index.

Several scenarios may necessitate the use of the **REINDEX** command. For instance, if an index becomes corrupted due to a system failure or application error, rebuilding the index ensures its proper functioning and improves query performance. Additionally, if an index contains many empty or nearly empty pages, it can slow down query processing due to increased disk I/O. Rebuilding the index with the **REINDEX** command consolidates these empty pages, thereby enhancing overall database performance.

Another scenario involves changing a storage parameter for an index. For example, increasing the fill factor of an index to reduce fragmentation and boost performance. In such cases, the **REINDEX** command rebuilds the index with the new storage parameter, ensuring it is optimized for the current database configuration.

It is important to note that the **REINDEX** command can be time-consuming, particularly for large indexes. Therefore, scheduling this operation during off-peak hours is advisable to minimize its impact on database performance.

Here are some examples of how to use the **REINDEX** command in PostgreSQL:

- **Recreate a single index:**
  `REINDEX INDEX index_name;`
- **Recreate all indexes in a table:**
  `REINDEX TABLE table_name;`
- **Recreate all indexes in a schema:**
  `REINDEX SCHEMA public;`
- **Recreate all indexes in a database:**
  `REINDEX DATABASE postgres;`
- **Recreate all indexes on system catalogs in a database:**
  `REINDEX SYSTEM postgres;`

Query Optimization Advantages

- **Query execution time:** Optimizing a query can greatly improve its execution time, thus leading to a quicker response time and an improved overall database performance.
- **Reduced server load:** When queries are optimized, they reduce the load on the server, thus preventing performance bottlenecks and improving the scalability of the database.
- **Improved database efficiency:** The efficiency of queries can be improved when they are optimized, thus reducing resource consumption and costs.

Tools to identify and address query performance bottlenecks include:

- **pgAdmin and DBeaver:** These are popular open-source management tools for PostgreSQL that provide a graphical interface for managing and optimizing databases. They are also used for analyzing and optimizing query performance.
- **pg_stat_statements:** This is a PostgreSQL extension that provides detailed statistics on query performance, including the number of times a query has been executed and the total time it has taken to execute.
- **pg_top:** This is a command-line tool that provides real-time statistics on the performance of a PostgreSQL server, including information on the most active queries and their resource usage.

Maintenance Tasks

Database maintenance is essential to ensure the integrity, performance, and reliability of a database system. Issues such as data corruption, inefficient queries, and system slowness are prevented from running maintenance tasks. Maintenance tasks thus improve database performance over time. Regularly maintaining the PostgreSQL database is crucial, especially with larger and complex databases.

VACUUM, **ANALYZE**, and **REINDEX** are some of the performance improvement maintenance tasks in PostgreSQL.

VACUUM

In PostgreSQL, the process of removing old or deleted data (dead tuples) that is no longer needed from the database is called vacuuming. The space held up by dead tuples is reclaimed when the bloated tables are vacuumed. Vacuum coupled with analyze, helps keep the database statistics up to date, thus improving query performance. Regular vacuuming of databases using the **VACUUM** command keeps them running smoothly by preventing database bloat caused by the Multi-version concurrency control (MVCC) architecture of PostgreSQL.

The reason why **VACUUM** is so important is that PostgreSQL allows concurrent access to the database by keeping multiple versions (MVCC) of rows. Whenever rows in a table are updated or deleted, the old versions of the rows are not physically removed immediately from the table, but rather become **dead tuples** and are no longer visible to new transactions. As more of these dead tuples accumulate, they occupy space, thus leading to what is called **table bloat**. This thus degrades database performance. The occupied space is marked as reusable, meaning that, though the dead tuples no longer exist on the table, the space they occupy is still reserved for them and cannot be used for any other purposes.

VACUUM is used to free up or reclaim the occupied space to be reused or returned to the OS by removing these dead tuples. **VACUUM** makes space available for new data inserts or updates. This ensures that the database does not become unnecessarily large and inefficient. This operation thus reduces the database size and improves its performance. Regularly vacuuming the database will ensure its storage space is utilized efficiently.

VACUUM can be executed with the **ANALYZE** keyword, which helps in freeing up space and updating statistics of the database simultaneously, thus optimizing the database.

When a database is vacuumed regularly, it prevents Transaction ID (XID) Wraparound. In PostgreSQL, every transaction is assigned a transaction ID called XID. The limit of the number of transaction IDs in PostgreSQL is **2 billion**. When this limit is reached, old transaction IDs need to be "*frozen*" to prevent what is called wraparound. Regular vacuuming ensures this process happens correctly and avoids corruption or shutdowns. A detailed explanation will be provided in subsequent sections.

The Visibility Map (VM), which tracks pages in a table that does not contain dead tuples, is maintained by vacuuming the PostgreSQL database. This helps in speeding up index-only scans.

There are two different types of vacuuming processes in PostgreSQL, namely, **VACUUM** and **VACUUM FULL**. There is a significant difference between them.

The **VACUUM** command **removes dead tuples** but does not release the unused space back to the operating system. Instead, it marks the space as available for future use by new rows. The syntax is **VACUUM table_name**.

It can run concurrently with other database operations, making it less disruptive and non-blocking, and does not cause any downtime.

VACUUM FULL reclaims the unused space and returns it to the operating system. It is more aggressive than **VACUUM** by holding exclusive locks on a table before vacuuming. It physically shrinks the table by compacting it. Due to its Blocking operation, it is not suitable to run on large tables in production, as the locking can be very disruptive for production operations. It is typically used when tables have become very bloated or significant space needs to be reclaimed.

The syntax is `VACUUM [FULL] [FREEZE] [VERBOSE] [ANALYZE] [table_name];`

The **FULL** option rewrites the entire table to reclaim all unused space, which is returned to the operating system. Transaction ID wraparound is prevented by using the **FREEZE** option. The **VERBOSE** option reviews details about the **VACUUM** process when it is running. The **ANALYZE** option updates PostgreSQL statistics for the query planner to produce a good execution plan while vacuuming the table.

Note that **VACUUM** or **VACUUM FULL** can be executed on an entire database or single tables. For the entire database, these commands are executed without specifying the table name.

Entire database vacuuming:

```
VACUUM;
VACUUM FULL;
```

For individual tables:

```
VACUUM table_name;
VACUUM FULL table_name;
VACUUM ANALYZE table_name;   This will update the statistics for
the query planner.
VACUUM VERBOSE table_name;   Will show details of the vacuuming
process.
```

pg_repack

In contrast to **`VACUUM FULL`**, **`pg_repack`** offers a more efficient and non-disruptive alternative for eliminating database bloat and reclaiming disk space. While both tools share the goal of optimizing storage, **`pg_repack`** minimizes downtime by requiring only a brief lock during the final phase when the old table is replaced with the new one. This makes it highly suitable for production environments where reducing service interruptions is critical. Moreover, **`pg_repack`** efficiently reorganizes data and rebuilds indexes, enhancing query performance, which is something **`VACUUM FULL`** does not achieve as effectively. However, **`pg_repack`** requires additional disk space and specific prerequisites, such as a primary key or unique index, which are not mandatory for **`VACUUM FULL`**.

`pg_repack` is a PostgreSQL extension designed to address bloat caused by frequent updates and deletes while allowing normal operations to continue. It achieves this by creating a temporary log table to track changes during the process, ensuring data consistency. The tool generates a new, optimized version of the table, applies logged changes, and performs a quick final swap to replace the old table. This efficient, minimal downtime approach makes **`pg_repack`** an excellent choice for maintaining performance and reclaiming space in large databases. Its key benefits include flexibility to target specific tables or indexes, online operation with reduced downtime, and improvements in disk usage and query efficiency. Despite its advantages, it is essential to test **`pg_repack`** in a controlled environment and ensure sufficient resources are available to mitigate potential performance impacts during the process.

Autovacuum

In PostgreSQL, autovacuum is a feature that automatically performs vacuuming in the background, depending on the configuration and workload of the database. It is lightweight, thus no impact on the performance of the database. The following parameters are used to configure autovacuum in PostgreSQL: **`autovacuum_vacuum_scale_factor`**, **`autovacuum_vacuum_threshold`**, and **`autovacuum_analyze_scale_factor`**:

These parameters control the threshold at which the autovacuum starts, the number of dead rows that must be present to trigger a vacuum, and the fraction of the table that must be vacuumed. Autovacuum reduces the need for frequent manual vacuuming, though after large updates or deletes, manual vacuuming is still very necessary.

Based on table activities such as inserts, updates, and deletes, the autovacuum runs automatically. Autovacuum parameters are configured in **`postgresql.conf`** file.

Consequences of not Running `VACUUM` Regularly

- **Table Bloat**: When vacuuming is not regular, tables grow unnecessarily large and become bloated, thus making queries slower as they consume more disk space.

- **Slower Queries**: Query performance is degraded, and queries become slow as dead tuples are still scanned during query execution.
- **Transaction ID Wraparound**: When vacuuming is not efficient, the database can shut down or experience corruption due to transaction IDs reaching their limit, causing transaction ID Wraparound
- **Index Bloat**: Not having efficient vacuuming can also cause bloated indexes, which thus make index-based queries inefficient.

Very important to note is that VACUUMING is a vital maintenance tool for PostgreSQL that helps clean up dead tuples, prevent transaction ID wraparound, and optimize database performance and efficiency. Though Autovacuum is used to automate much of the vacuuming process, manual `VACUUM` is still very much needed to be performed by database administrators for databases with large datasets. Manual vacuuming operations are used to reclaim space that has been occupied by dead tuples. We can manually vacuum specific tables or all the tables in the database.

For large datasets, autovacuum parameters should be aggressive. Choosing between manual vacuum and autovacuum depends on database resources available and specific requirements.

Reindexing is another maintenance tool that updates the indexes to reflect changes in the data. As earlier discussed, indexing is the process of creating a data structure that facilitates the database for data retrieval. In PostgreSQL, indexes are created on specific columns of a table and are used to speed up the execution of queries that filter or sort the data based on those columns.

Advantages of regular maintenance tasks:

- Improved performance: Regular maintenance tasks help to keep the database running smoothly and can improve the performance of queries by keeping data and indexes up to date.
- Reduced disk space usage: Regular vacuuming can help to reduce disk space usage by removing old, deleted, or unneeded data.
- Increased data integrity: Regular maintenance tasks can help to ensure the integrity of the data by removing inconsistencies and errors.

Reporting and Logging

One method to accomplish effective maintenance and troubleshooting is through logging. When an application operates, it generates logs containing brief information statements. PostgreSQL provides several configuration parameters for logging that can aid in maintaining the database. These parameters include:

- **log_statement**: Controls the types of queries logged into log files. Options include **none**, **ddl**, **mod**, and all. DDL queries involve creating, altering, or dropping tables, while MOD queries pertain to inserting, updating, deleting, truncating, and similar operations on a table.
- **logging_collector**: Captures all log messages sent to the standard error logger and redirects them to a log file. This ensures that all log messages are recorded, even those not appearing in the syslog.
- **log_checkpoints**: Logs checkpoints and restart points to the server log, including the number of buffers written and the time taken to write these buffers, which helps in better understanding and debugging.
- **log_min_error_statement**: Sets the minimum logging level for SQL queries generating errors to be logged into the system log. Values include **DEBUG1**, **DEBUG2**, **INFO**, **NOTICE**, **WARNING**, **ERROR**, **FATAL**, and so on.
- **log_line_prefix**: Defines the format of the database logs, enhancing readability.
- **log_lock_waits**: Identifies slow performance and lock waits caused by locking delays.

pgaudit

The **pgaudit** extension is a powerful tool for logging and auditing database activities in PostgreSQL. It enhances the native PostgreSQL logging capabilities by providing detailed session and object-level audit logs. This is particularly useful for organizations that need to comply with regulatory requirements, such as government, financial, or ISO certifications. **pgaudit** captures detailed information about database operations, including the user who performed the action, the specific objects affected, and the type of operation (for example, **SELECT**, **INSERT**, **UPDATE**, **DELETE**).

Key benefits of **pgaudit** include its ability to provide a comprehensive audit trail, making it easier to track and analyze database activities. It can be configured to log specific actions, users, or objects, offering a high degree of customization. Additionally, it redacts sensitive information, such as cleartext passwords, from the logs to enhance security. However, it is important to monitor storage consumption, as detailed logging can generate large volumes of data. **pgaudit** is an invaluable tool for maintaining accountability and security in PostgreSQL environments.

Optimizing Database Parameters for Maximum Efficiency

The goal of PostgreSQL performance optimization is to enhance the efficiency of a PostgreSQL database system by adjusting its configurations and implementing best practices to identify and resolve bottlenecks, improve query speed, and maximize database throughput and responsiveness.

Key areas include:

- **Configuration Parameter Tuning:** This involves modifying variables such as memory allocation, disk I/O settings, and concurrent connections based on specific hardware and requirements.
- **Query Optimization:** This includes analyzing query execution plans, identifying slow queries, and optimizing them through appropriate indexing techniques, query rewriting, or utilizing advanced features such as partial indexes or materialized views.
- **Hardware Optimization:** Ensuring that the CPU, memory, and storage components meet the performance requirements of the database workload is essential.
- **Statistics and Monitoring:** By enabling and analyzing performance statistics and using monitoring tools such as Grafana, Datadog, and Foglight Monitoring and Management, you can identify bottlenecks and track query performance over time.
- **Index Tuning:** PostgreSQL offers various types of indexes, including B-tree, hash, and generalized inverted indexes (GIN/GiST). Selecting the appropriate index type, creating composite indexes, and regularly analyzing and reindexing the database can significantly improve query speed.
- **Schema Design:** Evaluating the database schema design and making adjustments such as partitioning large tables, eliminating redundant data, and denormalizing tables for frequently accessed information can enhance performance.
- **Connection Pooling:** Minimizing connection overhead and improving response times for frequently accessed data can be achieved by implementing connection pooling and caching strategies.
- **Load Balancing and Replication:** Scaling the database system horizontally by distributing the workload across multiple servers using techniques such as connection pooling, read replicas, or a primary replica replication setup.
- **Memory Management:** Since PostgreSQL's performance heavily relies on efficient memory usage, monitoring and adjusting key memory-related

configuration parameters such as **`shared_buffers`**, **`work_mem`**, and **`effective_cache_size`** can have a significant impact on performance.

PostgreSQL performance optimization is an ongoing process involving monitoring and adjustments to maintain high-performing databases. Staying up to date with PostgreSQL releases and improvements can also contribute to this goal.

Importance of PostgreSQL Performance Tuning

The performance of a PostgreSQL database greatly influences the overall effectiveness of an application. Here's why having a high-performing database is crucial for businesses:

- **Responsiveness:** Slow queries and data retrieval can lead to frustrating delays, impacting user experience. A well-performing database ensures quick response times, meeting user expectations, and providing seamless interactions.
- **Throughput:** The speed at which the database processes and serves queries directly affects the application's throughput. A high-performance database reduces query execution time, allowing the application to handle more concurrent requests and deliver data faster.
- **Scalability:** As an application grows, the database must scale to handle more data and user loads. A well-performing PostgreSQL database can manage increased workloads, ensuring responsiveness and performance even under heavy usage.
- **Efficient Resource Utilization:** Optimizing system resources such as CPU, memory, and disk I/O maintains functionality while minimizing hardware requirements, resulting in cost savings and potentially reducing cloud expenses.
- **Data Integrity:** Ensuring data reliability and efficiency is critical. Features such as ACID compliance, transaction management, and strong error handling safeguard against data corruption or loss, which is vital for businesses relying on precise and consistent data for decision-making.
- **Competitive Advantage:** A high-performing database that consistently delivers excellent application performance can enhance user retention rates and customer satisfaction, providing a competitive edge for businesses.

Understanding the Impact of Queries on PostgreSQL Performance

While optimizing PostgreSQL server configuration enhances performance, database developers must also ensure they write efficient queries. Queries that perform full table scans where indexes could be used, or that involve heavy joins or costly aggregate operations, can still lead to poor performance even with optimized database parameters. It is crucial to prioritize performance when writing database queries.

However, database parameters are equally important, so let's examine the eight that hold the greatest potential for enhancing performance.

Tunable PostgreSQL Parameters

The following are some PostgreSQL parameters that can be adjusted for improved performance based on your system and specific workload:

- **`shared_buffers`:** PostgreSQL uses its own buffer along with the kernel-buffered I/O, resulting in data being stored in memory twice—first in PostgreSQL's buffer and then in the kernel buffer. This double buffering involves PostgreSQL's **`shared_buffers`**, which is a critical tunable parameter for most operating systems. It defines the amount of dedicated memory PostgreSQL will use for caching. The default value is set low, but modern machines often benefit from increasing this value. Typically, it is recommended to set **`shared_buffers`** to 25% of total machine RAM, but this can be adjusted based on benchmarking and workload requirements.
- **`wal_buffers`:** This parameter defines the buffer size for PostgreSQL's Write-Ahead Log (WAL). By default, it is set to 16MB. Increasing this value can improve performance, especially in environments with many concurrent connections.
- **`effective_cache_size`:** This provides an estimate of the memory available for disk caching, guiding the optimizer on the cache size. It does not allocate actual memory but helps the planner decide whether to use certain indexes. Setting a large value is generally beneficial.
- **`work_mem`:** Used for complex sorting operations, increasing **`work_mem`** can improve performance by enabling in-memory sorting, which is faster than disk-based sorting. This parameter is a per-user and per-sort operation, so it is recommended to adjust it at the session level to avoid excessive memory usage.
- **`maintenance_work_mem`:** This setting is used for maintenance tasks such as **`VACUUM`**, **`RESTORE`**, **`CREATE INDEX`**, **`ADD FOREIGN KEY`**, and **`ALTER TABLE`**. Increasing maintenance_work_mem can speed up these operations.

- **`synchronous_commit`**: This parameter determines whether commits wait for WAL to be written to disk before returning a success status to the client. Disabling **`synchronous_commit`** can improve performance at the cost of reliability, as data might be lost if a crash occurs before the WAL is flushed.
- **`checkpoint_timeout` and `checkpoint_completion_target`**: These parameters control how often and how long checkpoints occur. **`checkpoint_timeout`** sets the interval between checkpoints, while **`checkpoint_completion_target`** defines the fraction of time between checkpoints dedicated to completing a checkpoint. Adjusting these settings can balance crash recovery time and performance.

There are additional parameters that can be tuned for better performance, though they may have less impact than those highlighted. It is important to remember that not all parameters are relevant for every application type. Tuning PostgreSQL parameters must be tailored to the specific needs of the application and the operating system it runs on.

Table Bloat (Dead Tuples)

Table bloat in PostgreSQL occurs when tables accumulate unused or dead rows, leading to inefficient disk space usage and slower query performance. The primary cause of table bloat is PostgreSQL's Multi-Version Concurrency Control (MVCC) system, which allows multiple versions of a row to exist simultaneously. Here are the main factors contributing to table bloat:

Causes of Table Bloat in PostgreSQL

- **Multi-Version Concurrency Control (MVCC):**
  - PostgreSQL uses MVCC to allow concurrent transactions without locking entire tables. When a row is updated or deleted, its old version remains in the table until it is cleaned up, known as dead tuples.
  - Over time, as updates and deletes increase, the number of dead tuples grows. Although not visible to active transactions, they still consume storage space, resulting in table bloat.
- **Frequent Updates or Deletes:**
  - When a row is updated, PostgreSQL creates a new version while keeping the old one until all referencing transactions are complete.
  - Similarly, deleted rows are marked as "**`dead`**" but not immediately removed.
  - If the **`VACUUM`** process is not run regularly or effectively, these dead rows accumulate, causing bloat.

- **Long-Running Transactions:**
  - PostgreSQL tracks row visibility using transaction IDs (XIDs). If a transaction remains open for a long time, dead tuples cannot be reused because the database does not know if the transaction still needs them.
  - Long-running transactions delay the cleanup of dead tuples, leading to table bloat.
- **Ineffective `VACUUM` Process:**
  - PostgreSQL relies on **`VACUUM`** to clean up dead tuples. If **`VACUUM`** does not run frequently enough or is misconfigured, dead tuples will not be removed promptly, contributing to bloat.
  - The autovacuum feature automatically runs **`VACUUM`**, but if thresholds are too high or the system is heavily loaded, the autovacuum may not run often enough to keep up with dead tuples.
- **Lack of or Delayed Autovacuum:**
  - Autovacuum is designed to clean up dead tuples automatically. However, in high-transaction environments or large databases, autovacuum might lag behind or be sub-optimally configured.
  - If autovacuum does not trigger frequently enough due to high thresholds, dead tuples build up, leading to bloat.
- **Table Growth Without `VACUUM FULL`:** Regular **`VACUUM`** marks dead rows for reuse but does not physically reduce table size. For tables with significant deletions or updates, **`VACUUM FULL`** is needed to reclaim disk space. Without it, the table's physical size remains large despite containing less data.
- **Inefficient Indexing:**
  - Indexes can also bloat similarly to tables. When rows are updated or deleted, corresponding index entries remain until **`VACUUM`** cleans them up.
  - Over time, index bloat can slow queries and consume additional disk space, contributing to inefficiency.
- **Unbalanced Workloads:** High-frequency inserts, updates, and deletes without proper maintenance (for example, **`VACUUM`**, **`ANALYZE`**, **`REINDEX`**) can cause more bloat. Tables frequently updated or deleted are more prone to bloat compared to read-heavy tables.
- **TOAST Tables:**
  - PostgreSQL stores large columns (for example, **`TEXT`**, **`BYTEA`**) using **`TOAST`** (The Oversized-Attribute Storage Technique).
  - **`TOAST`** tables can bloat if large object columns are frequently updated or deleted without corresponding vacuuming, contributing to overall table bloat.

Effects of Table Bloat

- **Increased Disk Space Usage:** Bloat increases table sizes on disk, consuming more storage than necessary.
- **Slower Query Performance:** Query performance degrades as PostgreSQL scans more data (including dead tuples) than is needed.
- **Inefficient Indexes:** Bloated indexes slow down lookups and queries relying on indexed columns.
- **Increased I/O:** Bloat leads to increased disk I/O as PostgreSQL reads and writes more data than necessary, impacting overall system performance.

Identifying Table Bloat

You can identify table bloat by comparing live rows (active rows) with table size. PostgreSQL provides tools and queries for detecting bloated tables:

- **`pg_stat_user_tables`:** This system view provides information about tables, including the count of live and dead tuples.

  ```
  SELECT relname, n_live_tup, n_dead_tup
  FROM pg_stat_user_tables
  WHERE n_dead_tup > 0;
  ```

- **`pgstattuple` extension:** This extension provides detailed information about table bloat.

  ```
  CREATE EXTENSION pgstattuple;
  SELECT * FROM pgstattuple('your_table_name');
  ```

- **`pg_freespacemap`:** This extension shows free space in tables and indexes.

Preventing and Reducing Table Bloat

- **VACUUM Frequently:**
  - Run **`VACUUM`** regularly to remove dead tuples and prevent bloat. You can run it manually or rely on an autovacuum.
  - Ensure autovacuum settings are tuned to match the workload. Lower autovacuum thresholds for heavily updated or deleted tables.
- **Tune Autovacuum:**
  - Adjust autovacuum settings in **`postgresql.conf`** to run more frequently or aggressively on tables prone to bloat.
  - Key settings to adjust include:
    - `autovacuum_vacuum_threshold`
    - `autovacuum_vacuum_scale_factor`

        - `autovacuum_naptime`

- **`VACUUM FULL` for Severe Bloat:**
    - Run **`VACUUM FULL`** to reclaim space by physically shrinking tables. Useful for severe bloat but locks the table while running, so schedule it during low-traffic periods.

```
VACUUM FULL table_name;
```

- **Rebuild Indexes:**
    - Use **`REINDEX`** to rebuild bloated indexes. Index bloat can be as problematic as table bloat, slowing down queries relying on those indexes.

```
REINDEX TABLE table_name;
```

- **Partitioning:**
    - Partition large tables so that older, less frequently accessed data can be vacuumed separately, reducing bloat in frequently updated partitions.
- **Avoid Long-Running Transactions:**
    - Keep transactions as short as possible to prevent dead tuples from being vacuumed and reclaimed.
- **Archiving or Deleting Old Data:**
    - Regularly archive or delete old or unnecessary data to maintain manageable table sizes.

Transaction ID Wraparound

In PostgreSQL, Transaction ID (XID) Wraparound occurs when transaction IDs (XIDs) used to track database transactions reach their maximum possible value, potentially leading to data corruption or system failure if not managed properly. PostgreSQL uses 32-bit transaction IDs, meaning there is a finite limit to the number of transactions before the counter resets or "*wraps around.*"

Understanding Transaction ID:

- Transaction ID (XID) is a 32-bit integer assigned to each transaction in PostgreSQL.
- It tracks the relative timing of transactions, helping PostgreSQL determine the visibility of rows to specific transactions.
- Since XIDs are 32-bit, they can represent around 4.2 billion (2^32) transactions. After reaching this limit, the XID counter resets to zero.

Reason XID Wraparound is Problematic

PostgreSQL uses MVCC to manage database changes, with each row storing metadata that includes the XIDs of the transactions that inserted or deleted the row. When the XID counter reaches its maximum value (4.2 billion transactions), it resets to zero, creating an issue. PostgreSQL relies on the relative age of XIDs to decide which transactions and rows are visible. If old XIDs are mistaken for new ones post-wraparound, it can lead to data corruption or incorrect visibility checks.

How PostgreSQL Prevents XID Wraparound: To mitigate this, PostgreSQL employs a process called freezing, which makes older transactions irrelevant for visibility checks.

Key Concepts in Preventing XID Wraparound:

- **Frozen XIDs:**
  - PostgreSQL replaces very old transaction IDs with a special frozen XID value.
  - A frozen row is treated as if created by an infinitely old transaction, meaning it is always visible to any current or future transaction.
  - Freezing is crucial for rows with very old XIDs, preventing wraparound-related issues.
- **`VACUUM` and Autovacuum:**
  - The **`VACUUM`** process marks tuples (rows) as "**`frozen`**" when their XIDs age sufficiently.
  - Autovacuum is a background process that automatically vacuums tables to prevent XID wraparound.
  - The process freezes long-unmodified rows, allowing PostgreSQL to disregard their original XID values.
- **The `age()` Function:**
  - PostgreSQL uses the **`age()`** function to monitor the relative age of transaction IDs.
  - When the age of the oldest XID in a table nears 2 billion transactions (half of the 4.2 billion limits), PostgreSQL triggers an autovacuum to start freezing tuples, preventing wraparound.
- **`vacuum_freeze_min_age` and `vacuum_freeze_table_age`:**
  - **`vacuum_freeze_min_age`**: Controls the minimum age at which tuples are frozen during a **`VACUUM`** operation.

- **vacuum_freeze_table_age**: Sets the threshold for how old a table's oldest transaction ID can be before autovacuum is triggered to freeze its tuples.

XID Wraparound Prevention Process

- **Normal Operation:**
  - Each new transaction receives a new XID.
  - PostgreSQL tracks the age of all transaction IDs.
- **Autovacuum Kicks In:**
  - As transaction ages grow, PostgreSQL periodically runs the **VACUUM** process (or autovacuum).
  - **VACUUM** reclaims dead tuples and "**freezes**" old stable tuples.
- **Freezing:**
  - When rows in a table reach a certain age (as defined by the **vacuum_freeze_min_age** parameter), **VACUUM** marks their XIDs as frozen.
  - A frozen XID is treated as if the row was inserted by an ancient transaction, always visible to any current or future transaction.
- **Preventing Wraparound Failure:**
  - When the age of the oldest unfrozen XID in a table approaches the wraparound point (usually 2 billion transactions), PostgreSQL forces an autovacuum freeze to handle old XIDs before reaching the critical wraparound limit.

Consequences of Ignoring XID Wraparound

- **Transaction Failures:** PostgreSQL will stop accepting new transactions to avoid potential data corruption if it detects transaction IDs nearing the wraparound point and old rows have not been vacuumed/frozen.
- **Data Corruption:** Without proper freezing, PostgreSQL may misinterpret old rows as part of newer or ongoing transactions, leading to incorrect query results or data inconsistencies.
- **Database Shutdown:** If PostgreSQL gets too close to the wraparound limit, it can force a shutdown to prevent data loss or corruption.

Monitoring XID Age and Wraparound Risk

You can monitor the age of transaction IDs and assess wraparound risk with these queries:

- **Checking the age of XIDs:**

```
SELECT datname, age(datfrozenxid) AS xid_age
FROM pg_database;
```

This query shows the age of the oldest transaction (in terms of XIDs) in each database. If the age approaches 2 billion, action is needed to prevent wraparound.

- **Checking per-table XID status:**

```
SELECT relname, age(relfrozenxid) AS xid_age
FROM pg_class
WHERE relkind = 'r'
ORDER BY age(relfrozenxid) DESC;
```

This shows the age of the oldest unfrozen XID for each table, sorted by age. Tables with very old XIDs should be vacuumed to prevent wraparound.

Avoiding XID Wraparound Issues

- **Ensure the Autovacuum is Running:**
    - Verify that autovacuum is enabled and properly configured, as it is the primary mechanism for preventing XID wraparound.
    - Check autovacuum settings, such as:
        - **`autovacuum_vacuum_scale_factor`**: Determines how much of the table needs to change before triggering autovacuum.
        - **`vacuum_freeze_min_age`**: Controls how old a tuple must be before it is frozen.
        - **`autovacuum_freeze_max_age`**: When the transaction age of a table reaches this limit, an autovacuum is forced to prevent wraparound.
- **Run Manual `VACUUM` if Necessary:**
    - If tables experience many updates or deletes, or the autovacuum cannot keep up, manually running **`VACUUM FREEZE`** might be necessary.

      ```
      VACUUM FREEZE table_name;
      ```
- **Monitor XID Age Regularly:**
    - Use the **`pg_database`** and **`pg_class`** system views to monitor the age of XIDs and ensure they remain far from the wraparound limit.
- **Handle Long-Running Transactions Carefully:**
    - Long-running transactions can hinder vacuuming and freezing, as old rows might still be visible to those transactions.
    - Keep transaction lifetimes short to allow the vacuum to perform effectively.

Visibility Map

In PostgreSQL, the Visibility Map (VM) is a specialized data structure that optimizes the VACUUM process and enhances the performance of index-only scans. The VM tracks whether all tuples (rows) on a specific page of a table are visible to all active transactions. This enables PostgreSQL to skip unnecessary operations during **`VACUUM`** and perform faster index-only scans when possible.

Key Functions of the Visibility Map

- **Optimizing VACUUM:**
  - The VM indicates whether a page contains any dead tuples (deleted or updated rows that need cleanup).
  - If a page is marked as "**`all visible`**," the **`VACUUM`** process can skip it entirely since there are no dead tuples to remove. This increases **`VACUUM`** efficiency, especially for large tables, by avoiding unnecessary scans.
- **Index-Only Scans:**
  - The VM helps optimize index-only scans, allowing PostgreSQL to return query results using only index data without reading the corresponding table (heap) pages.
  - For an index-only scan to be possible, PostgreSQL must ensure that all tuples in the relevant heap pages are visible to all transactions (no row visibility checks needed). The VM provides this information.
  - If a page is marked "**`all visible`**" in the VM, PostgreSQL can skip fetching actual data from the table and return results based on the index alone.

Working of the Visibility Map

- The VM is a bitmap where each bit corresponds to a page (8 KB block) in a table.
- Each bit indicates whether all tuples on that page are visible to every transaction.
  - If a bit is set (marked as "**`all visible`**"), PostgreSQL knows that no further visibility checks are needed for that page.
  - If the bit is clear (not marked), it means some tuples on the page might not be visible to all transactions, necessitating a scan if required.
- The VM is stored as a separate file alongside the table (heap) data and is updated by the **`VACUUM`** process.

Updating the Visibility Map

- **During `VACUUM`:**
  - When **`VACUUM`** processes a table, it identifies pages containing only live tuples (visible to all active transactions) and updates the VM to mark those pages as "**`all visible`**."
  - If dead tuples are found on a page, **`VACUUM`** cleans them up, but the page will not be marked "**`all visible`**" until the next **`VACUUM`**, assuming no further changes occur.
- **When Rows are Modified:**
  - When a page is modified (for example, rows inserted, updated, or deleted), the bit in the VM for that page is cleared. This signals that the page may contain rows requiring further visibility checks, and it cannot be skipped by future **`VACUUM`** operations or index-only scans.
  - The bit will be set again after the next successful **`VACUUM`** run if no further changes have been made to the page.

Example Use Case: `VACUUM` Optimization

- Suppose you have a large table with 1 million pages, but only 100,000 have been modified recently (updated or deleted rows).
- The VM will mark the 900,000 unmodified pages as "**`all visible`**."
- When **`VACUUM`** runs, it only needs to scan the 100,000 modified pages, significantly reducing work and speeding up the process.

Example Use Case: Index-Only Scan

- For a query like:
  `SELECT id FROM users WHERE id > 1000;`
- If an index exists on the id column and the corresponding pages in the table are marked "**`all visible`**" in the VM, PostgreSQL can perform an index-only scan.
- Instead of fetching actual rows from the table, PostgreSQL uses the index alone to satisfy the query, improving performance by avoiding disk I/O.

Key Benefits of the Visibility Map

- **Faster `VACUUM`:**
  - By allowing **`VACUUM`** to skip pages known to contain only live tuples, the VM reduces the time and I/O resources needed for maintenance tasks.

- **Improved Index-Only Scans:**
  - Index-only scans can bypass heap access and retrieve data directly from indexes when the VM shows that all tuples in the relevant heap pages are visible.
  - This significantly reduces disk I/O and improves query performance, especially in read-heavy workloads.

Limitations and Considerations

- **Visibility Map is Approximate:** The VM provides a broad indication of tuple visibility at the page level but does not store tuple-level visibility information. If even a single tuple on a page is not visible to all transactions, the entire page will be marked as not "`all visible`."
- **Needs `VACUUM` to be Effective:** For the VM to remain accurate, `VACUUM` (or autovacuum) must run periodically. Without `VACUUM`, the map will not be updated, impacting both index-only scans and `VACUUM` efficiency.
- **Disk Space Overhead:** The VM is stored in a separate file on disk, introducing a small storage overhead (approximately 1 byte per page of the table).

Practical Section

For us to see all the above in action, we need to create a table with millions of rows and simulate index fragmentation and dead tuples situation in PostgreSQL.

Before the following steps, we first need to create a database and connect to it.

```
postgres=# \l
postgres=# CREATE DATABASE payments;
CREATE DATABASE
postgres=# \l
                              List of databases
   Name    |  Owner   | Encoding |   Collate   |    Ctype    |   Access privileges
-----------+----------+----------+-------------+-------------+-----------------------
 payments  | postgres | UTF8     | en_GB.UTF-8 | en_GB.UTF-8 |
 postgres  | postgres | UTF8     | en_GB.UTF-8 | en_GB.UTF-8 |
 template0 | postgres | UTF8     | en_GB.UTF-8 | en_GB.UTF-8 | =c/postgres          +
           |          |          |             |             | postgres=CTc/postgres
 template1 | postgres | UTF8     | en_GB.UTF-8 | en_GB.UTF-8 | =c/postgres          +
           |          |          |             |             | postgres=CTc/postgres
(4 rows)

postgres=# \c payments
You are now connected to database "payments" as user "postgres".
payments=#
```

Figure 9.1: *Create a new database and connect to it*

To generate a table with a million rows and simulate both index fragmentation and dead tuples in PostgreSQL, follow these steps:

1. **Create a Table**:

```
CREATE TABLE providers (
    id SERIAL PRIMARY KEY,
    data TEXT
);
```

2. **Insert a Million Rows**:

```
INSERT INTO providers (data)
SELECT 'Row ' || generate_series(1, 1000000);
```

```
payments=# CREATE TABLE providers (
    id SERIAL PRIMARY KEY,
    data TEXT
);
CREATE TABLE
payments=#
payments=# INSERT INTO providers (data)
SELECT 'Row ' || generate_series(1, 1000000);
INSERT 0 1000000
payments=#
```

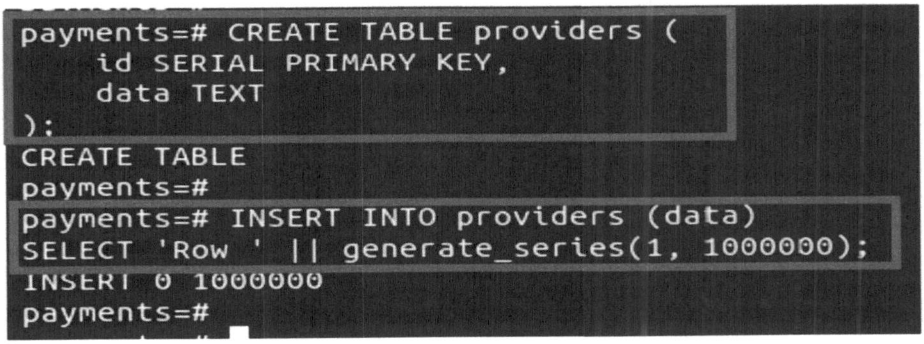

Figure 9.2: *Create a table and insert 1 million rows of data into it*

3. **Create an Index**:

```
CREATE INDEX idx_ providers_data ON providers(data);
```

```
payments=#
payments=# CREATE INDEX idx_providers_data ON providers(data);
CREATE INDEX
payments=#
```

Figure 9.3: *Create an index on the table*

4. **Simulate Index Fragmentation**: Perform multiple updates and deletes to fragment the index.

```
-- Update rows to create fragmentation
UPDATE providers SET data = data || ' updated'
WHERE id % 2 = 0;
-- Delete some rows to create more fragmentation and dead tuples
DELETE FROM providers WHERE id % 10 = 0;
```

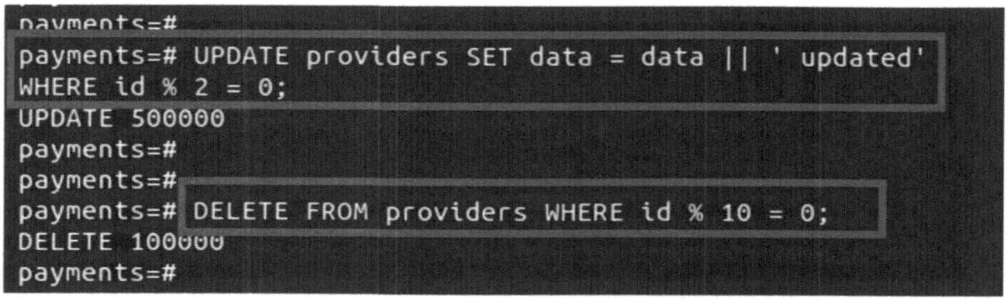

```
payments=#
payments=# UPDATE providers SET data = data || ' updated'
WHERE id % 2 = 0;
UPDATE 500000
payments=#
payments=#
payments=# DELETE FROM providers WHERE id % 10 = 0;
DELETE 100000
payments=#
```

Figure 9.4: *Update multiple times to simulate index fragmentation*

```
UPDATE providers SET data = data || ' updated'   WHERE id % 2 = 0;
UPDATE providers SET data = data || ' updated' WHERE id % 4 = 0;
UPDATE providers SET data = data || ' updated' WHERE id % 3 = 0;
```

```
payments=# UPDATE providers SET data = data || ' updated'
WHERE id % 2 = 0;
UPDATE 400000
payments=# UPDATE providers SET data = data || ' updated'
WHERE id % 4 = 0;
UPDATE 200000
payments=# UPDATE providers SET data = data || ' updated'
WHERE id % 3 = 0;
UPDATE 300000
```

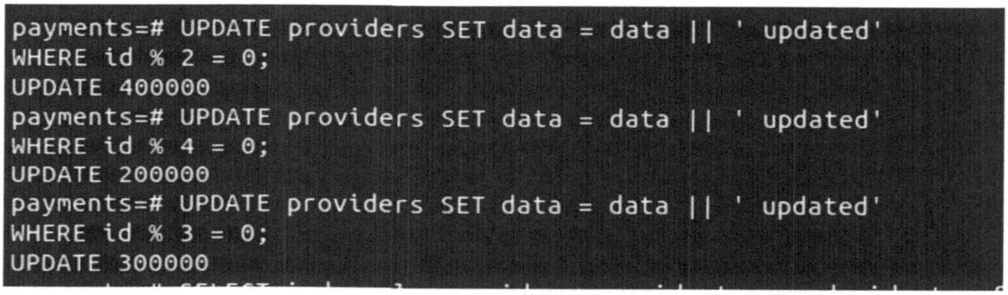

Figure 9.5: *More of the updates*

5. **Check for Dead Tuples**: Use the **pg_stat_user_tables** view to check the number of live and dead tuples.

```
SELECT relname, n_live_tup, n_dead_tup
FROM pg_stat_user_tables
WHERE relname = 'providers';
```

```
payments=# SELECT relname, n_live_tup, n_dead_tup
FROM pg_stat_user_tables
WHERE relname = 'providers';
  relname   | n_live_tup | n_dead_tup
------------+------------+------------
 providers  |     900000 |     100000
(1 row)

payments=#
```

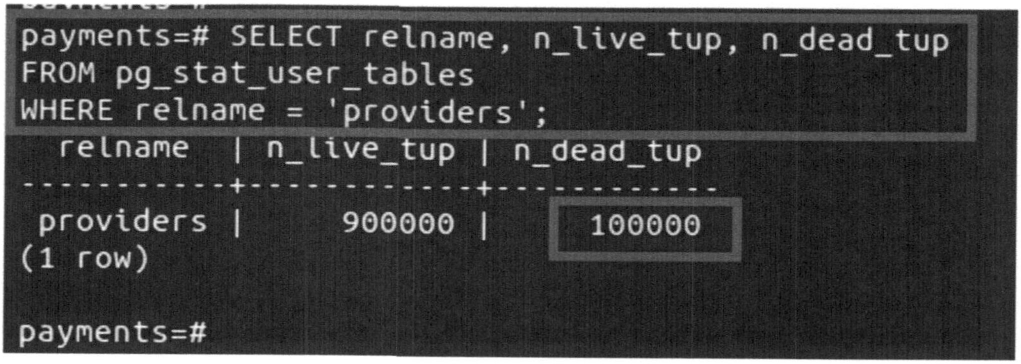

Figure 9.6: *Check the amount of dead tuples*

6. **Check Index Fragmentation**: Use the **pg_stat_all_indexes** view to check index statistics.

```
SELECT indexrelname, idx_scan, idx_tup_read, idx_tup_fetch
FROM pg_stat_all_indexes
WHERE relname = 'providers';
```

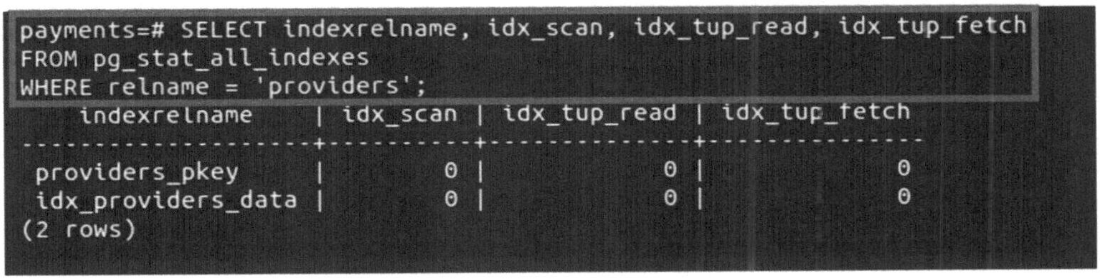

```
payments=# SELECT indexrelname, idx_scan, idx_tup_read, idx_tup_fetch
FROM pg_stat_all_indexes
WHERE relname = 'providers';
    indexrelname     | idx_scan | idx_tup_read | idx_tup_fetch
---------------------+----------+--------------+---------------
 providers_pkey      |        0 |            0 |             0
 idx_providers_data  |        0 |            0 |             0
(2 rows)
```

Figure 9.7: *Check index usage*

```
CREATE EXTENSION pgstattuple;
SELECT * FROM pgstatindex('idx_providers_data');
\x
SELECT * FROM pgstatindex('idx_providers_data');
```

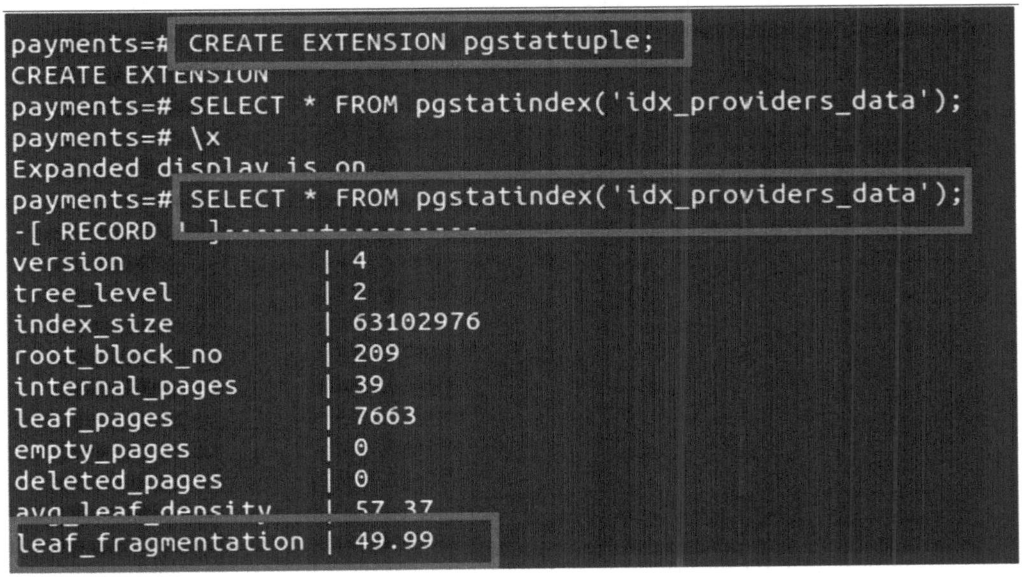

```
payments=# CREATE EXTENSION pgstattuple;
CREATE EXTENSION
payments=# SELECT * FROM pgstatindex('idx_providers_data');
payments=# \x
Expanded display is on.
payments=# SELECT * FROM pgstatindex('idx_providers_data');
-[ RECORD 1 ]------+---------
version            | 4
tree_level         | 2
index_size         | 63102976
root_block_no      | 209
internal_pages     | 39
leaf_pages         | 7663
empty_pages        | 0
deleted_pages      | 0
avg_leaf_density   | 57.37
leaf_fragmentation | 49.99
```

Figure 9.8: *Check the percentage of index fragmentation*

7. **Run VACUUM**: Execute the **VACUUM** command to clean up dead tuples.

```
VACUUM providers;
```

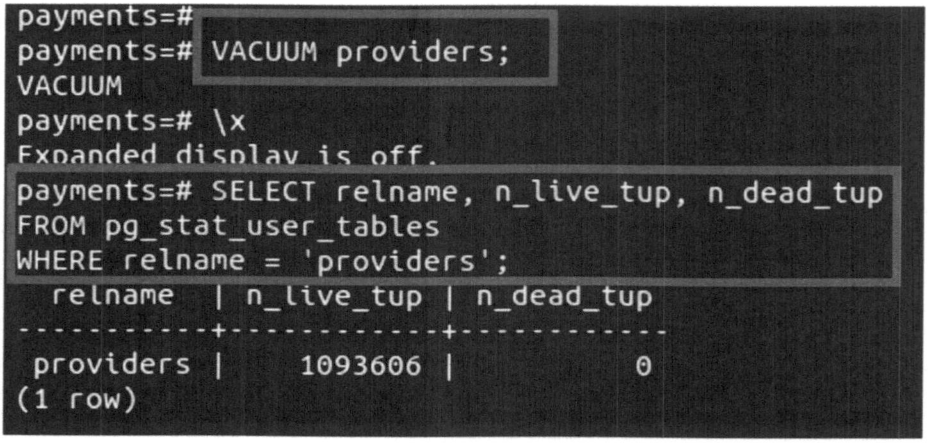

Figure 9.9: *Vacuum Providers table to clean the dead tuples*

8. **Run `REINDEX`**: Reindex the fragmented index.

```
REINDEX INDEX idx_ providers_data;
```

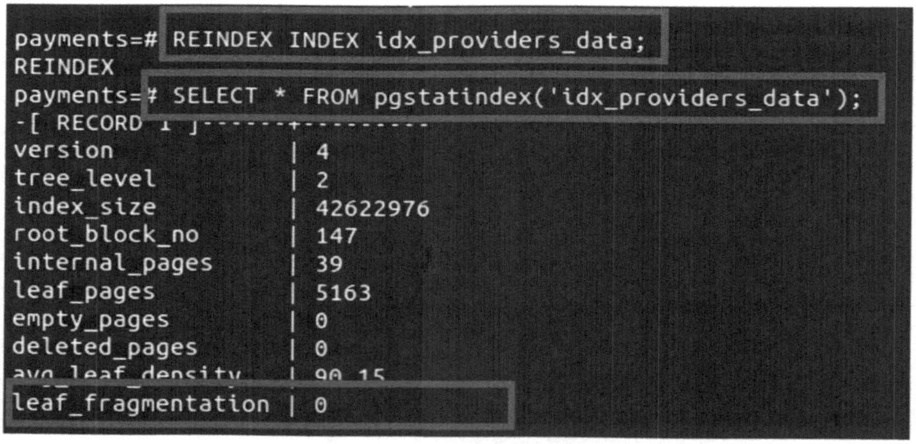

Figure 9.10: *Reindex to rebuild the fragmented indexes*

9. **Check Statistics Again**: Check the number of live and dead tuples as well as index statistics again after running **VACUUM** and **REINDEX**.

```
-- Check live and dead tuples again
SELECT relname, n_live_tup, n_dead_tup
FROM pg_stat_user_tables
WHERE relname = 'providers';

-- Check index statistics again
SELECT indexrelname, idx_scan, idx_tup_read, idx_tup_fetch
FROM pg_stat_all_indexes
WHERE relname = 'providers';
```

Here is a summary of the process:

- **Step 1**: Create a table named providers with columns id and data.
- **Step 2**: Insert one million rows using the **`generate_series`** function.
- **Step 3**: Create an index on the data column.
- **Step 4**: Perform updates and deletes to fragment the index and create dead tuples.
- **Step 5**: Check the number of live and dead tuples to see the current state.
- **Step 6**: Check index statistics to evaluate the level of fragmentation.
- **Step 7**: Run the **`VACUUM`** command to clean up dead tuples.
- **Step 8**: Run the **`REINDEX`** command to rebuild the fragmented index.
- **Step 9**: Check statistics again to observe the changes after running **`VACUUM`** and **`REINDEX`**.

Following these steps will help you simulate a situation with both index fragmentation and dead tuples in PostgreSQL and see how maintenance commands such as **`VACUUM`** and **`REINDEX`** can improve performance.

Locks in PostgreSQL

Locking in PostgreSQL is a critical concept that plays a central role in maintaining data consistency and ensuring concurrent access to the database. However, when not managed effectively, locking can lead to performance bottlenecks such as **deadlocks** and **table blocking**, which directly impact database efficiency.

Types of Locks in PostgreSQL

PostgreSQL uses various lock types to control access to tables, rows, and other database objects:

Row-Level Locks

- Row-level locks are lightweight and operate on individual rows within a table. These locks are useful for ensuring that only specific rows are affected during concurrent transactions, minimizing contention on other rows.
- Examples include **`SELECT FOR UPDATE`** and **`SELECT FOR SHARE`**.
  - **`SELECT FOR UPDATE`**: Locks selected rows for updates, preventing other transactions from modifying or locking them until the current transaction completes.
  - **`SELECT FOR SHARE`**: Allows other transactions to read the rows but prevents them from acquiring exclusive locks (for example, updates or deletions).

- These locks are ideal for fine-grained control, ensuring better concurrency compared to full table locks.

Table Locks

- Table locks apply to entire tables and are often used for operations such as schema changes (for example, `ALTER TABLE`) or maintenance tasks (for example, `VACUUM FULL`).
- They are more restrictive and can cause blocking if held by long-running queries, as other transactions are unable to access the table until the lock is released.
- PostgreSQL supports several types of table locks, including:
    - `ACCESS EXCLUSIVE`: Blocks all other operations, including `SELECTs`. Used for operations such as `DROP TABLE`.
    - `ACCESS SHARE`: Allows other transactions to read the table but blocks schema modifications.
    - `SHARE ROW EXCLUSIVE`: Used for operations such as `CREATE INDEX`, blocking other modifications but allowing `SELECT` queries.

Shared and Exclusive Locks

- **Shared Locks:**
    - Allow multiple transactions to hold the same lock on a resource simultaneously, as long as they are performing non-conflicting operations (e.g., reading data).
    - These locks block write operations but allow concurrent reads.
    - **Example**: A shared lock is acquired during a `SELECT` query to prevent changes to the table's structure while data is being read.
- **Exclusive Locks:**
    - Grant exclusive access to a resource, blocking all other transactions from reading or writing to it.
    - Exclusive locks are typically used for operations that alter data or the structure of a table.
    - **Example**: `INSERT`, `UPDATE`, or `DELETE` operations acquire exclusive locks on the modified rows.

Lock Modes in PostgreSQL

PostgreSQL provides several lock modes for fine-grained control:

- **Row Share**: Used by `SELECT FOR UPDATE` or `SELECT FOR SHARE`, allowing concurrent access for other queries that do not modify the locked rows.

- **Row Exclusive**: Acquired by **INSERT**, **UPDATE**, or **DELETE** queries on specific rows.
- **Share**: Allows concurrent queries to read data or create indexes but prevents data modification.
- **Share Update Exclusive**: Blocks vacuum operations but allows regular **SELECT**, **INSERT**, or **UPDATE** queries.
- **Access Exclusive**: The most restrictive lock, used by operations such as **ALTER TABLE** or **DROP TABLE**.

Deadlocks

In PostgreSQL, **deadlocks** occur when two or more transactions are unable to proceed because they are waiting on each other to release locks on resources they both need. For instance, if Transaction A locks Resource 1 and then tries to lock Resource 2, while Transaction B locks Resource 2 and tries to lock Resource 1, a cycle is created where both transactions are stuck. Deadlocks can severely impact database operations by halting transactions and delaying processes. However, PostgreSQL is equipped with a **deadlock detection mechanism** that identifies such situations and resolves them by forcefully aborting one of the transactions. While this mechanism helps keep the database functional, the aborted transaction may result in incomplete operations, leading to additional work to retry or handle the failure.

In PostgreSQL, **deadlocks** occur when two or more transactions are unable to proceed because they are waiting on each other to release locks on resources they both need. For instance, if Transaction A locks Resource 1 and then tries to lock Resource 2, while Transaction B locks Resource 2 and tries to lock Resource 1, a cycle is created where both transactions are stuck. Deadlocks can severely impact database operations by halting transactions and delaying processes. However, PostgreSQL is equipped with a **deadlock detection mechanism** that identifies such situations and resolves them by forcefully aborting one of the transactions. While this mechanism helps keep the database functional, the aborted transaction may result in incomplete operations, leading to additional work to retry or handle the failure.

How to Prevent Deadlocks

To minimize the likelihood of deadlocks, there are several key strategies:

- **Consistent Locking Order**: Always acquire locks in the same sequence across all transactions. For example, if multiple transactions need to lock Resource 1 first and then Resource 2, maintaining this order ensures that circular dependencies do not occur. Inconsistent locking order is one of the most common causes of deadlocks.

- **Short Transactions**: Keep transactions brief and efficient to limit the duration for which locks are held. The longer a transaction holds a lock, the greater the chances of contention with other transactions, increasing the risk of deadlocks.
- **Avoid Explicit Locks**: Explicit locking constructs such as `SELECT FOR UPDATE` should only be used when absolutely necessary. Often, PostgreSQL's internal concurrency mechanisms handle resource management effectively without the need for explicit locks. Relying too heavily on explicit locks can exacerbate the risk of deadlocks.
- **Index Optimization**: Optimize indexes to improve query performance, as faster queries reduce the time locks are held, thereby minimizing contention.
- **Regular Monitoring and Analysis**: Use PostgreSQL tools such as `pg_stat_activity` and `pg_locks` to monitor active locks and transactions. Proactively analyzing these can help identify patterns or problematic queries that frequently cause deadlocks.

Conclusion

In this chapter, we have delved into the Vacuum, Analyze, and Reindex commands in PostgreSQL and their significance in optimizing database performance. These commands are indispensable for maintaining a healthy database, freeing up storage space, and enhancing query execution times. By regularly executing these commands and adhering to best practices, you can ensure that your PostgreSQL database operates efficiently, providing fast and reliable access to your data.

Table bloat in PostgreSQL primarily arises from dead tuples accumulated due to the MVCC system and inadequate vacuuming. Frequent updates, deletions, long-running transactions, and inefficient autovacuum configurations exacerbate this issue. Regular maintenance tasks such as running `VACUUM`, `VACUUM FULL`, and `REINDEX`, along with proper tuning of autovacuum settings, are vital for preventing and reducing bloat, thereby ensuring optimal database performance.

Transaction ID (XID) Wraparound in PostgreSQL is a critical issue that occurs when the transaction ID counter reaches its 32-bit limit and resets. PostgreSQL manages this by freezing old XIDs, primarily during the `VACUUM` process (or autovacuum). To prevent wraparound, it is essential to run `VACUUM` regularly and monitor the age of transactions. Proper tuning of autovacuum settings and the use of monitoring tools can help maintain your PostgreSQL database's integrity and prevent wraparound issues.

The Visibility Map in PostgreSQL is an essential optimization tool for both **VACUUM** operations and index-only scans. By tracking which pages in a table contain tuples visible to all transactions, the Visibility Map allows PostgreSQL to skip unnecessary table scans during maintenance tasks and execute faster, more efficient queries. Regular **VACUUM** (or autovacuum) runs are crucial for keeping the Visibility Map up to date and fully leveraging its benefits.

In the next chapter, we will be looking into the different types of functions in PostgreSQL, which include Aggregate Functions, SQL Functions, PL/pgSQL Functions, Trigger Functions, Window Functions, Set Returning Functions, and String Functions.

CHAPTER 10
Functions in PostgreSQL

Introduction

In this chapter, learners will understand the different types of functions in PostgreSQL and their importance in managing the database. Functions in PostgreSQL are essential tools that encapsulate SQL logic, enabling the reuse of code, improving modularity, and simplifying database management. They can accept parameters, perform operations, and return results. Let us dive into the different types of functions in PostgreSQL, providing in-depth explanations and examples using the **`dvdrental`** sample database. Please refer to *Chapter 2, Installing and initializing PostgreSQL*, on how to install the **`dvdrental`** sample database.

Structure

In this chapter, the types of functions in PostgreSQL will be covered:

- Aggregate Functions
- SQL Functions
- PL/pgSQL Functions
- Trigger Functions
- Window Functions
- Set Returning Functions
- String Functions

Aggregate Functions in PostgreSQL

Aggregate functions in PostgreSQL perform calculations on a set of rows and return a single value. These functions are essential for summarizing and analyzing data, helping you derive meaningful insights from large datasets. Here are some of the most commonly used aggregate functions: **`SUM()`**, **`AVG()`**, **`MAX()`**, **`MIN()`**, and **`COUNT()`**.

Let us explore each function in detail using the **`dvdrental`** sample database, which includes tables such as payment, rental, film, and customer.

`SUM()` Function

The **`SUM()`** function adds up all the values in a specified column.

Example: Suppose you want to calculate the total amount of payments made by customers. The syntax is:

```
SELECT SUM(amount) AS total_payments
FROM payment;
```

```
dvdrental=#
dvdrental=# Select * from public.payment limit 4;
 payment_id | customer_id | staff_id | rental_id | amount |        payment_date
------------+-------------+----------+-----------+--------+----------------------------
      17503 |         341 |        2 |      1520 |   7.99 | 2007-02-15 22:25:46.996577
      17504 |         341 |        1 |      1778 |   1.99 | 2007-02-16 17:23:14.996577
      17505 |         341 |        1 |      1849 |   7.99 | 2007-02-16 22:41:45.996577
      17506 |         341 |        2 |      2829 |   2.99 | 2007-02-19 19:39:56.996577
(4 rows)

dvdrental=# SELECT SUM(amount) AS total_payments
FROM payment;
 total_payments
----------------
       61312.04
(1 row)

dvdrental=#
```

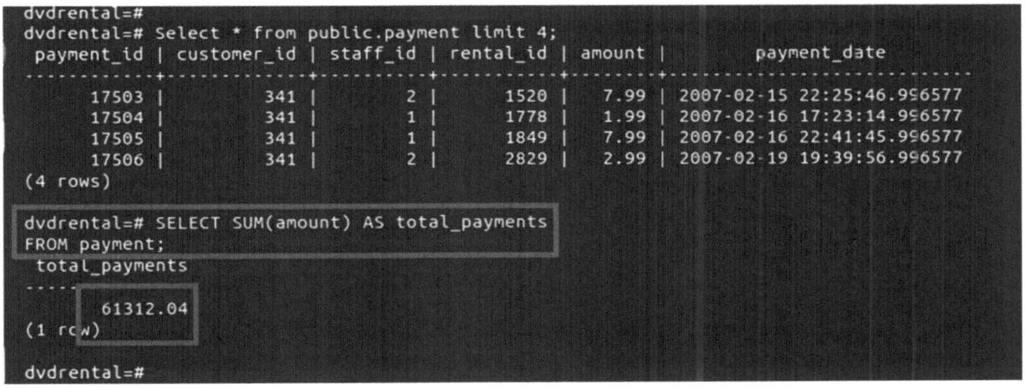

Figure 10.1: *Shows SELECT on payment and a sum function on amount column in payments table*

The first query above is just to see the amount column in the payment table. The second query is the sum function on the amount column. Note that the Select query is a limit of 4 rows.

Explanation: This query sums up all the values in the amount column of the payment table and returns the total amount of payments.

`AVG()` and `ROUND()` Functions

The **`AVG()`** function calculates the average value of a specified column.

Example: Suppose you want to find the average payment amount.

```
SELECT AVG(amount) AS average_payment
FROM payment;
```

```
dvdrental=#
dvdrental=# SELECT AVG(amount) AS average_payment
FROM payment;
  average_payment

 4.2006056453822965
(1 row)

dvdrental=#
```

Figure 10.2: *Average (AVG) function on amount column*

To ensure the result of your query displays the average payment amount to two decimal places, you can use the **ROUND** function. Here is the modified query:

```
SELECT ROUND(AVG(amount), 2) AS average_payment
FROM payment;
```

```
dvdrental=# SELECT ROUND(AVG(amount), 2) AS average_payment
FROM payment;
 average_payment
-----------------
            4.20
(1 row)

dvdrental=#
```

Figure 10.3: *Result of AVG function rounded to 2 decimal places*

Explanation:

- **SELECT Statement:**
  - This part of the query is used to specify the columns of data you want to retrieve from a database.
- **AVG(amount):**
  - **AVG()** is an aggregate function that calculates the average value of the specified column. In this case, it calculates the average of the amount column from the payment table.
- **ROUND(AVG(amount), 2):**
  - **ROUND()** is a function that rounds a number to a specified number of decimal places.
  - In this query, **ROUND()** is used to round the result of **AVG(amount)** to 2 decimal places.

- This ensures the average payment amount is displayed with two decimal points for better readability and precision.

- **AS average_payment:**
  - **AS** is a keyword used to provide an alias to the column or result. It renames the result of the expression for better readability in the output.
  - In this query, the result of **ROUND(AVG(amount), 2)** is given the alias **average_payment**, so it will appear as a column named **average_payment** in the result set.
- **FROM payment:**
  - This part specifies the table from which to retrieve the data. In this case, it is the payment table.

The above query calculates the average value of the amount column from the payment table, rounds that average to two decimal places, and names the resulting column average_payment. This provides a clear, formatted result that shows the average payment amount with two decimal places.

MAX() Function

The **MAX()** function returns the maximum value in a specified column.

Example: Suppose you want to find the highest payment amount made by any customer.

```
SELECT MAX(amount) AS highest_payment
FROM payment;
```

```
dvdrental=#
dvdrental=# SELECT MAX(amount) AS highest_payment
FROM payment;
 highest_payment
-----------------
           11.99
(1 row)

dvdrental=#
```

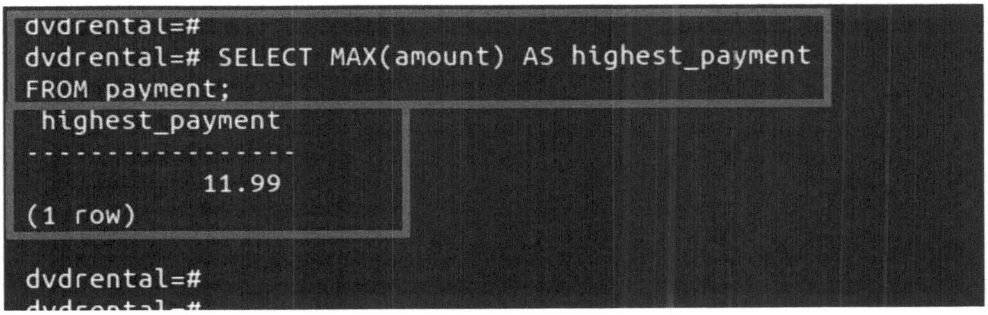

***Figure 10.4:** The Max function to find the maximum amount*

Explanation: This query retrieves the highest value in the amount column of the payment table and returns it as the highest payment amount.

MIN() Function

The **MIN()** function returns the minimum value in a specified column.

Example: Suppose you want to find the smallest payment amount made by any customer.

```
SELECT MIN(amount) AS smallest_payment
FROM payment;
```

```
dvdrental=# SELECT MIN(amount) AS smallest_payment
FROM payment;
 smallest_payment
------------------
             0.00
(1 row)

dvdrental=#
dvdrental=#
```

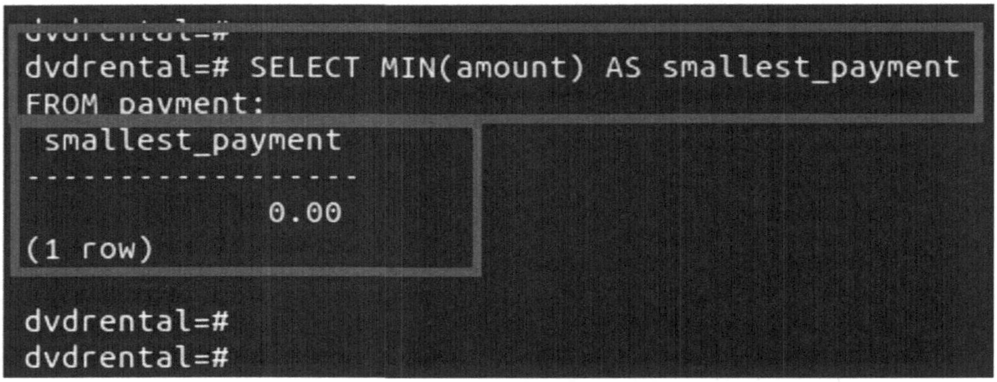

***Figure 10.5:** Min function to find the minimum or smallest amount*

Explanation: This query retrieves the smallest value in the amount column of the payment table and returns it as the smallest payment amount.

COUNT() Function

The **COUNT()** function counts the number of rows that match a specified condition.

Example: Suppose you want to count the total number of payments made.

```
SELECT COUNT(*) AS total_payments
FROM payment;
```

```
dvdrental=# SELECT COUNT(*) AS total_payments
FROM payment;
 total_payments
----------------
          14596
(1 row)

dvdrental=#
```

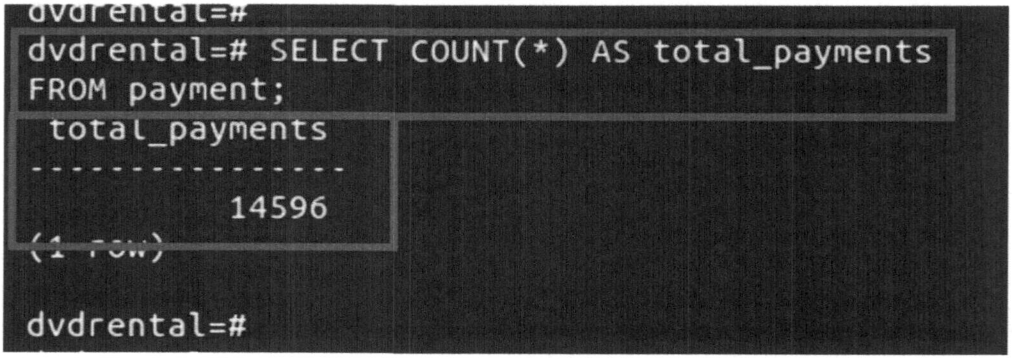

***Figure 10.6:** Count function to find the total number of rows in the payment table*

Explanation: This query counts the total number of rows in the payment table and returns the count as the total number of payments made.

Combining Aggregate Functions with GROUP BY

Aggregate functions are often used in conjunction with the **GROUP BY** clause to summarize data by categories. Let us see an example:

Example: Suppose you want to calculate the total amount of payments and the number of payments made by each customer.

```
SELECT customer_id, SUM(amount) AS total_payments, COUNT(*) AS
number_of_payments
FROM payment
GROUP BY customer_id;
```

```
dvdrental=# SELECT customer_id, SUM(amount) AS total_payments, COUNT(*) AS number_of_payments
FROM payment
GROUP BY customer_id;

 customer_id | total_payments | number_of_payments
-------------+----------------+--------------------
         184 |          80.80 |                 20
          87 |         137.72 |                 28
         477 |         106.79 |                 21
         273 |         130.72 |                 28
         550 |         151.69 |                 31
          51 |         123.70 |                 30
         394 |          77.80 |                 20
         272 |          65.87 |                 13
          70 |          75.83 |                 17
         190 |         102.75 |                 25
         350 |          63.79 |                 21
         539 |          84.80 |                 20
         554 |          95.80 |                 20
         278 |          71.79 |                 21
         424 |         109.71 |                 29
         406 |         121.69 |                 31
```

Figure 10.7: *Sum function with the GROUP BY condition*

- **`customer_id`**: This is the column by which the results are grouped.
- **`SUM(amount) AS total_payments`**: Calculates the total payment amount for each customer.
- **`COUNT(*) AS number_of_payments`**: Counts the number of payments made by each customer.
- **`GROUP BY customer_id`**: Groups the results by the **`customer_id`** column.

Practical Applications of Aggregate Functions

- **Financial Reporting**: Calculating total revenue, average sales, and other financial metrics.
- **Customer Analytics**: Understanding customer behavior by analyzing the frequency and number of transactions.
- **Performance Metrics**: Measuring key performance indicators (KPIs) like maximum and minimum values for various metrics.

- **Data Summarization**: Summarizing large datasets to extract meaningful insights and trends.

By understanding and effectively using aggregate functions, you can perform powerful data analysis and derive valuable insights from your PostgreSQL databases. These functions are foundational tools for any database administrator or data analyst, enabling efficient and comprehensive data management.

Aggregate functions perform a calculation on a set of rows and return a single row. They are used to summarize data. It is used to calculate the average value in a set of values.

SQL Functions

SQL functions are written in plain SQL. They are useful for straightforward tasks, such as data retrieval and simple calculations.

Example: Create a function to get the total payment amount from the payment table:

```
CREATE FUNCTION get_total_payment() RETURNS DECIMAL AS $$
BEGIN
    RETURN (SELECT SUM(amount) FROM payment);
END;
$$ LANGUAGE plpgsql;
```

```
dvdrental=# CREATE FUNCTION get_total_payment() RETURNS DECIMAL AS $$
BEGIN
    RETURN (SELECT SUM(amount) FROM payment);
END;
$$ LANGUAGE plpgsql;
CREATE FUNCTION
dvdrental=#
```

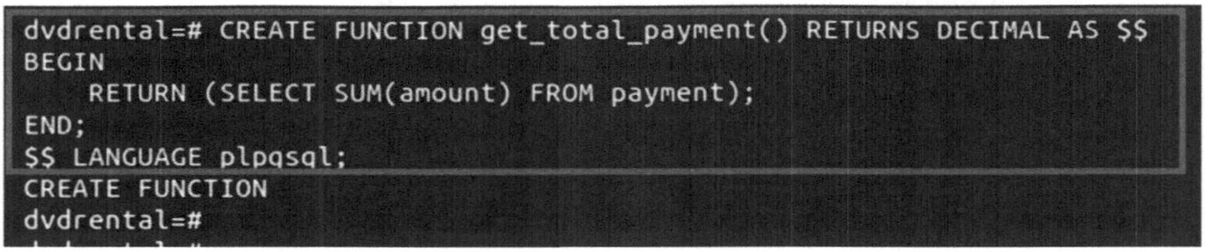

Figure 10.8: *A simple function that returns the sum of the amount*

- **`CREATE FUNCTION get_total_payment():`**
  - This part of the statement is used to define a new function named **`get_total_payment`**.
  - The empty parentheses **`()`** indicate that this function does not take any parameters.
- **`RETURNS DECIMAL:`**
  - This specifies that the function will return a value of the **`DECIMAL`** data type.
  - It means the result of the function will be a decimal number.
- **`AS $$:`**
  - The AS $$ syntax is used to define the body of the function using dollar-

quoted string constants. This allows for writing the function body in a way that can include quotes and other special characters without requiring them to be escaped.

  - It also signifies the start of the function body.

- **`BEGIN ... END;:`**
  - This block encloses the main logic of the function. The function body starts with **`BEGIN`** and ends with **`END;`**.

- **`RETURN (SELECT SUM(amount) FROM payment);:`**
  - Inside the function body, this line performs the main task.
  - The **`RETURN`** statement is used to specify the value that the function will return when it is called.
  - **`SELECT SUM(amount) FROM payment;`** is a subquery that calculates the total sum of the amount column from the payment table.
  - The result of this subquery, which is the total payment amount, is returned by the function.

- **`$$ LANGUAGE plpgsql;:`**
  - This closes the function body and specifies that the function is written in the PL/pgSQL procedural language, which is used for writing functions and trigger procedures in PostgreSQL.

The **`get_total_payment`** function is a user-defined function in PostgreSQL that calculates and returns the total sum of the amount column from the payment table. When you call this function, it executes the **`SELECT SUM(amount) FROM payment`** query, sums up all the amounts in the payment table, and returns that total as a decimal number.

Shown below is how to call the function.

```
SELECT get_total_payment();
```

```
dvdrental=# SELECT get_total_payment();
 get_total_payment
-------------------
          61312.04
(1 row)

dvdrental=#
```

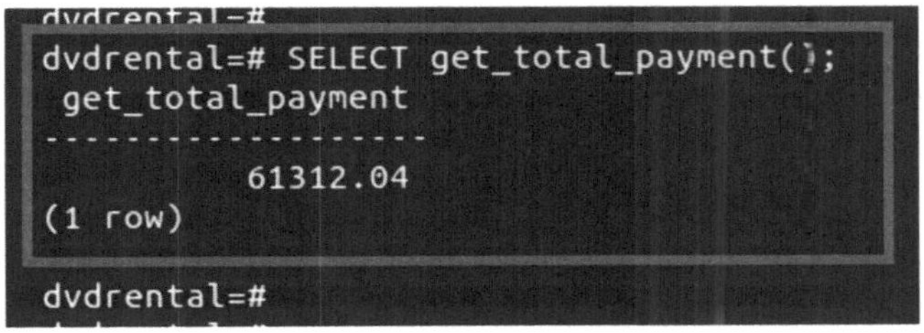

***Figure 10.9**: How to call the function*

This function sums all the payment amounts in the payment table and returns the total.

PL/pgSQL Functions

Procedural Language/PostgreSQL (PL/pgSQL) is a procedural language supported by PostgreSQL that allows you to write functions and procedures with more complex operations. It extends SQL by adding control structures (such as loops and conditionals), exception handling, and more, making it a powerful tool for database developers and administrators.

Example: Create a function to calculate the total payments made by a specific customer:

```
CREATE FUNCTION get_customer_payment(customer_id INT) RETURNS
DECIMAL AS $$
DECLARE
    total_payment DECIMAL;
BEGIN
    SELECT SUM(amount) INTO total_payment
    FROM payment
    WHERE payment.customer_id = get_customer_payment.customer_id;
    RETURN total_payment;
END;
$$ LANGUAGE plpgsql;
```

```
dvdrental=#
dvdrental=# CREATE FUNCTION get_customer_payment(customer_id INT) RETURNS DECIMAL AS $$
DECLARE
    total_payment DECIMAL;
BEGIN
    SELECT SUM(amount) INTO total_payment
    FROM payment
    WHERE payment.customer_id = get_customer_payment.customer_id;
    RETURN total_payment;
END;
$$ LANGUAGE plpgsql;
CREATE FUNCTION
dvdrental=#
dvdrental=#
```

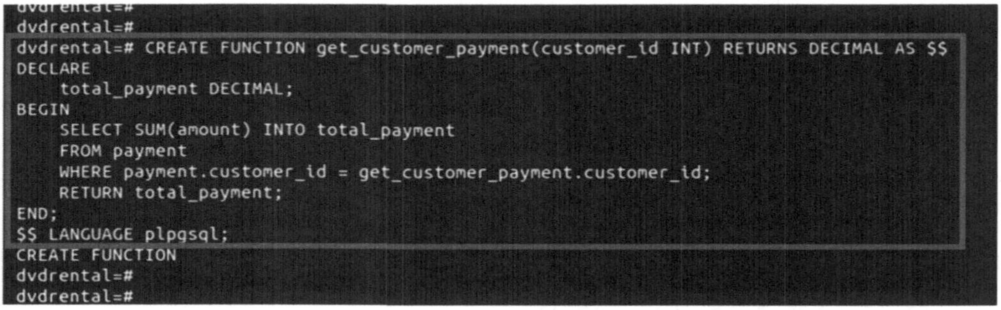

Figure 10.10: *How to create a store procedure that takes one parameter*

- The **CREATE FUNCTION get_customer_payment(customer_id INT)** statement creates a function named **get_customer_payment** that takes one parameter, **customer_id**, of type INT.
- The **RETURNS DECIMAL** specifies that the function will return a value of the **DECIMAL** data type.
- The **DECLARE** keyword introduces the declaration section where variables used within the function are defined.
- The **total_payment DECIMAL** declares a variable named **total_payment** of type **DECIMAL**, which will store the sum of payments for the specified customer.

- The **BEGIN ... END;** encloses the main logic of the function.
- The **SELECT SUM(amount) INTO total_payment** statement calculates the sum of the amount column from the payment table and stores the result into the **total_payment** variable.
- The **FROM payment WHERE payment.customer_id = get_customer_payment.customer_id** filters the records in the payment table to include only those where the **customer_id** matches the function's input parameter.
- The **payment.customer_id** refers to the **customer_id** column in the payment table.
- The **get_customer_payment.customer_id** refers to the input parameter of the function.
- The **RETURN total_payment** returns the value of **total_payment** to the caller of the function.
- The **$$** denotes the start and end of the function body using dollar-quoted string constants.
- The **LANGUAGE plpgsql** specifies that the function is written in the PL/pgSQL procedural language.

The following query is used to call the function.

```
SELECT get_customer_payment(1);
```

```
dvdrental=# SELECT get_customer_payment(1);
 get_customer_payment
----------------------
               114.70
(1 row)

dvdrental=#
```

Figure 10.11: *Calling a function called get_customer_payment*

This function calculates the total payments made by the customer with ID 1.

Trigger Functions

Trigger functions in PostgreSQL are a powerful feature that allows you to automatically perform specific actions in response to certain events on a table. Trigger functions are special procedures that are executed automatically in response to defined events on a table. These events include:

- **INSERT**: When a new row is inserted into the table.
- **UPDATE**: When an existing row is updated.

- **DELETE**: When a row is deleted from the table.

A trigger function is defined using the **CREATE FUNCTION** statement in PL/pgSQL or other procedural languages supported by PostgreSQL. The function contains the logic that should be executed when the trigger event occurs. Once the trigger function is defined, you create a trigger that specifies the event (**INSERT**, **UPDATE**, **DELETE**) and the timing (**BEFORE** or **AFTER**) when the function should be executed.

We are going to create a trigger function to update a timestamp when a row is modified in the rental table:

```
CREATE FUNCTION update_rental_timestamp() RETURNS TRIGGER AS $$
BEGIN
    NEW.last_update = NOW();
    RETURN NEW;
END;
$$ LANGUAGE plpgsql;
```

```
dvdrental=# CREATE FUNCTION update_rental_timestamp() RETURNS TRIGGER AS $$
BEGIN
    NEW.last_update = NOW();
    RETURN NEW;
END;
$$ LANGUAGE plpgsql;
CREATE FUNCTION
dvdrental=#
```

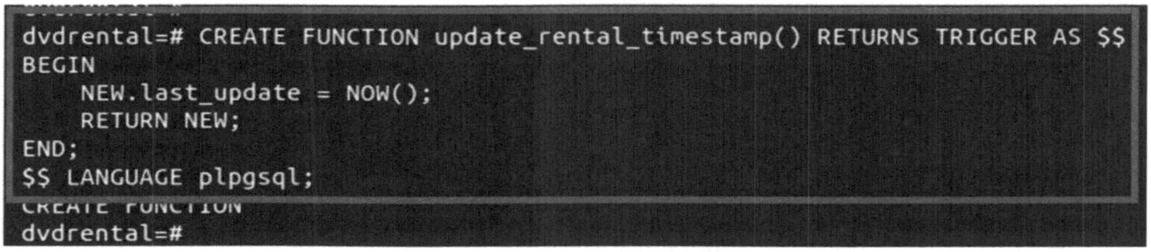

Figure 10.12: *Creating a function for the trigger*

Next is to create the trigger to fire when there is an update on a table.

```
CREATE TRIGGER update_rental_timestamp_trigger
BEFORE UPDATE ON rental
FOR EACH ROW
EXECUTE FUNCTION update_rental_timestamp();
```

```
dvdrental=#
dvdrental=# CREATE TRIGGER update_rental_timestamp_trigger
BEFORE UPDATE ON rental
FOR EACH ROW
EXECUTE FUNCTION update_rental_timestamp();
CREATE TRIGGER
dvdrental=#
```

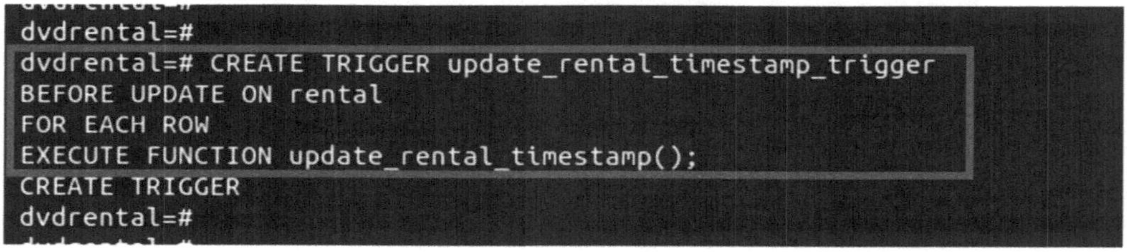

Figure 10.13: *Creating a trigger*

The next function automatically updates the **last_update** column to the current timestamp whenever a row in the rental table is updated.

```
SELECT tgname, tgrelid::regclass
FROM pg_trigger
WHERE tgname = 'update_rental_timestamp_trigger';
```

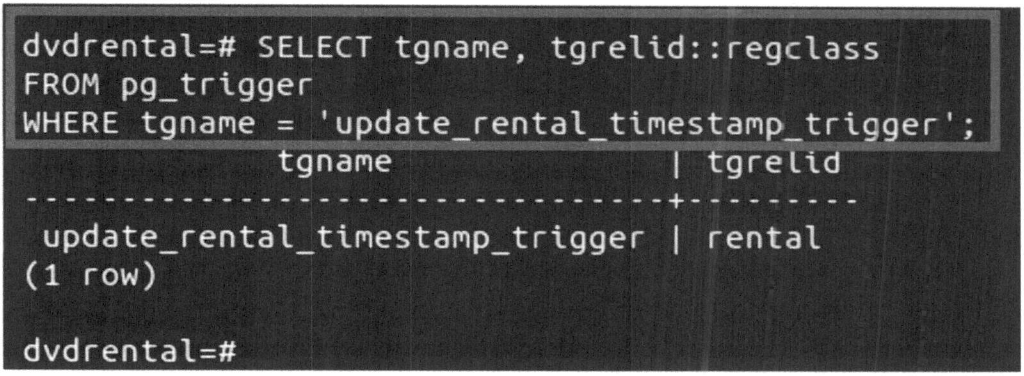

```
dvdrental=# SELECT tgname, tgrelid::regclass
FROM pg_trigger
WHERE tgname = 'update_rental_timestamp_trigger';
             tgname              | tgrelid
---------------------------------+---------
 update_rental_timestamp_trigger | rental
(1 row)

dvdrental=#
```

Figure 10.14: *Checking that the trigger is created*

```
-- Update a row in the rental table
UPDATE rental
SET return_date = '2024-12-24 10:00:00'
WHERE rental_id = 2;

-- Verify the last_update column
SELECT *
FROM rental
WHERE rental_id = 2;
```

```
dvdrental=#
dvdrental=# Select * from rental limit 10;
 rental_id |     rental_date     | inventory_id | customer_id |     return_date     | staff_id |     last_update
-----------+---------------------+--------------+-------------+---------------------+----------+---------------------
         2 | 2005-05-24 22:54:33 |         1525 |         459 | 2005-05-28 19:40:33 |        1 | 2006-02-16 02:30:53
         3 | 2005-05-24 23:03:39 |         1711 |         408 | 2005-06-01 22:12:39 |        1 | 2006-02-16 02:30:53
         4 | 2005-05-24 23:04:41 |         2452 |         333 | 2005-06-03 01:43:41 |        2 | 2006-02-16 02:30:53
         5 | 2005-05-24 23:05:21 |         2079 |         222 | 2005-06-02 04:33:21 |        1 | 2006-02-16 02:30:53
         6 | 2005-05-24 23:08:07 |         2792 |         549 | 2005-05-27 01:32:07 |        1 | 2006-02-16 02:30:53
         7 | 2005-05-24 23:11:53 |         3995 |         269 | 2005-05-29 20:34:53 |        2 | 2006-02-16 02:30:53
         8 | 2005-05-24 23:31:46 |         2346 |         239 | 2005-05-27 23:33:46 |        2 | 2006-02-16 02:30:53
         9 | 2005-05-25 00:00:40 |         2580 |         126 | 2005-05-28 00:22:40 |        1 | 2006-02-16 02:30:53
        10 | 2005-05-25 00:02:21 |         1824 |         399 | 2005-05-31 22:44:21 |        2 | 2006-02-16 02:30:53
        11 | 2005-05-25 00:09:02 |         4443 |         142 | 2005-06-02 20:56:02 |        2 | 2006-02-16 02:30:53
(10 rows)

UPDATE rental
SET return_date = '2024-12-24 10:00:00'
WHERE rental_id = 2;

dvdrental=#
dvdrental=# -- Verify the last_update column
SELECT *
FROM rental
WHERE rental_id = 2;
 rental_id |     rental_date     | inventory_id | customer_id |     return_date     | staff_id |        last_update
-----------+---------------------+--------------+-------------+---------------------+----------+----------------------------
         2 | 2005-05-24 22:54:33 |         1525 |         459 | 2024-12-24 10:00:00 |        1 | 2024-12-07 11:55:20.935016
(1 row)

dvdrental=#
```

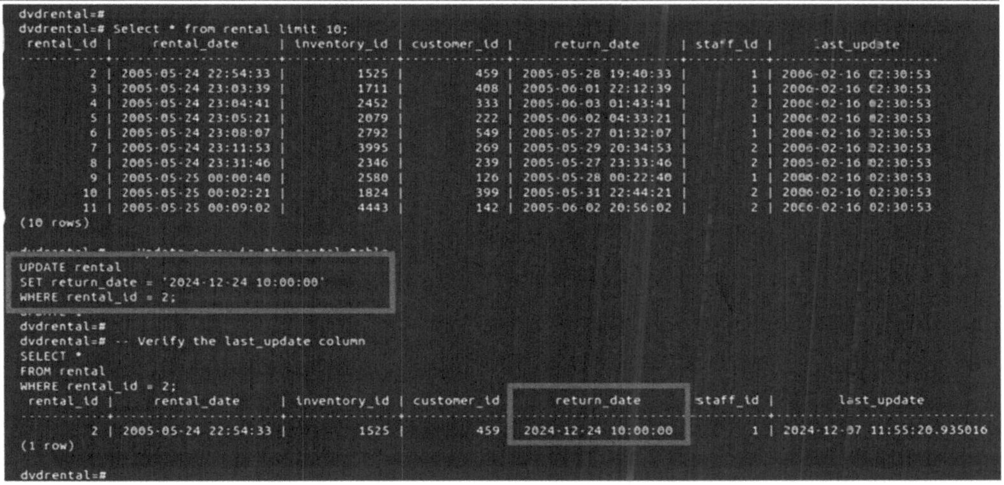

Figure 10.15: *Updating the return_date of the rental table where rental_id is 2*

Window Functions

Window functions in PostgreSQL are a powerful feature that allows you to perform calculations across a set of table rows that are somehow related to the current row. Unlike aggregate functions that return a single value for a group of rows, window functions retain the individual rows while still providing aggregate-like computations. It performs calculations across a set of table rows that are related to the current row.

Window functions compute values over a set of rows, known as the "**window**," which is defined by a **PARTITION BY** clause and an optional **ORDER BY** clause within a query. These functions are typically used to perform advanced analytics, such as running totals, moving averages, rank calculations, and more.

Syntax of Window Functions

The general syntax of a window function in PostgreSQL is:

```
function_name(expression) OVER (
    [PARTITION BY expression_list]
    [ORDER BY expression_list]
    [frame_clause]
)
```

- **function_name(expression)**: The window function and its argument.
- **OVER**: Specifies the window clause.
- **PARTITION BY**: Divides the result set into partitions to which the function is applied.
- **ORDER BY**: Defines the order of rows in each partition.
- **frame_clause**: Defines a subset of rows within the partition (optional).

The following are the common window functions:

- **ROW_NUMBER()**: Assigns a unique sequential integer to rows within a partition of the result set.
- **RANK()**: Assigns a rank to each row within a partition, with gaps for ties.
- **DENSE_RANK()**: Similar to **RANK()**, but without gaps between rank values.
- **SUM()**: Calculates the cumulative sum of a value in each row of a partition.
- **AVG()**: Computes the average of a set of values in each partition.
- **LEAD() and LAG()**: Accesses data from subsequent or preceding rows in the result set.

We are going to calculate the rank of films based on their rental rate:

```
SELECT title, rental_rate,
    RANK() OVER (ORDER BY rental_rate DESC) AS rank
FROM film;
```

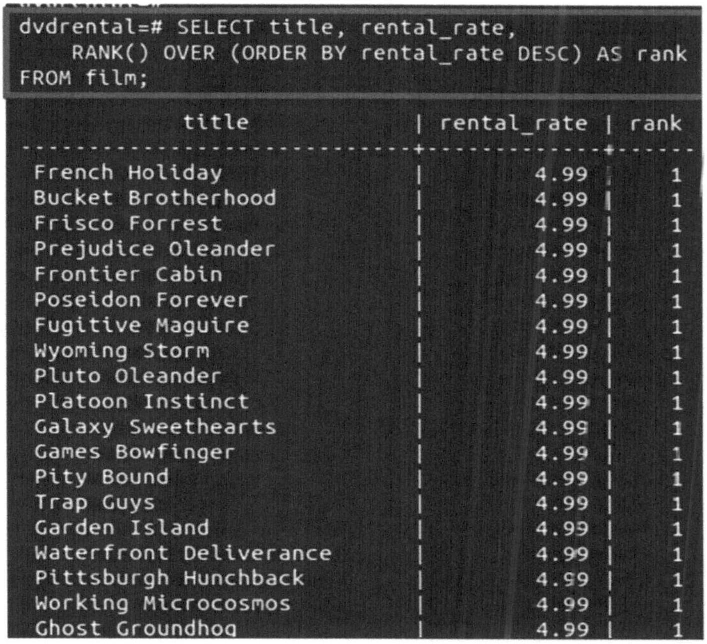

```
dvdrental=# SELECT title, rental_rate,
    RANK() OVER (ORDER BY rental_rate DESC) AS rank
FROM film;
            title             | rental_rate | rank
------------------------------+-------------+------
 French Holiday               |        4.99 |    1
 Bucket Brotherhood           |        4.99 |    1
 Frisco Forrest               |        4.99 |    1
 Prejudice Oleander           |        4.99 |    1
 Frontier Cabin               |        4.99 |    1
 Poseidon Forever             |        4.99 |    1
 Fugitive Maguire             |        4.99 |    1
 Wyoming Storm                |        4.99 |    1
 Pluto Oleander               |        4.99 |    1
 Platoon Instinct             |        4.99 |    1
 Galaxy Sweethearts           |        4.99 |    1
 Games Bowfinger              |        4.99 |    1
 Pity Bound                   |        4.99 |    1
 Trap Guys                    |        4.99 |    1
 Garden Island                |        4.99 |    1
 Waterfront Deliverance       |        4.99 |    1
 Pittsburgh Hunchback         |        4.99 |    1
 Working Microcosmos          |        4.99 |    1
 Ghost Groundhog              |        4.99 |    1
```

Figure 10.16: *Using the Windows function to calculate films' ranking*

The query above ranks films based on their rental rate, with the highest rental rate receiving the top rank.

Set Returning Functions

Set Returning Functions (SRFs) in PostgreSQL are powerful tools that allow functions to return a set of rows, much like a table. These functions are useful when you want to encapsulate complex queries or calculations and reuse them as if they were a table.

Set Returning Functions can be defined in SQL or PL/pgSQL and are typically used to return result sets that can be queried like a regular table. They are particularly useful for breaking down complex logic into reusable components.

Set returning functions return a set of rows, similar to a table.

The below query creates a function to return all films rented by a specific customer:

```
CREATE OR REPLACE FUNCTION get_rented_films(customer_id INT)
RETURNS TABLE (film_id INT, title VARCHAR) AS $$
BEGIN
    RETURN QUERY
    SELECT f.film_id, f.title
    FROM film f
```

```
    JOIN inventory i ON f.film_id = i.film_id
    JOIN rental r ON i.inventory_id = r.inventory_id
    WHERE r.customer_id = get_rented_films.customer_id;
END;
$$ LANGUAGE plpgsql;
```

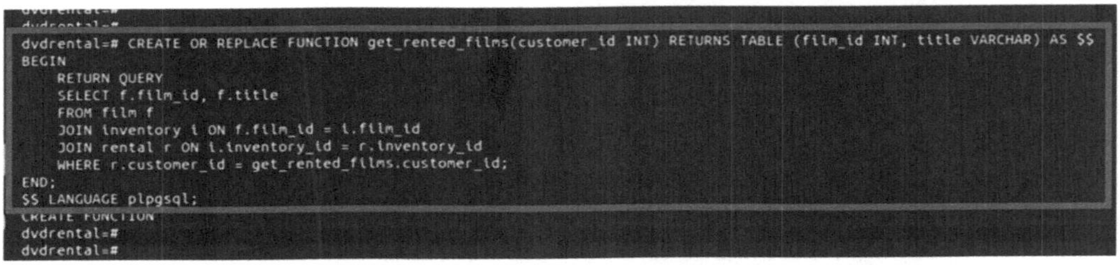

```
dvdrental=# CREATE OR REPLACE FUNCTION get_rented_films(customer_id INT) RETURNS TABLE (film_id INT, title VARCHAR) AS $$
BEGIN
    RETURN QUERY
    SELECT f.film_id, f.title
    FROM film f
    JOIN inventory i ON f.film_id = i.film_id
    JOIN rental r ON i.inventory_id = r.inventory_id
    WHERE r.customer_id = get_rented_films.customer_id;
END;
$$ LANGUAGE plpgsql;
CREATE FUNCTION
dvdrental=#
dvdrental=#
```

Figure 10.17: *How to create a set returning function*

Below, we are calling the set returning function.

```
SELECT * FROM get_rented_films(1);
```

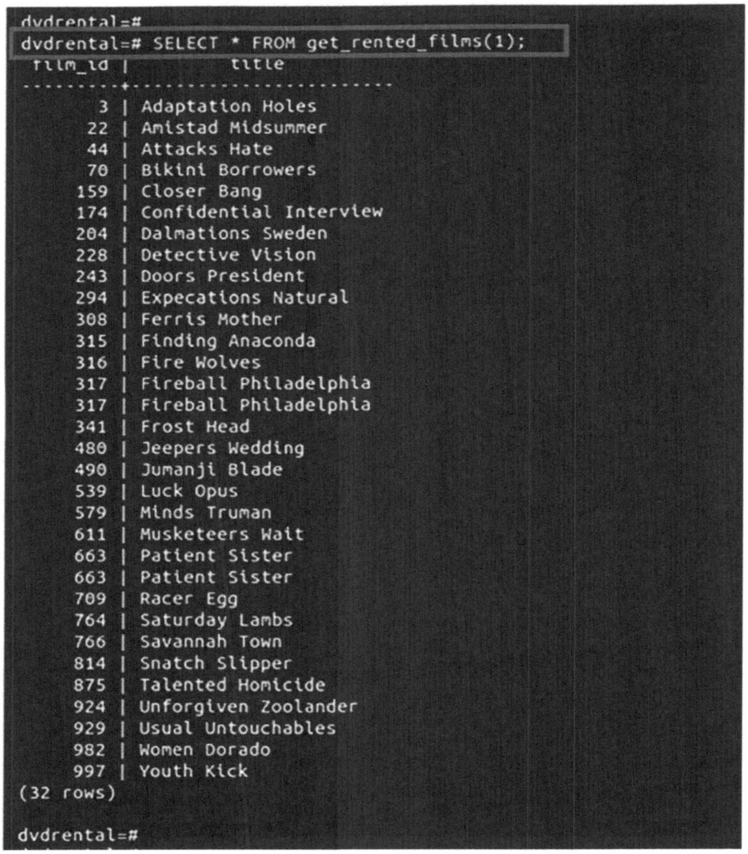

```
dvdrental=#
dvdrental=# SELECT * FROM get_rented_films(1);
 film_id |         title
---------+------------------------
       3 | Adaptation Holes
      22 | Amistad Midsummer
      44 | Attacks Hate
      70 | Bikini Borrowers
     159 | Closer Bang
     174 | Confidential Interview
     204 | Dalmations Sweden
     228 | Detective Vision
     243 | Doors President
     294 | Expecations Natural
     308 | Ferris Mother
     315 | Finding Anaconda
     316 | Fire Wolves
     317 | Fireball Philadelphia
     317 | Fireball Philadelphia
     341 | Frost Head
     480 | Jeepers Wedding
     490 | Jumanji Blade
     539 | Luck Opus
     579 | Minds Truman
     611 | Musketeers Wait
     663 | Patient Sister
     663 | Patient Sister
     709 | Racer Egg
     764 | Saturday Lambs
     766 | Savannah Town
     814 | Snatch Slipper
     875 | Talented Homicide
     924 | Unforgiven Zoolander
     929 | Usual Untouchables
     982 | Women Dorado
     997 | Youth Kick
(32 rows)

dvdrental=#
```

Figure 10.18: *SRFs function returns a set of all films rented by the customer with ID 1*

Practical Applications of Functions

- **Data Transformation**: Use functions to clean and transform data before inserting it into tables.
- **Custom Business Logic**: Implement complex business rules and calculations within the database.
- **Automation**: Automate repetitive tasks like updating records, generating reports, and performing backups.
- **Data Validation**: Use trigger functions to enforce data integrity and consistency.
- **Complex Queries**: Simplify complex queries by breaking them down into reusable functions.

String Functions

String functions in PostgreSQL allow you to manipulate and perform various operations on string data types. These functions are extremely useful for tasks such as formatting, searching, extracting, and modifying text. Here are some common string functions in PostgreSQL, with examples using the dvdrental sample database:

CONCAT

The **CONCAT** function concatenates two or more strings into one.

Example: Concatenating first and last names of customers:

```
SELECT CONCAT(first_name, ' ', last_name) AS full_name
FROM customer;
```

```
dvdrental=# SELECT CONCAT(first_name, ' ', last_name) AS full_name
FROM customer;

      full_name
-----------------------
 Jared Ely
 Mary Smith
 Patricia Johnson
 Linda Williams
 Barbara Jones
 Elizabeth Brown
 Jennifer Davis
 Maria Miller
 Susan Wilson
 Margaret Moore
 Dorothy Taylor
 Lisa Anderson
 Nancy Thomas
 Karen Jackson
 Betty White
 Helen Harris
 Sandra Martin
```

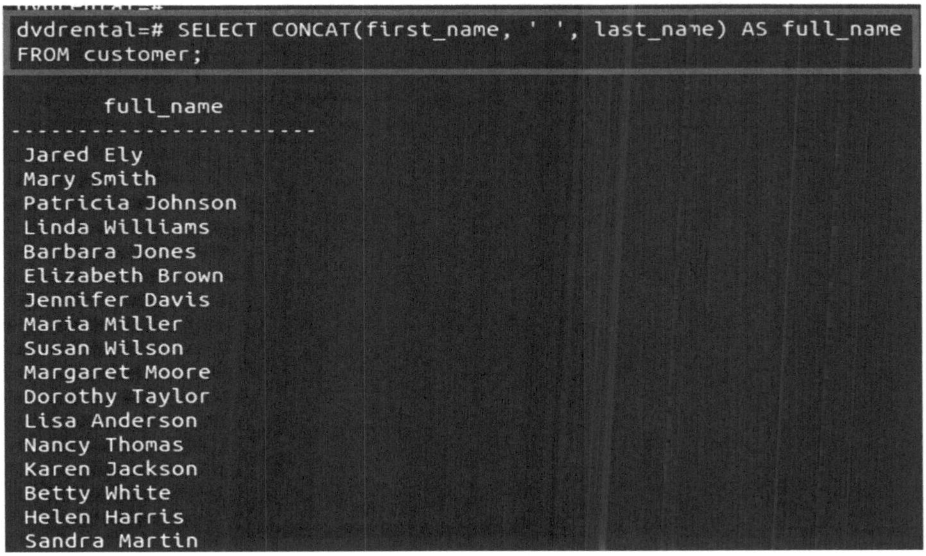

Figure 10.19: *The CONCAT function concatenates two words*

This query retrieves the **first_name** and **last_name** columns from the customer table and concatenates them with a space in between to form the **full_name**.

SUBSTRING

The **SUBSTRING** function extracts a part of a string based on specified positions.

Example: Extracting the first three characters of film titles:

```
SELECT title, SUBSTRING(title FROM 1 FOR 3) AS short_title
FROM film;
```

```
dvdrental=# SELECT title, SUBSTRING(title FROM 1 FOR 3) AS short_title
FROM film;
            title             | short_title
-----------------------------+-------------
 Chamber Italian             | Cha
 Grosse Wonderful            | Gro
 Airport Pollock             | Air
 Bright Encounters           | Bri
 Academy Dinosaur            | Aca
 Ace Goldfinger              | Ace
 Adaptation Holes            | Ada
 Affair Prejudice            | Aff
 African Egg                 | Afr
 Agent Truman                | Age
 Airplane Sierra             | Air
 Alabama Devil               | Ala
 Aladdin Calendar            | Ala
 Alamo Videotape             | Ala
 Alaska Phantom              | Ala
```

***Figure 10.20:** The SUBSTRING function that extracts part of a string*

This query retrieves the title of each film from the film table and extracts the first three characters to create a **short_title**.

LENGTH

The **LENGTH** function returns the number of characters in a string.

Example: Finding the length of each film title:

```
SELECT title, LENGTH(title) AS title_length
FROM film;
```

```
dvdrental=# SELECT title, LENGTH(title) AS title_length
FROM film;
          title           | title_length
--------------------------+--------------
 Chamber Italian          |           15
 Grosse Wonderful         |           16
 Airport Pollock          |           15
 Bright Encounters        |           17
 Academy Dinosaur         |           16
 Ace Goldfinger           |           14
 Adaptation Holes         |           16
 Affair Prejudice         |           16
 African Egg              |           11
 Agent Truman             |           12
 Airplane Sierra          |           15
 Alabama Devil            |           13
 Aladdin Calendar         |           16
 Alamo Videotape          |           15
 Alaska Phantom           |           14
 Date Speed               |           10
```

Figure 10.21: *The LENGTH function showing the length of film titles*

This query retrieves the title of each film from the film table and calculates the length of each title.

LOWER and UPPER

The **`LOWER`** function converts a string to lowercase, while the **`UPPER`** function converts a string to uppercase.

Example: Converting film titles to uppercase:

```
SELECT title, LOWER(title) AS title_lowercase
FROM film;
```

```
dvdrental=# SELECT title, LOWER(title) AS title_lowercase
FROM film;
          title            |      title_lowercase
---------------------------+--------------------------
 Chamber Italian           | chamber italian
 Grosse Wonderful          | grosse wonderful
 Airport Pollock           | airport pollock
 Bright Encounters         | bright encounters
 Academy Dinosaur          | academy dinosaur
 Ace Goldfinger            | ace goldfinger
 Adaptation Holes          | adaptation holes
 Affair Prejudice          | affair prejudice
 African Egg               | african egg
 Agent Truman              | agent truman
 Airplane Sierra           | airplane sierra
 Alabama Devil             | alabama devil
 Aladdin Calendar          | aladdin calendar
 Alamo Videotape           | alamo videotape
 Alaska Phantom            | alaska phantom
 Date Speed                | date speed
```

Figure 10.22: *The LOWER function that converts a string to lowercase*

```
SELECT title, UPPER(title) AS title_uppercase
FROM film;
```

```
dvdrental=# SELECT title, UPPER(title) AS title_uppercase
FROM film;
          title            |      title_uppercase
---------------------------+--------------------------
 Chamber Italian           | CHAMBER ITALIAN
 Grosse Wonderful          | GROSSE WONDERFUL
 Airport Pollock           | AIRPORT POLLOCK
 Bright Encounters         | BRIGHT ENCOUNTERS
 Academy Dinosaur          | ACADEMY DINOSAUR
 Ace Goldfinger            | ACE GOLDFINGER
 Adaptation Holes          | ADAPTATION HOLES
 Affair Prejudice          | AFFAIR PREJUDICE
 African Egg               | AFRICAN EGG
 Agent Truman              | AGENT TRUMAN
 Airplane Sierra           | AIRPLANE SIERRA
 Alabama Devil             | ALABAMA DEVIL
 Aladdin Calendar          | ALADDIN CALENDAR
 Alamo Videotape           | ALAMO VIDEOTAPE
 Alaska Phantom            | ALASKA PHANTOM
 Date Speed                | DATE SPEED
 Ali Forever               | ALI FOREVER
 Alice Fantasia            | ALICE FANTASIA
```

Figure 10.23: *The UPPER function that converts a string to uppercase*

This query retrieves the title of each film from the film table and converts each title to uppercase.

TRIM

The **TRIM** function in PostgreSQL is a versatile string manipulation function that helps clean up text by removing unwanted spaces. The **TRIM** function removes leading and trailing spaces from a string.

The **TRIM** function can remove:

- Leading spaces (spaces before the text).
- Trailing spaces (spaces after the text).
- Both leading and trailing spaces (default behavior).

We are going to be removing leading and trailing spaces from customer names:

```
SELECT TRIM(first_name) AS trimmed_first_name, TRIM(last_name) AS
trimmed_last_name
FROM customer;
```

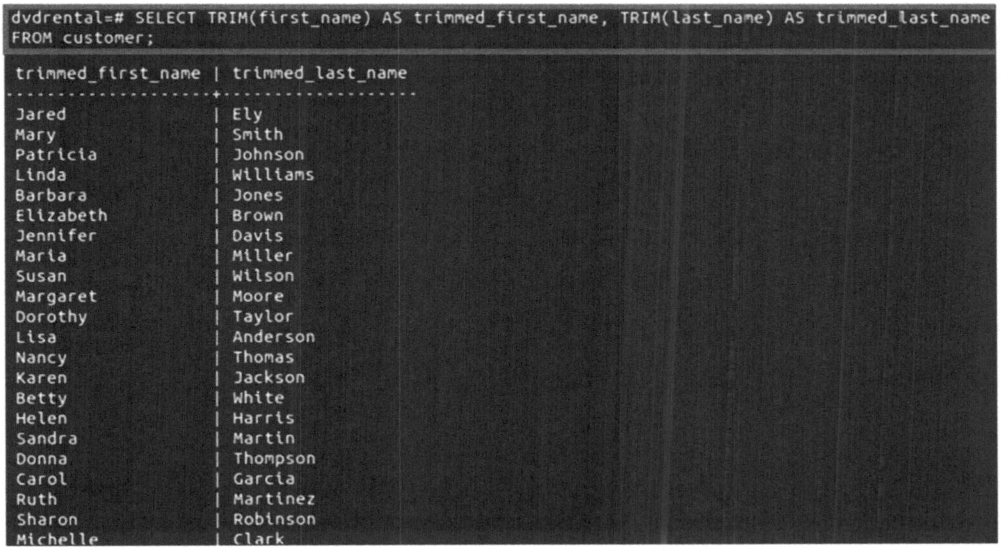

```
dvdrental=# SELECT TRIM(first_name) AS trimmed_first_name, TRIM(last_name) AS trimmed_last_name
FROM customer;

 trimmed_first_name | trimmed_last_name
--------------------+-------------------
 Jared              | Ely
 Mary               | Smith
 Patricia           | Johnson
 Linda              | Williams
 Barbara            | Jones
 Elizabeth          | Brown
 Jennifer           | Davis
 Maria              | Miller
 Susan              | Wilson
 Margaret           | Moore
 Dorothy            | Taylor
 Lisa               | Anderson
 Nancy              | Thomas
 Karen              | Jackson
 Betty              | White
 Helen              | Harris
 Sandra             | Martin
 Donna              | Thompson
 Carol              | Garcia
 Ruth               | Martinez
 Sharon             | Robinson
 Michelle           | Clark
```

Figure 10.24: *Trim function to remove leading and trailing spaces from customer names*

This query retrieves the **first_name** and **last_name** columns from the customer table and removes any leading or trailing spaces.

POSITION

The **POSITION** function returns the position of a substring within a string.

Example: Finding the position of the word **'Bright'** in film titles:

```
SELECT title, POSITION('Bright' IN title) AS bright_position
FROM film;
```

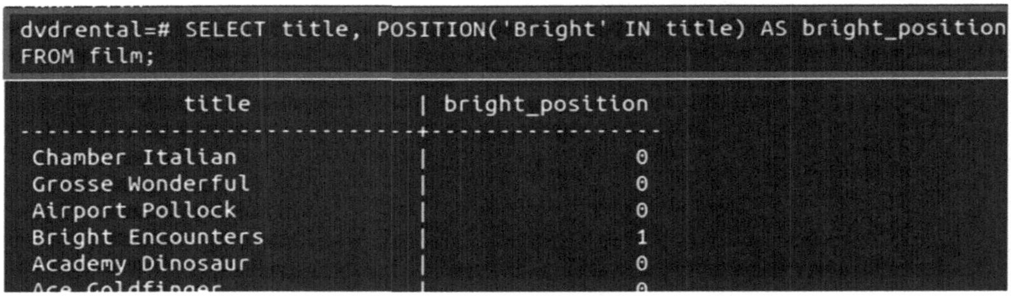

```
dvdrental=# SELECT title, POSITION('Bright' IN title) AS bright_position
FROM film;
         title              | bright_position
----------------------------+-----------------
 Chamber Italian            |               0
 Grosse Wonderful           |               0
 Airport Pollock            |               0
 Bright Encounters          |               1
 Academy Dinosaur           |               0
```

Figure 10.25: *The POSITION function returns the position of a substring*

This query retrieves the title of each film from the film table and finds the position of the substring **'Bright'** within each title.

REPLACE

The **REPLACE** function in PostgreSQL is used to replace all occurrences of a specified substring within a string with another substring. It is particularly useful for cleaning up or transforming data within the database.

REPLACE(string, from_substring, to_substring)

- **string**: The original string where you want to replace the substring.
- **from_substring**: The substring that you want to replace.
- **to_substring**: The substring that will replace the from_substring.

We are going to be replacing **'English'** with **'Eng'** in language table:

```
SELECT name, REPLACE(name, 'English', 'Eng') AS updated_name
FROM language;
```

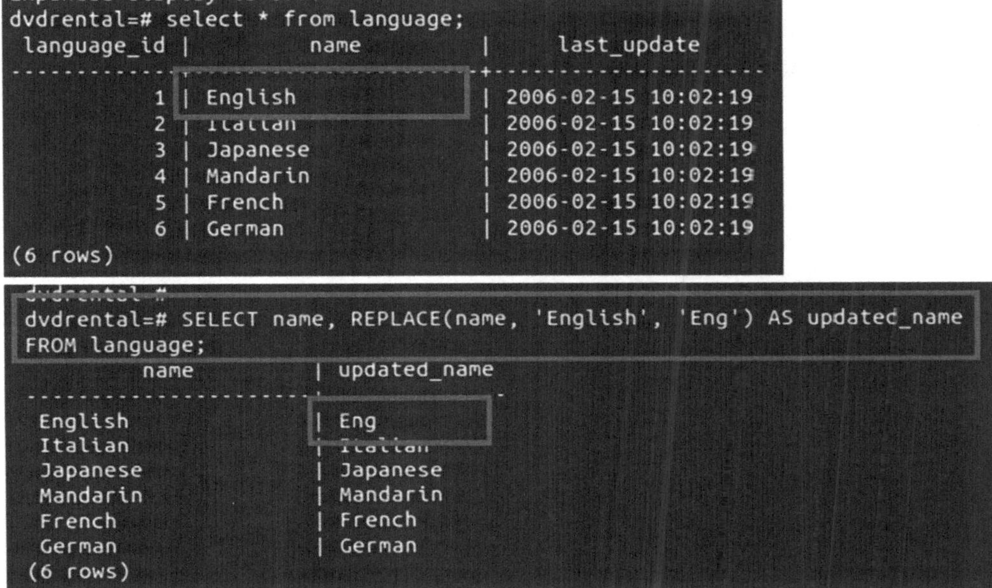

***Figure 10.26:** Replace function on a string*

This query retrieves the title of each film from the language table and replaces every occurrence of the word **'English'** with the word **'Eng'**.

INITCAP

The **INITCAP** function converts the first letter of each word in a string to uppercase and the rest to lowercase.

Example: Converting film titles to title case:

```
SELECT title, INITCAP(title) AS title_case
FROM film;
```

```
dvdrental=# SELECT title, INITCAP(title) AS title_case
FROM film;
```

```
          title           |        title_case
--------------------------+-------------------------
Chamber Italian           | Chamber Italian
Grosse Wonderful          | Grosse Wonderful
Airport Pollock           | Airport Pollock
Bright Encounters         | Bright Encounters
Academy Dinosaur          | Academy Dinosaur
Ace Goldfinger            | Ace Goldfinger
Adaptation Holes          | Adaptation Holes
Affair Prejudice          | Affair Prejudice
African Egg               | African Egg
Agent Truman              | Agent Truman
Airplane Sierra           | Airplane Sierra
Alabama Devil             | Alabama Devil
Aladdin Calendar          | Aladdin Calendar
Alamo Videotape           | Alamo Videotape
Alaska Phantom            | Alaska Phantom
```

Figure 10.27: *The INITCAP function, which converts the first letter of each word in a string*

This query retrieves the title of each film from the film table and converts each title to title case, where the first letter of each word is uppercase.

```
dvdrental=#
dvdrental=# SELECT INITCAP('hello world') AS capitalized_string;
 capitalized_string
--------------------
 Hello World
(1 row)
```

Figure 10.28: *Another use case of the INITCAP function*

By using these string functions, you can effectively manage and manipulate textual data in your PostgreSQL databases

Conclusion

In this chapter, we learned the different types of functions in PostgreSQL to enhance the efficiency and functionality of the database management and operations. Each type of function has its unique applications and strengths, allowing us to choose the best tool for the task at hand. We used the **`AVG()`** to calculate the average, the **`COUNT()`** function to perform a count, the **`SUM()`** function to calculate the total of values, the **`MIN()`** function to get the minimum value in a set of values, the **`MAX()`** function to get the maximum value in a set of values.

In the next chapter, we will look at Table joins, sub queries, and views in PostgreSQL.

Chapter 11
Demystifying Table Joints, Subqueries, CTE, and Views

Introduction

In this chapter, learners will understand the different types of joins, subqueries, and, most importantly, will understand CTE, views, and materialized views. In PostgreSQL, table joins are used to combine records from two or more tables in a database based on a related column between them. This process allows you to perform queries that span multiple tables and return results that incorporate data from these related tables.

Structure

In this chapter, we will cover the following topics:

- `INNER JOIN`
- `LEFT JOIN` (Left Outer Join)
- `RIGHT JOIN` (Right Outer Join)
- `FULL JOIN` (Full Outer Join)
- `CROSS JOIN`
- `NATURAL JOIN`
- Subqueries
- Common Table Expression (CTE)
- Views
- Materialized Views
- Task

INNER JOIN

An **INNER JOIN** returns rows when there is a match in both tables, combining records that have matching values in the specified columns. This type of join retrieves only the records with matching data in both tables, resulting in a combined dataset that includes rows meeting the specified condition.

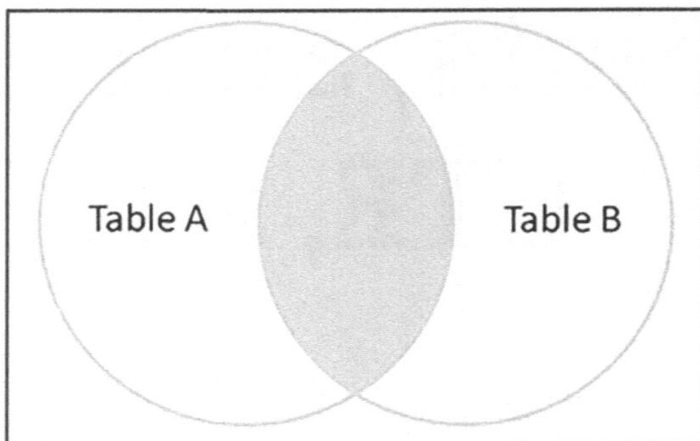

***Figure 11.1:** Diagrammatic representation of inner join between tables A and B*

Syntax:

```
SELECT tableA.column1, tableA.column2, tableB.column1, ...
FROM tableA
INNER JOIN tableB
ON tableA.matching_column = tableB.matching_column;
```

For our example, we are going to retrieve the rental id from the rental table alongside the first name and last name of the customers who made the rentals from the customer table.

```
dvdrental=# SELECT r.rental_id, c.first_name, c.last_name
FROM rental r
INNER JOIN customer c ON r.customer_id = c.customer_id;

 rental_id | first_name  |  last_name
-----------+-------------+-------------
         3 | Manuel      | Murrell
         4 | Andrew      | Purdy
         5 | Delores     | Hansen
         6 | Nelson      | Christenson
         7 | Cassandra   | Walters
         8 | Minnie      | Romero
         9 | Ellen       | Simpson
        10 | Danny       | Isom
        11 | April       | Burns
        12 | Deanna      | Byrd
        13 | Raymond     | Mcwhorter
        14 | Theodore    | Culp
        15 | Ronald      | Weiner
```

***Figure 11.2:** Inner join between two tables (rental and customer)*

In this example, the query returns the first and last names of customers along with their respective rental ids, but only for those customers who have a matching customer id in the rental table.

```
dvdrental=# SELECT f.title, c.first_name, c.last_name
FROM film f
INNER JOIN inventory i ON f.film_id = i.film_id
INNER JOIN rental r ON i.inventory_id = r.inventory_id
INNER JOIN customer c ON r.customer_id = c.customer_id;
            title             | first_name  |  last_name
------------------------------+-------------+-------------
 Graduate Lord                | Manuel      | Murrell
 Love Suicides                | Andrew      | Purdy
 Idols Snatchers              | Delores     | Hansen
 Mystic Truman                | Nelson      | Christenson
 Swarm Gold                   | Cassandra   | Walters
 Lawless Vision               | Minnie      | Romero
 Matrix Snowman               | Ellen       | Simpson
 Hanging Deep                 | Danny       | Isom
 Whale Bikini                 | April       | Burns
 Games Bowfinger              | Deanna      | Byrd
 King Evolution               | Raymond     | Mcwhorter
 Monterey Labyrinth           | Theodore    | Culp
 Pelican Comforts             | Ronald      | Weiner
 Boogie Amelie                | Steven      | Curley
 Contact Anonymous            | Isaac       | Oglesby
 Roman Punk                   | Ruth        | Martinez
 Hollow Jeopardy              | Ronnie      | Ricketts
```

Figure 11.3: *Inner join between four tables (film, inventory, rental, and customer)*

LEFT JOIN (Left Outer Join)

A **LEFT JOIN** returns all rows from the left table and the matched rows from the right table. If there is no match, the result is NULL on the side of the right table. This join is useful when you want to retrieve all records from the left table, regardless of whether there is a match in the right table.

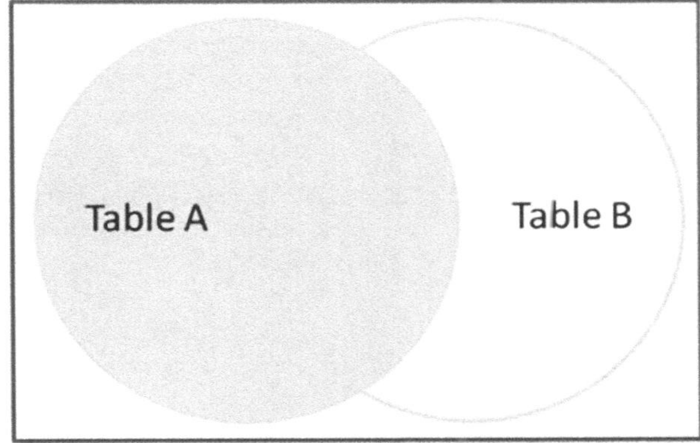

Figure 11.4: *Diagrammatic representation of Left Join*

This query retrieves all customer names and their payment amounts, if any. If a customer has not made any payments, the amount field will be **NULL**.

```
dvdrental=# SELECT c.first_name, c.last_name, p.amount
FROM customer c
LEFT JOIN payment p ON c.customer_id = p.customer_id;
```

```
 first_name  |  last_name   | amount
-------------+--------------+--------
 Peter       | Menard       |   7.99
 Peter       | Menard       |   1.99
 Peter       | Menard       |   7.99
 Peter       | Menard       |   2.99
 Peter       | Menard       |   7.99
 Peter       | Menard       |   5.99
 Harold      | Martino      |   5.99
 Harold      | Martino      |   5.99
 Harold      | Martino      |   2.99
 Douglas     | Graf         |   4.99
 Douglas     | Graf         |   6.99
 Douglas     | Graf         |   0.99
```

Figure 11.5: *Left Join between customer and payment tables*

This query returns all customer names and their corresponding rented films, if any. If a customer has not rented any films, the title field will be **NULL**.

```
dvdrental=# SELECT c.first_name, c.last_name, f.title
FROM customer c
LEFT JOIN rental r ON c.customer_id = r.customer_id
LEFT JOIN inventory i ON r.inventory_id = i.inventory_id
LEFT JOIN film f ON i.film_id = f.film_id;
```

```
 first_name  |  last_name   |            title
-------------+--------------+-----------------------------
 Manuel      | Murrell      | Graduate Lord
 Andrew      | Purdy        | Love Suicides
 Delores     | Hansen       | Idols Snatchers
 Nelson      | Christenson  | Mystic Truman
 Cassandra   | Walters      | Swarm Gold
 Minnie      | Romero       | Lawless Vision
 Ellen       | Simpson      | Matrix Snowman
 Danny       | Isom         | Hanging Deep
 April       | Burns        | Whale Bikini
 Deanna      | Byrd         | Games Bowfinger
 Raymond     | Mcwhorter    | King Evolution
 Theodore    | Culp         | Monterey Labyrinth
```

Figure 11.6: *Left Join involving four tables (customer, rental, inventory, and film)*

RIGHT JOIN (Right Outer Join)

A **RIGHT JOIN** returns all rows from the right table and the matched rows from the left table. If there is no match, the result is NULL on the side of the left table. This join is less common but is useful when you want to retrieve all records from the right table, regardless of whether there is a match in the left table.

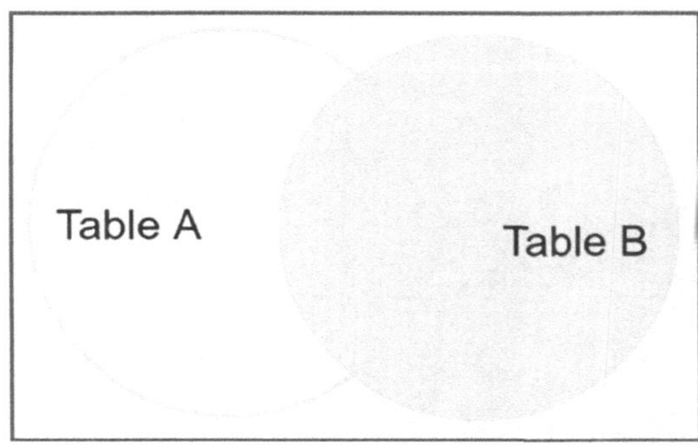

Figure 11.7: *Diagrammatic representation of Right Join*

Syntax:

```
SELECT TableA.name, TableB.id
FROM TableA
RIGHT JOIN TableB
ON TableA.matching_column = TableB.matching_column;
```

Suppose we want to list all rental records and the customers who made them, including rentals that might not have customer information.

```
dvdrental=# SELECT r.rental_id, c.first_name, c.last_name
FROM rental r
RIGHT JOIN customer c ON r.customer_id = c.customer_id;

 rental_id | first_name  |  last_name
-----------+-------------+-------------
         3 | Manuel      | Murrell
         4 | Andrew      | Purdy
         5 | Delores     | Hansen
         6 | Nelson      | Christenson
         7 | Cassandra   | Walters
         8 | Minnie      | Romero
         9 | Ellen       | Simpson
        10 | Danny       | Isom
        11 | April       | Burns
        12 | Deanna      | Byrd
        13 | Raymond     | Mcwhorter
        14 | Theodore    | Culp
        15 | Ronald      | Weiner
```

Figure 11.8: *Right Join between rental and customer tables*

This query returns all rental IDs and customer names. If a rental record does not have customer information, the **first_name** and **last_name** fields will be **NULL**.

Let us list all payments and the names of the staff members who processed them, including payments that might not have staff information:

```
dvdrental=# SELECT p.payment_id, s.first_name, s.last_name
FROM payment p
RIGHT JOIN staff s ON p.staff_id = s.staff_id;
```

```
 payment_id | first_name | last_name
------------+------------+-----------
      17503 | Jon        | Stephens
      17504 | Mike       | Hillyer
      17505 | Mike       | Hillyer
      17506 | Jon        | Stephens
      17507 | Jon        | Stephens
      17508 | Mike       | Hillyer
      17509 | Jon        | Stephens
      17510 | Mike       | Hillyer
      17511 | Mike       | Hillyer
      17512 | Jon        | Stephens
      17513 | Mike       | Hillyer
```

Figure 11.9: *Right Join between payment and staff tables*

This query returns all payment IDs and the names of the staff who processed them. If a payment record does not have staff information, the **first_name** and **last_name** fields will be **NULL**.

FULL JOIN (Full Outer Join)

A **FULL JOIN** returns all rows when there is a match in either left or right table. It includes all records from both tables and fills in **NULL**s for missing matches on either side. This join is useful when you want a complete set of records from both tables.

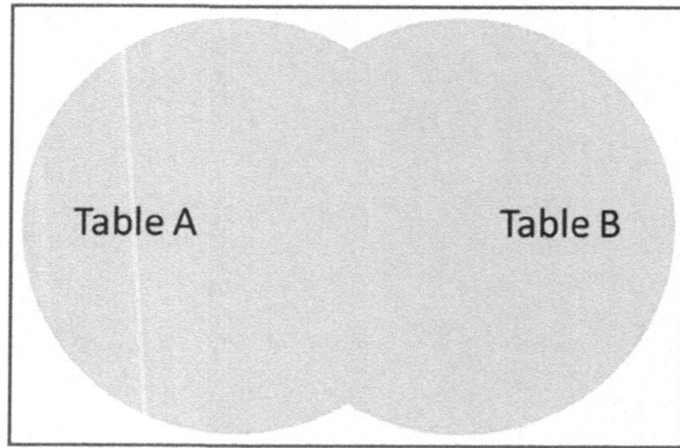

Figure 11.10: *Diagrammatic representation of Full Join*

Let us list all categories and films within those categories, including categories that might not have films and films that might not have categories.

```
dvdrental=# SELECT c.name AS category_name, f.title AS film_title
FROM category c
FULL JOIN film_category fc ON c.category_id = fc.category_id
FULL JOIN film f ON fc.film_id = f.film_id;
```

```
 category_name |          film_title
---------------+-----------------------------
 Documentary   | Academy Dinosaur
 Horror        | Ace Goldfinger
 Documentary   | Adaptation Holes
 Horror        | Affair Prejudice
 Family        | African Egg
 Foreign       | Agent Truman
 Comedy        | Airplane Sierra
 Horror        | Airport Pollock
 Horror        | Alabama Devil
 Sports        | Aladdin Calendar
 Foreign       | Alamo Videotape
```

Figure 11.11: *Full Join between three tables*

This query returns all category names and film titles. If a category does not have any films or if a film does not belong to any category, the corresponding fields will be NULL.

Explanation of the query:

a. **`SELECT c.name AS category_name, f.title AS film_title:`**
   - This part of the query specifies the columns to be selected and included in the result set.
   - **`c.name`** is selected from the category table and is aliased as **`category_name`**.
   - **`f.title`** is selected from the film table and is aliased as **`film_title`**.

b. **`FROM category c:`**
   - This specifies the category table and gives it an alias **`c`** to be used in the query.

c. **`FULL JOIN film_category fc ON c.category_id = fc.category_id:`**
   - A **`FULL JOIN`** is performed between the category table (c) and the **`film_category table (fc)`**.
   - The join is based on the **`category_id`** column, which must be present in both tables.

   - A **FULL JOIN** returns all records when there is a match in either left (category) or right (**film_category**) table. It includes rows that do not have matching rows in either table, filling with **NULL** where applicable.

d. **FULL JOIN film f ON fc.film_id = f.film_id:**

   - Another **FULL JOIN** is performed between the result of the previous join (**category** and **film_category**) and the film table (**f**).
   - The join is based on the **film_id** column, which must be present in both the film_category table (**fc**) and the film table (**f**).
   - This **FULL JOIN** again returns all records when there is a match in either left or right table, including rows that do not have matching rows in either table, filling with **NULL** where applicable.

CROSS JOIN

A **CROSS JOIN** returns the Cartesian product of the two tables, meaning it returns all possible combinations of rows from the two tables. This join is used when you need all combinations of rows from the tables involved.

Suppose we want to pair each customer with each film:

```
dvdrental=# SELECT c.first_name, c.last_name, f.title
FROM customer c
CROSS JOIN film f;
```

```
 first_name  |  last_name   |          title
-------------+--------------+-------------------------
 Jared       | Ely          | Chamber Italian
 Mary        | Smith        | Chamber Italian
 Patricia    | Johnson      | Chamber Italian
 Linda       | Williams     | Chamber Italian
 Barbara     | Jones        | Chamber Italian
 Elizabeth   | Brown        | Chamber Italian
 Jennifer    | Davis        | Chamber Italian
 Maria       | Miller       | Chamber Italian
 Susan       | Wilson       | Chamber Italian
 Margaret    | Moore        | Chamber Italian
 Dorothy     | Taylor       | Chamber Italian
 Lisa        | Anderson     | Chamber Italian
```

Figure 11.12: *Cross Join between customer and film tables*

This query generates all possible combinations of customers and film titles, producing a Cartesian product of the customer and film tables.

NATURAL JOIN

A **NATURAL JOIN** automatically joins tables based on columns with the same names in both tables. It is a shorthand for an **INNER JOIN** on all columns with matching names. However, it is important to use this join type carefully to avoid unexpected results.

Let us join the customer and payment tables to list all customers and the payment amounts they have made, assuming both tables have a common column.

```
dvdrental=# SELECT customer.first_name, customer.last_name, payment.amount
FROM customer
NATURAL JOIN payment;
```

```
 first_name  |  last_name    | amount
-------------+---------------+--------
 Peter       | Menard        |   7.99
 Peter       | Menard        |   1.99
 Peter       | Menard        |   7.99
 Peter       | Menard        |   2.99
 Peter       | Menard        |   7.99
 Peter       | Menard        |   5.99
 Harold      | Martino       |   5.99
 Harold      | Martino       |   5.99
 Harold      | Martino       |   2.99
 Douglas     | Graf          |   4.99
 Douglas     | Graf          |   6.99
 Douglas     | Graf          |   0.99
 Douglas     | Graf          |   0.99
 Douglas     | Graf          |   6.99
 Douglas     | Graf          |   8.99
```

Figure 11.13: *Natural Join between customer and payment tables*

This query above returns the first names, last names, and payment amounts for all customers who have made payments, where the **customer_id** matches in both tables.

Practical Considerations of joins

When using joins, it is essential to consider indexing the join columns to improve performance, especially for large datasets. Also, be mindful of the join type you choose, as it impacts the results and performance of your queries.

Subqueries

In PostgreSQL, a **subquery** (also known as an inner query or nested query) is a query embedded within another query. The results of the subquery are used by the outer query to further restrict or summarize the results. Subqueries can be placed in various parts of the SQL statement, including the **SELECT** clause, **FROM** clause, and **WHERE** clause.

Types of Subqueries

There are different types of subqueries that are presented in this section.

Single-row Subquery

It returns a single row and can be used in places where a single value is expected.

Suppose we want to find the customer who made the most payments:

```
dvdrental=#
dvdrental=# SELECT first_name, last_name
FROM customer
WHERE customer_id = (
    SELECT customer_id
    FROM payment
    GROUP BY customer_id
    ORDER BY SUM(amount) DESC
    LIMIT 1
);
 first_name | last_name
------------+-----------
 Eleanor    | Hunt
(1 row)
```

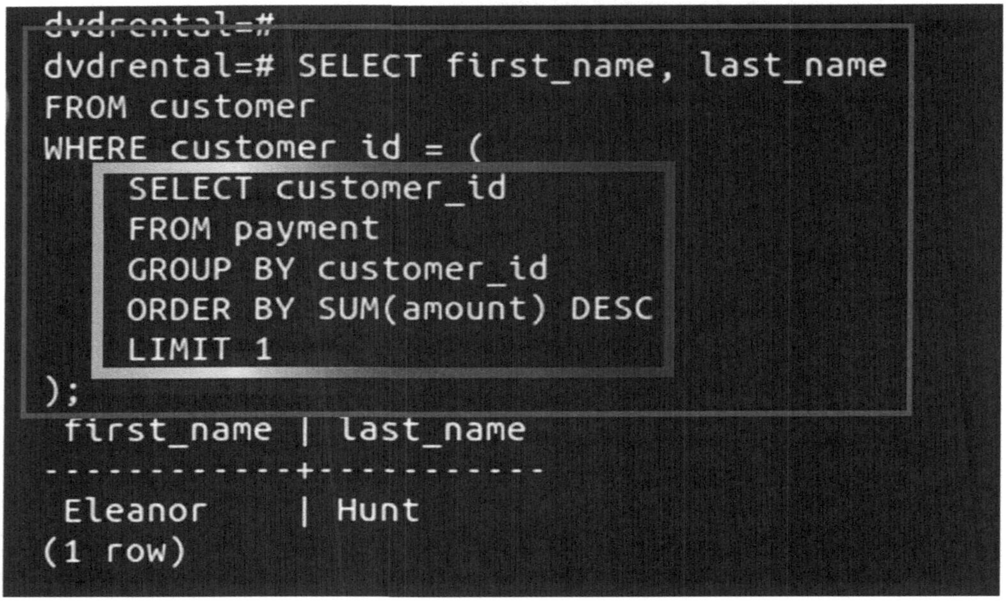

Figure 11.14: *Single-row Subquery*

Here, the subquery returns the **`customer_id`** of the customer who made the most payments, which is then used by the outer query to fetch the customer's name.

Multi-row Subquery

Returns multiple rows and is often used with **`IN`**, **`ANY`**, or **`ALL`** operators.

Suppose we want to list all films that have been rented by customers who have spent more than **`$100`**:

```
dvdrental=# SELECT title
FROM film
WHERE film_id IN (
    SELECT i.film_id
    FROM inventory i
    JOIN rental r ON i.inventory_id = r.inventory_id
    JOIN payment p ON r.rental_id = p.rental_id
    WHERE p.customer_id IN (
        SELECT p2.customer_id
        FROM payment p2
        GROUP BY p2.customer_id
        HAVING SUM(p2.amount) > 100
    )
);
```

```
           title
-----------------------------
 Chamber Italian
 Grosse Wonderful
 Airport Pollock
 Bright Encounters
 Academy Dinosaur
 Ace Goldfinger
 Adaptation Holes
```

Figure 11.15: *Nested subqueries that returns multiple rows*

This subquery first identifies customers who spent more than **`$100`** and then selects the films rented by these customers. An explanation of the different parts of the query is given below.

Main Query

The main query retrieves the titles of films from the film table where the **`film_id`** matches those found in the intermediate subquery.

- **`SELECT` title**: This part of the query selects the title of films from the film table.
- **`FROM` film**: This specifies the film table as the source of data.
- **`WHERE film_id IN (...)`**: This filters the films to include only those whose **`film_id`** is in the result set of the subquery.

Intermediate Subquery

The intermediate subquery finds the **`film_id`**s of films that were rented by the high-spending customers identified in the nested subquery.

It does this by joining the inventory, rental, and payment tables to link films to payments made by those customers.

- **`SELECT i.film_id`**: This part of the subquery selects the **`film_id`** from the inventory table.

- **`FROM inventory i`**: This specifies the inventory table and assigns it an alias **`i`**.
- **`JOIN rental r ON i.inventory_id = r.inventory_id`**: This joins the inventory table with the rental table based on the **`inventory_id`**.
- **`JOIN payment p ON r.rental_id = p.rental_id`**: This joins the rental table with the payment table based on the **`rental_id`**.
- **`WHERE p.customer_id IN (...)`**: This filters the results to include only those records where the **`customer_id`** is in the result set of the nested subquery.

Nested Subquery

The nested subquery calculates the total payment amount for each customer. It filters the customers to include only those who have spent more than **`$100`**.

- **`SELECT p2.customer_id`**: This part of the nested subquery selects the **`customer_id`** from the payment table.
- **`FROM payment p2`**: This specifies the payment table and assigns it an alias **`p2`**.
- **`GROUP BY p2.customer_id`**: This groups the results by **`customer_id`**.
- **`HAVING SUM(p2.amount) > 100`**: This filters the grouped results to include only those customers whose total payments exceed **`$100`**.

Scalar Subquery

Returns a single value and can be used wherever a single value is expected, such as in the **`SELECT`** clause.

Suppose we want to display the average rental duration alongside each **`rental id`** and **`date`**:

```
dvdrental=# SELECT rental_id, rental_date,
    (SELECT AVG(return_date - rental_date)
     FROM rental) AS avg_rental_duration
FROM rental;
```

```
 rental_id |     rental_date     |   avg_rental_duration
-----------+---------------------+------------------------
         3 | 2005-05-24 23:03:39 | 4 days 35:25:34.709476
         4 | 2005-05-24 23:04:41 | 4 days 35:25:34.709476
         5 | 2005-05-24 23:05:21 | 4 days 35:25:34.709476
         6 | 2005-05-24 23:08:07 | 4 days 35:25:34.709476
         7 | 2005-05-24 23:11:53 | 4 days 35:25:34.709476
         8 | 2005-05-24 23:31:46 | 4 days 35:25:34.709476
         9 | 2005-05-25 00:00:40 | 4 days 35:25:34.709476
        10 | 2005-05-25 00:02:21 | 4 days 35:25:34.709476
        11 | 2005-05-25 00:09:02 | 4 days 35:25:34.709476
        12 | 2005-05-25 00:19:27 | 4 days 35:25:34.709476
        13 | 2005-05-25 00:22:55 | 4 days 35:25:34.709476
        14 | 2005-05-25 00:31:15 | 4 days 35:25:34.709476
        15 | 2005-05-25 00:39:22 | 4 days 35:25:34.709476
        16 | 2005-05-25 00:43:11 | 4 days 35:25:34.709476
        17 | 2005-05-25 01:06:36 | 4 days 35:25:34.709476
```

Figure 11.16: *Scalar subquery*

The subquery calculates the average rental duration, which is then included in the result set of the outer query.

Correlated Subquery

It is a subquery that references columns from the outer query. Unlike a regular subquery, which is executed only once, a correlated subquery is evaluated once for each row processed by the outer query. This means that the subquery is dependent on the outer query and cannot be executed independently.

The correlated subquery relies on the outer query for its values, creating a relationship between the two queries.

Because of this dependency, the subquery is re-executed for each row of the outer query.

Since a correlated subquery is executed multiple times (once for each row of the outer query), it can be less efficient than a regular subquery.

Optimizing correlated subqueries may involve indexing and query restructuring.

Suppose we want to list all customers who have rented more films than the average number of films rented:

```
dvdrental=# SELECT first_name, last_name
FROM customer c
WHERE (
    SELECT COUNT(*)
    FROM rental r
    WHERE r.customer_id = c.customer_id
) > (
    SELECT AVG(rental_count)
    FROM (
        SELECT COUNT(*) AS rental_count
        FROM rental
        GROUP BY customer_id
    ) sub
);
```

```
 first_name  |  last_name
-------------+-------------
 Mary        | Smith
 Patricia    | Johnson
 Elizabeth   | Brown
 Jennifer    | Davis
 Maria       | Miller
 Nancy       | Thomas
 Karen       | Jackson
```

Figure 11.17: *Correlated Subquery*

The subquery counts the number of rentals for each customer and compares it to the average rental count.

Scenarios Where Subqueries are Applicable

In this section, we will look at some situations where subqueries are used.

- **Filtering Results:** Subqueries are often used in the `WHERE` clause to filter results based on complex conditions.

Suppose we want to list all customers who have rented at least one film:

```
dvdrental=# SELECT first_name, last_name
FROM customer
WHERE customer_id IN (
    SELECT customer_id
    FROM rental
);
```

```
 first_name  |  last_name
-------------+-------------
 Jared       | Ely
 Mary        | Smith
 Patricia    | Johnson
 Linda       | Williams
 Barbara     | Jones
 Elizabeth   | Brown
 Jennifer    | Davis
 Maria       | Miller
 Susan       | Wilson
 Margaret    | Moore
 Dorothy     | Taylor
```

Figure 11.18: *Using subqueries to filter results*

- **Calculating Aggregate Values:** Subqueries can be used to calculate aggregate values and include them in the result set.

Let us display each customer and their total spending:

```
dvdrental=# SELECT first_name, last_name,
    (SELECT SUM(amount)
     FROM payment
     WHERE customer_id = c.customer_id) AS total_spending
FROM customer c;
```

```
 first_name  |  last_name   | total_spending
-------------+--------------+----------------
 Jared       | Ely          |          81.83
 Mary        | Smith        |         114.70
 Patricia    | Johnson      |         123.74
 Linda       | Williams     |         130.76
 Barbara     | Jones        |          81.78
 Elizabeth   | Brown        |         134.65
 Jennifer    | Davis        |          84.75
 Maria       | Miller       |         130.72
 Susan       | Wilson       |          85.77
```

Figure 11.19: *Using subqueries in Calculating Aggregate Values*

Common Table Expression (CTE)

A **Common Table Expression (CTE)** in PostgreSQL is a temporary result set that you can reference within a `SELECT`, `INSERT`, `UPDATE`, or `DELETE` statement. CTEs are defined using the `WITH` keyword and provide a way to structure complex queries in a more readable and maintainable manner. They are particularly useful for breaking down complicated queries into simpler, more understandable parts and can be used multiple times within the same SQL statement.

Scenarios Where CTEs are Applicable

In this section, we will look at some situations where CTEs are used.

- **Improving Query Readability**: CTEs can break down complex queries into simpler, named components, making the overall query easier to read and understand.
- **Eliminating Duplicate Code**: By defining a CTE once and referencing it multiple times, you can avoid repeating the same subquery logic throughout your SQL statement.
- **Working with Recursive Data**: Recursive CTEs are particularly useful for querying hierarchical or recursive data, such as organizational structures or adjacency lists.

Types of CTEs

In this section, we will look at the different types of CTEs.

- **Non-Recursive CTE:** A standard CTE used for creating a temporary result set from a simple query.
- **Recursive CTE:** A CTE that refers to itself. It is used for queries that involve hierarchical or recursive relationships, such as organizational charts or adjacency lists.

Suppose we want to find the customer who made the most payments. Instead of using a subquery, we can use a CTE to improve query readability:

```
dvdrental=# WITH top_customer AS (
    SELECT customer_id, SUM(amount) AS total_amount
    FROM payment
    GROUP BY customer_id
    ORDER BY total_amount DESC
    LIMIT 1
)
SELECT first_name, last_name
FROM customer
WHERE customer_id = (SELECT customer_id FROM top_customer);
 first_name | last_name
------------+-----------
 Eleanor    | Hunt
(1 row)
```

Figure 11.20: *Using CTE to find the customer who made the most payments*

In this CTE example, we define **top_customer** to calculate the customer with the highest total payments and then use it in the main query to retrieve the customer's name. The different parts of the query are explained below:

- **WITH top_customer AS (...):** This part of the query defines a CTE named **top_customer**.
- **SELECT customer_id**, **SUM(amount) AS total_amount:** This selects **customer_id** and the total amount they have paid, calculated using **SUM(amount)**.
- **FROM payment:** This specifies the payment table as the source of data.
- **GROUP BY customer_id:** This groups the results by **customer_id** to aggregate payments for each customer.
- **ORDER BY total_amount DESC:** This orders the grouped results in descending order by the total amount paid, so the customer who paid the most is at the top.
- **LIMIT 1:** This limits the results to the top customer (the one with the highest total amount paid).

Filtering Results

Suppose we want to list all films rented by customers who have spent more than **$100**. We resolved this problem statement earlier on using subqueries, but let us see how to resolve it using CTE:

```
dvdrental=# WITH high_spenders AS (
    SELECT customer_id
    FROM payment
    GROUP BY customer_id
    HAVING SUM(amount) > 100
),
films_rented AS (
    SELECT film_id
    FROM inventory i
    JOIN rental r ON i.inventory_id = r.inventory_id
    JOIN payment p ON r.rental_id = p.rental_id
    WHERE p.customer_id IN (SELECT customer_id FROM high_spenders)
)
SELECT title
FROM film
WHERE film_id IN (SELECT film_id FROM films_rented);
```

```
           title
-----------------------------
 Chamber Italian
 Grosse Wonderful
 Airport Pollock
 Bright Encounters
 Academy Dinosaur
 Ace Goldfinger
 Adaptation Holes
```

Figure 11.21: *Using CTE to list all films rented by customers who have spent more than $100*

In this CTE example, we define **high_spenders** to find customers who spent more than **$100** and **films_rented** to find films rented by these customers, then use these CTEs in the main query to retrieve the film titles.

Suppose we want to list all films along with the number of times they were rented:

```
dvdrental=# WITH rental_counts AS (
    SELECT film_id, COUNT(*) AS rental_count
    FROM inventory i
    JOIN rental r ON i.inventory_id = r.inventory_id
    GROUP BY film_id
)
SELECT f.title, rc.rental_count
FROM film f
JOIN rental_counts rc ON f.film_id = rc.film_id;
```

```
           title             | rental_count
-----------------------------+--------------
 Chamber Italian             |           13
 Grosse Wonderful            |            8
 Airport Pollock             |           18
 Bright Encounters           |           13
 Academy Dinosaur            |           23
 Ace Goldfinger              |            7
 Adaptation Holes            |           12
 Affair Prejudice            |           23
 African Egg                 |           12
```

Figure 11.22: *Using CTE to list all films along with the number of times rented*

In this CTE example, we define **`rental_counts`** to calculate the number of rentals for each film, then use it in the main query to join with the film table and retrieve the titles and rental counts.

Views

In PostgreSQL, a **`view`** is a virtual table representing the result of a stored query. Unlike a regular table, a view does not store data physically. Instead, it dynamically retrieves data from the base tables whenever it is queried. Views can simplify complex queries, enhance security by restricting access to certain data, and provide a level of abstraction.

Types of Views

In this section, we will look at different types of views.

- **Simple View:** A simple view is based on a single table and does not contain any complex logic like joins or subqueries. It typically selects a subset of columns or rows from the base table.

```
dvdrental=# CREATE VIEW customer_names AS
SELECT customer_id, first_name, last_name
FROM customer;
CREATE VIEW
```

Figure 11.23: *How to create a view from a base table*

The view **`customer_names`** provides a simplified representation of the customer table, focusing only on the customer identifiers and their names.

It abstracts the underlying table structure, allowing users to query the view without needing to understand the details of the customer table.

This can be particularly useful in complex databases with many tables and relationships.

Explanation of the query above:

- **`CREATE VIEW customer_names AS`**:
    - This part of the query is creating a **`view`** named **`customer_names`**.
    - A view is a virtual table based on the result of a **`SELECT`** query. It does not store data physically but provides a way to simplify and reuse complex queries.
    - In this case, the **`view`** will be named **`customer_names`**.

- **`SELECT customer_id, first_name, last_name`**:
  - This **`SELECT`** statement defines the columns that will be included in the view.
  - It selects the **`customer_id`**, **`first_name`**, and **`last_name columns`** from the customer table.
- **`FROM customer`**:
  - This specifies the customer table as the source of data for the view.

Instead of writing the same **`SELECT`** query repeatedly to fetch customer names and **`IDs`**, you can use this view to streamline your queries.

Views can be used to restrict access to specific columns or rows in a table. By providing access to the **`customer_names`** view, you can limit users from accessing other sensitive information in the customer table, such as addresses or payment details.

Once the view **`customer_names`** is created, you can query it like a regular table:

```
dvdrental=# SELECT customer_id, first_name, last_name
FROM customer_names;
dvdrental=#
dvdrental=#
dvdrental=# SELECT * FROM customer_names;
dvdrental=#
```

```
 customer_id | first_name  |  last_name
-------------+-------------+--------------
         524 | Jared       | Ely
           1 | Mary        | Smith
           2 | Patricia    | Johnson
           3 | Linda       | Williams
           4 | Barbara     | Jones
           5 | Elizabeth   | Brown
           6 | Jennifer    | Davis
           7 | Maria       | Miller
           8 | Susan       | Wilson
           9 | Margaret    | Moore
          10 | Dorothy     | Taylor
          11 | Lisa        | Anderson
          12 | Nancy       | Thomas
          13 | Karen       | Jackson
```

Figure 11.24: *Querying data from a view rather than a base table*

This query will return the **`customer_id`**, **`first_name`**, and **`last_name`** for all customers.

- **Complex View**: A complex view can be based on multiple tables and include joins, subqueries, and various expressions.

  It can present a more comprehensive dataset derived from multiple sources.

```
dvdrental=# CREATE VIEW rental_info_vw AS
SELECT c.first_name, c.last_name, f.title, r.rental_date
FROM customer c
INNER JOIN rental r ON c.customer_id = r.customer_id
INNER JOIN inventory i ON r.inventory_id = i.inventory_id
INNER JOIN film f ON i.film_id = f.film_id;
CREATE VIEW
```

Figure 11.25: *Creating a complex view with multiple tables and joins*

The **rental_info_vw** view combines data from the **customer**, **rental**, **inventory**, and film tables to present information about rentals along with customer names and film titles.

```
dvdrental=# SELECT * FROM rental_info_vw;
 first_name | last_name   | title              |     rental_date
------------+-------------+--------------------+---------------------
 Manuel     | Murrell     | Graduate Lord      | 2005-05-24 23:03:39
 Andrew     | Purdy       | Love Suicides      | 2005-05-24 23:04:41
 Delores    | Hansen      | Idols Snatchers    | 2005-05-24 23:05:21
 Nelson     | Christenson | Mystic Truman      | 2005-05-24 23:08:07
 Cassandra  | Walters     | Swarm Gold         | 2005-05-24 23:11:53
 Minnie     | Romero      | Lawless Vision     | 2005-05-24 23:31:46
 Ellen      | Simpson     | Matrix Snowman     | 2005-05-25 00:00:40
 Danny      | Isom        | Hanging Deep       | 2005-05-25 00:02:21
 April      | Burns       | Whale Bikini       | 2005-05-25 00:09:02
 Deanna     | Byrd        | Games Bowfinger    | 2005-05-25 00:19:27
```

Figure 11.26: *Querying data from a complex view created in Figure 11.24*

- **Updatable View**: An updatable view allows you to perform insert, update, or delete operations, which are then reflected in the underlying base table(s).

Not all views are updatable. A view must meet specific criteria (for example, it should not include complex joins or subqueries).

```
dvdrental=# CREATE VIEW customer_details AS
SELECT customer_id, first_name, last_name, email
FROM customer;
CREATE VIEW
```

```
dvdrental=# SELECT *
FROM  customer_details
WHERE customer_id = 1;
 customer_id | first_name | last_name |              email
-------------+------------+-----------+----------------------------------
           1 | Mary       | Smith     | mary.smith@sakilacustomer.org
(1 row)
```

Figure 11.27: *Creating an Updatable View*

You can update customer emails through the **`customer_details`** view:

```
dvdrental=# UPDATE customer_details
SET email = 'newemail@example.com'
WHERE customer_id = 1;
UPDATE 1
dvdrental=#
dvdrental=# SELECT *
FROM  customer_details
WHERE customer_id = 1;
 customer_id | first_name | last_name |        email
-------------+------------+-----------+----------------------
           1 | Mary       | Smith     | newemail@example.com
(1 row)

dvdrental=# SELECT *
FROM  customer
WHERE customer_id = 1;
 customer_id | store_id | first_name | last_name |        email         |
-------------+----------+------------+-----------+----------------------+-
           1 |        1 | Mary       | Smith     | newemail@example.com |
(1 row)
```

Figure 11.28: *Result of updating an updatable view on base table*

Scenarios Where Views are Applicable

- **Data Abstraction**
    - Views can abstract complex queries, making them easier to use.
    - For example, instead of writing a complex join query repeatedly, you can create a view that simplifies the process:
- **Security**
    - Views can limit access to specific columns or rows, enhancing security.

   - Suppose you want to grant access to customer service representatives to view customer names and emails but not their addresses or payment details:
- **Simplifying Maintenance**
   - Using views can simplify database maintenance by centralizing complex logic in one place.
   - If business logic changes, you can update the view definition rather than altering multiple queries scattered throughout the application.

Materialized View

A **materialized view** is different from a standard view in that it stores the result of the query physically. This can improve performance for complex queries because the result set is precomputed and stored, rather than recalculated each time the view is accessed. However, the data in a materialized view must be refreshed manually or on a schedule to reflect changes in the underlying base tables.

Creating a Materialized View

Let us create an example of a materialized view using our **dvdrental** database:

Suppose we want to create a materialized view that stores the aggregated data of film rentals:

```
dvdrental=# CREATE MATERIALIZED VIEW film_rental_summary AS
SELECT f.film_id, f.title, COUNT(r.rental_id) AS rental_count
FROM film f
LEFT JOIN inventory i ON f.film_id = i.film_id
LEFT JOIN rental r ON i.inventory_id = r.inventory_id
GROUP BY f.film_id, f.title;
SELECT 1000
```

```
dvdrental=# SELECT * FROM film_rental_summary;
```

```
 film_id |            title            | rental_count
---------+-----------------------------+--------------
     652 | Pajama Jawbreaker           |           14
     273 | Effect Gladiator            |           25
      51 | Balloon Homeward            |           23
     951 | Voyage Legally              |           28
      70 | Bikini Borrowers            |           17
     839 | Stallion Sundance           |            7
     350 | Garden Island               |           20
     758 | Saints Bride                |            8
     539 | Luck Opus                   |           10
```

Figure 11.29: *Creating and using a Materialized View*

In this example, the materialized view **film_rental_summary** precomputes and stores the number of rentals for each film.

Refreshing a Materialized View

To keep the data in a materialized view up-to-date, you need to refresh it manually or on a schedule:

```
dvdrental=# REFRESH MATERIALIZED VIEW film_rental_summary;
REFRESH MATERIALIZED VIEW
```

Figure 11.30: *How to refresh a Materialized view*

Nested Materialized Views

A **Nested Materialized View** refers to a materialized view that depends on other materialized views as its source. Essentially, one materialized view is built on top of another. This is useful for structuring complex queries efficiently. It speeds up complex queries by precomputing results, and it simplifies handling of large datasets with hierarchical data views.

If we want to analyze rental data, then we need to create multiple levels of aggregation. We will first create a materialized view, and then a nested materialized view.

Step 1: Create a Materialized View for total rentals by film:

```
CREATE MATERIALIZED VIEW film_rental_totals AS
SELECT f.film_id, f.title, COUNT(r.rental_id) AS total_rentals
FROM film f
JOIN inventory i ON f.film_id = i.film_id
JOIN rental r ON i.inventory_id = r.inventory_id
GROUP BY f.film_id, f.title;
```

```
dvdrental=# CREATE MATERIALIZED VIEW film_rental_totals AS
SELECT f.film_id, f.title, COUNT(r.rental_id) AS total_rentals
FROM film f
JOIN inventory i ON f.film_id = i.film_id
JOIN rental r ON i.inventory_id = r.inventory_id
GROUP BY f.film_id, f.title;
SELECT 958
dvdrental=#
```

Figure 11.31: *Create a based materialized view for the nested view below*

The above materialized view, **film_rental_totals**, calculates the total number of rentals for each film.

Step 2: Create a Nested Materialized View for top rented films in each category:

```
CREATE MATERIALIZED VIEW category_top_films AS
SELECT c.name AS category_name, frt.title, frt.total_rentals
FROM film_rental_totals frt
JOIN film_category fc ON frt.film_id = fc.film_id
JOIN category c ON fc.category_id = c.category_id
WHERE frt.total_rentals > 25
ORDER BY c.name, frt.total_rentals DESC;
```

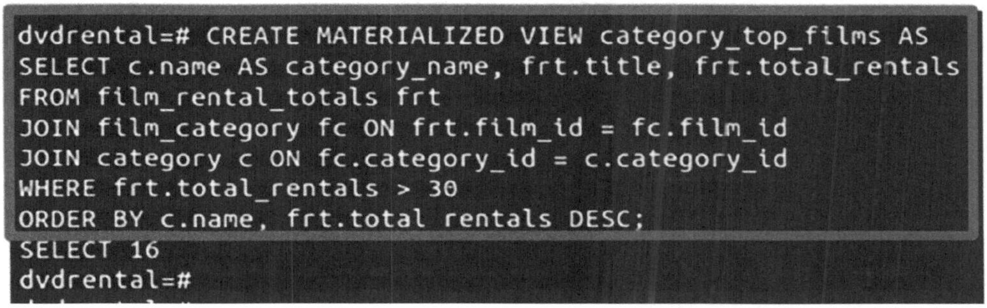

```
dvdrental=# CREATE MATERIALIZED VIEW category_top_films AS
SELECT c.name AS category_name, frt.title, frt.total_rentals
FROM film_rental_totals frt
JOIN film_category fc ON frt.film_id = fc.film_id
JOIN category c ON fc.category_id = c.category_id
WHERE frt.total_rentals > 30
ORDER BY c.name, frt.total_rentals DESC;
SELECT 16
dvdrental=#
```

***Figure 11.32**: Create a nested materialized view from the based materialized view above*

This second materialized view, **category_top_films**, filters the top-performing films in each category based on total rentals.

This query uses **film_rental_totals** (created in Step 1) as its base. By reusing precomputed data, it avoids rejoining and recalculating the rental counts. The **film_category** table connects films (**film_id**) to their categories (**category_id**), while the category table provides the name of each category. The **WHERE frt.total_rentals > 30** condition ensures that only films with more than 30 rentals are included, focusing on high-performing titles. The **ORDER BY c.name, frt.total_rentals DESC** clause ensures the output is sorted by category name first and within each category by rental counts in descending order. This materialized view provides a precomputed snapshot of the top-performing films by category, making it quick and efficient to retrieve this data later.

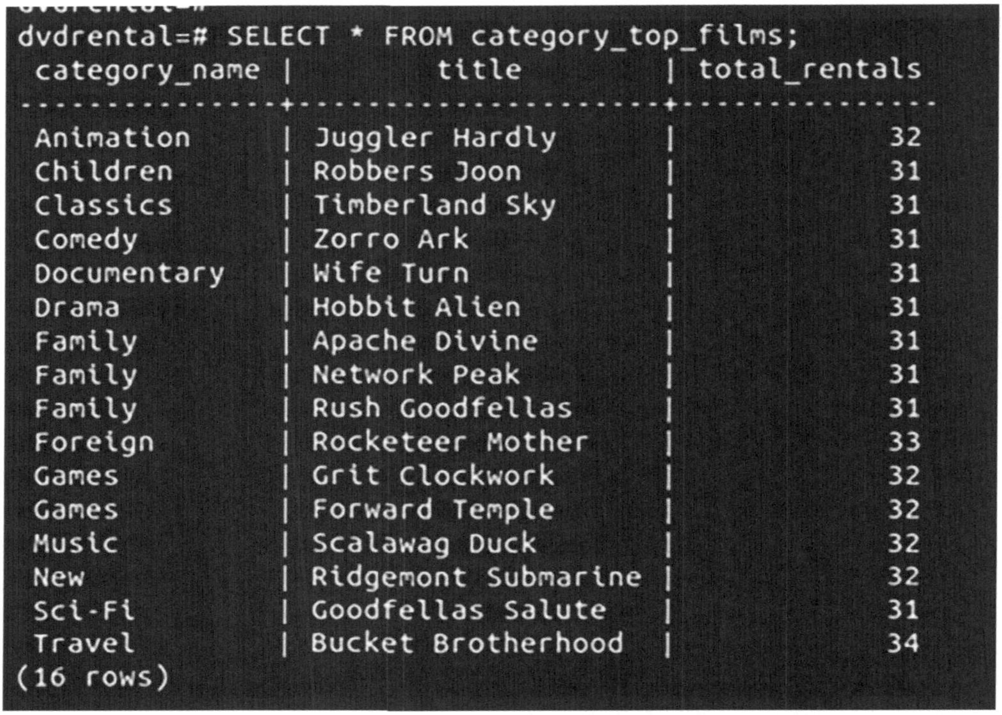

```
dvdrental=# SELECT * FROM category_top_films;
 category_name |        title        | total_rentals
---------------+---------------------+---------------
 Animation     | Juggler Hardly      |            32
 Children      | Robbers Joon        |            31
 Classics      | Timberland Sky      |            31
 Comedy        | Zorro Ark           |            31
 Documentary   | Wife Turn           |            31
 Drama         | Hobbit Alien        |            31
 Family        | Apache Divine       |            31
 Family        | Network Peak        |            31
 Family        | Rush Goodfellas     |            31
 Foreign       | Rocketeer Mother    |            33
 Games         | Grit Clockwork      |            32
 Games         | Forward Temple      |            32
 Music         | Scalawag Duck       |            32
 New           | Ridgemont Submarine |            32
 Sci-Fi        | Goodfellas Salute   |            31
 Travel        | Bucket Brotherhood  |            34
(16 rows)
```

Figure 11.33: *A select query on nested materialized view (category_top_films)*

Defaults

In PostgreSQL and other RDBMS, a default value is the value automatically assigned to a column when a row is inserted without explicitly specifying a value for that column.

Imagine you want to enhance the customer table to ensure new rows have default values for certain columns:

```
ALTER TABLE customer
ADD COLUMN preferred_language TEXT DEFAULT 'English',
ADD COLUMN account_status TEXT DEFAULT 'active',
ADD COLUMN signup_date TIMESTAMP DEFAULT CURRENT_TIMESTAMP;
```

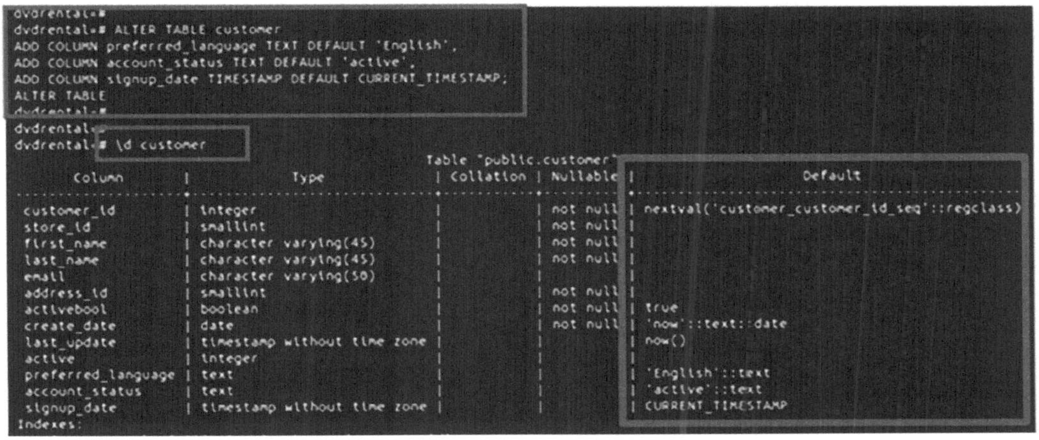

Figure 11.34: Adding new columns with default values

If no value is provided for **preferred_language**, it will default to **'English'**. If no value is given for **account_status**, it will default to **'active'**.

signup_date will automatically use the current timestamp for new rows.

```
INSERT INTO customer (store_id, address_id, first_name, last_name, email)
VALUES (1, 101, 'John', 'Doe', 'john.doe@example.com');
select * from customer  where first_name ='John' AND last_name='Doe';
```

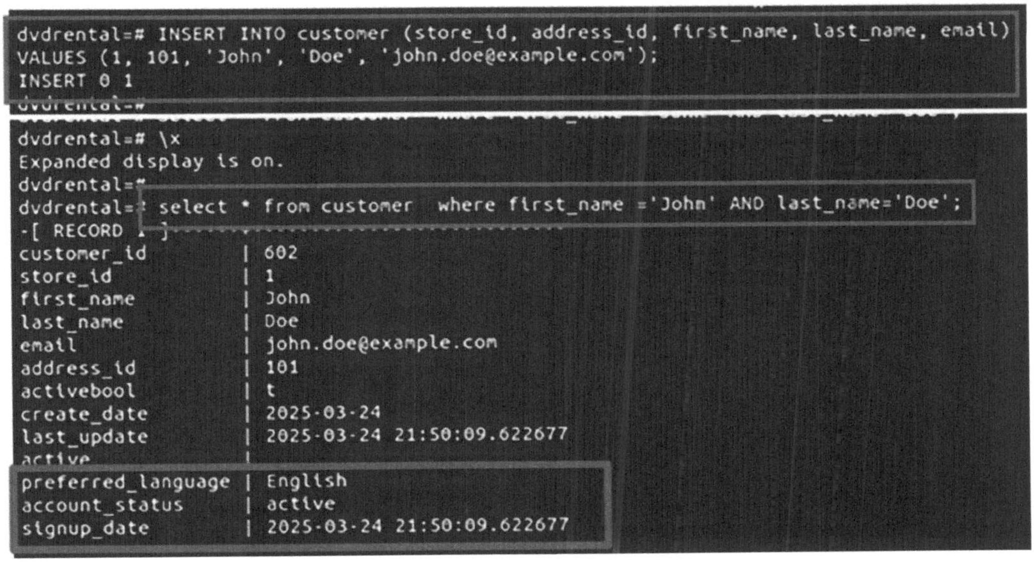

Figure 11.35: Insert into table and check default values for new columns

The resulting row above has:

- `preferred_language = 'English'`
- `account_status = 'active'`
- `signup_date = current timestamp`

Default values ensure consistency by preventing columns such as **`preferred_language`** or **`account_status`** from being null, reducing inconsistencies. They simplify data entry by removing the need to explicitly specify common values, easing application logic. Additionally, they help prevent errors or omissions during insertion by automatically assigning predefined values.

Task

Possible Technical Interview Questions

This section provides interview questions to help solidify the learning of this chapter.

- **Basic Join Types:**
  - What are the different types of joins in SQL?
  - Explain the difference between an **`INNER JOIN`** and a **`LEFT JOIN`**.
  - How does a **`RIGHT JOIN`** differ from a **`FULL JOIN`**?
- **Practical Applications:**
  - Write a query to find all customers who have rented a film, including the film titles.
  - How would you retrieve a list of all films along with their categories using a join?
- **Performance Considerations:**
  - What are some performance implications of using joins in SQL?
  - How can indexing affect the performance of join operations?

Subqueries

- **Basic Understanding:**
  - What is a subquery in SQL, and where can it be used?
  - Explain the difference between a nested subquery and a correlated subquery.
- **Practical Applications**:
  - Write a query using a subquery to find customers who have not rented any films.

    - How would you use a subquery to retrieve the top 5 customers by total payment amount?
- **Complex Scenarios:**
    - Explain how you would use a subquery to calculate the average rental duration for all customers.
    - Write a query with a subquery to list all films that were rented by customers who have spent more than `$200`.

Common Table Expressions (CTEs)

- **Basic Understanding:**
    - What is a CTE, and how is it different from a subquery?
    - Explain the syntax of a CTE.
- **Practical Applications:**
    - Write a query using a CTE to find the total number of films rented by each customer.
    - How would you use a recursive CTE to generate a sequence of rental dates for a specific customer?
- **Advanced Usage:**
    - Explain how CTEs can improve the readability and maintainability of complex queries.
    - Write a query using multiple CTEs to calculate the average payment amount per customer and then use it to find customers who exceed this average.
- **Views**
    - What is a view in SQL, and how is it different from a table?
    - Explain the syntax for creating a view.
- **Practical Applications:**
    - Write a query to create a view that shows the names and rental counts of all customers.
    - How can views be used to enhance security in a database?
- **Advanced Usage:**
    - What are materialized views, and how do they differ from standard views?
    - Explain a scenario where you would use a materialized view over a standard view.

Conclusion

In this chapter, we learned about the different types of joins, and by mastering these different types of joins, learners will be able to construct complex queries that can handle a wide variety of data retrieval scenarios in PostgreSQL.

As a PostgreSQL database administrator or developer, utilizing subqueries effectively can enhance the flexibility and power of your SQL queries in PostgreSQL.

By using CTEs, you can enhance the clarity and maintainability of your SQL queries, making complex data manipulations more manageable.

The `CREATE VIEW` query creates a view that simplifies and abstracts access to information from the base table, enhancing readability, maintainability, and security of SQL queries.

Materialized views are particularly useful when dealing with large datasets and complex queries that do not need to be run in real-time, thus improving query performance significantly.

In the next chapter, we will look at cloud computing, such as creating and using AWS EC2 instances, RDS instances, storage, and migration.

CHAPTER 12
Cloud Computing (AWS)

Introduction

Amazon Web Services (AWS) is a cloud computing platform that offers a wide array of services, including storage, databases, networking, machine learning, and analytics. Companies and developers use AWS to host websites, build applications, and manage large datasets, among other tasks.

In this chapter, learners will understand how to create an account in AWS, create EC2 instances, RDS instances, Install PostgreSQL in EC2 and RDS, increase storage and change instance type, Create S3 bucket and upload files into it, migrate PostgreSQL from on-prem to the cloud (EC2 instance), migrate from EC2 to RDS using AWS database migration services (DMS), configure monitoring and learn the basic metrics. Learners will become proficient in working in the cloud.

Structure

In this chapter, we will cover the following topics:

- Creating an AWS Account
- Launching an EC2 Instance
- Install PostgreSQL on EC2 Instances
- Amazon Relational Database Service (RDS)
- S3 Bucket
- Task

Creating an AWS Account

Creating an AWS account is a simple process that lets you access Amazon's vast array of cloud services. The following is a step-by-step guide for you to create your account:

1. **Go to the AWS Website**: Open your browser and navigate to **`aws.amazon.com`**.

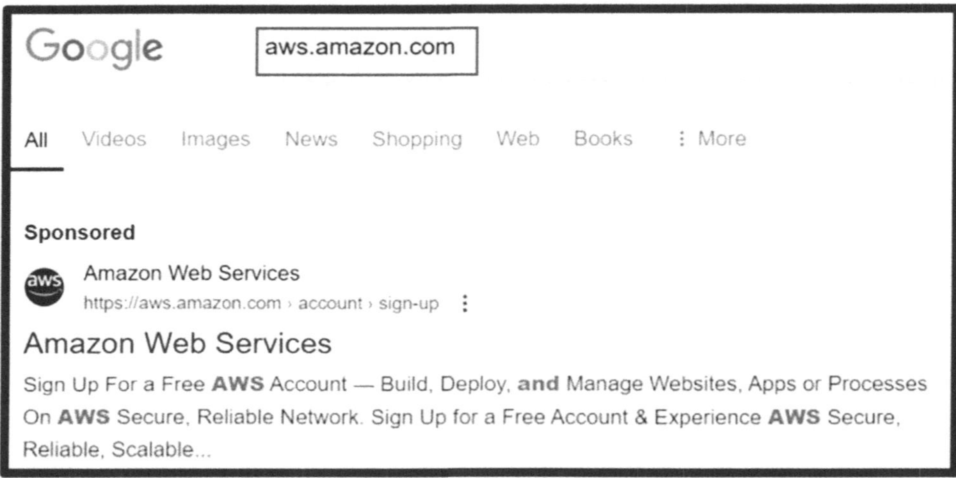

***Figure 12.1:** Locating the aws web services site*

2. **Start the Sign-Up Process**: Click on the "**`Create an AWS Account`**" button.

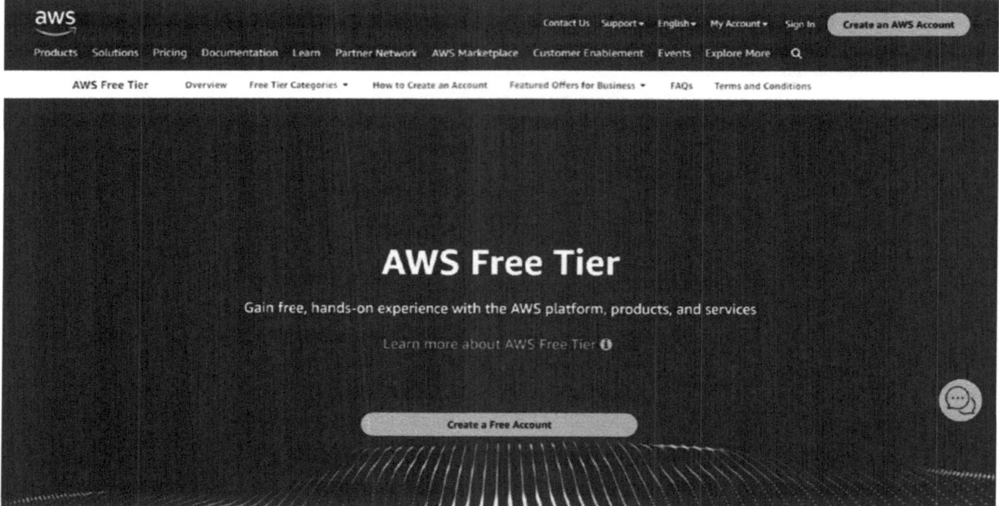

***Figure 12.2:** How to create an AWS account*

3. **Enter Your Email and Set a Password**: Provide your email address and choose a secure password.

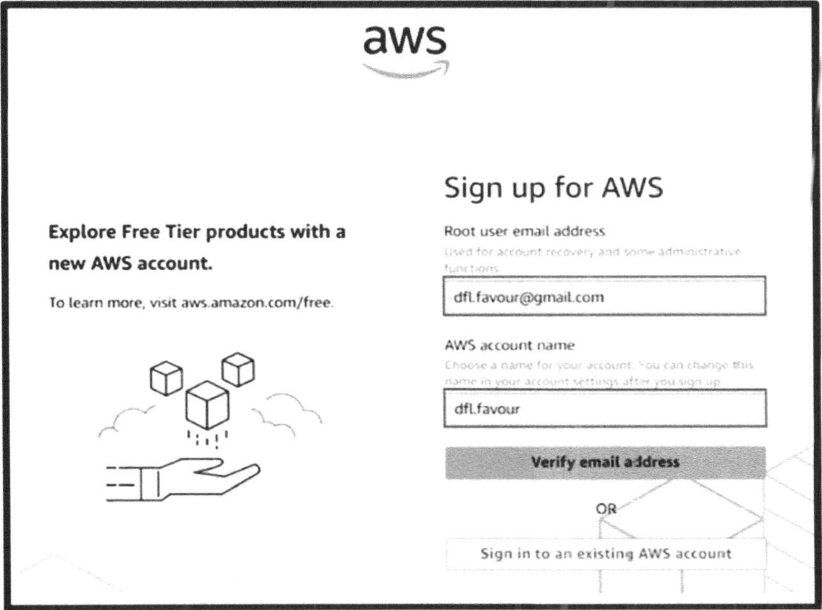

Figure 12.3: *Shows how to enter credentials for signup*

Figure 12.4: *Enter a verification token*

In the next screen, you will provide the account password for the signup account.

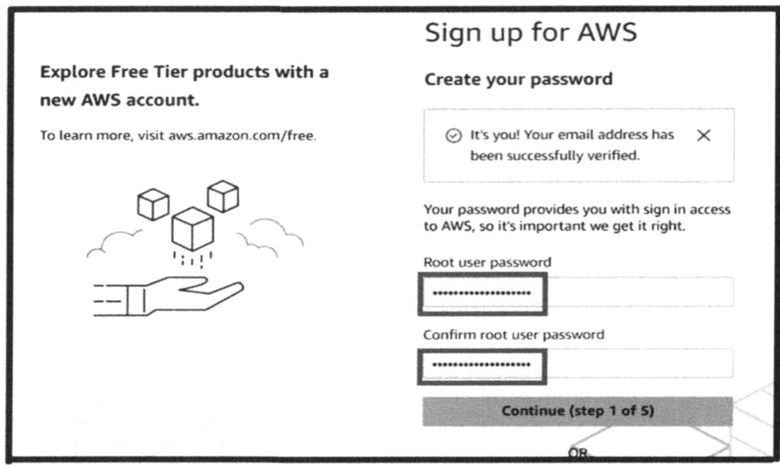

Figure 12.5: *Providing a password for the signup account*

4. **Provide Your Contact Information**: Fill in your name, phone number, and address.
5. **Add Payment Details**: Enter your credit or debit card information. AWS uses this to verify your identity, even if you plan to use the free tier.

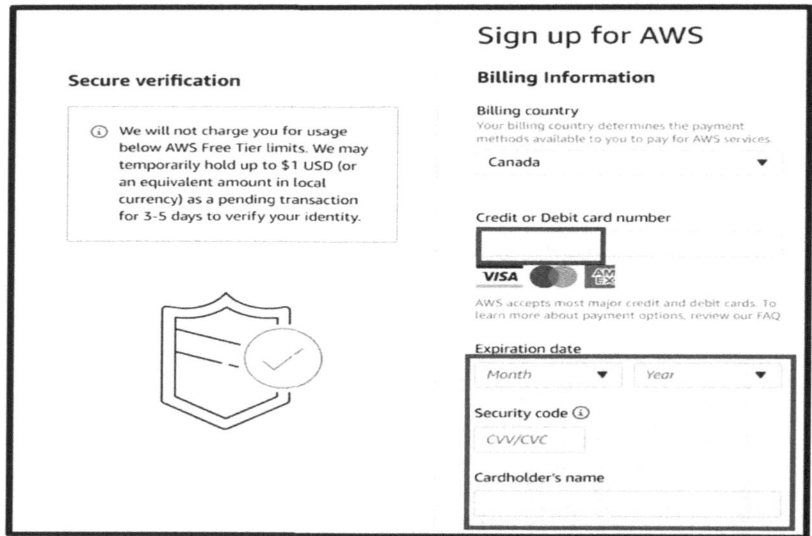

Figure 12.6: *Provide card information for verification purposes by AWS*

6. **Verify Your Identity**: AWS will either call or send you an SMS to verify your identity.

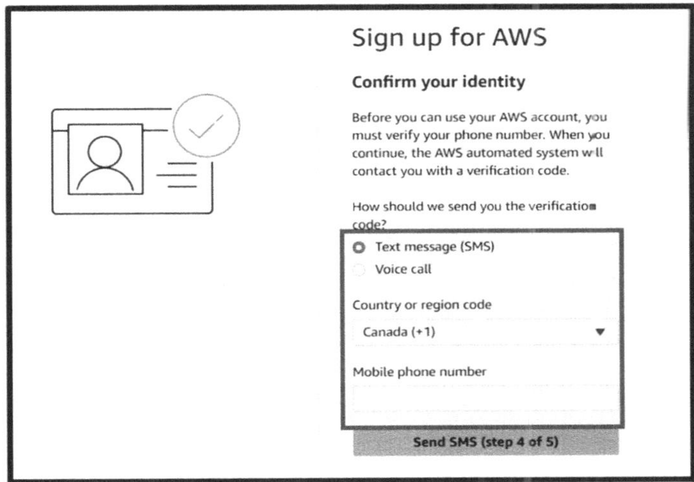

Figure 12.7: *Enter phone number for verification purposes*

7. **Select a Support Plan**: Choose a support plan that suits your needs. The **`Basic Support Plan`** is free, but you can opt for more comprehensive plans if needed.

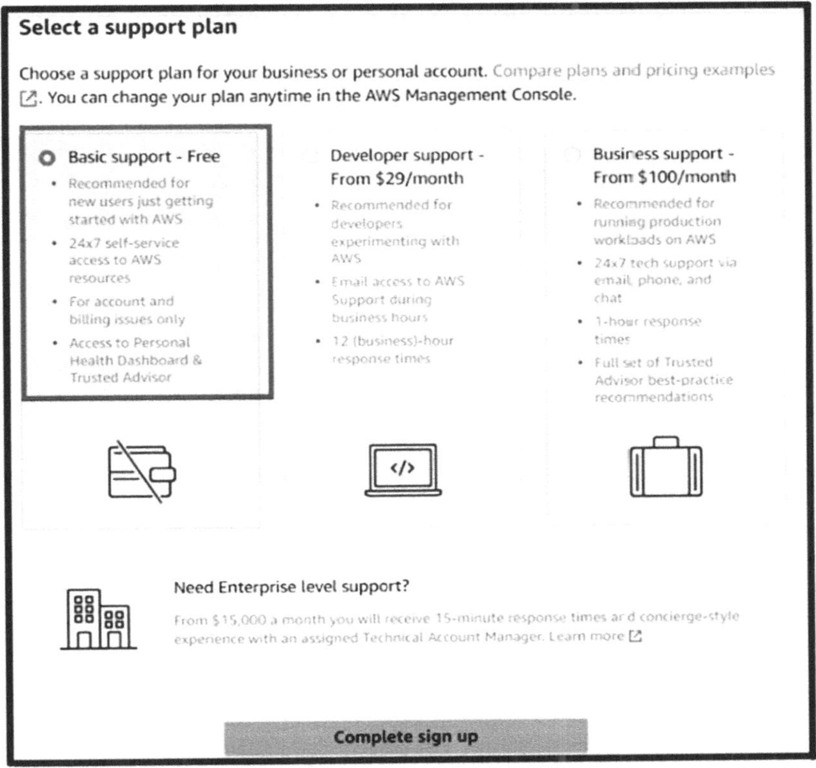

Figure 12.8: *Choosing a plan and completing the signup process*

8. **Complete the Registration**: Review your details and complete the sign-up process.

Figure 12.9: *You have successfully signed up*

9. **Access the AWS Management Console**: Once your account is set up, you can log in to the AWS Management Console and start exploring the services available. First, you will receive an email from Amazon such as the following one:

Figure 12.10*: An email from AWS confirming your registration*

That is it! You are now ready to use AWS and explore its myriads of cloud computing services as follows.

Launching an EC2 Instance

Amazon Elastic Compute Cloud (EC2) is a web service that provides scalable computing capacity in the cloud. It allows users to rent virtual servers, known as EC2 instances, to run their applications. This flexibility eliminates the need for investing in physical hardware, making it easier to scale resources up or down based on demand.

Follow these steps to launch your EC2 instance:

1. **Sign in to the AWS Management Console**:
    a. Go to the AWS Management Console.
    b. Sign in with your AWS account credentials.

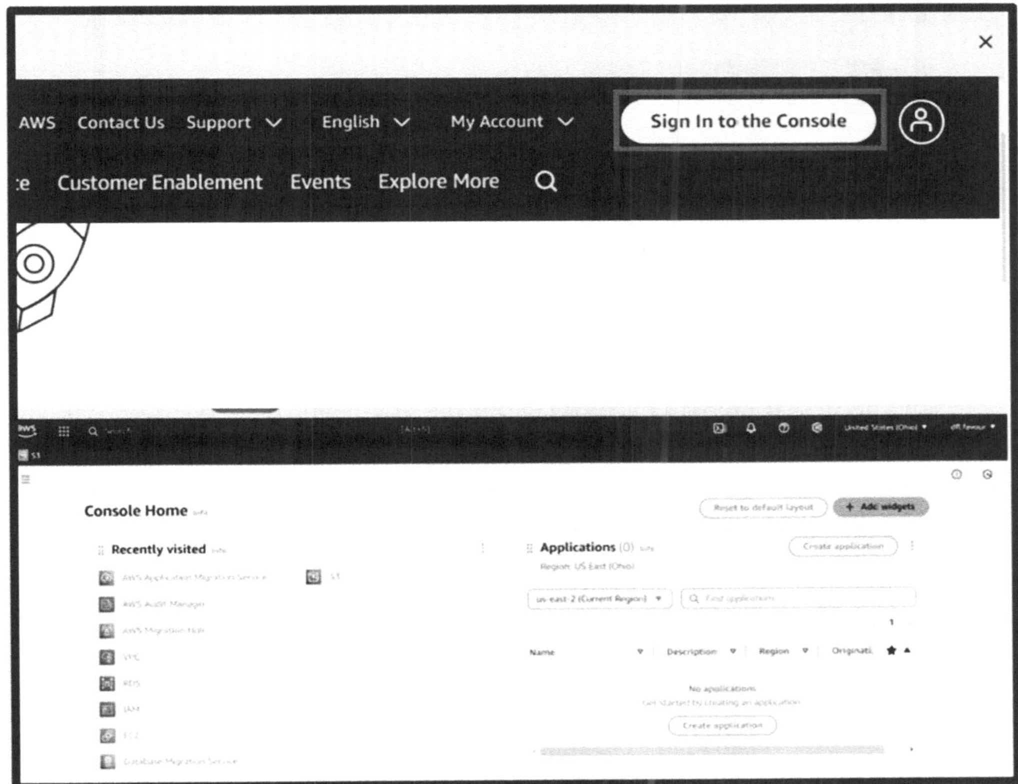

Figure 12.11: *Shows an AWS sign-in console*

2. **Navigate to the EC2 Dashboard**: In the AWS Management Console, select "**EC2**" from the "**Services**" menu to access the **EC2 Dashboard**.

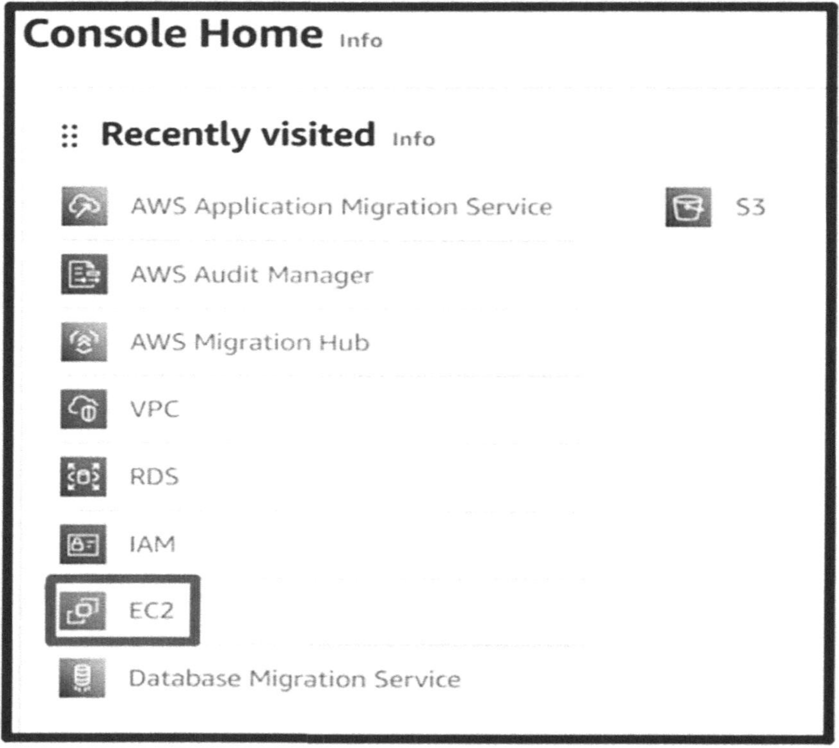

Figure 12.12: *Selecting EC2 from the dashboard if you recently selected it*

OR

For the first time, you have to search the **`EC2`** from the service search bar as shown in the following figure.

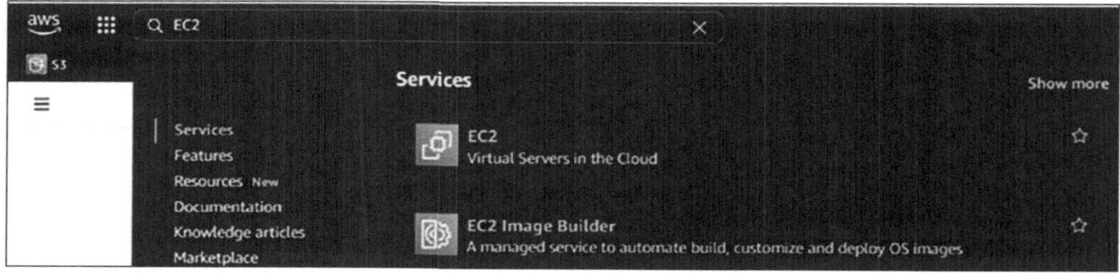

Figure 12.13: *Obtaining EC2 from the search bar*

3. **Launch Instance**: Click on the "**`Launch Instance`**" button to start the setup process.

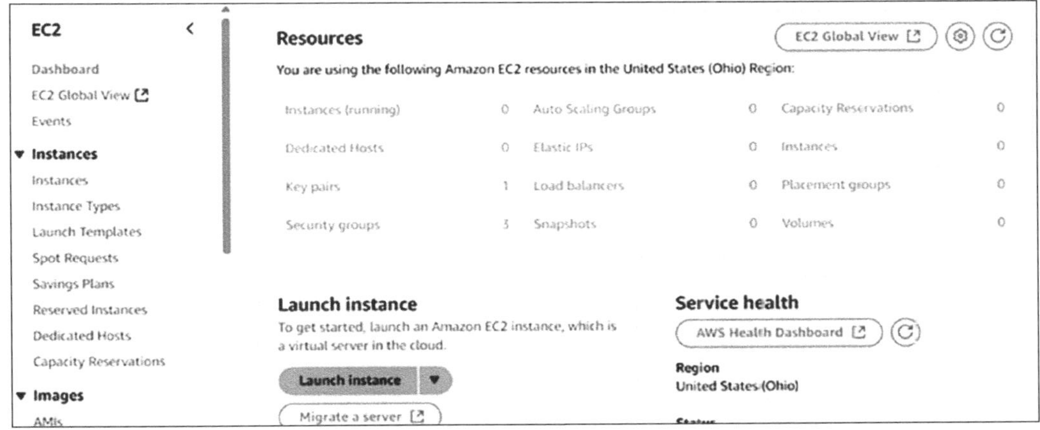

Figure 12.14: *Launching an EC2 instance*

Give your instance a name as shown in the following figure:

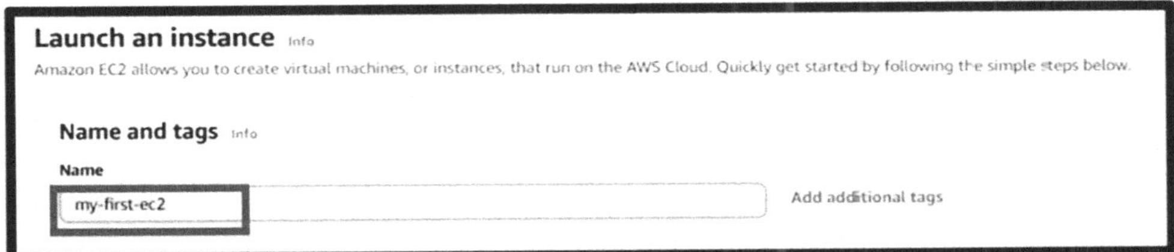

Figure 12.15: *Naming the EC2 instance*

4. **Choose an Amazon Machine Image (AMI)**: Select an AMI, which is a pre-configured template for your instance. You can choose from a range of Linux and Windows AMIs provided by AWS or the community.

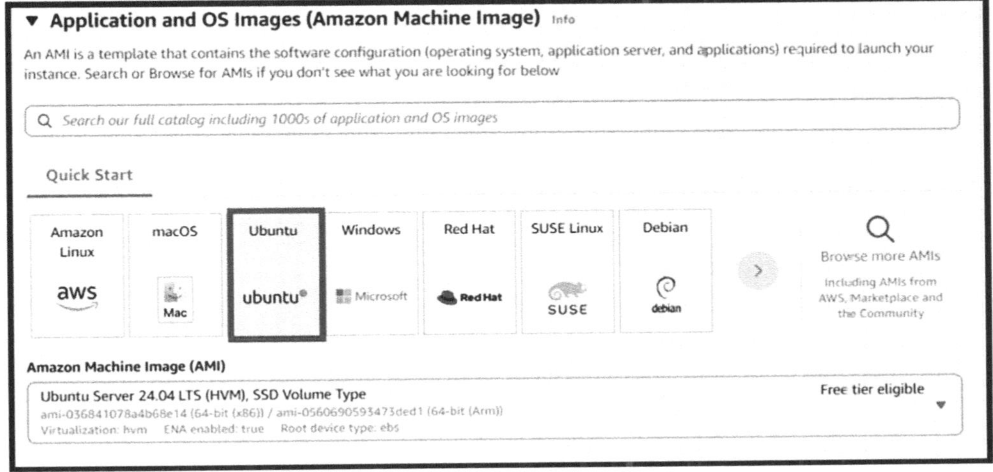

Figure 12.16: *Choosing Amazon machine image for the EC2 instance*

5. **Choose an Instance Type**: Select the instance type based on the amount of CPU, memory, storage, and networking capacity you need. AWS offers various instance types optimized for different use cases. For the purpose of learning, let us choose the free tier t2 micro from the drop-down menu.

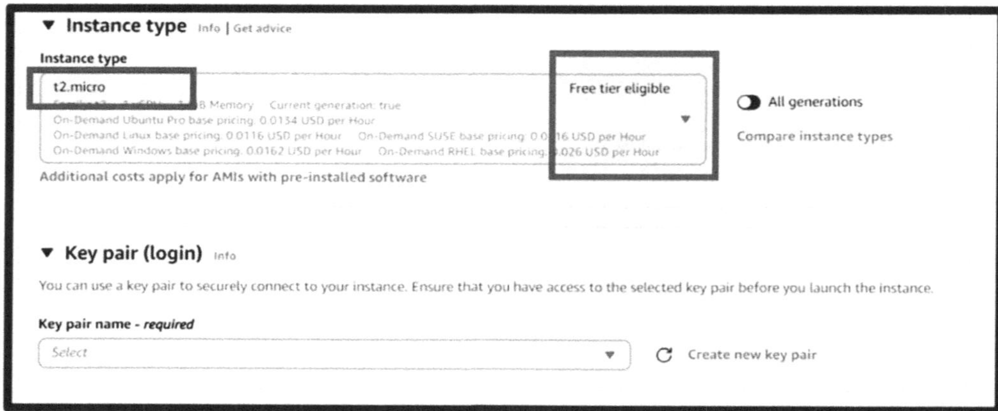

***Figure 12.17:** Choosing the instance type*

We also need to create a key pair. This must be kept as a sensitive file and not exposed to the public. In AWS, a key pair is a set of security credentials that you use to prove your identity when connecting to an instance. It consists of a public key and a private key.

- **Public Key:** This key is used to encrypt data. It is stored on the instance that you want to access.
- **Private Key:** This key is used to decrypt data that has been encrypted with the public key. You store this key securely and use it to connect to your instance.

As we have launched an instance, we need to specify a key pair to use for Secure Shell (SSH) access to the instance. The public key is embedded in the instance, and you use the private key to log in to the instance securely.

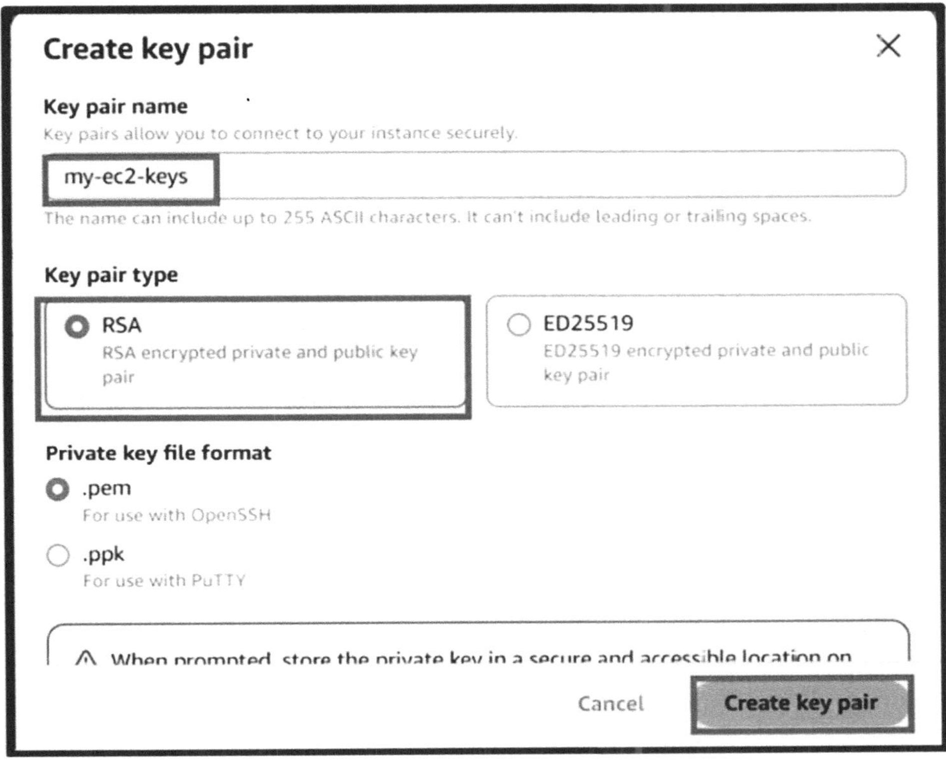

Figure 12.18: *Creating a key pair*

Once created, the key will appear in your downloads. Keep it safe.

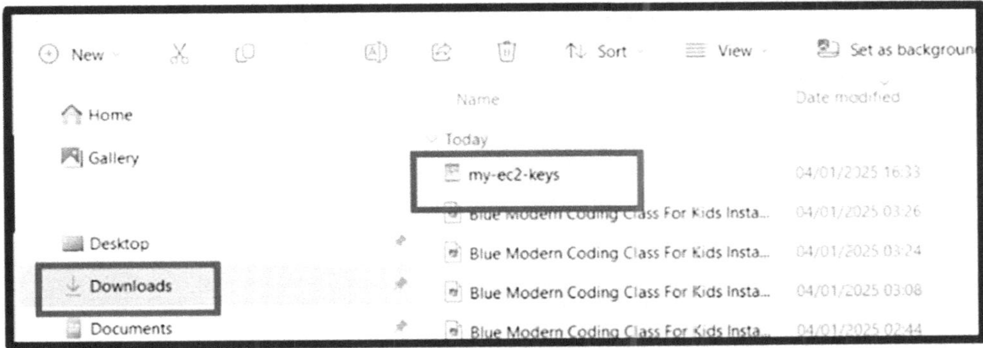

Figure 12.19: *The key pair in the downloads*

6. **Configure Instance Details**: Customize your instance settings, such as the number of instances and network settings. You can accept the default settings or modify them as per your requirements. We are going to keep the default settings.

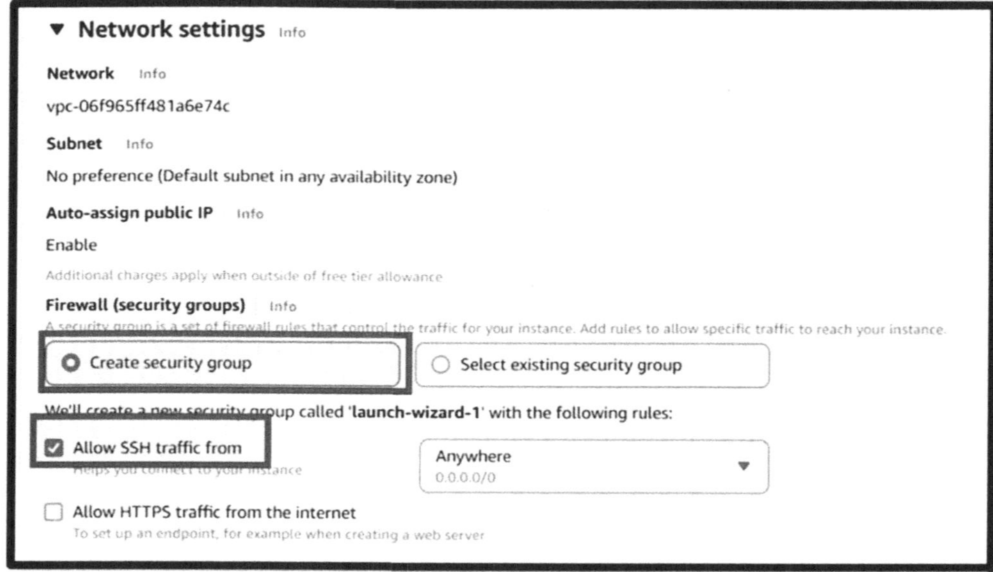

Figure 12.20: *Customizing the network settings*

7. **Configure Security Group**: Create a new security group or select an existing one to define the traffic rules for your instance. Ensure to configure appropriate rules for SSH (for Linux instances access as shown in *Figure* 12.20.

   We are going to assign 30GB of storage as shown in the following figure and then click on launch instance:

8. **Add Storage**: Specify the storage volumes for your instance. You can add additional storage volumes or use the default configuration.

Figure 12.21: *Allocate storage and launch the instance*

Review all your settings, and once satisfied, click "**Launch**." During the launch process, you will be prompted to create or select an existing key pair for SSH/RDP access to your instance if not already selected.

9. **Access Your Instance**: After launching, navigate to the **`EC2 Dashboard`** and check the status of your instance. Once it is running, you can connect to your instance using SSH (for Linux).

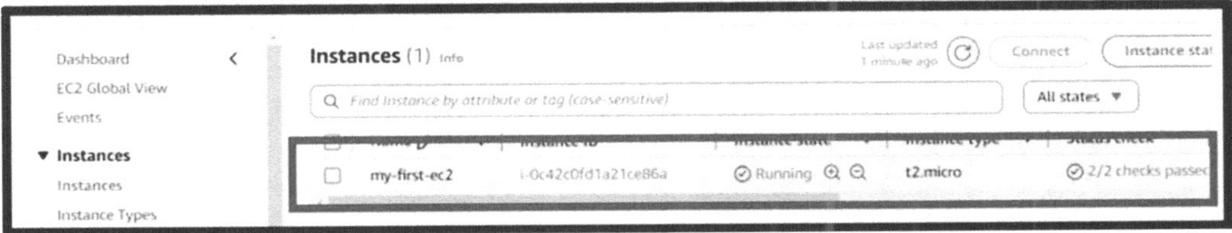

Figure 12.22: *Shows the status of our new instance as running*

Now we are ready to connect to our new EC2 instance. First, we need to create a **`.pem`** file on our virtual machine and copy the content of the key pair created in *Figure* 12.19 into it.

Figure 12.23: *Creating a key pair called my-ec2-keys.pem*

Grant the file read-only permissions with the command **`chmod 400 "my-ec2-keys.pem"`** as shown in the following figure.

```
jack@jack-virtual-machine: $
jack@jack-virtual-machine: $ chmod 400 "my-ec2-keys.pem"
jack@jack-virtual-machine: $
jack@jack-virtual-machine: $ ls
                              my-ec2-keys.pem
jack@jack-virtual-machine: $ ls -al
total 100
drwxr-x---  18 jack     jack      4096 Jan  4 12:16
drwxr-xr-x   4 root     root      4096 Jul 30  2023
drwxrwxr-x   2 jack     jack      4096 Jul 30  2023
-rw-------   1 jack     jack      1451 Feb  3  2024 .bash_history
-rw-r--r--   1 jack     jack       220 Jul  5  2023 .bash_logout
-rw-r--r--   1 jack     jack      3771 Jul  5  2023 .bashrc
drwx------  10 jack     jack      4096 Jul 30  2023
drwx------  12 jack     jack      4096 Dec 30  2023
drwxr-xr-x   3 jack     jack      4096 Dec 28  2023
drwxr-xr-x   2 jack     jack      4096 Jul  5  2023
drwxr-xr-x   2 jack     jack      4096 Jan 11  2024
drwx------   2 jack     jack      4096 Aug 31 06:38
-rw-------   1 jack     jack        20 Dec 28  2023 .lesshst
drwx------   3 jack     jack      4096 Jul  5  2023
-r--------   1 jack     jack      1679 Jan  4 12:16 my-ec2-keys.pem
drwxr-xr-x   2 jack     jack      4096 Jul  5  2023
```

Figure 12.24: *Granting the file called my-ec2-keys.pem read-only permissions*

Let us now connect to our new EC2 instance. First, select the instance and then click on **Connect** as shown in the following figure:

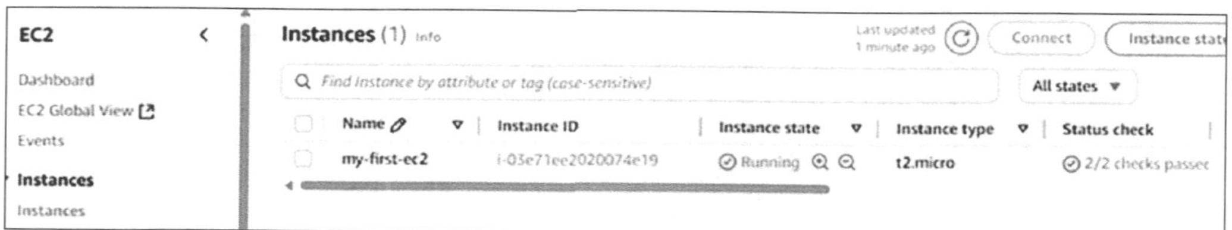

Figure 12.25: *Connecting to an EC2 instance process*

After you have given your key pair the right permissions, copy the connection string as shown in the following figure and run it in your virtual machine.

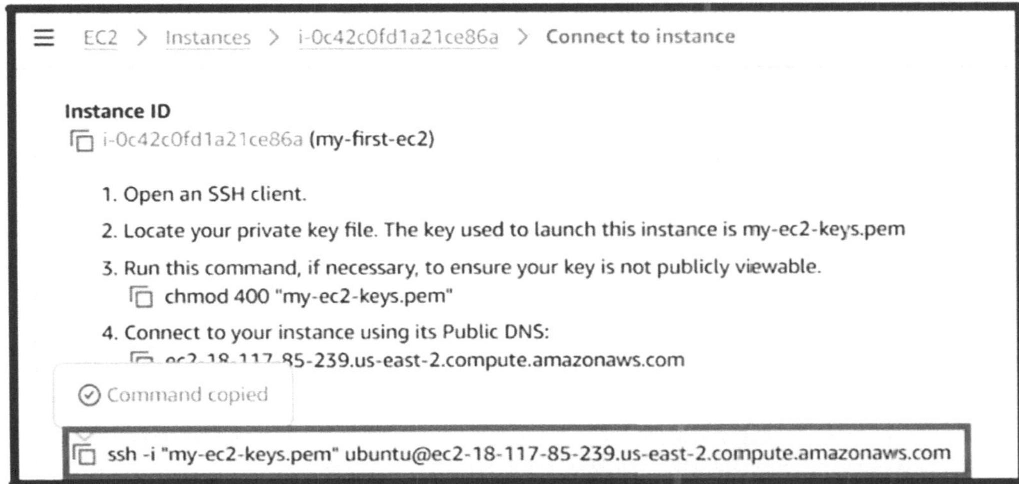

Figure 12.26: *Copy the connection string*

The connection string enables you to log in remotely to the new EC2 instance using the key pair we created in *Figure* 12.18.

```
jack@jack-virtual-machine $ ssh -i "my-ec2-keys.pem" ubuntu@ec2-18-117-85-239.us-east-2.compute.amazonaws.com
The authenticity of host
ED25519 key fingerprint is SHA256:f+enT5OPAwT2PHKTdjTOMDA+sbqQ8LXt4Kqcm+DXPVs.
This key is not known by any other names
Are you sure you want to continue connecting (yes/no/[fingerprint])? yes
Warning: Permanently added 'ec2-18-117-85-239.us-east-2.compute.amazonaws.com' (ED25519) to the list of known hosts.
Welcome to Ubuntu 24.04.1 LTS (GNU/Linux 6.8.0-1018-aws x86_64)

 * Documentation:  https://help.ubuntu.com
 * Management:     https://landscape.canonical.com
 * Support:        https://ubuntu.com/pro

 System information as of Sat Jan  4 17:23:11 UTC 2025
```

Figure 12.27: *Connecting to the new EC2 instance via a virtual machine*

The following screen shows that we have successfully connected to the new instance. You can see that the virtual machine sign-in name has changed from **`jack@jack-virtual-machine`** to **`ubuntu@ip-172-31-20-82`**. These details will be different for everyone.

```
ubuntu@ip-172-31-20-82: $
ubuntu@ip-172-31-20-82: $ hostname
ip-172-31-20-82
ubuntu@ip-172-31-20-82: $
ubuntu@ip-172-31-20-82: $
```

Figure 12.28: *Successfully sign in to the new EC2 instance*

Installing PostgreSQL on EC2 Instance

Setting up PostgreSQL on our new EC2 instance involves several steps.

1. **Connect** to the new EC2 instance using SSH as shown in *Figure 12.26*.

   `ssh -i "your-key-pair-file.pem" ubuntu@your-instance-public-dns.`

   Edit as per your instance connection details.

   ```
   ssh -i "my-ec2-keys.pem" ubuntu@ec2-18-117-85-239.us-east-2.com-
   pute.amazonaws.com
   ```

2. **Update Your System Packages**: Open your terminal and run:

   ```
   sudo apt update -y
   sudo apt upgrade -y
   ```

```
ubuntu@ip-172-31-20-82: $
ubuntu@ip-172-31-20-82: $ sudo apt update -y
Hit:1 http://us-east-2.ec2.archive.ubuntu.com/ubuntu noble InRelease
Get:2 http://us-east-2.ec2.archive.ubuntu.com/ubuntu noble-updates InRelease [126 kB]
Get:3 http://us-east-2.ec2.archive.ubuntu.com/ubuntu noble-backports InRelease [126 kB]
Get:4 http://us-east-2.ec2.archive.ubuntu.com/ubuntu noble/universe amd64 Packages [15.0 MB]
Get:5 http://security.ubuntu.com/ubuntu noble-security InRelease [126 kB]
Get:6 http://us-east-2.ec2.archive.ubuntu.com/ubuntu noble/universe Translation-en [5982 kB]
Get:7 http://security.ubuntu.com/ubuntu noble-security/main amd64 Packages [572 kB]
Get:8 http://us-east-2.ec2.archive.ubuntu.com/ubuntu noble/universe amd64 Components [3871 kB]
Get:9 http://us-east-2.ec2.archive.ubuntu.com/ubuntu noble/universe amd64 c-n-f Metadata [301 kB]
Get:10 http://us-east-2.ec2.archive.ubuntu.com/ubuntu noble/multiverse amd64 Packages [269 kB]
```

Figure 12.29: *Updating the new EC2 instance*

Next, run the upgrade command for new packages to be installed.

```
ubuntu@ip-172-31-20-82: $ sudo apt upgrade -y
Reading package lists... Done
Building dependency tree... Done
Reading state information... Done
Calculating upgrade... Done
The following NEW packages will be installed:
  linux-aws-headers-6.8.0-1021 linux-aws-tools-6.8.0-1021 linux-heade
The following packages will be upgraded:
  apport apport-core-dump-handler chrony cloud-init curl dmidecode fw
  libcurl4t64 libexpat1 libfwupd2 libglib2.0-0t64 libglib2.0-bin libg
  libmodule-scandeps-perl libopeniscsiusr libpackagekit-glib2-18 libp
  needrestart open-iscsi packagekit packagekit-tools python3-apport p
  ssh-import-id udisks2 vim vim-common vim-runtime vim-tiny xxd
58 upgraded, 6 newly installed, 0 to remove and 0 not upgraded.
```

Figure 12.30: *Upgrade the new EC2 instance*

3. **Install PostgreSQL** using the command:

   ```
   sudo apt install postgresql postgresql-contrib
   ```

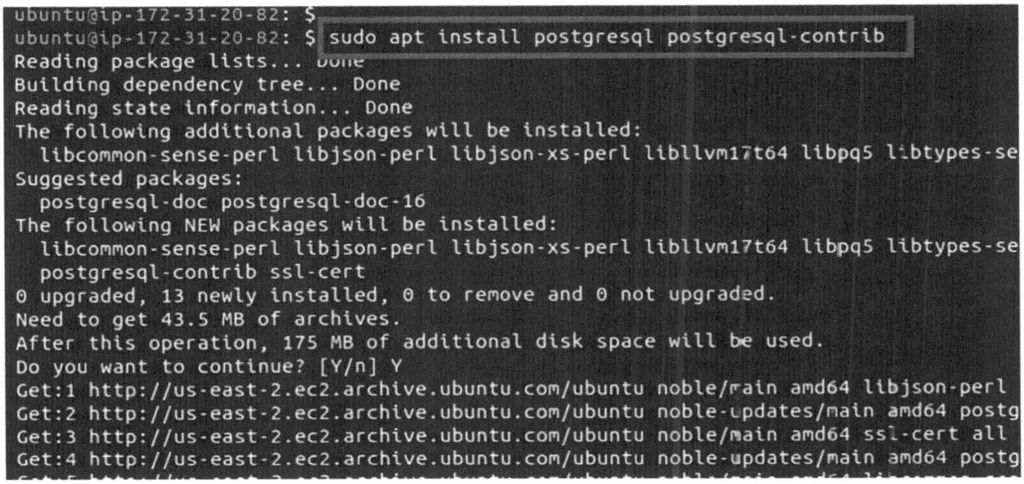

Figure 12.31: *Installing PostgreSQL onto the new EC2 instance*

4. **Verify Installation**: Check the status of PostgreSQL to ensure it is running:

```
sudo systemctl status postgresql
```

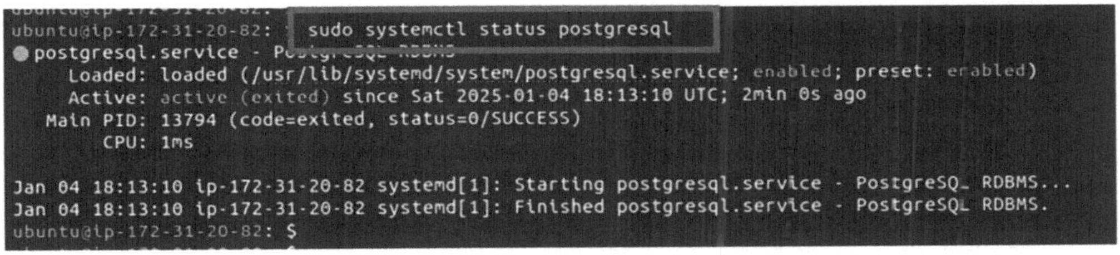

Figure 12.32: *Shows PostgreSQL is installed and active on the new EC2 instance*

5. **Access PostgreSQL**: Switch to the **postgres** user and access the PostgreSQL prompt:

```
sudo -i -u postgres
psql -U postgres
```

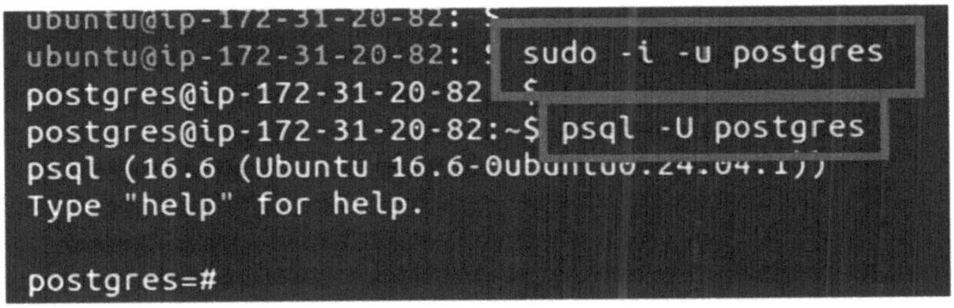

Figure 12.33: *Log in to the PostgreSQL prompt as user postgres*

6. **Create a Database and User**: Now that we are connected in the PostgreSQL prompt, let us create a new database and user in our EC2 instance.

```
CREATE DATABASE mydatabase;
CREATE USER myEC2user WITH ENCRYPTED PASSWORD 'mypassword';
GRANT ALL PRIVILEGES ON DATABASE mydatabase TO myEC2user;
```

```
postgres=# CREATE DATABASE mydatabase;
CREATE DATABASE
postgres=#
postgres=# CREATE USER myEC2user WITH ENCRYPTED PASSWORD 'mypassword';
CREATE ROLE
postgres=#
postgres=# GRANT ALL PRIVILEGES ON DATABASE mydatabase TO myEC2user;
GRANT
postgres=#
postgres=#
postgres=# \l
                                          List of databases
    Name     |  Owner   | Encoding | Locale Provider | Collate   |  Ctype    |
-------------+----------+----------+-----------------+-----------+-----------+
 mydatabase  | postgres | UTF8     | libc            | C.UTF-8   | C.UTF-8   |
             |          |          |                 |           |           |
             |          |          |                 |           |           |
 postgres    | postgres | UTF8     | libc            | C.UTF-8   | C.UTF-8   |
 template0   | postgres | UTF8     | libc            | C.UTF-8   | C.UTF-8   |
             |          |          |                 |           |           |
 template1   | postgres | UTF8     | libc            | C.UTF-8   | C.UTF-8   |
             |          |          |                 |           |           |
(4 rows)

postgres=# \du
                             List of roles
 Role name |                         Attributes
-----------+------------------------------------------------------------
 myec2user |
 postgres  | Superuser, Create role, Create DB, Replication, Bypass RLS
```

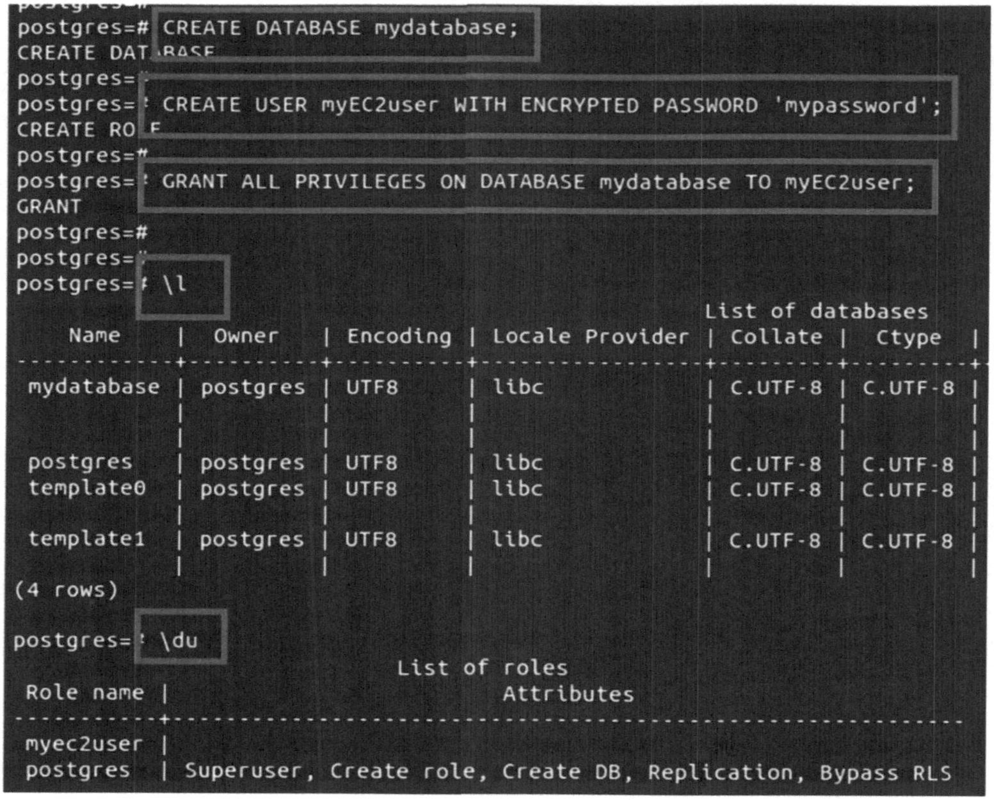

Figure 12.34: *Creating a new database and a user*

To exit the PostgreSQL prompt, we type **\q**, then switch back to the regular user by typing **exit**.

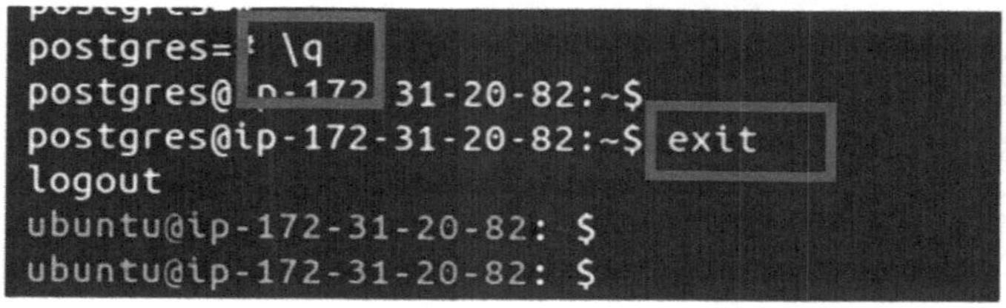

```
postgres=# \q
postgres@ip-172-31-20-82:~$
postgres@ip-172-31-20-82:~$ exit
logout
ubuntu@ip-172-31-20-82: $
ubuntu@ip-172-31-20-82: $
```

Figure 12.35: *Log out of the database*

We now have PostgreSQL installed and running on our Ubuntu EC2 instance.

Amazon RDS

Amazon Relational Database Service is a managed database service provided by AWS. It simplifies the process of setting up, operating, and scaling a relational database in the cloud.

RDS Instance

An RDS instance is a virtual machine in the cloud, dedicated to running your relational database, managed and maintained by AWS to offer reliability and efficiency.

- **Managed Service**: RDS takes care of routine database tasks such as provisioning, patching, backup, recovery, and scaling.
- **Supports Multiple Database Engines**: RDS supports popular databases such as PostgreSQL, MySQL, MariaDB, Oracle, SQL Server, and Amazon Aurora.
- **Scalable**: You can easily scale the compute and storage resources to meet your application needs without downtime.
- **Automated Backups**: RDS provides automated backups, database snapshots, and point-in-time recovery.
- **High Availability**: With Multi-AZ (Availability Zones) deployments, RDS ensures high availability and failover support.
- **Security**: RDS offers features such as network isolation using Amazon VPC, encryption at rest and in transit, and integration with AWS IAM for access management.

Key Features of RDS Instances

RDS simplifies database management tasks, making it easier for developers and DBAs to focus on application development. It is optimized for performance with features such as provisioned IOPS and read replicas for read-heavy workloads. It is cost-effective; thus, you pay only for the resources you use, with options for reserved instances to reduce costs. AWS handles maintenance and updates, ensuring your database runs on the latest and most secure software versions.

Creating an RDS instance for PostgreSQL

Creating an Amazon RDS instance for PostgreSQL is a straightforward process. The following step-by-step guide will help you to create an RDS instance:

1. **Log in to AWS Management Console**: Open the AWS Management Console and sign in with your credentials.

2. **Navigate to the RDS Dashboard**: In the AWS Management Console, select "RDS" from the services menu to access the RDS Dashboard.

Figure 12.36: *How to access the RDS dashboard*

3. **Create a New Database Instance**: Click on the "`Create Database`" button.

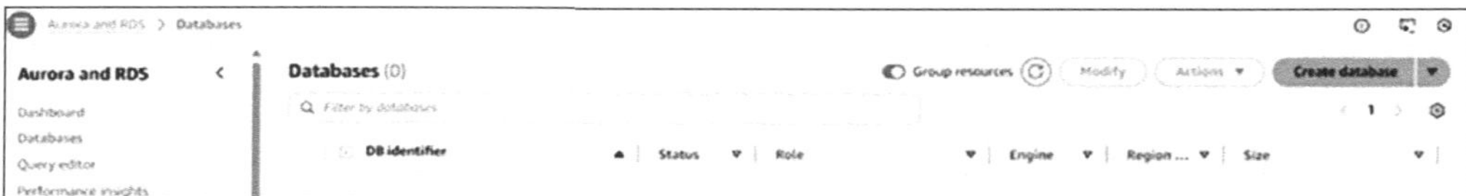

Figure 12.37: *Creating an RDS instance*

4. **Choose the Database Creation Method**: Select the "`Standard Create`" option to customize your database settings.

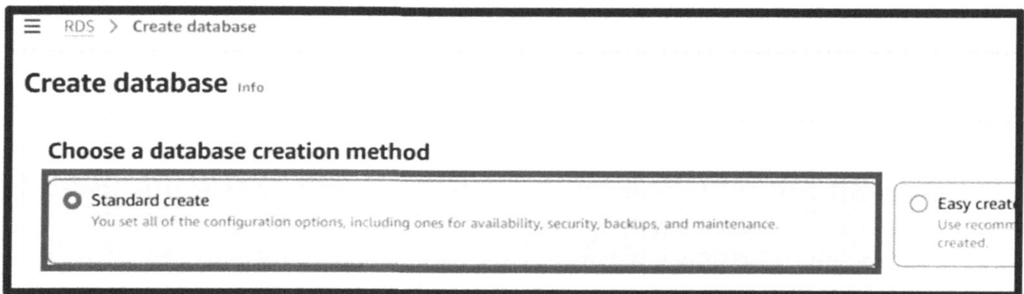

Figure 12.38: *Choosing the creation method for the instance*

5. **Select PostgreSQL as the Engine**: Choose PostgreSQL as the database engine and select the appropriate version.

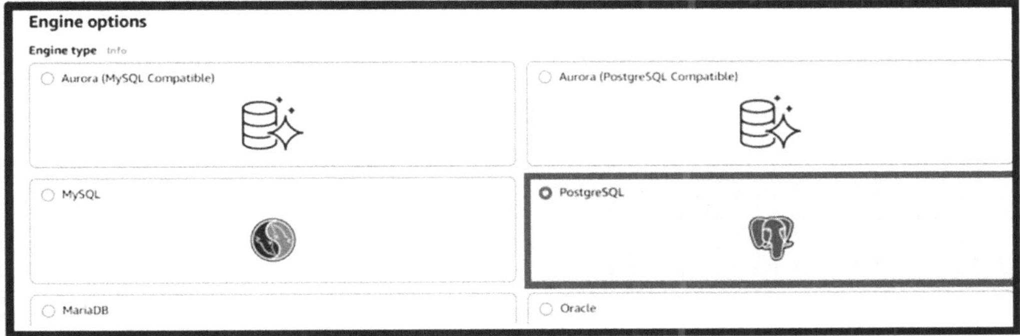

Figure 12.39: *Choose the PostgreSQL database engine for the new instance*

Next, we will choose the PostgreSQL version associated with the minor version, for example, **`16.6`**, as follows.

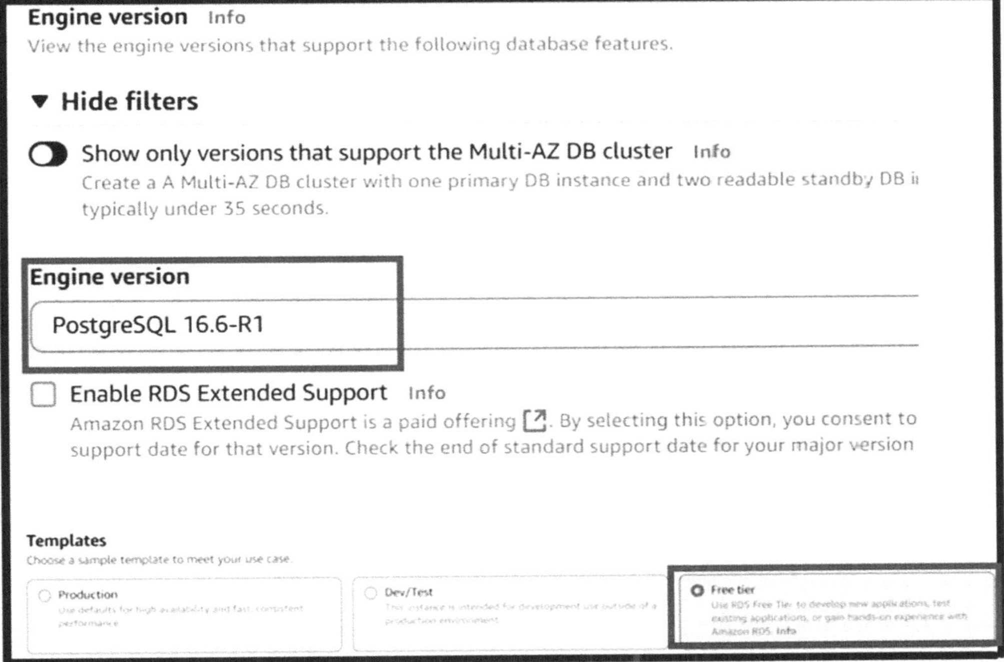

Figure 12.40: *Choose the PostgreSQL version and template to be used*

6. **Configure the DB Instance**: Specify the DB instance **`identifier`**, **`master username`**, and **`password`**.

Settings
DB instance identifier Info
my-database-instance
consecutive hyphens. Can't end with a hyphen.
▼ Credentials Settings
Master username Info
postgres
characters. The first character must be a letter.

Figure 12.41: *Choose the name of the instance and master username*

a. In the next screen, we will provide master username credentials.

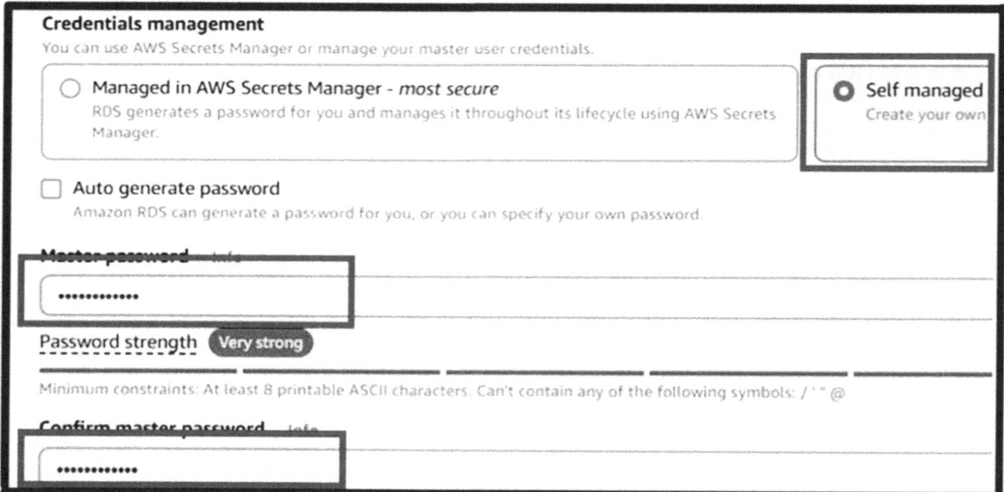

Figure 12.42: *Assign a password to the master username*

b. Choose the instance class (size) based on your performance needs.

Instance configuration
The DB instance configuration options below are limited to those supported by the engine that you selected above.
DB instance class Info
▼ Hide filters
Include previous generation classes
Burstable classes (includes t classes)
db.t3.micro
2 vCPUs 1 GiB RAM Network: Up to 2,085 Mbps

Figure 12.43: *Choose the instance class*

c. Set the allocated storage size.

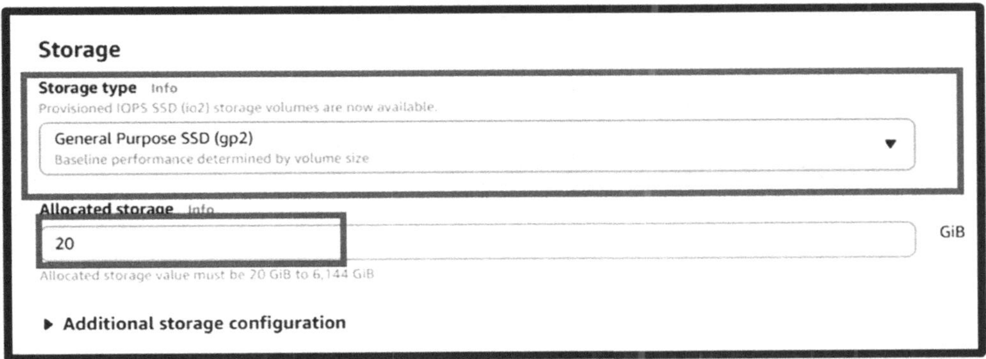

Figure 12.44: *Choose the storage type and allocate the storage size*

Next, we will configure the VPC, subnet group, and availability zone.

7. **Configure Advanced Settings**:
    a. Set the port number (default is **`5432`**).
    b. Choose the license model (default is "**`General Purpose`**").
    c. Configure backup and maintenance settings, such as automated backups and preferred backup windows.
8. **Configure Security Group**:
    a. Create or select a security group to control access to your database instance.

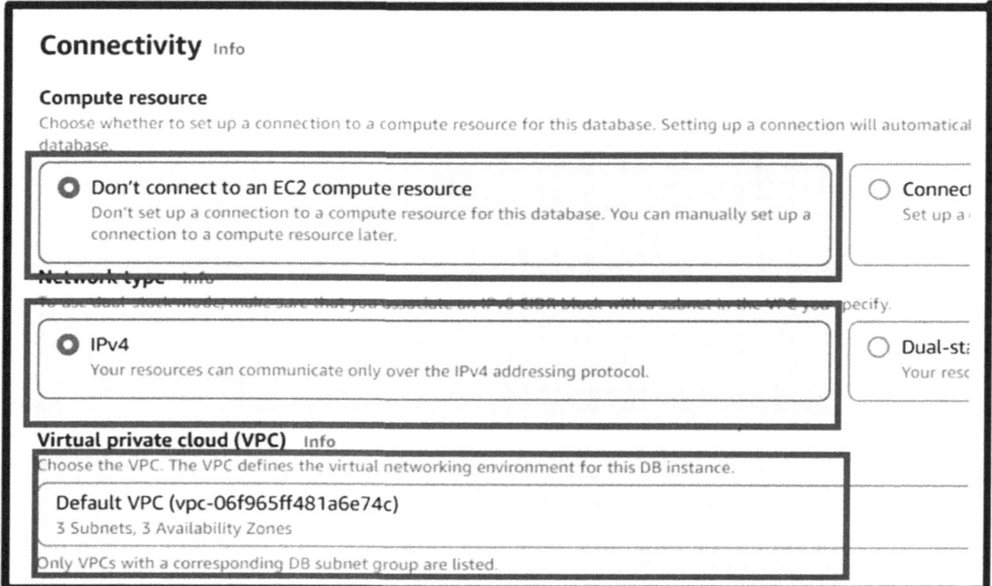

Figure 12.45: *Choose the default VPC*

Next, select the **`default`** subnet and allow for public access. Choose "**`create new`**" and give it a name.

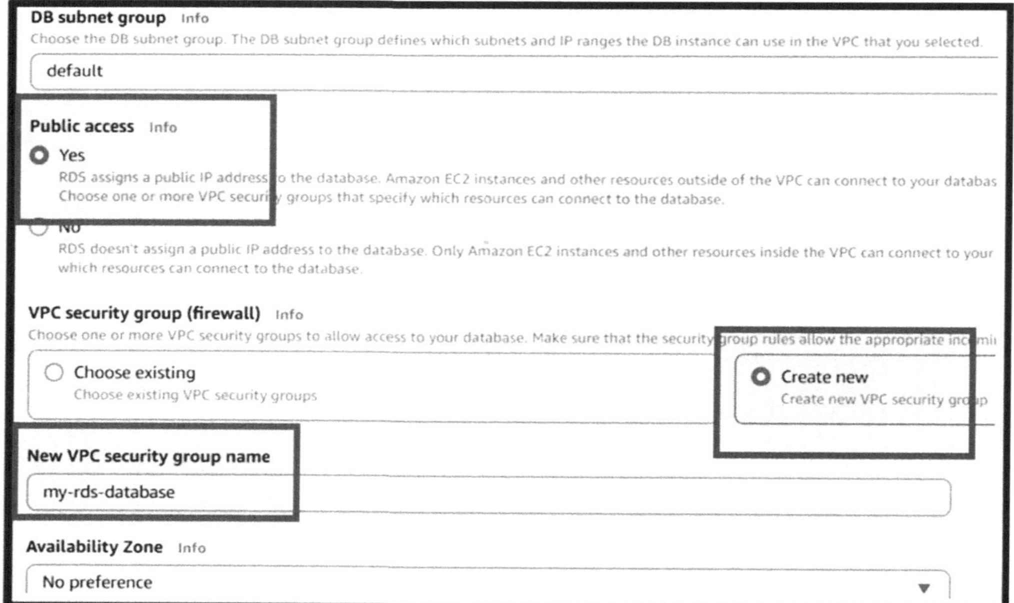

Figure 12.46: *Creating a new security group for the new RDS instance*

In the next section, we will choose the database authentication method and then check the monitoring box if you want the instance to be monitored via AWS Performance Insight.

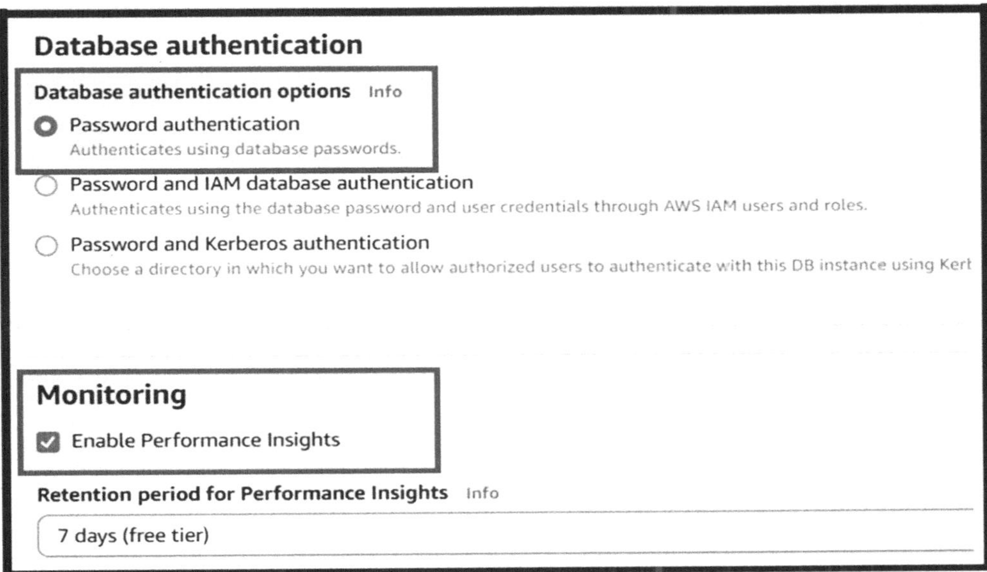

***Figure 12.47:** Choosing the database authentication method and monitoring*

9. **Review and Create**: Review all the settings and click "`Create Database`" to launch your RDS instance.

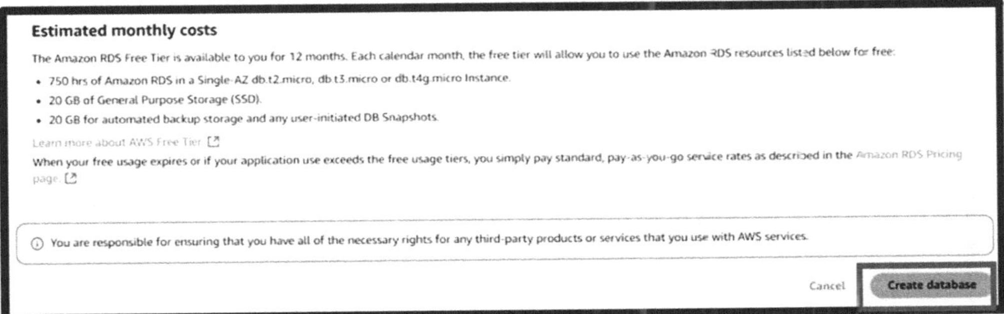

***Figure 12.48:** Create the database*

The RDS instance is being created:

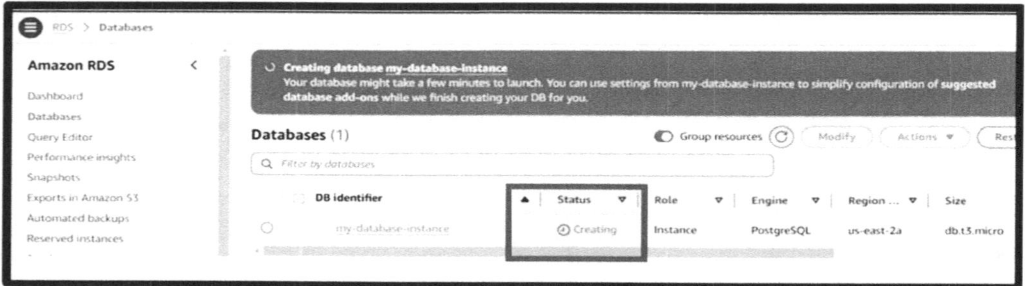

Figure 12.49: *Shows the status as creating*

10. **Connect to Your Database**: Our RDS instance is available and ready, connect to it using a PostgreSQL client such as **`psql`** or **`pgAdmin`**.

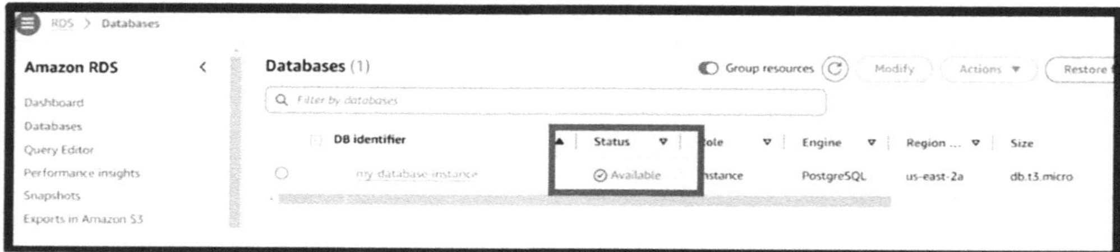

Figure 12.50: *Shows the status as available*

Connecting to a new RDS instance in AWS is a straightforward process. Here is a step-by-step guide to help you get connected:

Step-by-Step Guide to Connect to an RDS Instance

1. **Ensure Security Group Rules**: Make sure the security group associated with your RDS instance allows inbound traffic on the database port (default is **5432** for PostgreSQL).

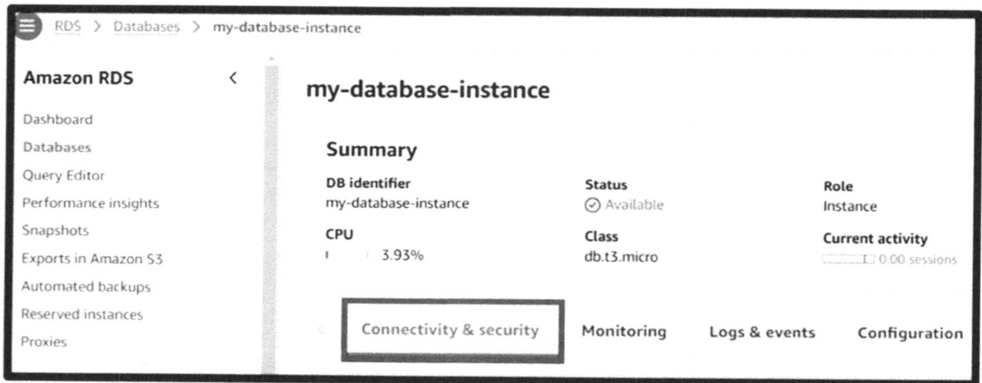

Figure 12.51: *Process of setting the inbound rules*

Scroll down to security group rules and select inbound.

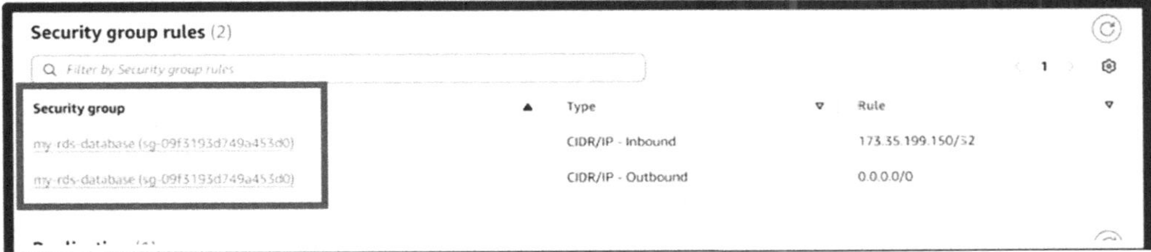

***Figure 12.52:** Select the security group to edit its inbound rules*

Choose the security group by clicking on it.

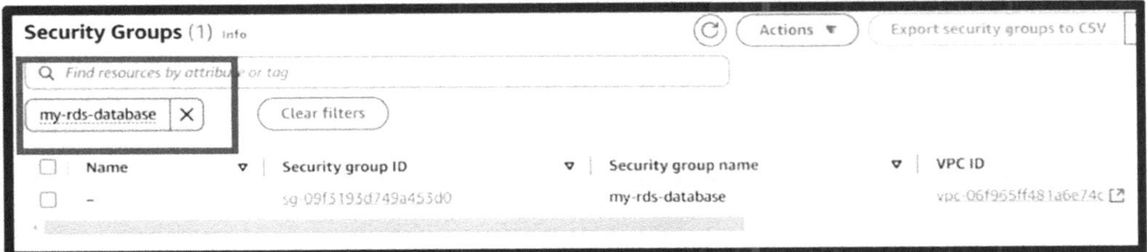

***Figure 12.53:** The security group is selected for editing*

Scroll down to Inbound rules and select edit inbound rules.

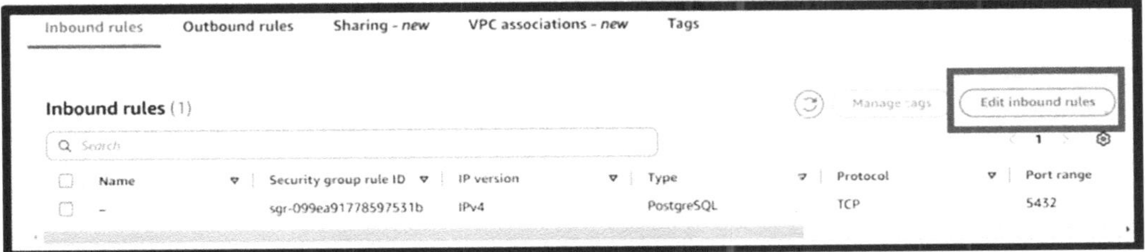

***Figure 12.54:** Editing the inbound rules*

Make sure your IP address or the IP addresses provided by your organization are able to connect to PostgreSQL via port **`5432`**, which is the default port. To obtain your IP address, go to the Google search engine and type "**`What is my IP?`**".

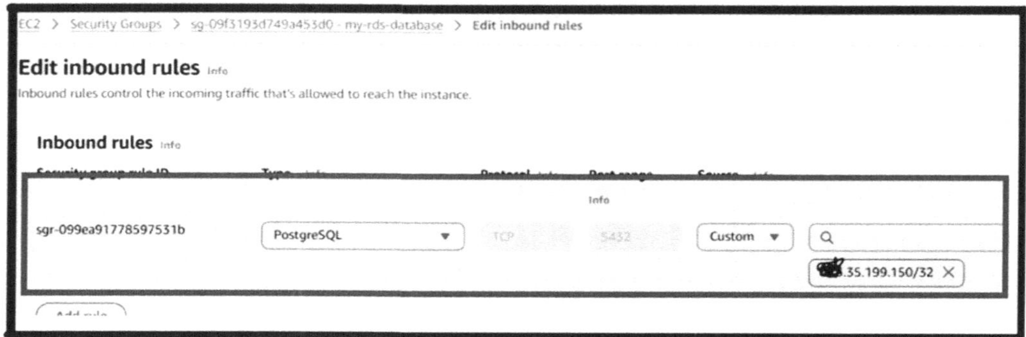

Figure 12.55: *Enable your IP address via inbound rules*

Add your IP address or your application's IP address to the inbound rules of the security group to allow access, as shown in the preceding figure.

2. **Connect Using a Database Client**: You can use a variety of database clients to connect, such as psql for **`PostgreSQL`**, **`pgAdmin`**, **`DBeaver`**, or any other SQL client. We will connect using psql via our Ubuntu virtual machine as we have been connecting to databases.

   In the **`AWS Management Console`**, go to the RDS Dashboard. Select your RDS instance and find the "**`Endpoint`**" information. This endpoint will be used to connect to your database.

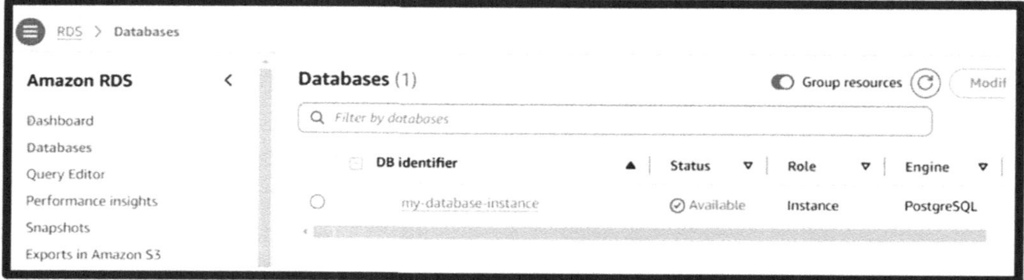

Figure 12.56: *Click on the instance to see the endpoint*

The following screenshot shows that the endpoint is under **`Connectivity & Security`**. Copy this endpoint.

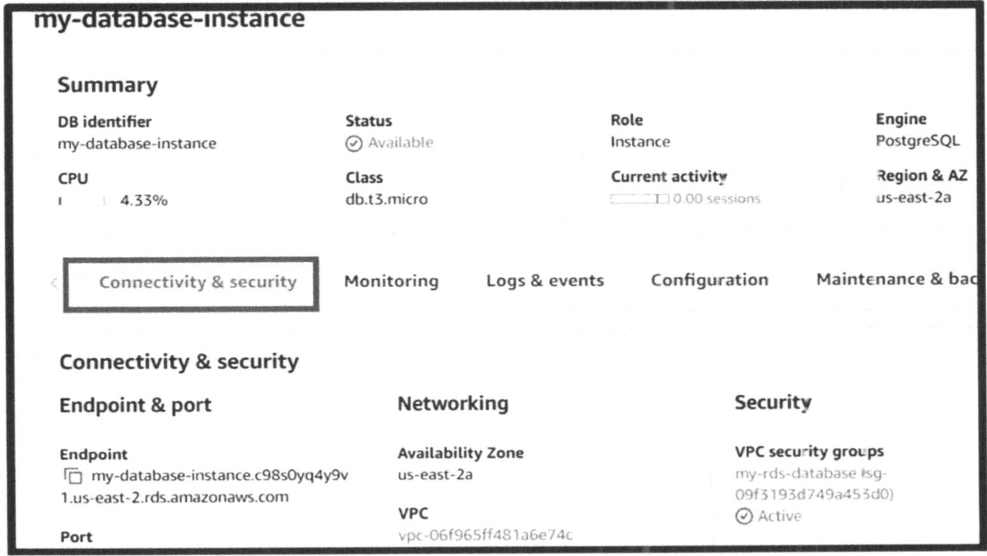

Figure 12.57: *Click on Connectivity and security to see the endpoint for connections*

7. **Connect to the RDS Instance**:

   a. Use the following command to connect:

   ```
   psql -h your-rds-endpoint -U your-master-username -d
   your-database-name
   ```

   b. Replace **`your-rds-endpoint`** with the RDS endpoint, **`your-master-username`** with the username you created, and your-database-name with the name of your database.

   ```
   psql -h my-database-instance.c98s0yq4y9v1.us-east-2.rds.
   amazonaws.com  -d postgres -p 5432 -U postgres
   ```

```
jack@jack-virtual-machine: $ psql -h my-database-instance.c98s0yq4y9v1.us-east-2.rds.amazonaws.com  -d postgres  p 5432 -U postgres
Password for user postgres:
psql (16.1 (Ubuntu 16.1-1.pgdg22.04+1), server 16.6)
SSL connection (protocol: TLSv1.3, cipher: TLS_AES_256_GCM_SHA384, compression: off)
Type "help" for help.

postgres=>
postgres=>
```

Figure 12.58: *Remote login using the endpoint and user postgres*

We are now connected to our new RDS instance. Let us create a database department inside using the **`CREATE DATABASE department;`** command.

```
postgres=> CREATE DATABASE department;
CREATE DATABASE
postgres=>
postgres=> \l
                                        List of databases
    Name    |  Owner   | Encoding | Locale Provider |   Collate   |    Ctype    | I
------------+----------+----------+-----------------+-------------+-------------+---
 department | postgres | UTF8     | libc            | en_US.UTF-8 | en_US.UTF-8 |
 postgres   | postgres | UTF8     | libc            | en_US.UTF-8 | en_US.UTF-8 |
 rdsadmin   | rdsadmin | UTF8     | libc            | en_US.UTF-8 | en_US.UTF-8 |
 template0  | rdsadmin | UTF8     | libc            | en_US.UTF-8 | en_US.UTF-8 |
            |          |          |                 |             |             |
 template1  | postgres | UTF8     | libc            | en_US.UTF-8 | en_US.UTF-8 |
            |          |          |                 |             |             |
(5 rows)
```

Figure 12.59: *Creating a new database in the new RDS instance*

AWS S3 Bucket

It is a fundamental storage unit within Amazon Simple Storage Service (Amazon S3). It is like a digital container for your data, allowing you to store and organize files in the cloud. S3 stores data as objects, which include the file itself and any associated metadata. Each bucket has a unique name across all of AWS, which helps in identifying it. An S3 bucket can handle virtually unlimited amounts of data, from tiny files to massive datasets. Designed for 99.999999999% durability and 99.99% availability, ensuring your data is safe and accessible. You can set granular permissions to control who can access your data, using mechanisms such as bucket policies and access control lists (ACLs). It is commonly used for backups, archival storage, data lakes, and even hosting static websites.

Creating an S3 Bucket

Creating an S3 bucket and uploading files to it is straightforward.

1. **Sign in to AWS Management Console**: Go to the AWS Management Console and sign in with your credentials.
2. **Open Amazon S3 Console**: In the AWS Management Console, search for "`S3`" and select `Amazon S3`.

Figure 12.60: Obtaining the S3 bucket dashboard for the first time

3. **Create a Bucket**: Click on the **`Create Bucket`** button.

Figure 12.61: Create a new bucket

4. **Enter Bucket Details**: Provide a unique bucket name (for example, **`my-s3-bucket`**), choose a region, and configure other settings as needed.

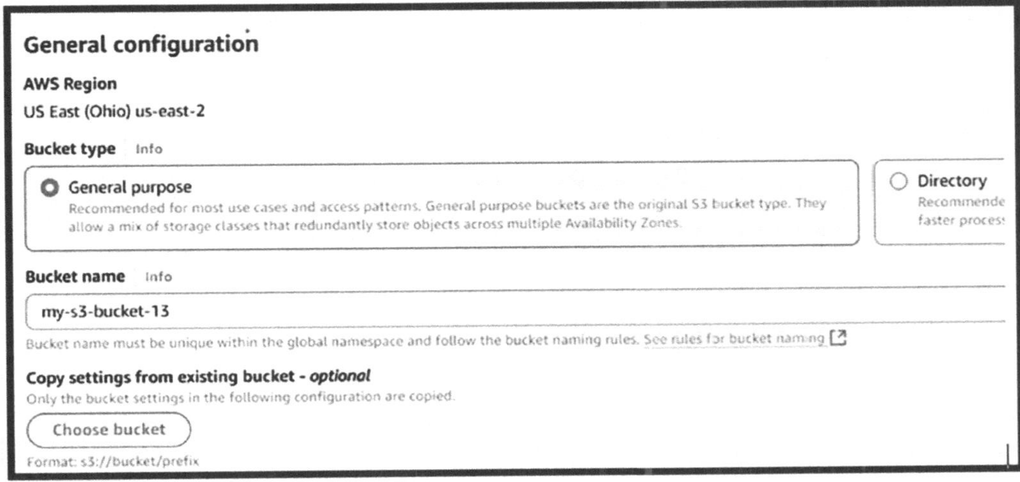

Figure 12.62: Choosing the bucket type and name

5. **Configure Permissions**: Set permissions and policies for the bucket. By default, the bucket owner has full control, and public access is blocked.

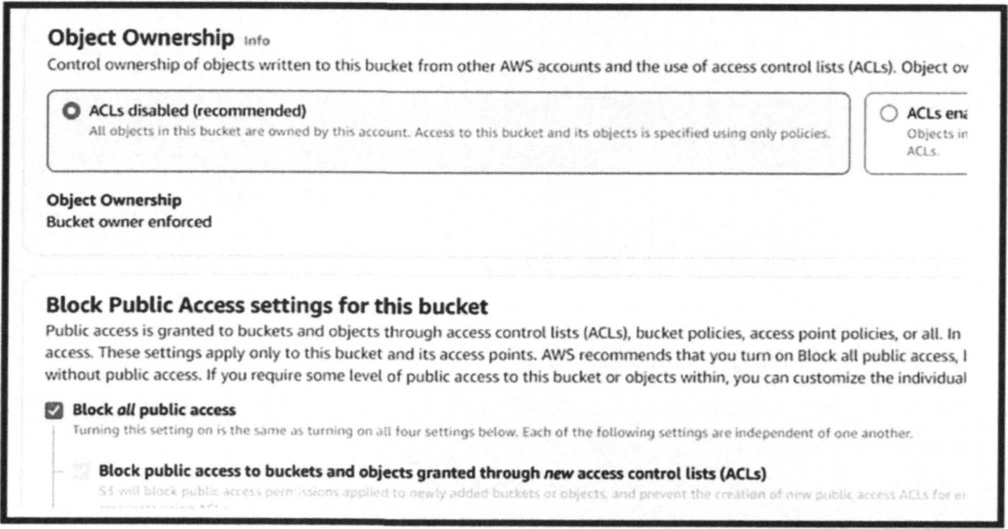

Figure 12.63: *Setting the bucket permissions and policy*

6. **Create the Bucket**: Click **`Create`** to finalize the bucket creation.

   Leave other settings as default. Scroll to the bottom and click on **`Create bucket`**.

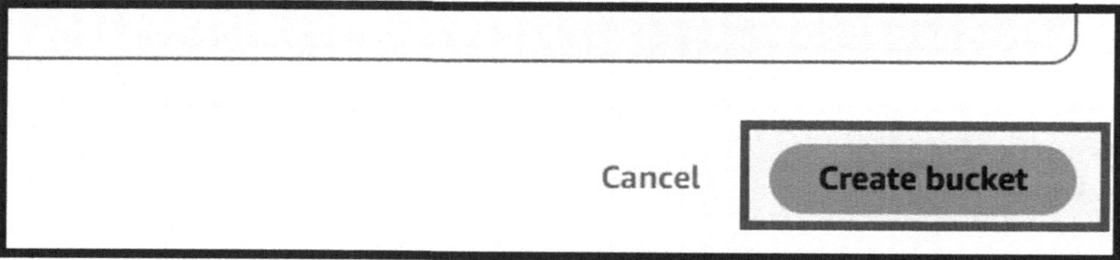

Figure 12.64: *Creating the bucket*

Our bucket is created as follows.

Next, we are going to upload files into our s3 bucket.

1. **Navigate to Your Bucket**: Go back to the Amazon S3 console and select the bucket you just created.

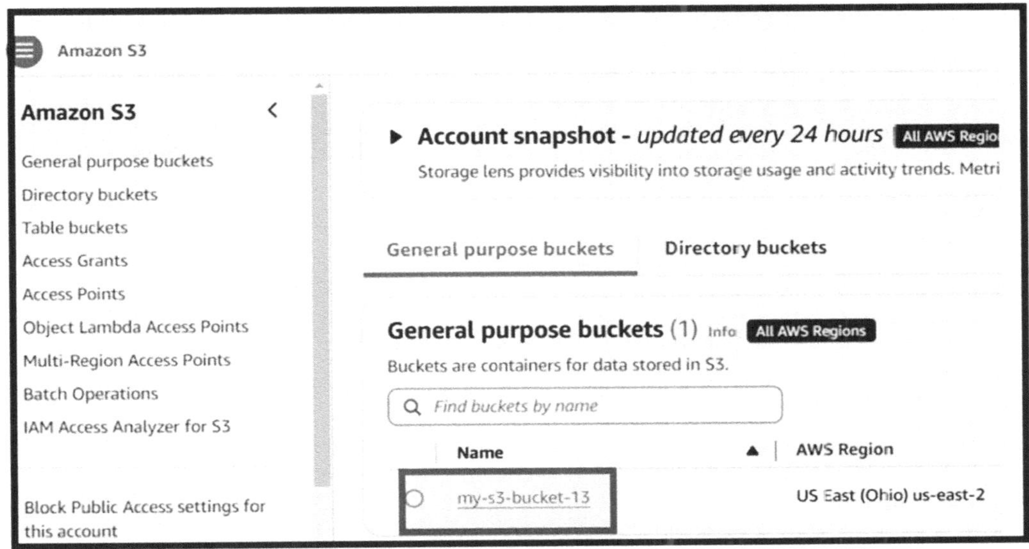

Figure 12.65: *Bucket is created*

Next, we are going to upload a file into the bucket.

2. **Upload Files**: Click on the **`Upload`** button.

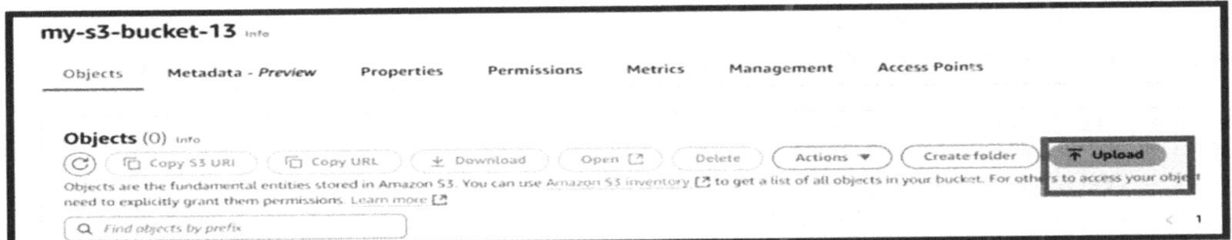

Figure 12.66: *Uploading a file into the bucket*

3. **Select Files**: Drag and drop files into the upload area or click **`Add files`** to select files from your computer. Let us upload the **`my-ec2-keys`** we created earlier. You can upload any file of your choice.

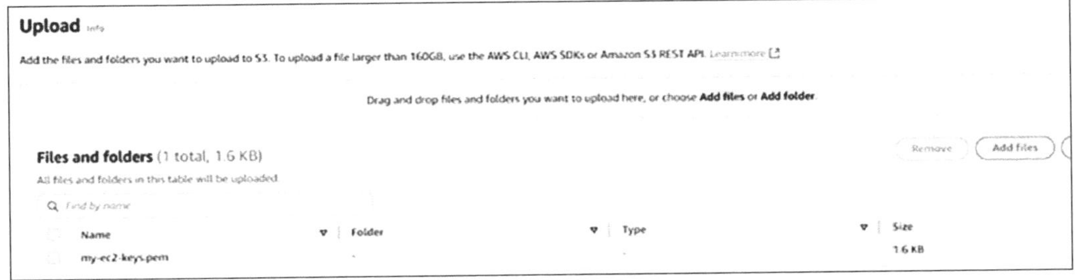

Figure 12.67: *Choosing a file to be uploaded to the bucket*

4. **Complete Upload**: Once the file is selected, scroll down and click **`Upload`** to start the upload process.

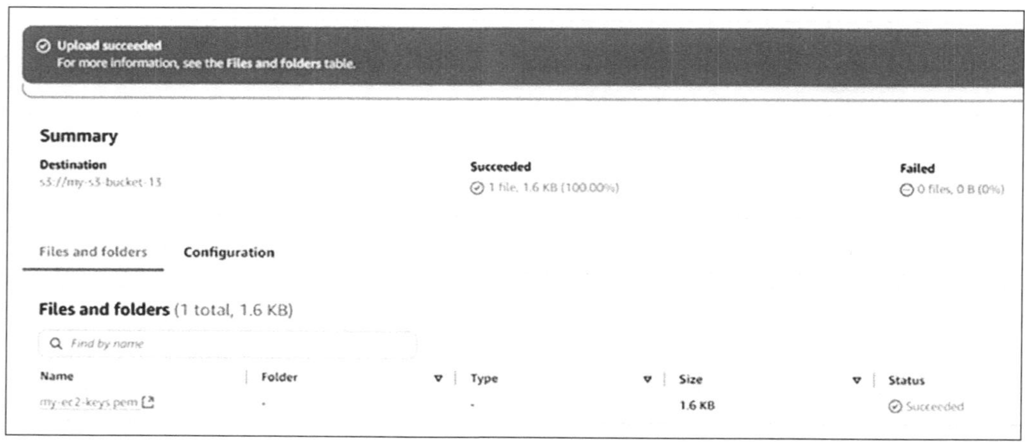

Figure 12.68: *File uploaded successfully into the S3 bucket*

For us to upload using the AWS command line (CLI), we first need to create an IAM user, so we can use the credentials.

Task

Technical Interview Questions

The following are some technical interview questions for the preceding AWS topics.

1. What are the steps to create an AWS account?
2. How do you secure your AWS account?
3. What is the role of IAM in AWS account management?
4. How do you launch an EC2 instance using the AWS Management Console?
5. What are the different instance types available in EC2?
6. How do you configure security groups for EC2 instances?
7. What are the key features of Amazon RDS?
8. How do you create and manage an RDS instance?
9. What are the different types of backups available in RDS?
10. How do you create an S3 bucket?
11. What are the best practices for securing S3 buckets?
12. How do you manage access control for S3 buckets?

Conclusion

In this chapter, learners have gained foundational knowledge and hands-on experience with key AWS services and functionalities. The journey began with creating an AWS account, the essential first step toward accessing the extensive AWS ecosystem. Learners then explored how to launch and configure EC2 instances, providing the compute power necessary for hosting applications and databases. This was followed by a deep dive into installing PostgreSQL on EC2 instances, demonstrating how to configure and manage relational databases directly on virtual servers.

Additionally, learners explored Amazon Relational Database Service (RDS), understanding how to use this managed service to simplify database setup and administration while ensuring scalability and durability. The chapter also introduced S3 buckets, showcasing their role in secure, scalable, and cost-effective storage of data and objects. By mastering these essential topics, learners are equipped with the knowledge and skills to build, manage, and integrate cloud-based resources, laying the groundwork for more advanced AWS applications and services.

In the next chapter, we will look at creating AWS IAM users and AWS migration.

CHAPTER 13
IAM Users and AWS Migration

Introduction

In this chapter, learners will explore the essentials of AWS Identity and Access Management (IAM) and the role it plays in securing access to AWS resources. The chapter begins by introducing IAM as a foundational service that allows administrators to create and manage users, groups, and roles with tailored permissions for authorized access. It emphasizes the importance of managing IAM users—individual accounts within an AWS environment—during the migration of workloads, data, and applications. Learners will delve into creating users, assigning policies, and leveraging roles for temporary access, along with strategies to ensure seamless integration of access controls during migrations.

Additionally, the chapter introduces the concept of AWS Migration, highlighting the transformative potential of moving workloads and applications to the cloud. It covers the comprehensive process of planning, strategizing, and executing migrations while minimizing downtime and ensuring operational continuity. Learners will explore methodologies and tools such as AWS Migration Hub and AWS Database Migration Service, as well as best practices for rehosting, replatforming, and refactoring workloads. The chapter underscores the opportunities that AWS Migration provides for scalability, cost efficiency, and modernization in a cloud-driven world.

Structure

In this chapter, we will cover the following topics:

- IAM User
    - Assign Permissions to IAM User
- AWS Database Migration Service

- Creating Source and Target Endpoints
    - Create a Replication Instance
    - Create a Migration Task
- Task

IAM User

An Identity and Access Management (**IAM**) user in AWS is an entity that you create to represent a person or application that interacts with AWS. Each IAM user has a unique identity within your AWS account and can have permissions assigned to allow access to various AWS resources.

Each IAM user has their own credentials, which can include a password for the AWS Management Console and access keys for programmatic access using the AWS CLI, SDKs, or APIs.

You can assign specific permissions to each IAM user using IAM policies, allowing you to follow the principle of least privilege and grant only the necessary access. IAM users can be added to IAM groups, which allow you to manage permissions for multiple users collectively.

You can enable MFA for IAM users to add an extra layer of security for accessing the AWS Management Console or making API requests.

Use Cases for IAM Users:

- **Employees**: Create IAM users for your employees to securely access AWS resources and perform their job functions.
- **Applications**: Assigning IAM users to applications or services that need to interact with AWS resources programmatically.
- **Contractors and Partners**: Granting temporary access to contractors or partners without sharing your root account credentials.

Create the IAM User:

1. Navigate to the IAM console (*Figure* 13.1).
2. Click on `Users` in the left-hand menu and then `Add user` (*Figure* 13.2).
3. Enter a username (for example, developer-user) and select the access type (AWS Management Console, programmatic access, or both).

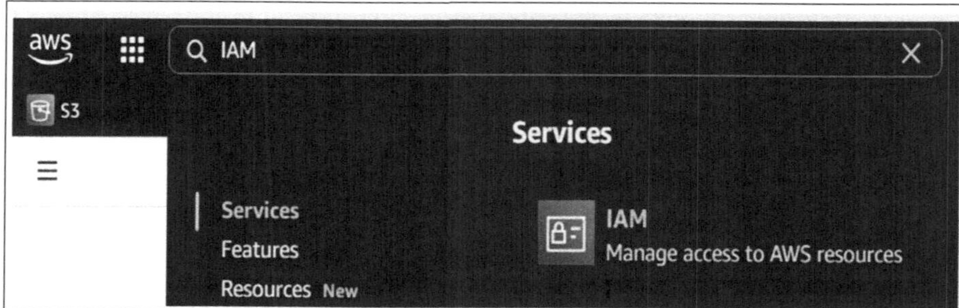

Figure 13.1: *Searching the IAM services for the first time*

The next screen shows when the user creation starts.

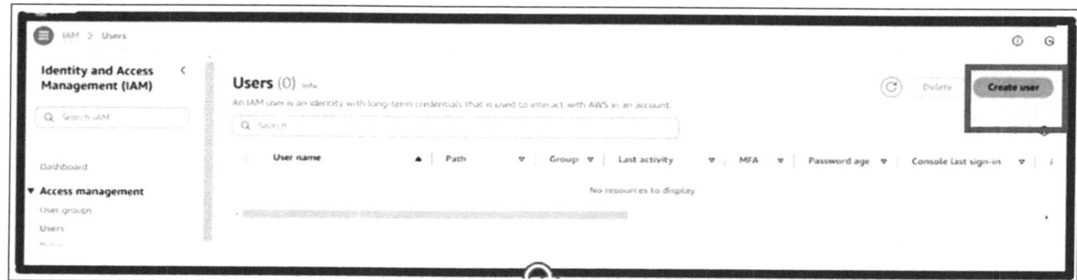

Figure 13.2: *Creating an IAM user*

In the next section, we will specify the IAM user details.

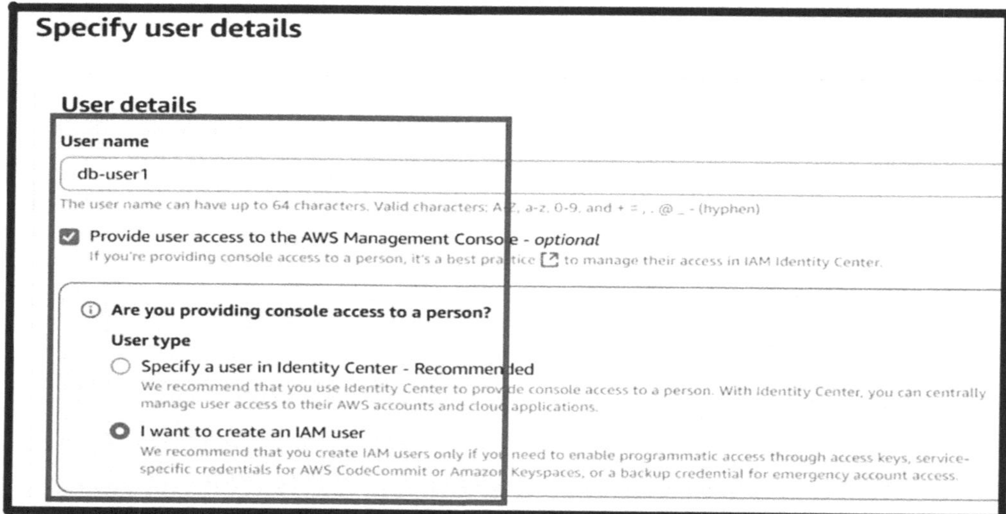

Figure 13.3: *Assign the IAM user details*

4. **Assign Permissions**: Attach an appropriate policy (for example, **`AmazonS3FullAccess`** if the user needs access to S3).

Figure 13.4: *Attaching policies to the IAM user*

5. Review and create the user.

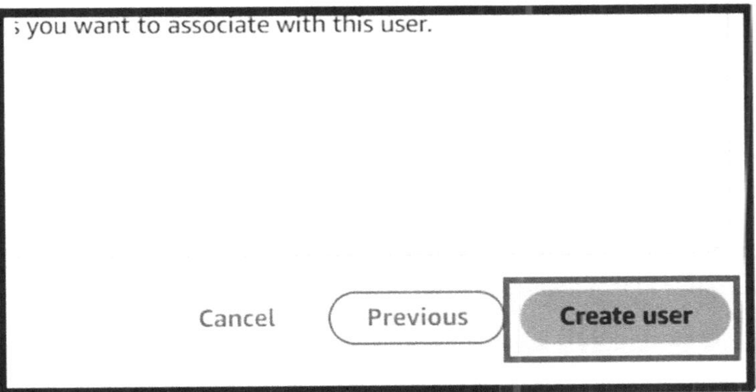

Figure 13.5: *Create the IAM user*

Now that the IAM user is created, click on it and select **`Create access key`**.

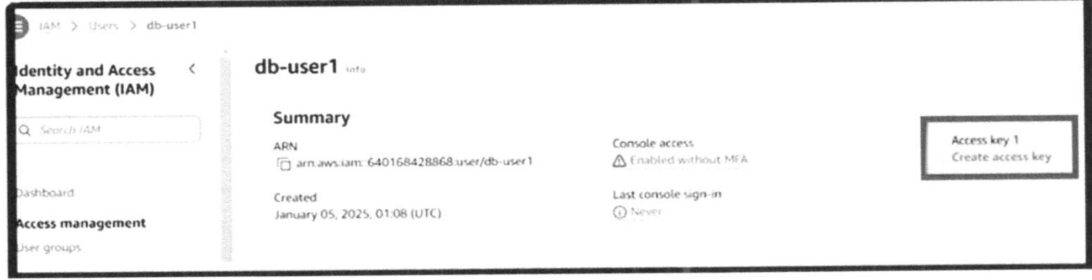

Figure 13.6: *Create an access key for the IAM user*

On the next screen, select the first option.

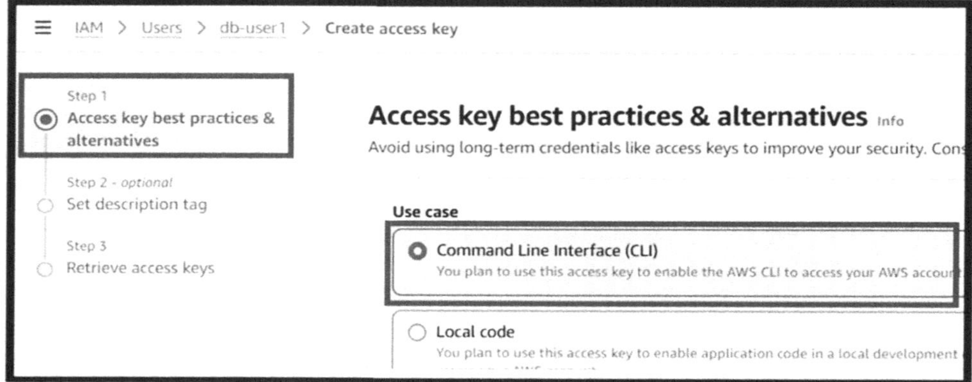

Figure 13.7: *Choosing the options for the access key*

Acknowledge and click on **Next**.

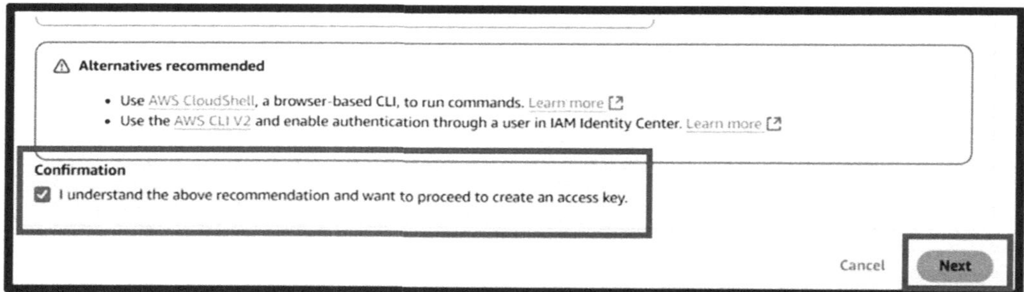

Figure 13.8: *Acknowledge the recommendations for the access key*

Click on the **Create access key** as shown in the following figure.

side the access key.

rotate this access key confidently later.

and: _ . : / = + - @

Cancel Previous Create access key

Figure 13.9: *Create the access key*

5. **Download Credentials**: Save the access key ID, secret access key, and any other relevant credentials provided. Click on the **`Download .csv file`**.

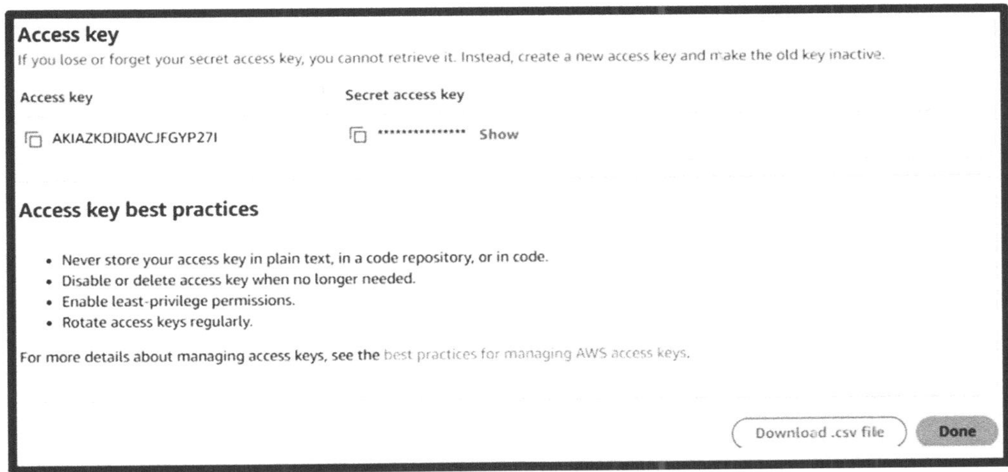

***Figure 13.10**: Download and save the access key*

Using AWS CLI for Upload

Now that we have created the IAM user and have the AWS credentials, we can use the AWS Command Line Interface (CLI) to upload files: Let us use the access key ID and secret access key to configure our AWS CLI using the **`aws configure`** command. Run the command and provide the access key and secret access keys as shown in the following figure.

```
postgres=> \q
jack@jack-virtual-machine: $ aws configure
AWS Access Key ID [****************3GVT]: AKIAZKDIDAVCJFGYP27I
AWS Secret Access Key [****************iXje]: VMyX13I9tlhNExj2yabiflvsKeC
Default region name [None]:
Default output format [None]:
jack@jack-virtual-machine: $
```

***Figure 13.11**: Configure the AWS CLI with the access key*

1. **Navigate to Your Bucket Directory**: Use the **`cd`** command to navigate to the directory containing your files.

2. **Upload Files**: Run the following command:

   ```
   aws s3 cp file.txt s3://my-s3-bucket-13/
   ```

   Replace **`file.txt`** with the name of your file and my-s3-bucket with your bucket name.

These steps should help you get started with creating an S3 bucket and uploading files to it.

Migration

AWS migration in the context of PostgreSQL refers to the process of moving a PostgreSQL database to Amazon Web Services (AWS). This can involve transferring data, applications, and configurations to AWS-managed services, such as Amazon RDS for PostgreSQL or Amazon Aurora PostgreSQL-Compatible Edition.

Key Steps in AWS Migration for PostgreSQL:

1. **Assessment**: Evaluate the current PostgreSQL environment and determine the migration strategy.
2. **Planning**: Develop a detailed migration plan, including timelines and resources.
3. **Migration Execution**: Use AWS tools such as AWS Database Migration Service (DMS) to move data and applications.
4. **Testing and Validation**: Ensure everything works correctly in the new AWS environment.
5. **Optimization**: Fine-tune the new environment for performance and cost efficiency.

Benefits of AWS Migration for PostgreSQL:

- **Scalability**: Easily scale resources up or down based on demand.
- **Cost Efficiency**: Pay only for what you use, potentially reducing costs.
- **Security**: Leverage AWS's robust security infrastructure and compliance standards.
- **Flexibility**: Access a wide range of services and tools to innovate and optimize operations.

Tools and Services for AWS Migration:

- **AWS Database Migration Service (DMS)**: Supports homogeneous migrations, data encryption, and more.
- **`pg_dump` and `pg_restore`**: PostgreSQL utilities for exporting and importing data.
- **Logical Replication**: For minimal downtime migrations.
- **COPY Command**: For migrating filtered table data.

AWS Database Migration Service

For the purpose of this learning, we are going to use the database migration service (DMS) tool to migrate the **dvdrental database** from an EC2 instance to an RDS instance. The EC2 instance is the source, and the RDS instance is the target.

Here is a step-by-step guide on using AWS Database Migration Service (DMS) to transfer the dvdrental PostgreSQL database from an EC2 instance to an RDS instance:

Prerequisites:

- **AWS DMS Set Up**: Ensure AWS DMS is properly set up in your AWS environment.
- **Source EC2 Instance**: Have the PostgreSQL database (dvdrental) running on an EC2 instance.

  Since we already have our EC2 instance, we are going to connect to the EC2 instance and add the Postgres user into the sudoers file. Following that, we will enable firewalls.

```
sudo vi /etc/sudoers
      postgres             All=(All)              NOPASS-
WD:ALL    #Add this line into the sudoers file
```

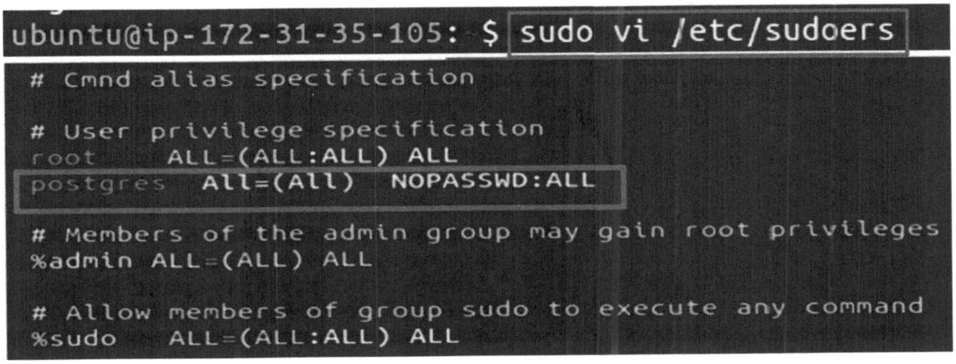

```
ubuntu@ip-172-31-35-105: $ sudo vi /etc/sudoers
# Cmnd alias specification

# User privilege specification
root    ALL=(ALL:ALL) ALL
postgres  All=(All)  NOPASSWD:ALL

# Members of the admin group may gain root privileges
%admin ALL=(ALL) ALL

# Allow members of group sudo to execute any command
%sudo   ALL=(ALL:ALL) ALL
```

Figure 13.12: *Adding the postgres user in the sudoers file*

Adding the line **postgres All=(All) NOPASSWD:ALL** to the sudoers file gives the **postgres** user the ability to run any command as any user without needing to enter a password.

Next, we are going to enable the firewall using the following commands:

```
sudo ufw enable    # Enter y, when prompted
     sudo ufw allow ssh
     sudo ufw allow 5432
```

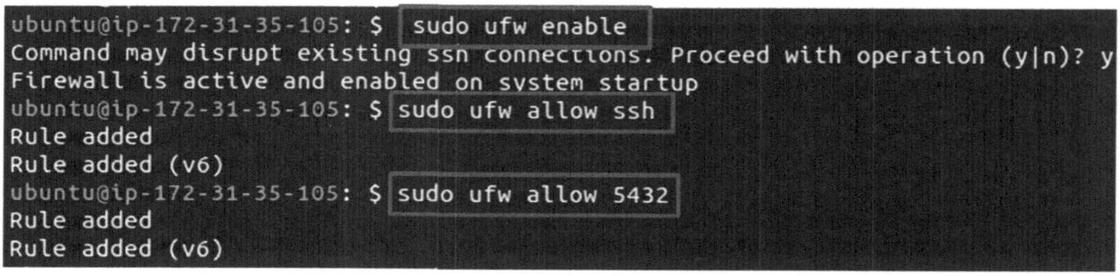

```
ubuntu@ip-172-31-35-105: $ sudo ufw enable
Command may disrupt existing ssh connections. Proceed with operation (y|n)? y
Firewall is active and enabled on system startup
ubuntu@ip-172-31-35-105: $ sudo ufw allow ssh
Rule added
Rule added (v6)
ubuntu@ip-172-31-35-105: $ sudo ufw allow 5432
Rule added
Rule added (v6)
```

Figure 13.13: *Enabling the firewall*

Enabling and configuring a firewall is essential for securing your system and network.

A firewall acts as a barrier between your computer and potentially malicious traffic from the internet. It helps prevent unauthorized access and cyberattacks. By configuring specific rules, you can control which services and ports are accessible, reducing the risk of exploitation. Firewalls can log network activity, allowing you to monitor and analyze any suspicious behavior.

Next, we are going to switch to user **postgres** and then allow all IP addresses using the following commands. We will then assign a password to user **postgres** once we log in to the database.

```
 sudo su - postgres
psql -U postgres
ALTER SYSTEM SET listen_addresses='*'; #Or specify who can
connect with its ip addresses.
```

Note that in an organization, this will never use the wildcard (*).

```
ALTER USER postgres password 'password';  #To give user postgres
password
```

```
ubuntu@ip-172-31-35-105: $ sudo su - postgres
postgres@ip-172-31-35-105:~$
postgres@ip-172-31-35-105:~$ psql
psql (16.9 (Ubuntu 16.9-1.pgdg24.04+1))
Type "help" for help.

postgres=#
postgres=# ALTER SYSTEM SET listen_addresses='*';
ALTER SYSTEM
postgres=#
postgres=# ALTER USER postgres password 'password';
ALTER ROLE
postgres=#
```

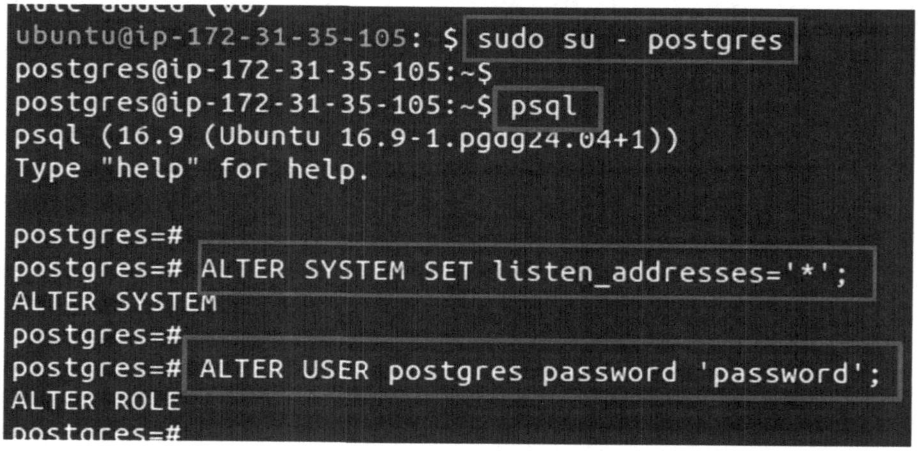

Figure 13.14: *Enabling all IP addresses access through listen_addresses*

Next, let us create a **database dvdrental**, table actor, and insert some data as follows.

```
create database dvdrental;
grant connect on database dvdrental to postgres;
grant usage on schema public to postgres;
create table actor (actor_id int, first_name varchar(30), last_
name varchar(30));
insert into actor values (34,'lemy','zao'),(67,'divine','zao'),
                    (90,'nash','zao'),(100,'jack','zao');
```

```
postgres=#
postgres=# create database dvdrental;
CREATE DATABASE
postgres=#
postgres=# grant connect on database dvdrental to postgres;
GRANT
postgres=#
postgres=# grant usage on schema public to postgres;
GRANT
postgres=#
postgres=# create table actor (actor_id int, first_name varchar(30) last_name varchar(30));
CREATE TABLE
postgres=#
postgres=# insert into actor values (34,'lemy','zao'),(67,'divine','zao'),
                    (90,'nash','zao'),(100,'jack','zao');
INSERT 0 4
postgres=#
```

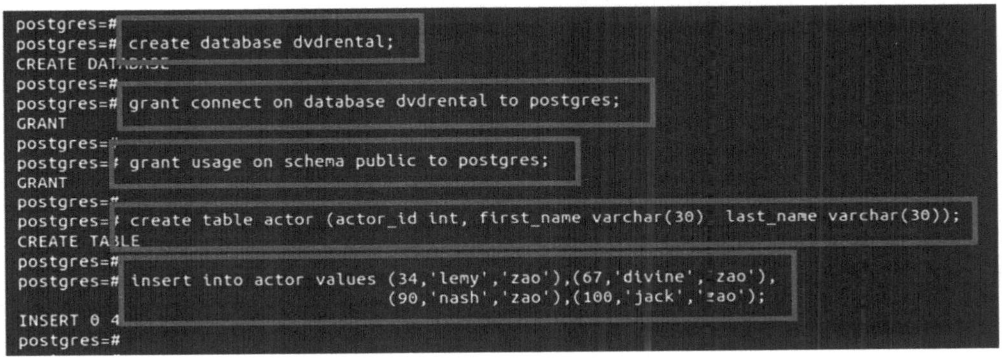

Figure 13.15: *Create a database and a table*

Remember to restart this system for the changes to be applied. Make sure you quit from the database and then exit from **postgres** user as shown below. Vi into the **postgresql.conf** file and change the listen address to * .

```
ps -ef | grep postgres  # To get the path to the configuration file
vi /etc/postgresql/16/main/postgresql.conf
```

Change the **listen_addresses** to * before you restart.

After saving with **:wq!** Which is to save and quit from the vi editor, we need to restart the database for the changes to take effect.

```
exit
sudo systemctl restart postgresql
```

Or

```
sudo systemctl start postgresql@16-main
```

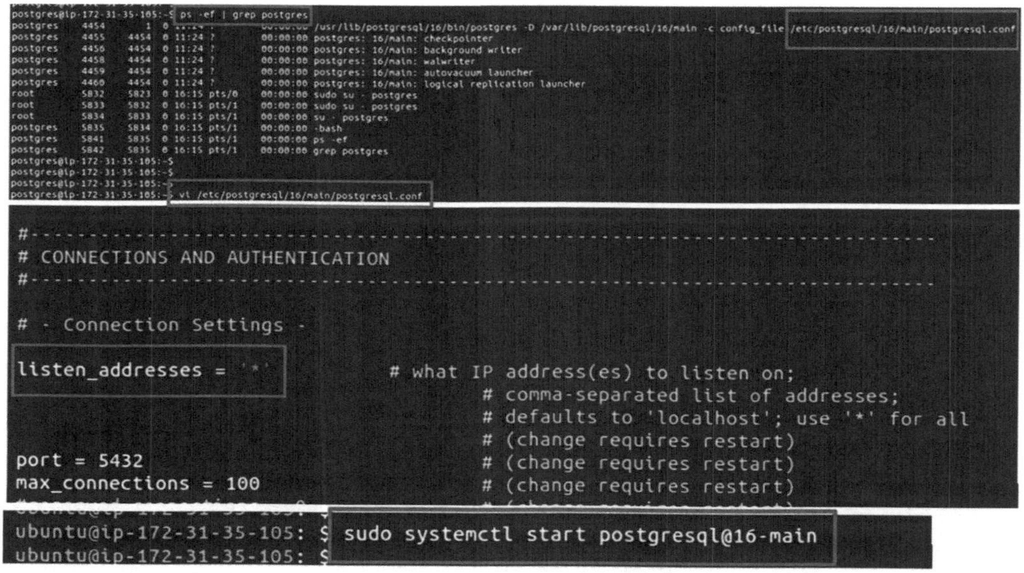

Figure 13.16: *Restarting the server*

Target RDS Instance: We have an RDS PostgreSQL instance ready for migration. This is our target instance.

The next step is to create a migration user in both our EC2 instance and the **rds** instance. Authenticate the user via **pg_hba.conf** and grant the user usage on the public schema and connectivity to the **dvdrental** databases. Remember, we have already created the **dvdrental** database in EC2.

```
CREATE USER dms_user WITH LOGIN PASSWORD 'strongpassword';
GRANT CONNECT ON DATABASE postgres TO dms_user;
GRANT USAGE ON SCHEMA public TO dms_user;
GRANT SELECT ON ALL TABLES IN SCHEMA public TO dms_user;
ALTER DEFAULT PRIVILEGES IN SCHEMA public GRANT SELECT ON TABLES
TO dms_user;
CREATE DATABASE dvdrental;
GRANT CONNECT ON DATABASE dvdrental TO dms_user;
```

```
jack@jack-virtual-machine: $ psql -h my-database-instance.c98s0yq4y9v1.us-east-2.rds.amazonaws.com -U postgres -d postgres -W
Password:
psql (16.1 (Ubuntu 16.1-1.pgdg22.04+1), server 16.9)
SSL connection (protocol: TLSv1.3, cipher: TLS_AES_256_GCM_SHA384, compression: off)
Type "help" for help.

postgres=>
postgres=> CREATE USER dms_user WITH LOGIN PASSWORD 'strongpassword';
GRANT CONNECT ON DATABASE postgres TO dms_user;
GRANT USAGE ON SCHEMA public TO dms_user;
GRANT SELECT ON ALL TABLES IN SCHEMA public TO dms_user;
ALTER DEFAULT PRIVILEGES IN SCHEMA public GRANT SELECT ON TABLES TO dms_user;
CREATE ROLE
GRANT
GRANT
GRANT
ALTER DEFAULT PRIVILEGES
postgres=>
postgres=> CREATE DATABASE dvdrental;
CREATE DATABASE
postgres=> GRANT CONNECT ON DATABASE dvdrental TO dms_user;
GRANT
postgres=>
```

Figure 13.17: *Creating a migration user and a dvdrental database in our rds instance*

The next step is creating a migration user in our EC2 instance and granting user privileges.

```
ubuntu@ip-172-31-35-105: $ sudo su - postgres
postgres@ip-172-31-35-105:~$ psql
psql (16.9 (Ubuntu 16.9-1.pgdg24.04+1))
Type "help" for help.

postgres=#
postgres=# CREATE USER dms_user WITH LOGIN PASSWORD 'strongpassword';
GRANT CONNECT ON DATABASE postgres TO dms_user;
GRANT USAGE ON SCHEMA public TO dms_user;
GRANT SELECT ON ALL TABLES IN SCHEMA public TO dms_user;
ALTER DEFAULT PRIVILEGES IN SCHEMA public GRANT SELECT ON TABLES TO dms_user;
CREATE DATABASE dvdrental;
GRANT CONNECT ON DATABASE dvdrental TO dms_user;
CREATE ROLE
GRANT
GRANT
GRANT
ALTER DEFAULT PRIVILEGES
ERROR:  database "dvdrental" already exists
GRANT
postgres=# GRANT CONNECT ON DATABASE dvdrental TO dms_user;
GRANT
postgres=#
```

Figure 13.18: *Creating a migration user in our EC2 instance*

Steps to Transfer PostgreSQL Database from EC2 to RDS

1. **Configure Source and Target Endpoints**:

   a. **Source Endpoint**: Open the AWS DMS console.

Figure 13.19: *Accessing the AWS DMS Service*

b. Go to the Migrate or replicate pane, then to **Endpoints,** and create a new endpoint.

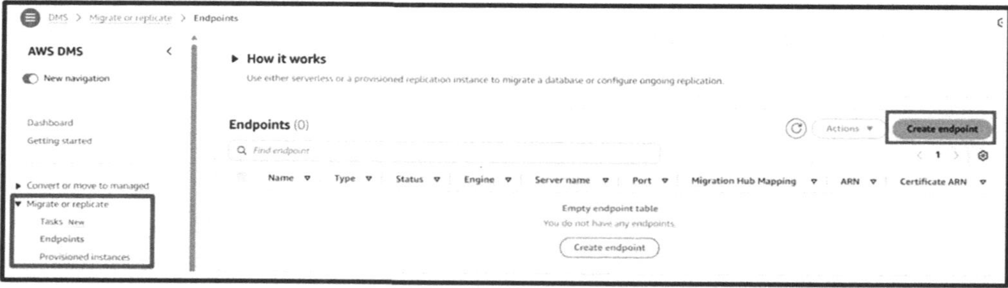

Figure 13.20: *Start to create a new endpoint*

c. Set the endpoint type to **`Source`** and give it a unique name.

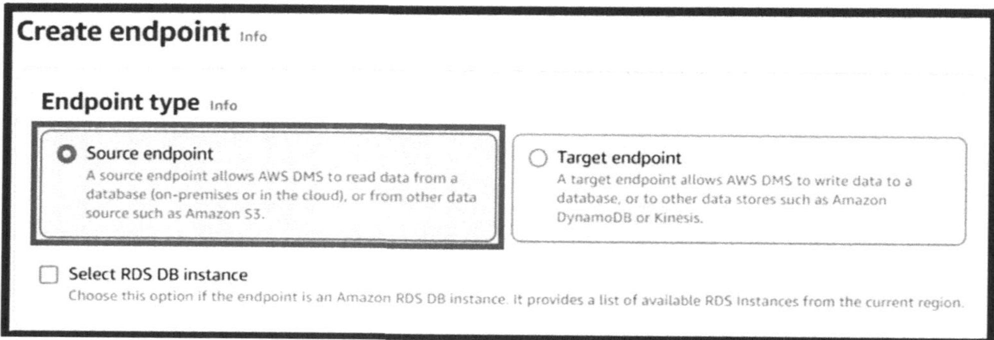

Figure 13.21: *Choosing the endpoint type*

2. In the next screen, we will give the endpoint a name and then choose the engine type, which in this case is PostgreSQL.

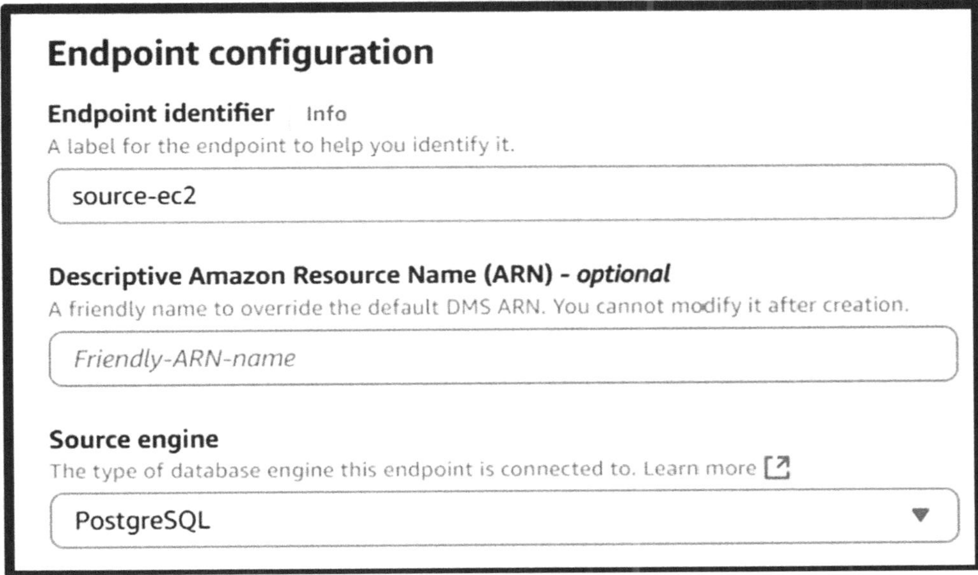

Figure 13.22: *Naming the endpoint and choosing the engine*

a. Provide the connection details for your PostgreSQL database running on the EC2 instance.

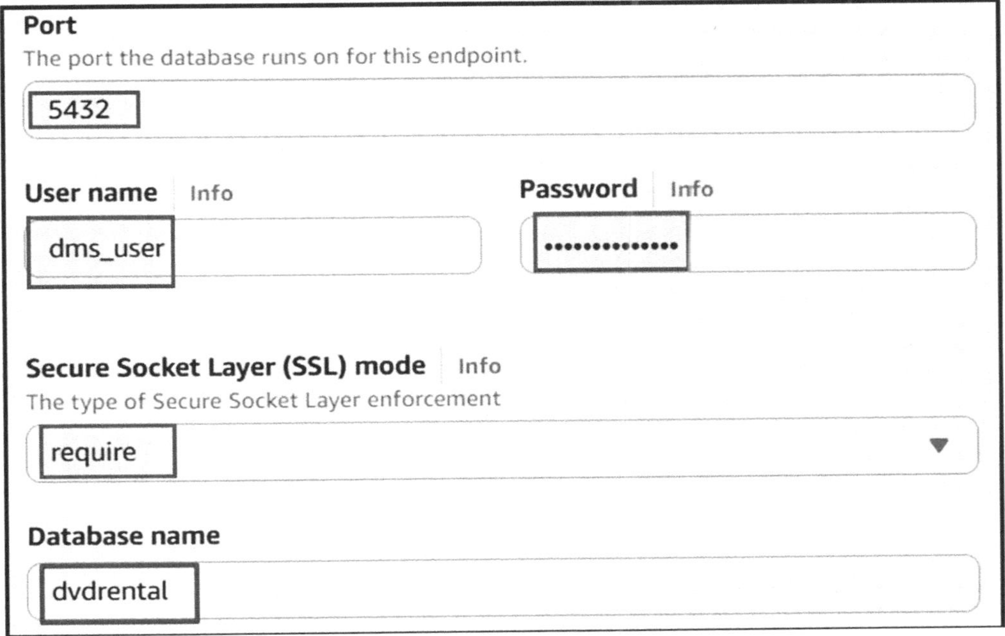

Figure 13.23: *Providing the server's name and user credentials*

Note, it is very important you get the right server name. To get the correct server name, click on the EC2 instance we created in *Figure 12.25*, and copy the public DNS in the top right corner.

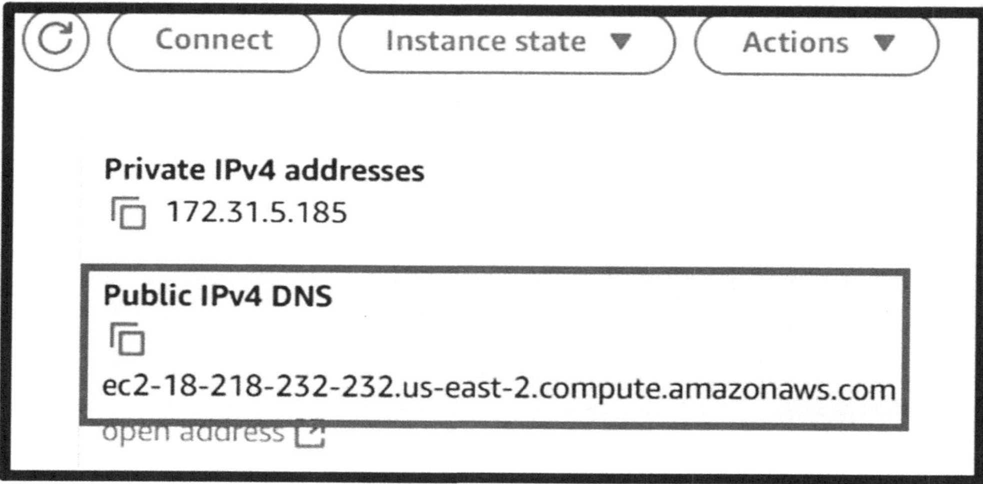

Figure 13.24: *Getting the server name from EC2's Public DNS*

3. In the next screen, we will be creating the source endpoint. Note that the database name is the one we created inside the EC2 instance in *Figure 13.15*.

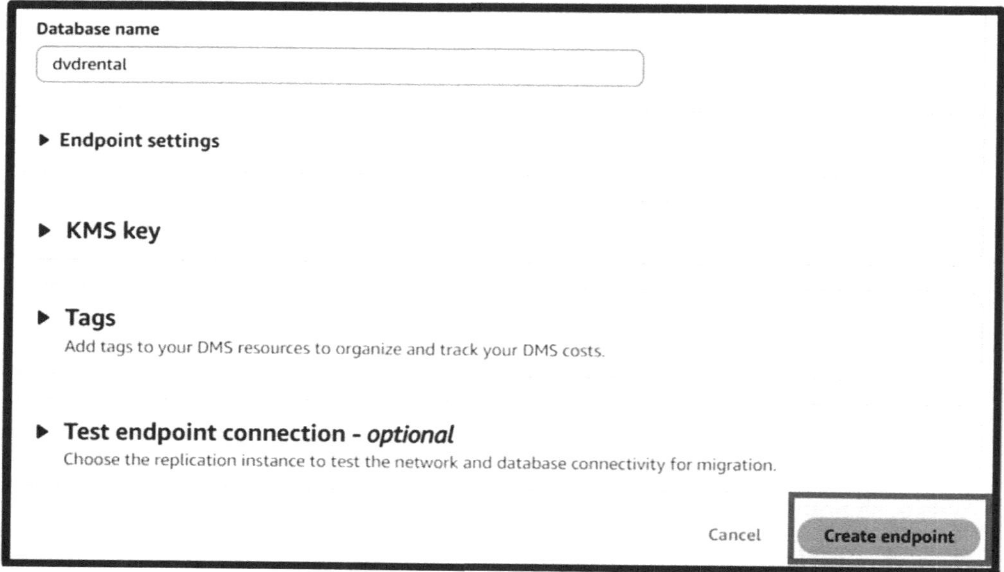

Figure 13.25: *Name the database and create the endpoint*

a. **Target Endpoint**: Create another endpoint and set the endpoint type to **Target**. Check the `Select RDS DB Instance` button and from the dropdown choose your RDS instance.

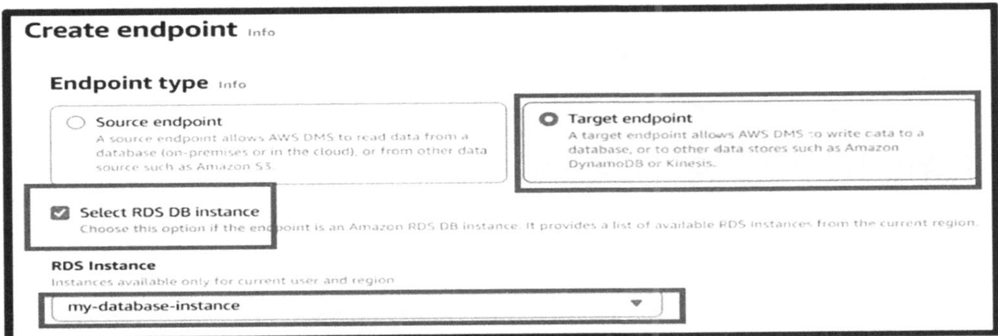

Figure 13.26: *Creating the target endpoint*

4. In the next screen, we will provide the endpoint name and choose the engine and the server name. By selecting Provide access information manually, some of the information, such as server name, will be pre-populated.

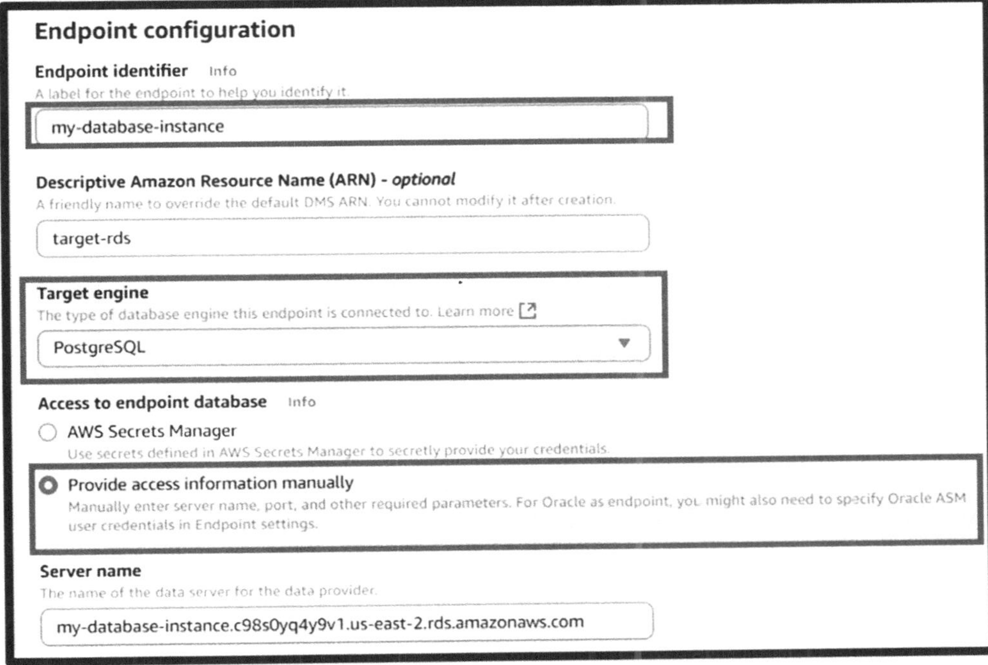

Figure 13.27: *Configuring the target endpoint*

a. Next, we will configure the port number and the user credentials. Note that the password is the one assigned to user **postgres** when we were creating the RDS instance. The **postgres** database is the default database created when you install PostgreSQL. The secure socket layer (SSL) mode must be changed to require.

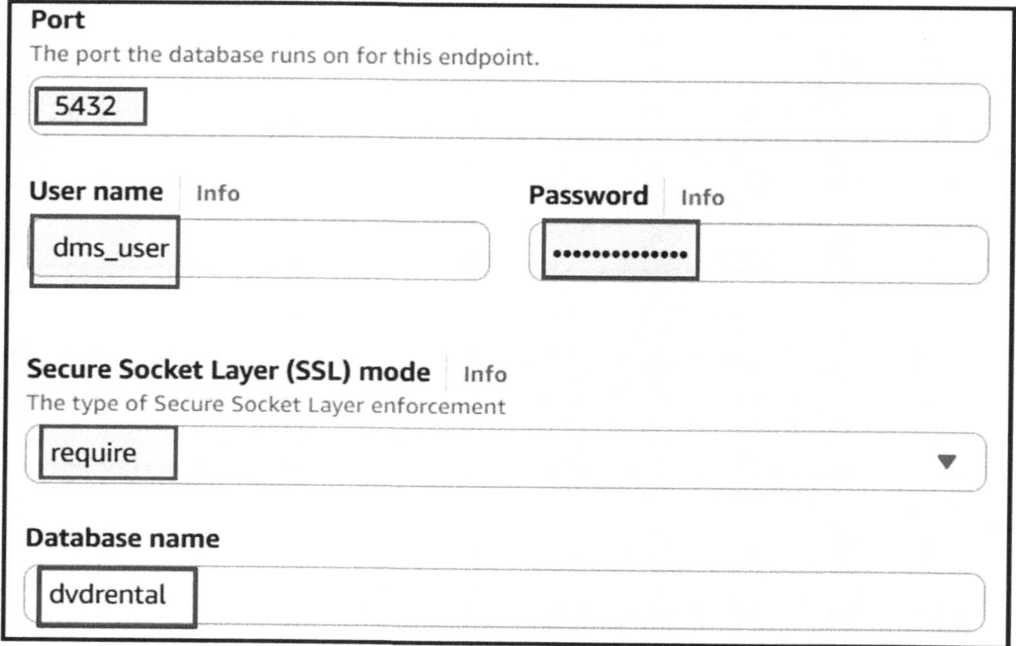

Figure 13.28: *Configuring the port number and user credentials*

b. Remember to connect to the **rds** instance and create the **dvdrental database**.

Now we have two endpoints created.

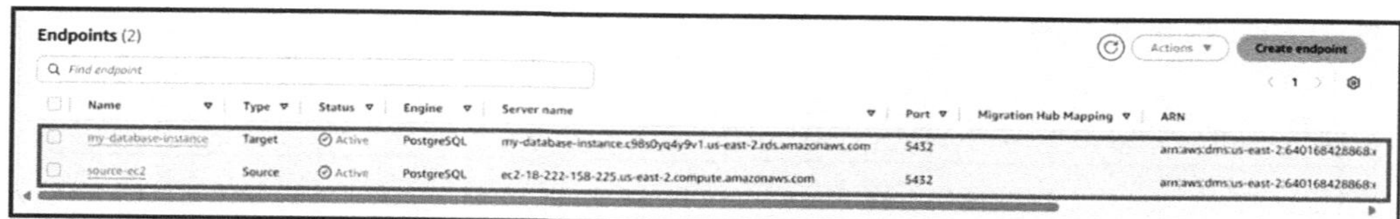

Figure 13.29: *Source and target endpoints available*

Before testing our endpoints, we need to first create the replication instance.

5. **Create a Replication Instance**:

   a. A **replication instance** in AWS Database Migration Service (DMS) is an essential component that facilitates the migration process of your database. It plays a crucial role in connecting to your source database, extracting the data, transforming it into a suitable format for the target database, and then loading the data into the target database.

   b. In the DMS console, go to Migrate or replicate, then to **`Provisioned instances`**, and create a new replication instance.

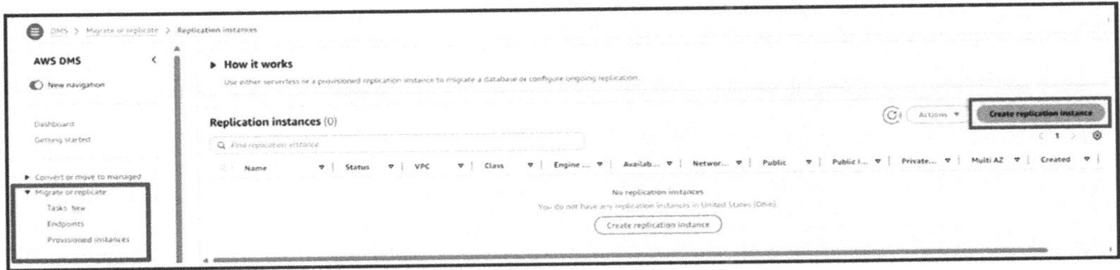

Figure 13.30: *Creating a replication instance*

The replication instance is responsible for managing the entire migration process, including ongoing replication tasks and initial data load tasks.

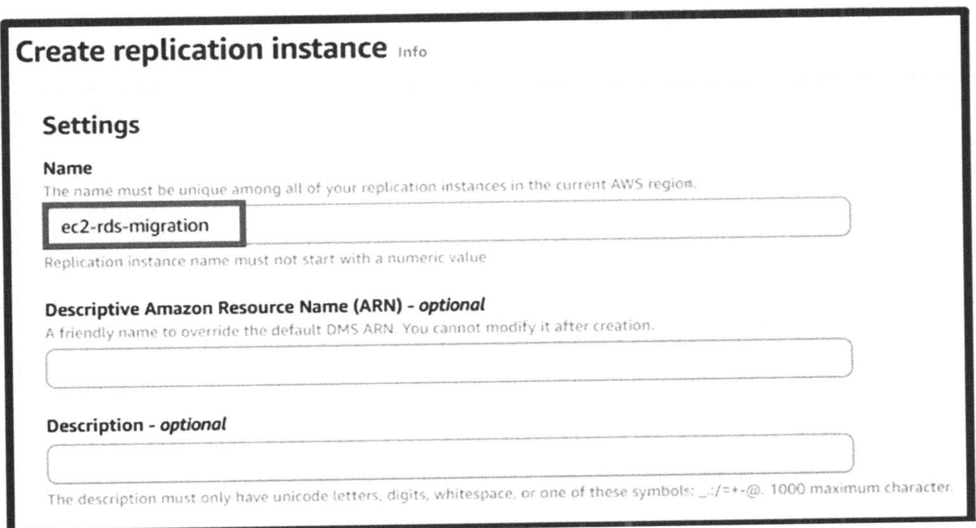

Figure 13.31: *Naming the replication instance*

Keep the rest of the configuration as default, and then create a new instance.

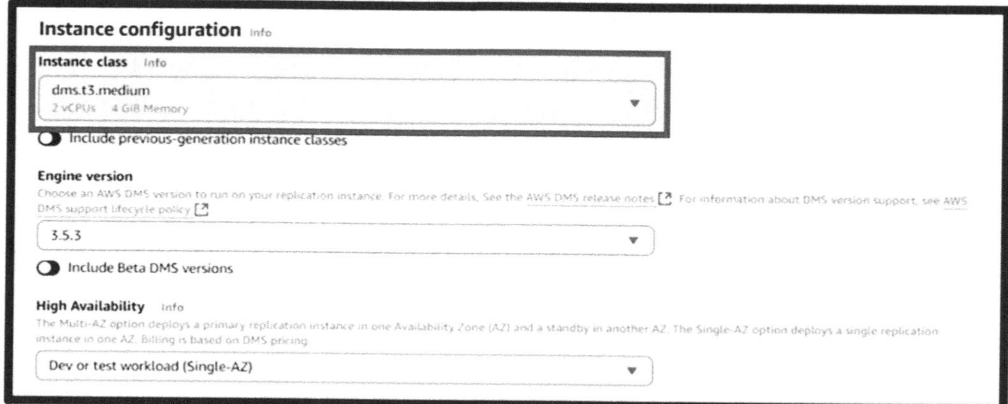

Figure 13.32: *Configuring the replication instance class*

In the next screen, select **`IPv4`** and public access. For the VPC, leave it as default.

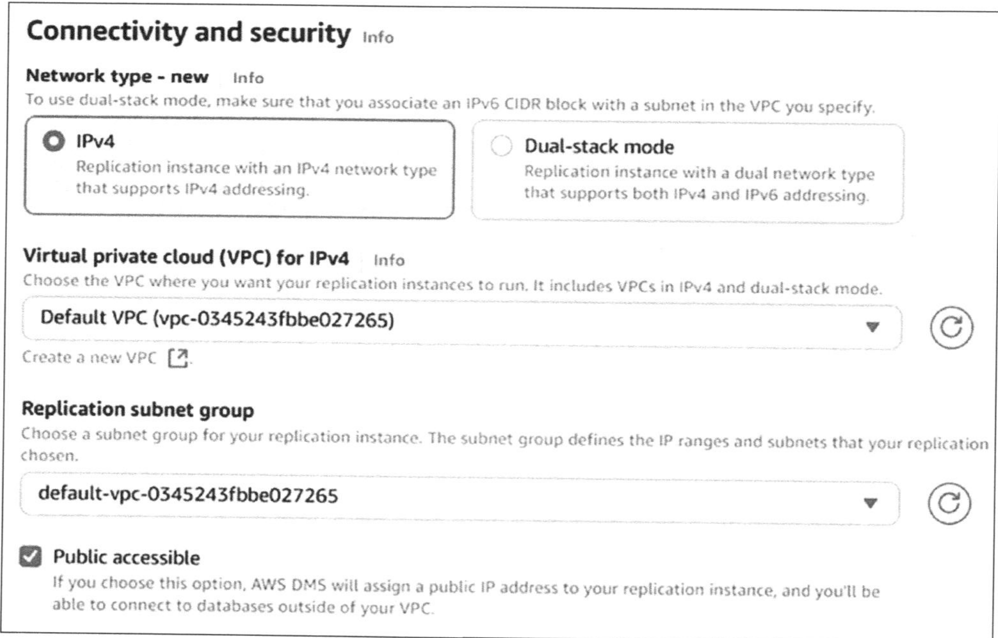

Figure 13.33: *Configuring the VPC of the replication instance*

The next screen shows that every other thing is left as default. Create the instance.

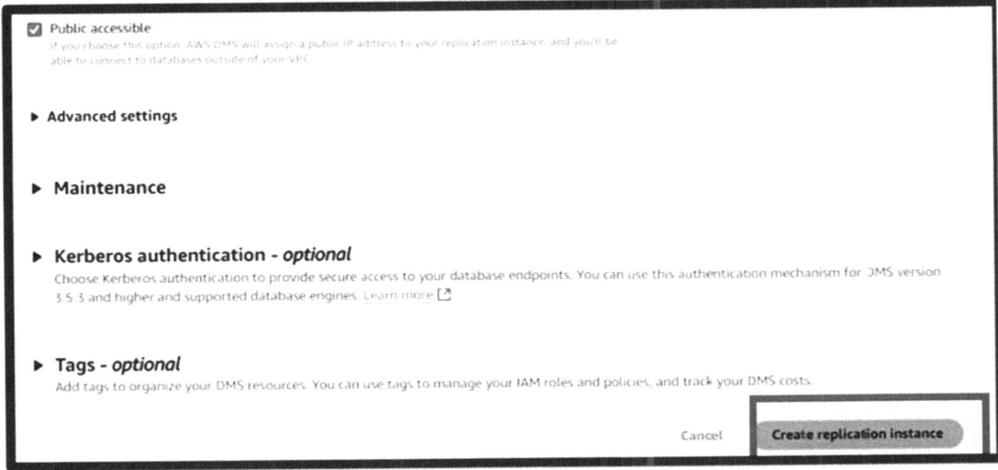

Figure 13.34: *Create the replication instance*

Ensure this instance can communicate with both your EC2 instance and RDS instance.

`#CONFIGURE THE SECURITY GROUPS` of EC2 and RDS to allow the IP address of the DMS replication instance private IP.

`--->` To get this IP `--->` Go to the DMS replication instance created in *Figure* 13.34 `-->` **`Overview details`** `-->` You will see the private IP there.

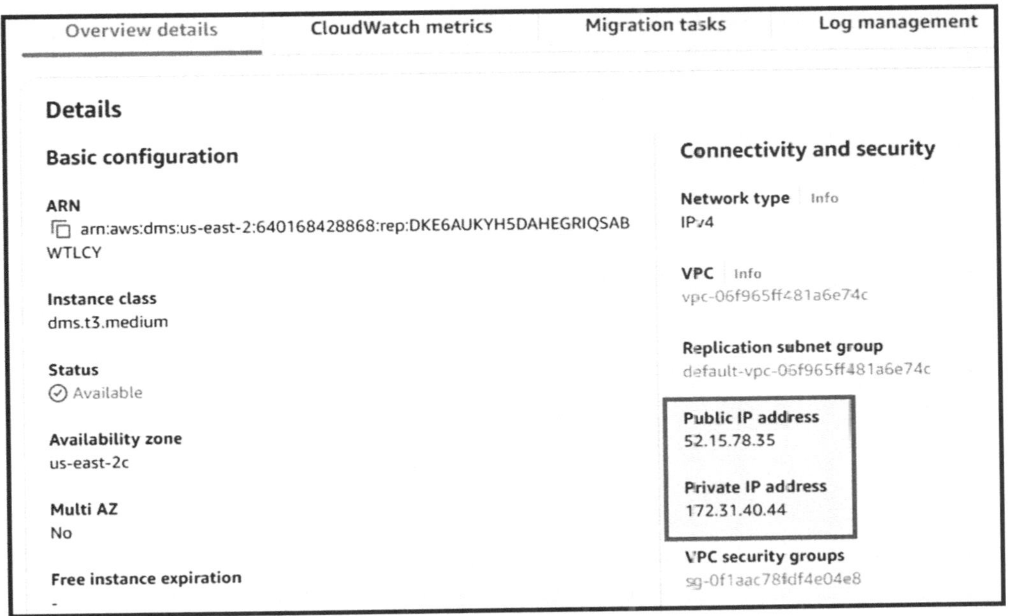

Figure 13.35: *Obtaining the IP addresses of the replication instance*

The private IP in my case is **`172.31.40.44`**.

Next, we will go to our **`EC2 instance`** created in *Figure 12.22* ---> **`Security`** --> Scroll down to **`Security groups`** ---> Edit **`Security groups`** ---> Edit **`inbound rules`** ---> **`Add rule`** -----> **`PostgreSQL`** ---> add IP of DMS replication instance

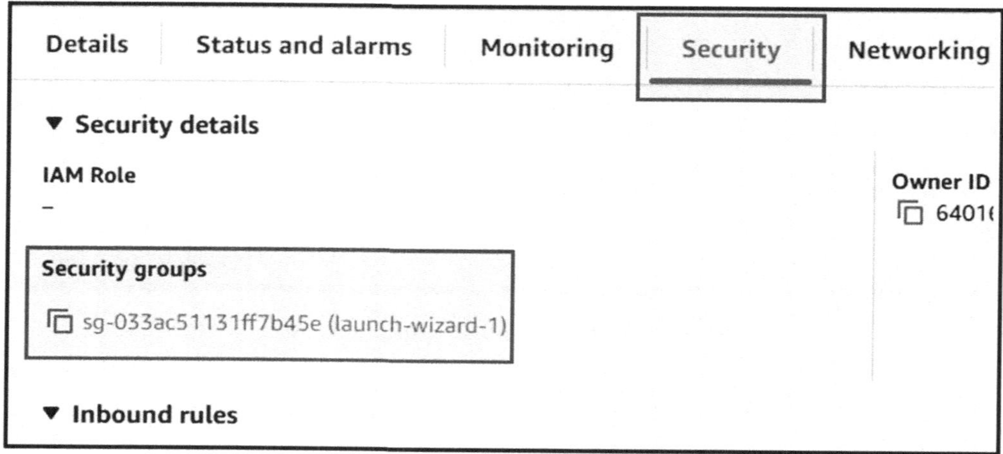

Figure 13.36: *Process of adding IP addresses of the replication instance in the EC2 instance inbound rule*

The next screen shows how the IP address of the replication instance is added to the inbound rule of the EC2 instance.

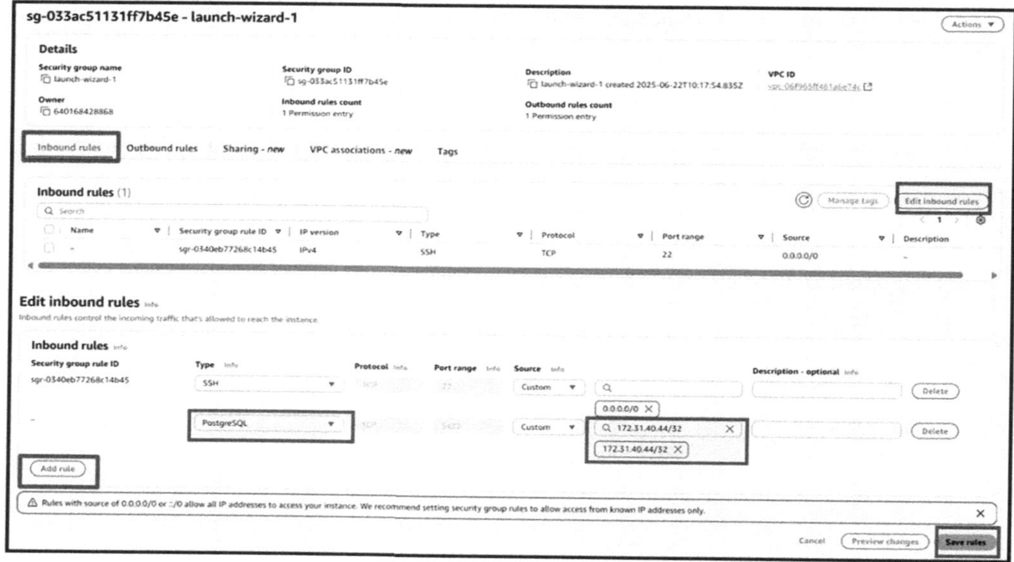

Figure 13.37: *Added the IP address of the replication instance in the EC2 instance inbound rule*

In the next screen, we are going to repeat the process, but this time for the RDS instance we created in *Figure 12.49*.

--> Go to **`RDS instance`** ---> **`Security groups`** ---> Edit **`Security groups`** ---> Edit **`Inbound rules`** ---> **`add rule`** -- ---> PostgreSQL ---> add IP of DMS replication instance

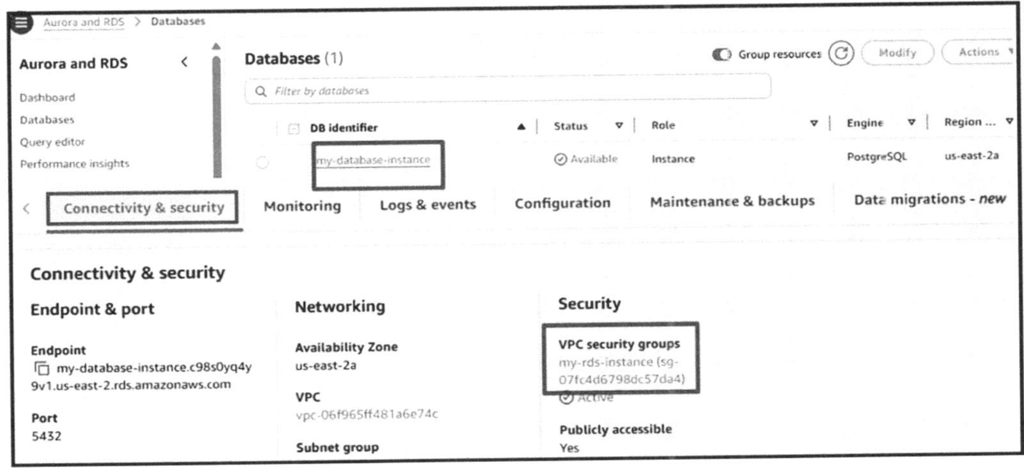

Figure 13.38: *Process of adding the IP address of the replication instance in the RDS instance inbound rule*

The next screen is locating the security group of the RDS instance.

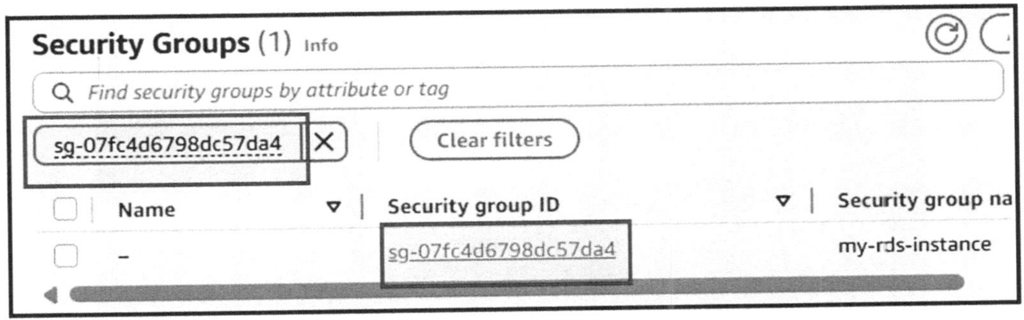

Figure 13.39: *Accessing the security group of the RDS instance*

In the next screen, we will add the IP address of the replication instance in the inbound rule of the RDS security group.

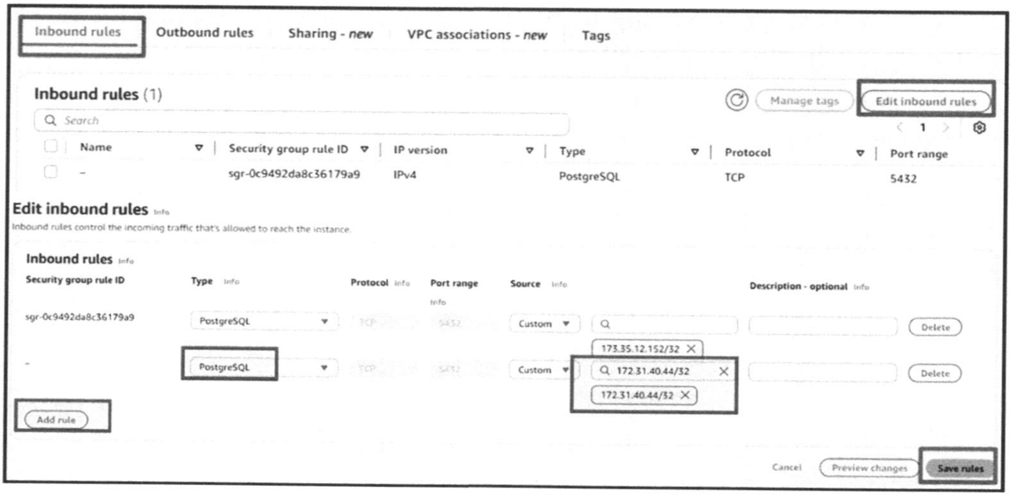

Figure 13.40: *Adding the replication instance IP into the security group of the RDS instance*

Now that the IP address of our DMS replication instance has been added to the EC2 and RDS security groups, we will authenticate the IP addresses of the replication instance, our EC2 server, and our local machine in the **`pg_hba.conf`** file on the EC2 instance. Vi into the **`pg_hba.conf`**.

Before we configure the last task, which is the migration task, let us first edit the **`pg_hba.conf`** file of our EC2 instance. We can either obtain this file from the OS level by running the **`ps -ef | grep postgres`** command or from within the database by running show **`config_file;`**

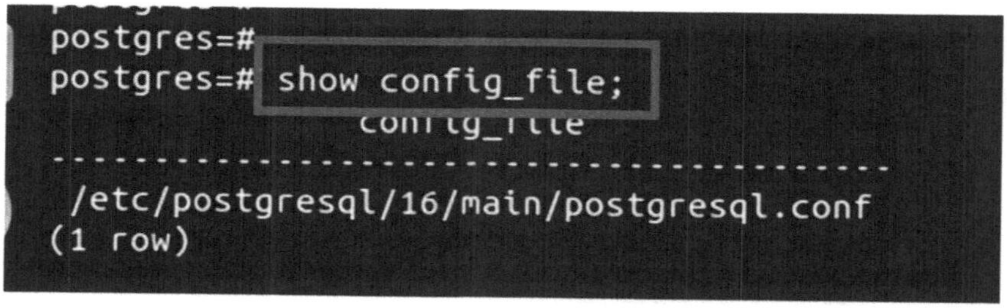

Figure 13.41: *Obtaining config files from within the database*

We change the directory to the location of the files as shown in the following screenshot.

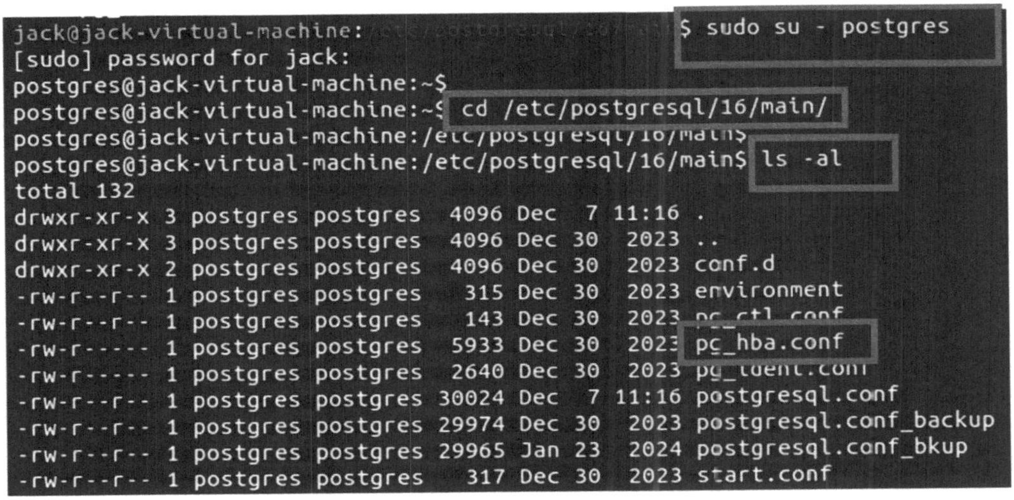

```
jack@jack-virtual-machine:                    $ sudo su - postgres
[sudo] password for jack:
postgres@jack-virtual-machine:~$
postgres@jack-virtual-machine:~$ cd /etc/postgresql/16/main/
postgres@jack-virtual-machine:/etc/postgresql/16/main$
postgres@jack-virtual-machine:/etc/postgresql/16/main$ ls -al
total 132
drwxr-xr-x 3 postgres postgres  4096 Dec  7 11:16 .
drwxr-xr-x 3 postgres postgres  4096 Dec 30  2023 ..
drwxr-xr-x 2 postgres postgres  4096 Dec 30  2023 conf.d
-rw-r--r-- 1 postgres postgres   315 Dec 30  2023 environment
-rw-r--r-- 1 postgres postgres   143 Dec 30  2023 pg_ctl.conf
-rw-r----- 1 postgres postgres  5933 Dec 30  2023 pg_hba.conf
-rw-r----- 1 postgres postgres  2640 Dec 30  2023 pg_ident.conf
-rw-r--r-- 1 postgres postgres 30024 Dec  7 11:16 postgresql.conf
-rw-r--r-- 1 postgres postgres 29974 Dec 30  2023 postgresql.conf_backup
-rw-r--r-- 1 postgres postgres 29965 Jan 23  2024 postgresql.conf_bkup
-rw-r--r-- 1 postgres postgres   317 Dec 30  2023 start.conf
```

Figure 13.42: *Locating the pg_hba.conf files*

To edit the file and add the following sentences, we vi into it first. We must add our local machine IP address and the replication instance IP address as follows.

```
host    all                          localmachine_IP/24        md5  #localmachine_IP get it by typing "what is my ip" in google
host    dvdrental        postgres        DMS_repication_instance_private_IP/32        md5
```

Replace the **localmachine_IP** with the actual IP and the **DMS_replication_instance_private** with the actual one as shown on the next screen.

Note that **dvdrental** is the name of the database we want to migrate from EC2 to RDS.

```
Vi /etc/postgresql/16/main/pg_hba.conf
host    all              all              172.31.35.105/32         md5
#Your EC2 ip address
host    all              all              172.31.0.0/16            md5
#VPC range like AWS EC2
host    all              all              173.35.12.152/24         md5
#local machine ip
host    dvdrental        dms_user         172.31.40.44/32          md5
#DMS user and replication instance ip.
```

Reload the system for the changes to take effect using the following command:

```
Sudo systemctl reload postgresql@16-main
```

Or

```
Sudo systemctl reload postgresql
```

```
postgres@ip-172-31-35-105:~$
postgres@ip-172-31-35-105:~$ vi /etc/postgresql/16/main/pg_hba.conf

# Database administrative login by Unix domain socket
local   all             postgres                                peer

# TYPE  DATABASE        USER            ADDRESS                 METHOD

# "local" is for Unix domain socket connections only
local   all             all                                     peer
# IPv4 local connections:
host    all             all             127.0.0.1/32            scram-sha-256
host    all             all             172.31.35.105/32        md5    #Your EC2
host    all             all             172.31.0.0/16           md5    #VPC range like AWS EC2
host    all             all             173.35.12.152/24        md5    #local machine ip
host    dvdrental       dms_user        172.31.40.44/32         md5    #DMS user and replication instance ip   #replication
# IPv6 local connections:
host    all             all             ::1/128                 scram-sha-256
# Allow replication connections from localhost, by a user with the
# replication privilege.
local   replication     all                                     peer
host    replication     all             127.0.0.1/32            scram-sha-256

ubuntu@ip-172-31-35-105: $ sudo systemctl reload postgresql@16-main
```

Figure 13.43: *Adding local machine IP address and replication instance IP into the pg_hba.conf file*

In the next screen, we are going to test the new endpoints. They must both be successful before we can migrate the task.

Next, we are going to test the endpoints. Select each endpoint shown in *Figure 13.26*. From the dropdown at the top right of the page, select "**`Test endpoint connection`**". It must have the status of successful before you can proceed with migration.

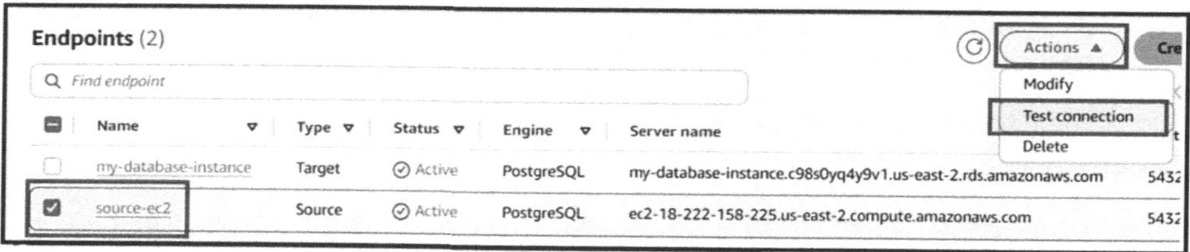

Figure 13.44: *Selecting the endpoint for testing*

Repeat the above process for the target endpoint as well.

Figure 13.45: *Test of the source endpoint*

The next screen is a test for the target endpoint.

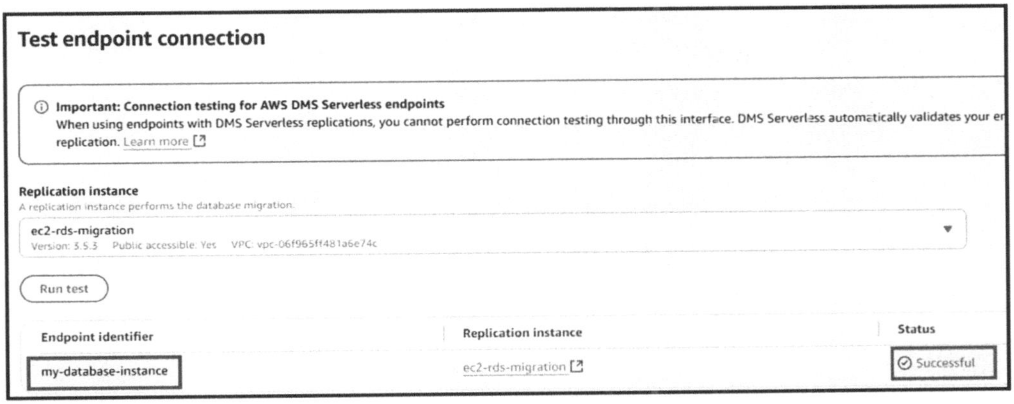

Figure 13.46: *Test of the target endpoint*

Our final step is to create a migration task.

Creating a Migration Task

A migration task in AWS Database Migration Service (DMS) is responsible for the actual data movement from the source database to the target database. There are three main types of migration tasks:

- **Full Load**: Transfers all the data from the source to the target database.
- **Change Data Capture (CDC)**: Continuously captures and applies changes from the source to the target.
- **Full Load and CDC**: Combines both full load and change data capture, ideal for minimal downtime migrations.

AWS DMS offers an option to perform a pre-migration assessment. This tool evaluates your task configuration and the source database to identify potential issues that could affect the migration process.

Once the migration task is running, you can monitor its progress through the AWS Management Console, CloudWatch metrics, and DMS-specific logs. You can track details such as the number of rows migrated, data throughput, and any errors encountered.

AWS DMS supports basic data transformation during the migration process. You can convert schema, apply filtering, and rename tables and columns.

After the migration is complete, you can use the data validation feature to compare the source and target databases, ensuring that all data has been accurately migrated.

Tasks can be paused, resumed, and stopped as needed. AWS DMS also provides options for handling network disruptions and other issues to ensure smooth migration.

1. In the **DMS** dashboard, go to **Tasks** under Migrate or Replicate, and create a new migration task.

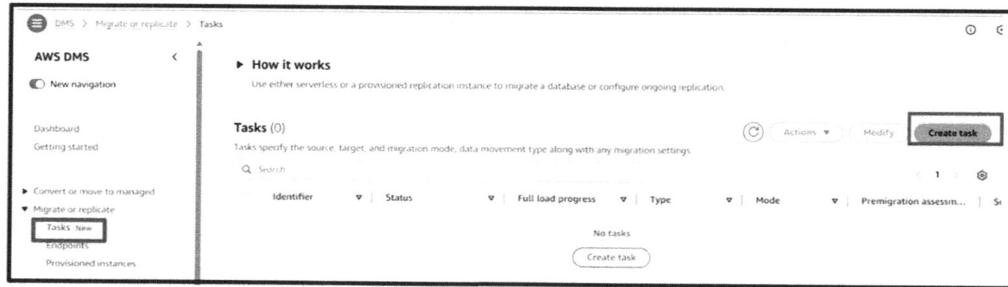

Figure 13.47: *Creating the Migration Task*

   a. Select your replication instance.

   b. Choose the source endpoint (EC2 PostgreSQL) and the target endpoint (RDS PostgreSQL).

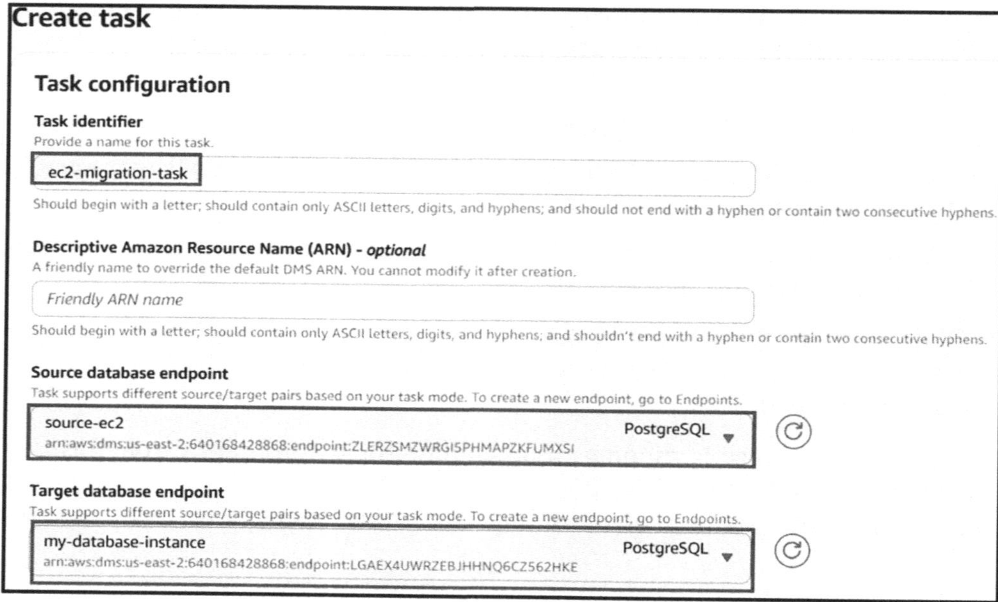

Figure 13.48: *Configuring the Migration Task*

2. In the next screen, we are going to choose the migration type and the replication instance.

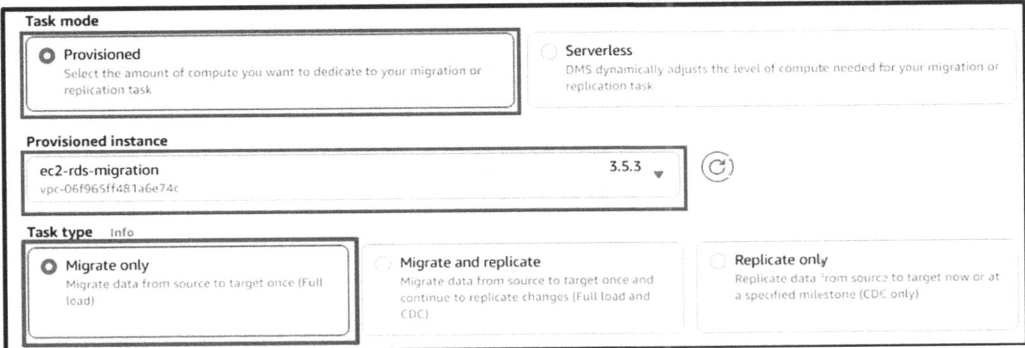

Figure 13.49: *Choosing the Migration type*

3. The next screen shows more settings for the migration task.

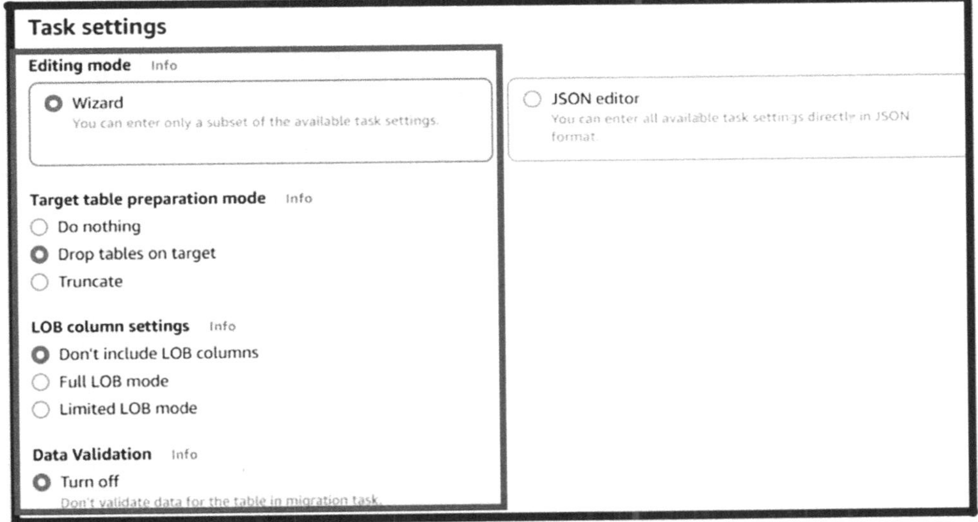

Figure 13.50: *The Migration task settings*

The following is shown under advanced task settings.

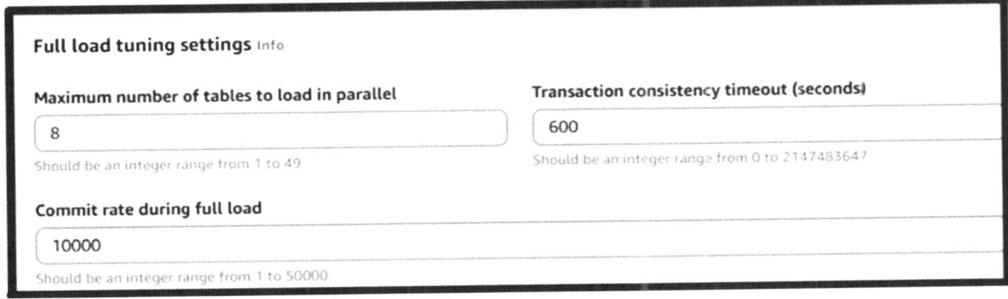

Figure 13.51: *The Migration task advance task settings*

Table mappings

Editing mode

Wizard
You can enter only a subset of the available task settings.

JSON editor
You can enter all available task settings directly in JSON format.

Specify at least one selection rule with an include action. After you do this, you can add one or more transformation rules.

▼ Selection rules

Choose the schema and/or tables you want to include with, or exclude from, your migration task.

Add new selection rule

Figure 13.52: *Adding the rule for migration*

4. Select the **`dvdrental`** database for migration.

   Under table mapping.

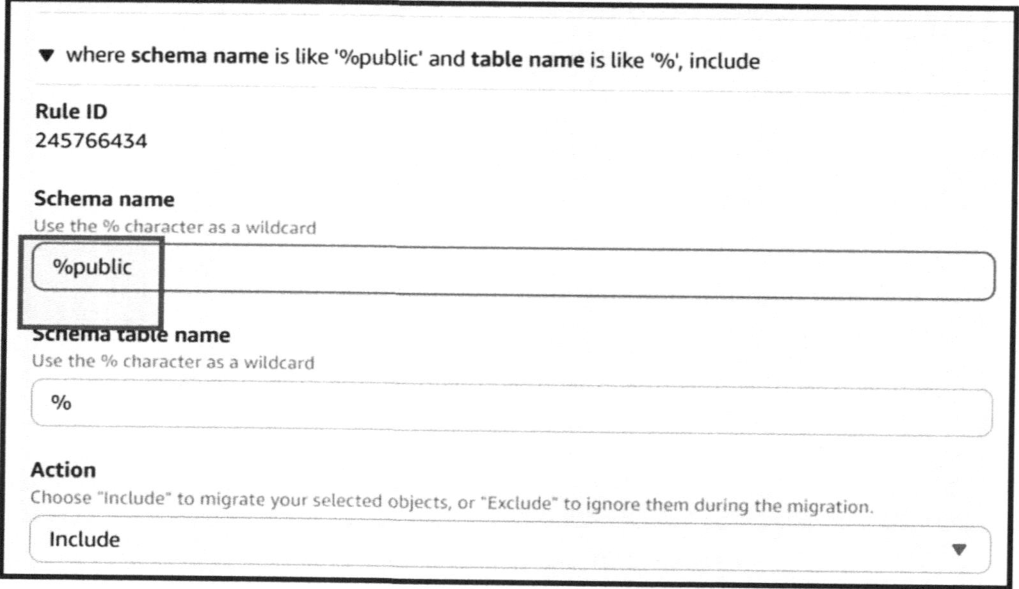

Figure 13.53: *Choosing the schema to migrate*

5. **Start the Migration Task**:

   a. Review the task settings and click **`Create task`**.

   b. Start the task and monitor its progress in the AWS DMS console.

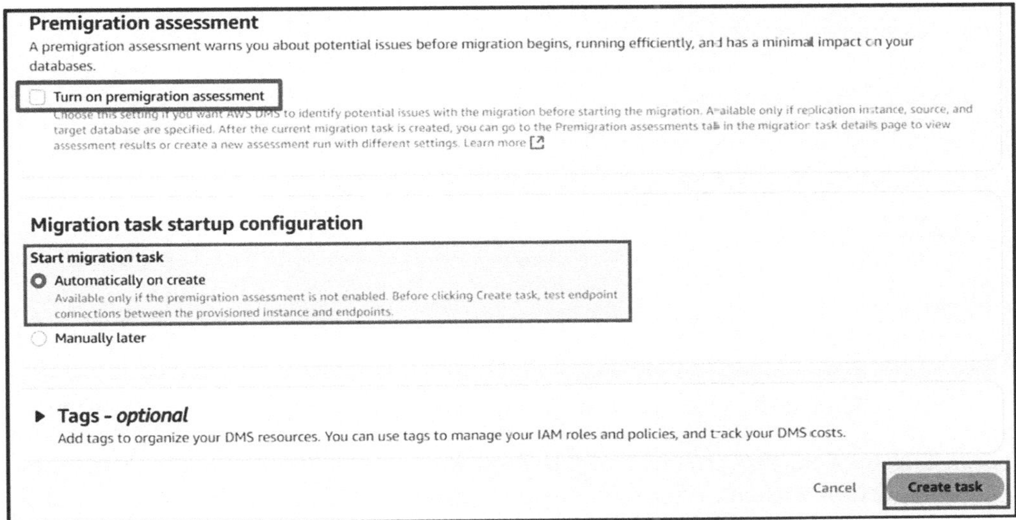

Figure 13.54: Create the migration task and run it automatically

6. The next screen shows that the task is being created.

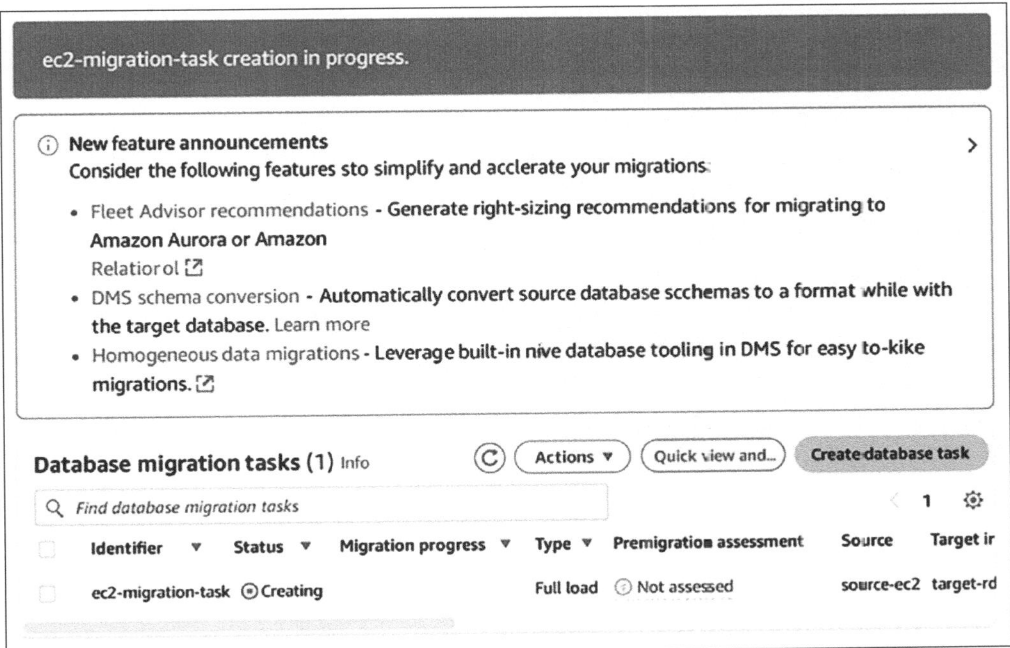

Figure 13.55: Migration task is being created

7. The next screen shows that the task has been created successfully, showing the source and target endpoints.

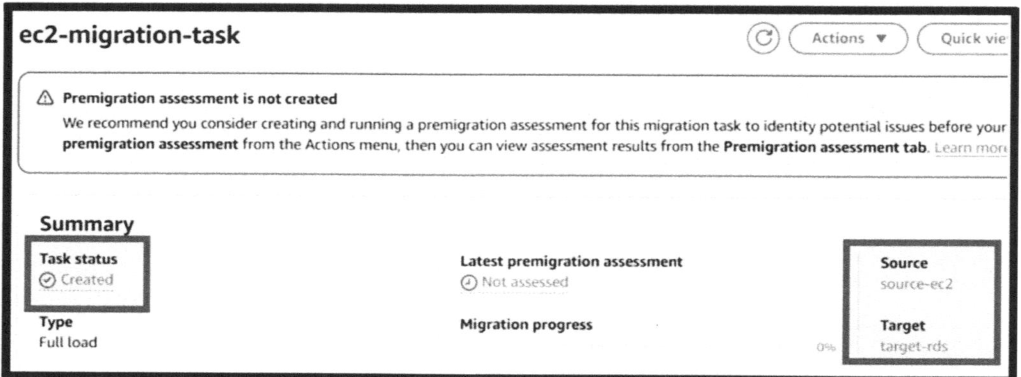

Figure 13.56: *Migration task successfully created and type is full load*

8. The next screen shows that the load was completed 100% which means the migration was successful.

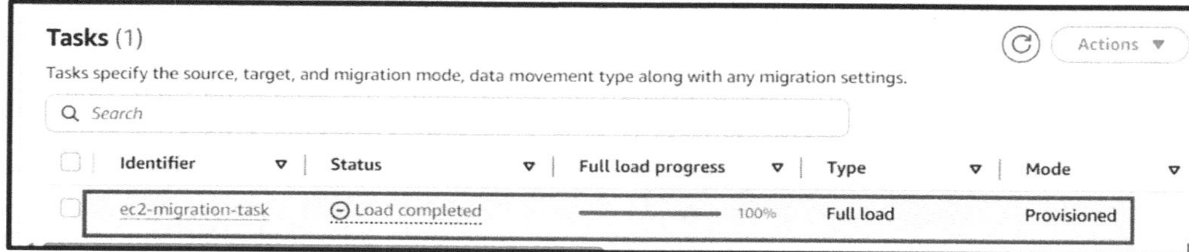

Figure 13.57: *Migration of the dvdrental database completed successfully*

9. Next is to connect to the RDS instance as shown in *Figure* 12.57, in *Chapter* 12, *Cloud Computing* (AWS), and check that the **dvdrental** database has migrated successfully.

```
psql -h my-database-instance.c98s0yq4y9v1.us-east-2.rds.
amazonaws.com -U postgres -d postgres -W
```

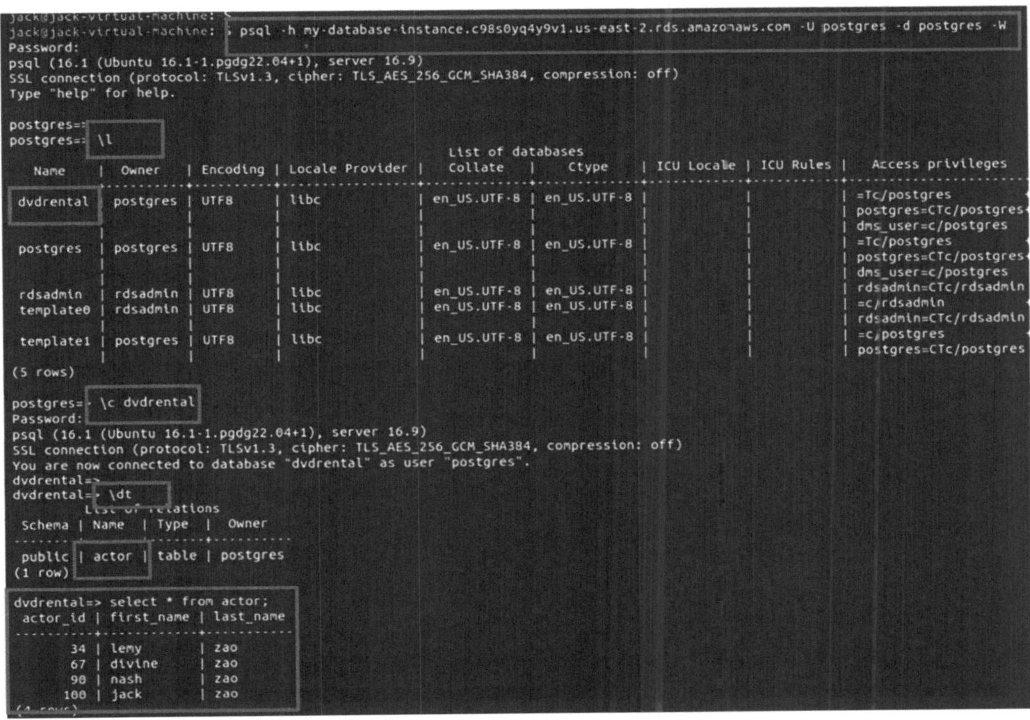

Figure 13.58: *RDS shows the migrated data*

AWS RDS Upgrade

Upgrading an Amazon RDS (Relational Database Service) instance involves updating the database engine version to a newer release. This process ensures that your database benefits from the latest features, performance improvements, bug fixes, and security patches. AWS RDS supports two types of upgrades: **minor version upgrades** and **major version upgrades.** In the following, we will look at the major version upgrade from PostgreSQL version 14 to version 16.

1. The first step is to log in to the RDS console and select the instance to upgrade, as shown.

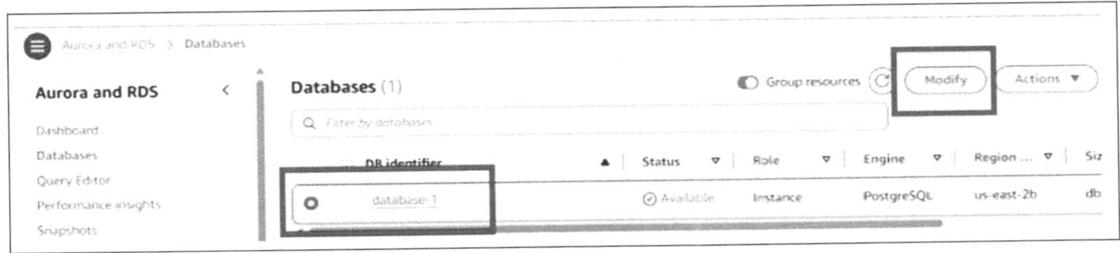

Figure 13.59: *Selecting the RDS instance for upgrade*

2. Once selected, click on the modify button. You will be shown the following screen with the current version.

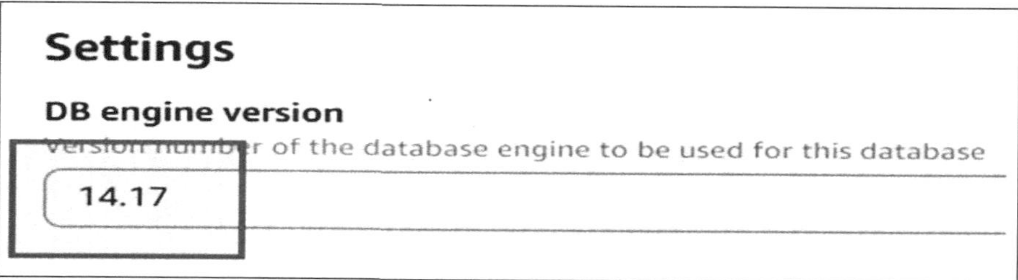

Figure 13.60: *Current PostgreSQL version*

3. Click on the dropdown menu to select the version of PostgreSQL you want to upgrade to.

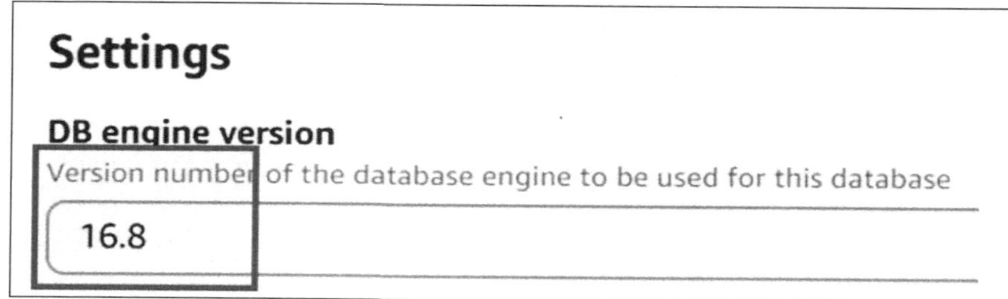

Figure 13.61: *PostgreSQL version to be upgraded to*

4. Next, scroll to the bottom and click on continue. You will then be shown a screen showing that it is going to be modified from Postgresql14 (14.7) to Postgresql16 (16.8).

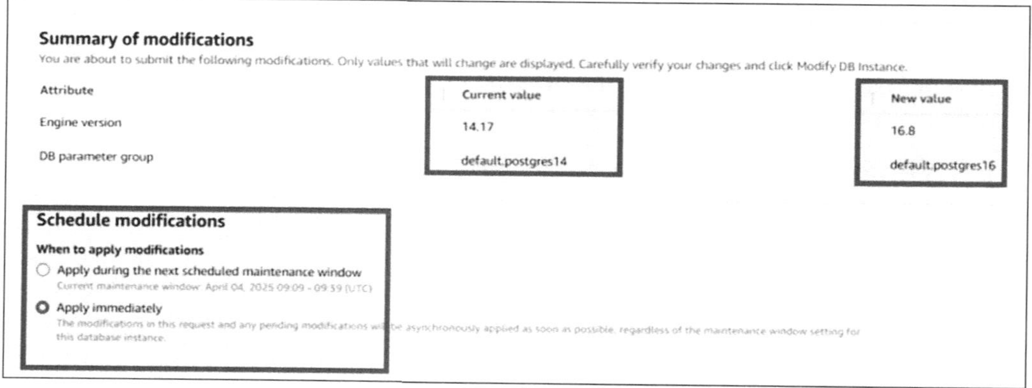

Figure 13.62: *Upgrade will be from PostgreSQL 14 to PostgreSQL 16*

Make sure to click on apply immediately for the upgrade to run now. Scroll to the end and click on modify database. The next screen shows that the system is upgrading.

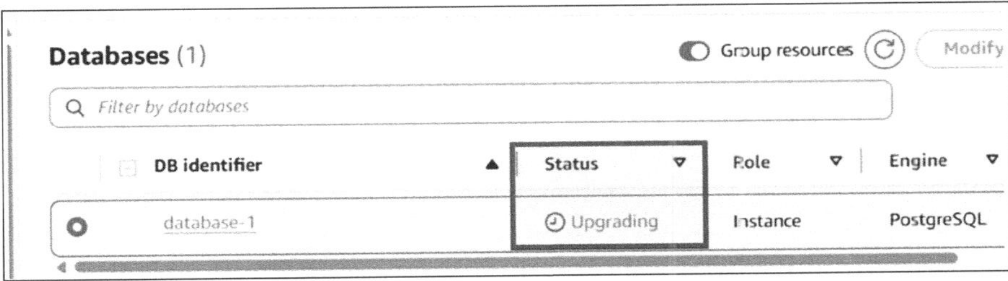

***Figure 13.63**: The RDS instance is upgrading from PostgreSQL 14 to PostgreSQL 16*

5. The next screen shows that the upgrade is done and the instance is available to be connected to.

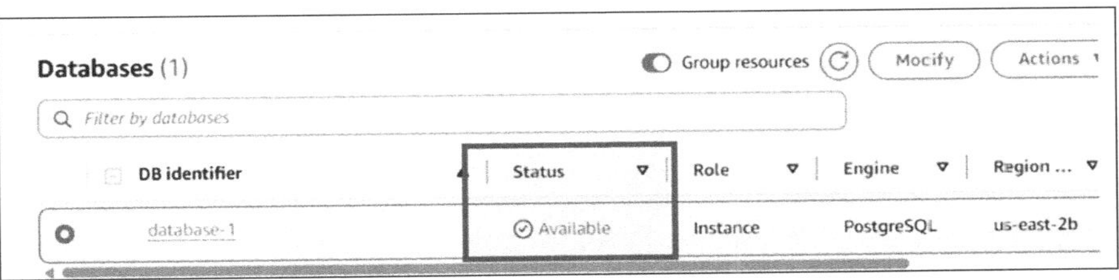

***Figure 13.64**: The RDS instance is unavailable after the upgrade is done*

6. Click on the RDS instance and then click on Configuration to see that it has been upgraded to PostgreSQL 16.8 by checking the **`Engine version`**.

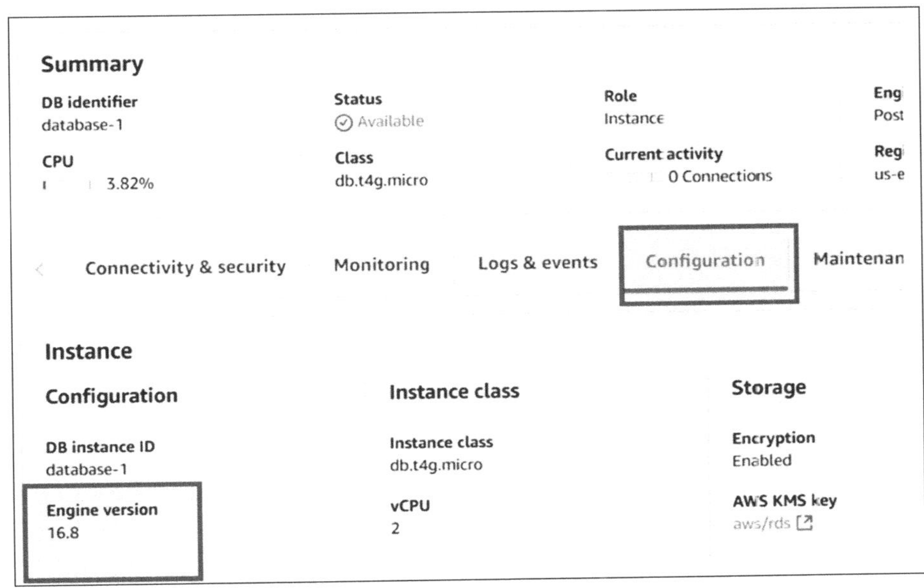

***Figure 13.65**: The RDS instance now has engine version as 16.8*

Task

Technical Interview Questions

1. What are the different types of policies that can be attached to an IAM user, and how do they affect permissions?
2. Describe the steps to create and attach a custom policy to an IAM user. What are some common mistakes to avoid when doing this?
3. How can you use IAM policies to enforce the principle of least privilege for users?
4. If an IAM user reports issues accessing a specific AWS resource, how would you troubleshoot and resolve the problem?
5. What are the key features of AWS Database Migration Service, and how does it simplify the migration process?
6. Which databases are supported by AWS DMS for migration, and what are the prerequisites for setting up the service?
7. How does AWS DMS ensure minimal downtime during database migration?
8. Can you explain the difference between continuous replication and one-time migration in AWS DMS?
9. What are the key parameters required to create source and target endpoints in AWS DMS?
10. How would you validate connectivity between the source and target endpoints before starting a migration task?
11. Describe the process of setting up SSL/TLS for secure endpoint connections in AWS DMS.
12. In what scenarios would you use an IP whitelist for source and target endpoints, and how would you configure this?
13. What is the purpose of a replication instance in AWS DMS, and how does its performance impact the migration process?
14. What considerations should you keep in mind when sizing a replication instance for database migration?
15. Explain the process of creating and configuring a replication instance in AWS DMS.
16. How would you monitor replication instance performance and troubleshoot issues during a migration?

17. What are the different types of migration tasks available in AWS DMS, and how do you select the appropriate type for a given scenario?
18. Can you walk through the steps to create a database migration task in AWS DMS? What configurations are critical for success?
19. How would you use transformation rules to modify data during migration?
20. What are some common migration task failures in AWS DMS, and how would you troubleshoot them?

Conclusion

This chapter has provided a comprehensive understanding of managing IAM users and their critical role in securely controlling access to AWS resources. By exploring how to assign permissions to IAM users, learners have gained insights into tailoring access levels to meet specific operational requirements while maintaining security best practices. These principles form the foundation for effectively managing AWS environments and preparing for seamless data migrations.

The chapter also delved into the AWS Database Migration Service (DMS), offering a step-by-step guide to creating source and target endpoints, setting up a replication instance, and initiating a migration task. These processes enable businesses to efficiently transfer databases to AWS with minimal downtime and disruption. By combining IAM user management with DMS capabilities, learners are now equipped with the knowledge to securely and effectively migrate and manage workloads in the AWS cloud, setting the stage for scalability and modernization.

Index

C

D

E

F

G

H

T

U

V